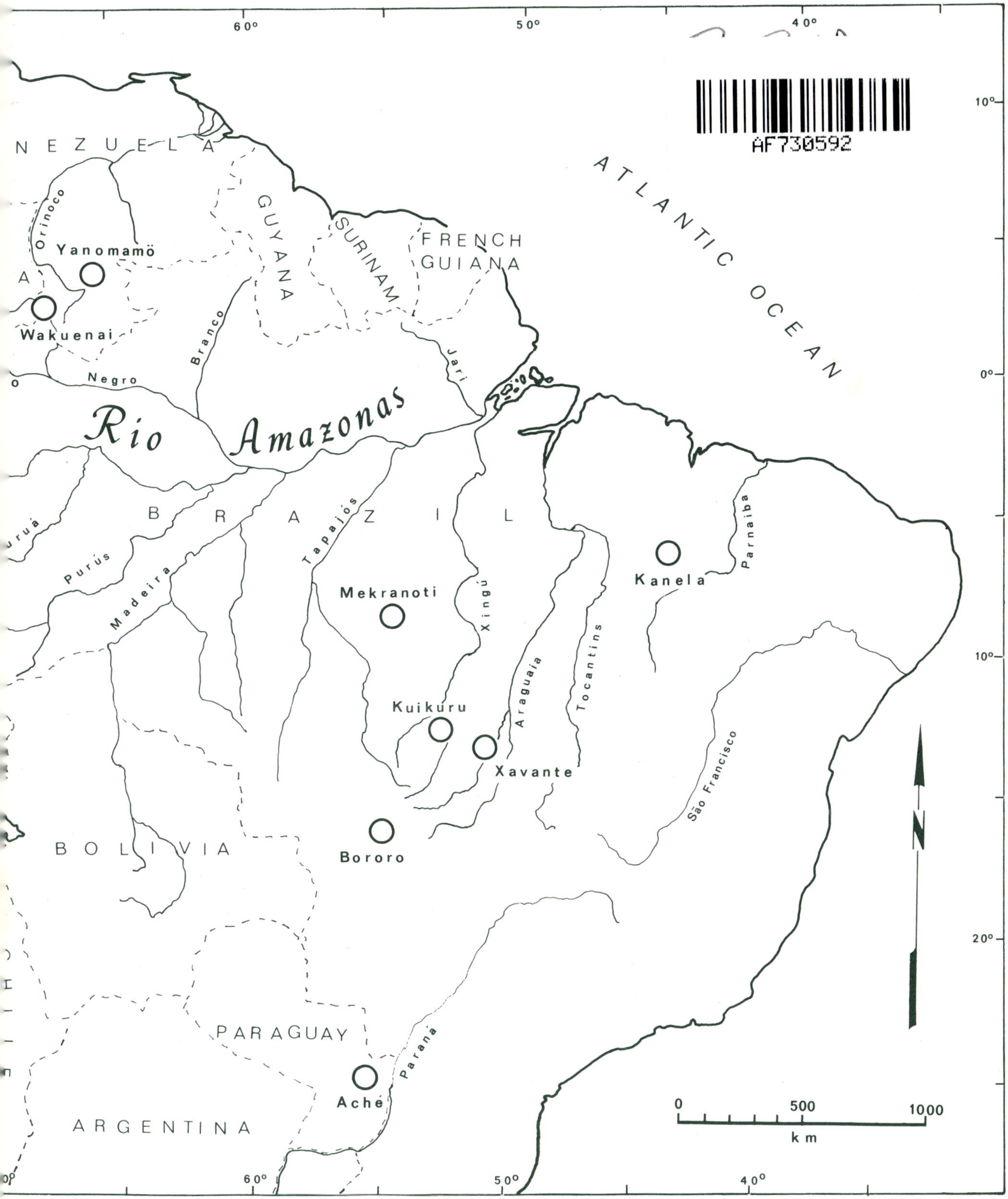
60°
50°
40°
10°
0°
10°
20°
NEZUELA
Orinoco
Yanomamö
Wakuenai
GUYANA
SURINAM
FRENCH
GUIANA
ATLANTIC OCEAN
Branco
Negro
Jari
Rio Amazonas
B R A Z I L
Purús
Madeira
Tapajós
Xingú
Araguaia
Tocantins
Parnaiba
São Francisco
Mekranoti
Kuikuru
Xavante
Kanela
Bororo
BOLIVIA
PARAGUAY
Paraná
Aché
ARGENTINA
0
500
1000
km
N
60°
50°
40°

Adaptive Responses of Native Amazonians

This is a volume in

STUDIES IN ANTHROPOLOGY

Under the consulting editorship of
E. A. Hammel, University of California, Berkeley

A complete list of titles appears at the end of this volume.

Adaptive Responses of Native Amazonians

EDITED BY

Raymond B. Hames
Department of Anthropology
University of Nebraska
Lincoln, Nebraska

William T. Vickers
Department of Sociology and Anthropology
Florida International University
Miami, Florida

1983

ACADEMIC PRESS
A Subsidiary of Harcourt Brace Jovanovich, Publishers
New York London
Paris San Diego San Francisco São Paulo Sydney Tokyo Toronto

ACADEMIC PRESS, INC.
111 Fifth Avenue, New York, New York 10003

United Kingdom Edition published by
ACADEMIC PRESS, INC. (LONDON) LTD.
24/28 Oval Road, London NW1 7DX

Library of Congress Cataloging in Publication Data

Main entry under title:

Adaptive responses of native Amazonians.

(Studies in anthropology)
Bibliography: p.
Includes index.
1. Indians of South America--Amazon River Valley--Social conditions--Addresses, essays, lectures.
2. Indians of South America--Amazon River Valley--Social life and customs--Addresses, essays, lectures.
3. Human ecology--Amazon River Valley--Addresses, essays, lectures. 4. Amazon River Valley--Social conditions--Addresses, essays, lectures. I. Hames, Raymond B. II. Vickers, William T. III. Series: Studies in anthropology (Academic Press)
F2230.1.S68A3 1982 305.8'98'0811 82-18399
ISBN 0-12-321250-2

PRINTED IN THE UNITED STATES OF AMERICA

83 84 85 86 9 8 7 6 5 4 3 2 1

Contents

IV. SETTLEMENT PATTERN

Contributors

Numbers in parentheses indicate the pages on which the authors' contributions begin.

Stephen Beckerman (269), Department of Anthropology, Pennsylvania State University, University Park, Pennsylvania 16802

Brent Berlin (301), Department of Anthropology, University of California, Berkeley, California 94720

Elois Ann Berlin (301), Division of Family Medicine, Department of Family, Community and Preventive Medicine, Stanford University, School of Medicine, Stanford, California 94305

Robert L. Carneiro (65), Department of Anthropology, American Museum of Natural History, New York, New York 10024

Darna L. Dufour[1] (329), The Institute of Physiology, The University, Glasgow, G12 8QQ, Scotland

Nancy M. Flowers (357), Department of Anthropology, City University of New York, New York, New York 10021

Daniel R. Gross (429), Department of Anthropology, Hunter College of the City University of New York, New York, New York 10021

Raymond B. Hames (1, 393), Department of Anthropology, University of Nebraska, Lincoln, Nebraska 68588

Kristen Hawkes (139), Department of Anthropology, University of Utah, Salt Lake City, Utah 84112

Jonathan Hill (113), Department of Anthropology, Indiana University, Bloomington, Indiana 47405

Kim Hill (139), Department of Anthropology, University of Utah, Salt Lake City, Utah 84112

[1]Present address: Department of Anthropology, University of Colorado, Boulder, Colorado 80309.

Allen Johnson (29), Department of Anthropology, University of California, Los Angeles, California 90024

Patricia M. Kelley (189), International Linguistics Center, Summer Institute of Linguistics, Dallas, Texas 75236

Emilio F. Moran (113), Department of Anthropology, Indiana University, Bloomington, Indiana 47405

Anthony Stocks (239), Department of Sociology, Anthropology, and Social Work, Idaho State University, Pocatello, Idaho 83209

William T. Vickers (1, 451), Department of Sociology and Anthropology, Florida International University, Miami, Florida 33199

Dennis Werner (225), Pos-Graduačao em Ciências Sociais, Universidade Federal de Santa Catarina, Florianópolis, Santa Catarina 88.000, Brazil

James A. Yost (189), International Linguistics Center, Summer Institute of Linguistics, Dallas, Texas 75236

Preface

Amazonia is a region that has always had an aura of mystery and intrigue about it. The sprawling expanses of forests, rivers, and savannas that form the Amazon Basin have frustrated would-be rulers of the Inca, Spanish, and Portuguese empires, as well as countless explorers searching for El Dorado, lost cities in the jungles, and human, animal, and plant exotica. Yet today Amazonia is no longer terra incognita, as eight nations strive to develop their holdings within the basin. Paralleling this process is an explosion of scientific research on Amazonia. This volume focuses on such research from the ecological perspective within anthropology.

The native societies of Amazonia hold a certain fascination for many anthropologists. Much of this interest stems from the endemic warfare of many Amazonian people and the associated traits of cannibalism, shrunken heads, and other forms of human trophy-taking, along with Amazonian societies' rich cosmologies, varied social structures, and apparent lack of state-level political organization. Because the shifting cultivation and foraging economies of Amazonian peoples bring them into intimate association with their environments, serious students of ethnology have often turned to ecological analysis when attempting to understand and explain the foundation of Amazonian cultural dynamics. At one time or another, soil quality, protein availability, population regulation, and sexual competition have all been proposed as key influences on native adaptation. Through the mid-1950s much of the debate concerning environmental influences on Amazonian societies was frankly speculative. But in 1957 Robert L. Carneiro completed a doctoral dissertation at the University of Michigan that has had seminal impact; his *Subsistence and Social Structure: An Ecological Study of the Kuikuru Indians* was the first community-based quantitative analysis of the subsistence ecology of an Amazonian people. Since that event, both the quantity and quality of problem-oriented field investigations into the human ecology of native Amazonian societies have been impressive. This volume is based on our belief that the ecological approach to the understanding of aboriginal Amazonia is indeed a powerful perspective, and on our desire to present a collection of empirical studies that are representative of recent trends within ecological anthropology. It was conceived in the course of informal conversations at the 1979 Annual Meeting of the American Anthropological Association in Cincinnati, Ohio. Both of us were concerned about the sometimes polemic nature of the debates concerning Amazonian human ecology. At the same time, we were encouraged by the new wave of empirical research being carried out by younger scholars; we believed that a source book featuring up-to-date research of this type would be most useful for specialists and for anyone else interested in the general fields of human ecology, South American ethnology, and tropical studies. This volume was prepared with these goals in mind.

The essays presented in this book are organized into sections dealing with the major modes of Amazonian subsistence (cultivation, and hunting and fishing), nutrition, and settlement pattern. These topical headings relate to major aspects of human articulation with the environment and represent a logical order that emerged from the papers submitted by our contributors. But the individual chapters frequently transcend any static categorization and make points that apply across many aspects of human ecological analysis. Because we wanted a source book that would represent original and innovative research, we did not attempt to straitjacket our contributors into a single ecological theory. All of the authors approach their subject matter from an ecological perspective, but the reader will find studies representing the major contending theories of cultural ecology, cultural materialism, evolutionary ecology, and ethnoecology. We believe that the proponents of each of these theories have something to say and that our readers should have the opportunity to evaluate their data as well as their interpretations.

As editors, we do not seek to promote mere eclecticism, but wish to engage our readers in an interactive process of exposure to data and analytical judgment and debate. As scientists, we do have our theoretical preferences; our arguments are made in the introductory chapter and in our individually authored chapters. As scholars, we believe that the general field of Amazonian studies has made great strides in recent years, and that ecologically oriented anthropologists have been among the most vitally active of all Amazonian researchers in presenting new data and in formulating and testing theory. Yet, all who work in this area agree that much remains to be done. This volume is in large measure dedicated to the evaluation of where we are and where we should be heading as we seek to understand the complex interrelationships between man, environment, and culture in Amazonia.

We would like to thank a number of people for their assistance in this project as well as those contributors who rapidly responded to our queries concerning their manuscripts. Raymond Hames would like to thank Nina Cuèllar for compiling and editing the consolidated bibliography, Ilene Hames for editorial advice, Dorothy McEwen for under-the-gun typing, and Peter Bleed, chairman of the Department of Anthropology of the University of Nebraska, for allocation of departmental resources to the project, and Allan Osborn for encouragement. A Summer Faculty Fellowship from the Research Council of the University of Nebraska greatly facilitated Raymond Hames's work on this volume. William Vickers would like to express his appreciation to William E. Carter and Charles Wagley, who encouraged him to pursue ecological and Amazonian studies, and those at Florida International University who gave support in various phases of his work on the manuscript. The latter include Mark B. Rosenberg, director of the Latin American and Caribbean Center, James A. Mau, dean of the College of Arts and Sciences, Anthony P. Maingot, chairman of the Department of Sociology and Anthropology, Michelle Lamarre, Betty Jo Sargent, and Linda Miller. We would like to thank the editors at Academic Press for their initial faith in the project and for their understanding and patience during the longer-than-anticipated process of bringing together this research collection.

CHAPTER 1

Introduction

Raymond B. Hames and William T. Vickers

Introduction

For centuries Amazonia has captured the imaginations of many throughout the world as a mysterious and forbidding jungle inhabited by untamed savages. Perceptions such as these were largely based on ignorance, and were fueled by stories of ferocious animals, cannibalistic tribes, female warriors, and fantastic plant forms. Without a doubt, these perceptions contained kernels of truth, for the Amazon was remote and held a rich variety of life forms and cultures that were unusual from a European, temperate-zone perspective. From a scientific point of view, however, there is nothing about the Amazonian environment that is potentially unknowable. As scientific studies in the region have increased, our knowledge of Amazonia has expanded dramatically. Yet we have only begun to scratch the surface in our attempts to understand this vast geographical and culture area. The purpose of this volume is to bring together a selection of recent studies that are representative of the ecological approach to the study of aboriginal human adaptation in Amazonia. It is based on our belief that the ecological approach offers powerful tools for the analysis of sociocultural phenomena. Recent decades have witnessed an ever-escalating debate among anthropologists and other specialists concerning the significance of man–nature relationships in lowland South America, but the level of this discussion has suffered from a paucity of rigorous studies based on carefully executed research, and from a scantiness of empirical data. This volume addresses these problems by providing a collection of

ISBN 0-12-321250-2

chapters that contribute a wealth of new data on human adaptation in the tropical forests of South America. The analyses of major human ecological topics include (1) availability of vital natural resources, and the strategies used to exploit them through hunting, fishing, and horticulture; (2) how such procurement strategies in turn influence nutrition, settlement pattern, and population dynamics; and (3) the complex interplay between ecological adaptation and social institutions. Although all of the contributors are agreed on the utility of an ecological perspective, they place differing emphases on specific theoretical approaches such as cultural ecology, cultural materialism, ethnoecology, and evolutionary ecology. Hence, many of our contributors disagree on certain theoretical assumptions and specific points of interpretation, but we feel that this is healthy because it represents the current state of research and model building.

Some Characteristics of Amazonian Ecosystems

Amazonia is defined by the world's greatest river system; a basin encompassing 6 million km^2 and supplying approximately 20% of the fluvial discharge on this planet. For purposes of anthropological analysis, *Amazonia* often is defined broadly to include adjacent lowland tropical areas (such as the Orinoco River basin) because of environmental and cultural similarities to the Amazon Basin proper. The chapters by Beckerman, by Hames, and by Hill and Hawkes in this volume deal with peoples who are technically outside of the Amazon drainage, but who are located in adjacent tropical lowland areas, so they are eminently relevant to the focus of the book.

The present-day land mass comprising Amazonia has a long and complex geological history. During the Archeozoic era it consisted of two land masses that were separate bodies within the Atlantic Ocean. These Archaic massifs are today referred to as the Guyana and Brazilian shields. During the Miocene, geological forces raised the Andes range to the west and a mediterranean sea was created between these new mountains and the more ancient massifs. Gradually the area occupied by this sea was reduced through sedimentation, and the continuing formation of the Andes in the west produced the eastward flowing rivers of the Amazon Basin. As a consequence of this history, the major geomorphic regions of Amazonia include the highly weathered and crystalline uplands of the Guyana and Brazilian shields (to the north and the south of the basin, respectively), the young and rugged slopes of the eastern Andes of the upper Amazon (in the western and southwestern portions of the basin), a lowland plateau of Tertiary lake bed running across the center of the basin (bisecting the

Guyana and Brazilian shields), and the *várzea*, or modern-day flood plain, which comprises an estimated 3% of the basin (Pires and Prance 1977).

Although the soils of the Amazon have often been characterized as *lateritic* (i.e., containing plinthite or "ironstone"), and as having low agricultural potential (e.g., see Meggers 1954), recent research indicates that this is an oversimplification. In an important paper published in 1974, Falesi identified no fewer than 20 soil types in the Brazilian Amazon (Falesi 1974:202). Wambeke also emphasizes the diversity of soils in the Amazon Basin and correlates soil types with the degree of weathering to which the materials have been exposed.

> Soils exhibit differing degrees of weathering due to the interaction of many environmental variables and the chemical composition of the rocks from which the soils derive . . . plinthite, often wrongly described as a major barrier to agricultural development in the Amazon, is in fact only found to be a problem in 2% of the soils. . . . The diversity of soils . . . is also landscape specific. Topography favors particular vegetation and soil types. The greatest diversity of soils and land use potentials can occur within an average of 89 meters of altitude differential between rivers and the plateaus they divide [1978:233].

Wambeke also emphasizes that it is necessary to conduct fine-grained surveys to determine fully the distribution of soil types because of the considerable variability in distributions at the local level (1978:234).

Cochrane and Sanchez (in press, cited in Hecht 1981) indicate that the major soil orders of tropical South America are Oxisols, Ultisols, and Entisols (covering an estimated 45.5, 29.4, and 14.9% of the Amazon, respectively). *Oxisols* have physical characteristics that are favorable for cultivation, but are acidic, low in nutrients, and may have toxic levels of aluminum compounds (Hecht 1981:70). *Ultisols* are similar to Oxisols but have less-favorable structures. *Entisols* are often alluvial and offer better nutrient status in many cases. Additional, but less common, soil orders are also found and some of these have good to excellent characteristics (e.g., see N. Smith 1980).

Sioli's well-known limnological work in the Amazon has contributed another set of ecological contrasts in which white-water, black-water, and clear-water rivers have been described as having different optical qualities, pH, flow patterns, and loads of suspended solids (1975a). These variations in riverine conditions are attributed to differences in the geomorphological characteristics of the catchment area within which each river originates. Such factors significantly influence the rates of primary and secondary production in these rivers, as well as other processes such as the rates of erosion and siltation.

White-water rivers such as the Amazon proper and some of its tributaries (e.g., the Napo) derive their designations from the sediments they carry. These sediments are eroded materials from the Andean headwaters and they contribute to the formation and maintenance of the *várzea* floodplain.

Such rivers are rich in nutrients, but their turbulence and opacity limit the primary production of phytoplankton. *Black-water rivers* such as the Rio Negro (see Hill and Moran, this volume) are said to originate in areas where white sand soils are covered by *caatinga* (a type of forest that is lower and more open than tropical rainforest). Although dark due to dissolved humic matter, these waters are relatively transparent due to a lack of inorganic suspended particles. They are also nutrient-poor and acidic (Sioli 1975a:284). *Clear-water rivers* such as the Tapajós and Xingú derive from forest-covered zones of flat or mild elevation where erosion is minimal. They are poor in nutrients, like the black-water rivers, but lack the humic coloration.

Goulding has argued that Sioli's classification system "will eventually have to be tested in terms of ecological and biogeographical consequences, which themselves will demand better and more extensive hydrochemical studies" (1980:17). He finds, for example, that there is little difference in the fish species present in the lower Rio Machado (clear-water) and the lower Rio Negro (black-water), and postulates that the most significant ecological distinction is that between the white-water and the nonwhite-water rivers.

The predominant morphological forms for Amazonian rivers are braided, meandering, and channelized (Goulding 1980:17). The Amazon proper is of the *braided* type, with many secondary channels, lenticular islands, and associated bodies of water that are periodically flooded or drained by the shifting water levels of the channel. Fittkau *et al.* (1975) found that phytoplankton production is highest in decanted white waters (due to improved light transmission) and in mixed waters that receive a white-water input. Clear waters have a moderate phytoplankton density due to their low nutrient content, but overall production is relatively high due to their light transmission properties. Black-water rivers have very low production due to their poor nutrient status and poor light transmission. *Channelized* rivers tend to be relatively young and fast flowing, and have more narrow floodplains cut through the Andes or the archaic Brazilian and Guyana massifs. *Meandering* rivers are often found in association with relatively wide floodplains that are often covered with swamp forest. Numerous cutoffs or oxbow lakes straddle these rivers and provide good fishing grounds (see Stocks, this volume), particularly during periods of low water when fish become trapped in them and the volume of water is reduced.

The Amazon Basin has one of the wettest rainfall regimes in the world. Its major source of water is the Atlantic Ocean, although it is estimated that over 50% of the rainfall derives from water that is recycled within the region (Marques *et al.* 1977, cited in Salati *et al.* 1978:202). Precipitation generally increases as one moves from east to west; at Marajó Island at the mouth of the Amazon the mean annual rainfall is 2600 mm, whereas it is 3600 mm in the Vaupés region. To the north of the equator, maximum rains occur around June, and to the south, around January. Salati *et al.* (1978:205) identify four hydraulic regimes for the Amazon. North of the equator, from the Atlantic to 60° west longitude, the period of heaviest rainfall is from November through June with somewhat elevated values due to the ocean's proximity. South of the equator, from the eastern extent of the Amazon Basin to 65° west longitude, the wet and dry seasons are reversed. In the western portions of the Amazon Basin (north and south of the equator), the annual rainfall increases and the wet and dry season variations are less pronounced.

Water levels on the mid- and lower Amazon are at their lowest in October, but begin rising in November and reach their peak in June (Sternberg 1975:22). The mean annual fluctuation is 10 m at Manaus, 6–7 m at Santarem (downstream from Manaus), and only 4 m near the mouth of the Amazon (Sioli 1975b, cited in N. Smith 1981a:11). This annual rising and falling of the river is of great ecological importance as it is the mechanism by which silt is deposited in the floodplain and nutrients are infused into the *várzea* lakes. This process of enrichment enhances both the floodplain's aquatic and terrestrial productivity.

Temperatures in the Amazon are remarkably stable throughout the annual cycle, and the warm conditions, high levels of solar radiation, and high humidity create ideal conditions for plant growth. As a consequence, Amazonia is one of the most floristically complex regions of the world. Pires (1978:611), for example, identified 224 plant species in 136 genera and 52 families in a 5-ha area of *terra firme* forest near Belém. Overall, it is estimated that the Amazon has at least 50,000 species of vascular plants (Myers 1979:23). The distributions of these plants are not homogeneous. Prance, for example, proposes eight vegetation types for Amazonia (1978:212):

1. Forest on *terra firme* (nonflooded) with eight subtypes
2. Forest on inundated areas with two subtypes further broken into six categories
3. Savanna on *terra firme* (nonflooded) with eight locational categories
4. *Várzea* savannas (flooded)
5. *Campina* (low vegetation on white sand)
6. Montane vegetation on the fringes of Amazonia
7. Littoral vegetation (coastal dunes, etc.)
8. River beaches

The various subtypes of *terra firme* forest are estimated to cover about 85% of the Amazon Basin. They include high forest (with high biomass), liana forest, low forest (with reduced biomass), *campina* or *caatinga* forest (on sandy soil), dry forest (in transitional areas), montane and cloud forest (in elevated areas), bamboo forests (such as in Acre), and miscellaneous types of low, sparse forest. So the Amazon is not only complex in terms of its floristic composition, but also in terms of its varied associations that are influenced by differences in rainfall, soils, drainage, elevation, and proximity to the sea.

South America as a whole has an extremely high diversity of animal forms and is notable for its large number of archaic endemic species. Gilmore (1950:348) attributes this endemism and richness to the continent's (1) large area, (2) diverse environments, (3) isolation during the Tertiary period, (4) good stock of animals prior to this isolation, and (5) peripheral position to the Holoarctic land mass. In the lowland tropics the arboreal and aquatic forms are particularly notable and those of Amazonia are the most diverse on the planet. The Amazon is also notable for its many edentates (sloths, armadillos, and anteaters), primates, and reptiles. For some time, zoologists have recognized that the tropical forest regions of the world tend to have a somewhat lower biomass of terrestrial mammals than temperate regions, and relatively few species of large ungulates (Bourliere 1963; Eisenberg and Thorington 1973). The fact that the tropical rainforest represents a climax association with the bulk of the biomass tied up in the wood of trees limits the availability of nutrients to terrestrial herbivores. Arboreal hervibores such as sloths and monkeys comprise much of the mammalian biomass. There is an abundance of rodent forms in Amazonia, and their densities may be comparable to rodent densities in temperate zones (Fleming 1971, cited in UNESCO 1978:163).

Most of the terrestrial mammals of Amazonia have solitary or paired habits. A notable exception is the white-lipped peccary (*Tayassu pecari*), which organizes into nomadic herds of a hundred or more individuals (Kiltie 1980). Monkeys display both family and troop types of social organization. The population densities and structures of many mammal and bird populations vary according to the temporal and spatial availabilities of food in the environment.

Only about half of the estimated 2500–3000 fish species of Amazonia have been described scientifically. Approximately 80% of the species are characins and siluroids (Goulding 1980). As with the terrestrial and arboreal animals, there are a number of archaic endemic forms, including the *paiche*, or *pirarucu* (*Arapaima gigas*), which reaches weights in excess of 100 kg. The interesting adaptations of many fish species to Amazonian conditions include feeding on the fruits and seeds of terrestrial plants that have fallen into the water, and seasonal migrations related to feeding and reproductive behavior (Goulding 1980). Overall, the fishes of Amazonia represent all of the major trophic levels since they include planktivores, frugivores, carnivores, and detritivores (Goulding 1980:30). Turtles (and their eggs), caimans, and manatees are additional aquatic animals that are significant sources of food for aboriginal peoples.

Research of recent years indicates that the ecology of the Amazon is far more heterogeneous than previously supposed. As we have just outlined, new work by soil scientists, botanists, zoologists, and other scientists is presenting us with new data. These data indicate that Amazonia really consists of a highly complex and variable series of habitat types whose outlines and dynamics are only beginning to be recognized and understood. This is one reason why Amazonia is such an exciting place for scientific investigation, and also why many researchers feel a strong sense of urgency in their work as the region undergoes development. It seems certain that as these studies continue, many of our assumptions and generalizations about Amazonian environments and cultures will be modified and improved upon.

For example, Denevan (1976) has proposed a classification of habitat types for analyzing aboriginal demographic patterns that is similar to, if somewhat simpler than, Prance's classification of vegetation types (described earlier). Denevan's seven human habitat types are Floodplain, Brazilian Coast, Lowland Savanna, Upland Forest, Upland Savanna, Lowland Forest, and Uninhabitable. Even Denevan's classification probably does not indicate the true degree of variability in Amazonian habitats. Furthermore, it is highly likely that specific Amazonian communities exploit a range of biotopes in much the same manner as Nietschmann has described for the Miskito of Nicaragua (1973). As Vayda and Rappaport (1968) point out, the proper starting point in the detailed analysis of human ecological adaptation is the community and the activities of its members within a defined space. It is within this context that anthropologists can survey resources, quantify inputs and outputs, study demographic trends, and investigate historical circumstances. As we describe subsequently, the paucity of such community-based ecological studies has been a weakness that has plagued the debate on the constraints on aboriginal adaptation in Amazonia. In a large measure this volume developed out of a sense that this gap needed to be addressed. However, we would like to stress that our goal is not to particularize Amazonian human ecology. The point of studying individual communities is not to show that each represents a unique repertoire of behaviors adapted to a unique complement of biotopes. Instead, community and regional studies are employed to test hypotheses concerning general adaptive processes in Amazonian societies.

Environmental Models of Amazonian Sociocultural Evolution

Ever since Julian Steward's masterful editorship of the *Handbook of South American Indians* (seven volumes, 1944–1949) the Amazonian habitat has been cited by cultural ecologists as a prime example of how the environment determines and limits sociocultural evolution. Lowie, one of Steward's collaborators on the handbook, stated in his summary chapter on the social and political organization of tropical forest tribes: "In the Tropical Forest area there is almost every type of social structure, yet the variations in economic activities are trifling" (Lowie 1949:331). Lowie had in mind the relationships among rather specific aspects of social organization such as postmarital residence rules and conceptions of descent. However, Steward's approach was much more general—that is, he focused on the relationship between (1) core elements of culture and levels of sociocultural integration, and (2) technology and gross environmental factors (Steward 1955:34). Steward (1949:677,698) suggested that environmentally richer riverine and coastal habitats permitted large and socially stratified settlements to form. These settlements could be integrated on a large scale due to the ease of water transportation. However, the tropical forest *terra firme* horticulturalists and the so-called marginal foot nomads of Amazonia could not evolve socially because either no surplus could be produced or food could be obtained only in limited amounts (Steward 1949:669).

Steward developed his position more fully in *Native Peoples of South America* (1959), authored with Faron. They divided the societies of the Amazon into three different sociocultural types, each associated with a particular environmental regime: (1) small single-lineage villages with populations less than 100 were located near small streams and away from large rivers; (2) large multilineage villages of more than 1000 inhabitants were found along the Amazon and its major tributaries; and finally, (3) the marginal foot nomads (Steward and Faron 1959:373,424) lived in small groups of 25–50 individuals, were organized as composite or family bands, and occupied "regions that were economically marginal, inferior, or relatively unproductive" (Steward and Faron 1959:374). Steward and Faron postulated that chiefdom or state-level societies could not easily evolve in Amazonia because of poor soils, sparse faunal resources, lack of efficient transportation, and isolation from the main centers of cultural invention (e.g., the Andean region). In their framework, poor soils led to shifting cultivation that did not allow dense, permanent, or aggregated populations. Furthermore, shifting cultivation did not allow a surplus of food to be produced to support a stratified society (Steward and Faron 1959:61). They did recognize that root crop cultivation was somewhat productive, but noted: "These crops, however, provide predominantly starchy foods, and while no adequate analysis has ever been made of the diet, there is little doubt that the people had a need for protein foods which were obtained principally by hunting and fishing" (Steward and Faron 1959:291). Interestingly, regarding the cultural and physiological determinants of behavior, they noted that "when cultivated foods provide a badly unbalanced diet for which hunting and fishing can compensate it is quite possible that the craving for meat, rather than mere tradition led men to hunt" (Steward and Faron 1959:292). Steward and Faron believed that the marginal foot nomads could not achieve even the modest levels of egalitarian organization of the *terra firme* populations because they "barely met minimum subsistence needs and often fell far short of them" (1959:60), and all of their time was spent in the food quest, which allowed them little time for nonsubsistence activities.

The paradigm developed by Steward has served as the basic model for nearly all subsequent ecological theorizing on Amazonian societies. As we shall show, Steward's rather vague model has been more fully developed by detailed ethnographic research and greater knowledge of neotropical ecology. However, his basic observations on soil and protein problems and their relation to *várzea* (or fluvial) and *terra firme* (or interfluvial) populations are still central. Steward's model forced him to assume much that was not known concerning subsistence adaptation in Amazonia. For him, low levels of sociocultural integration and low population density were proof of either an unproductive environment, inefficient technology, or both. Refinements of Steward's model, examined subsequently, still face this problem of scarce empirical research. Recent cultural ecological theory has tended to stress one of the two main components of Steward's model: poor soils (Meggers 1954, 1957, 1971; Oberg 1973; Roosevelt 1980) or protein problems (Carneiro 1970a; Gross 1975; Lathrap 1968; Ross 1978).

Oberg's (1955) explanation for differential cultural evolution in Amazonia emphasizes one of Steward's environmental factors—quality of soil for horticulture. Amazonian societies are divided into homogeneous and segmented tribes, which parallel Steward's lineal and multilineal tribes, respectively. He argues that good soil leads to the production of a surplus, which supports greater population aggregations and stratification, and which leads to an increase in leisure time for culture building (Oberg 1955). Additionally, in a rather haphazard fashion, he alludes to subsidiary factors, such as a diverse environment and a varied crop inventory (Oberg 1955), that promote aggregation and cultural evolution. These other factors, along with good soil, account for chiefdoms and higher levels of sociocultural integration outside of Amazonia.

Meggers takes up the theme of environmental determinism in a series of important publications (1954, 1957, 1971). In a much clearer fashion than Steward or Oberg, she divides Amazonia into two basic habitats, *várzea* and *terra firme* (Meggers 1957), and argues that silt-rich *várzea* soils enabled denser and more advanced societies to develop compared to *terra firme* soils. Her major theoretical contribution to Amazonian human ecology came in 1971 with the publication of *Amazonia: Man and Culture in a Counterfeit Paradise*. For the first time, the soil poverty hypothesis was systematically tested using higher quality data than Steward or Oberg had at their disposal. Data on social organization, village size, population density, and village permanence are cited to evaluate her thesis. Significantly, she uses the concepts of carrying capacity and environmental equilibrium to account for a number of cultural practices such as warfare, infanticide, and sorcery, which are argued to be adaptive responses to environmental problems (Meggers 1971:106). On occasion she implies that protein shortage may be an adaptive constraint (e.g., Meggers 1971:26–27), but its role never emerges clearly.

We are indebted to Carneiro for the first thoroughly modern cultural ecological study of an Amazonian society (1957) and the first critical and quantitative examination of the soil poverty hypothesis (1960, 1961). Carneiro (1960) developed a measure of agricultural carrying capacity to determine how large a permanent village could grow without irreversibly depleting soil resources. By doing so he was able to demonstrate forcefully that something other than agricultural soil limited the size, permanence, and hence cultural evolution of Amazonian societies. Carneiro later answered his own question in terms of the effects of protein needs (1970a) and circumscription (1970b). In his characteristic style Carneiro developed quantitative indexes of sedentariness and subsistence and tested the relationship on a sample of seven Amazonian societies. He concluded that a high dependence on fishing and agriculture led to larger and more permanent village aggregations whereas a high dependence on hunting and gathering led to smaller and more mobile villages. The rich fisheries found in large rivers emerged as the main factor that led to large and complex societies because it lessened dependence on dispersive hunting and gathering.

Early on, Steward (1949:665) remarked, "One has the impression, however, that vast regions of potentially arable land were never utilized and that the population may have been expanding at the time of conquest"; and "many other tribes similarly failed to achieve comparable size which their economy would have permitted." This observation presages Carneiro's (1970b) theory of environmental circumscription. One of the reasons tropical forest settlements in the *terra firme* habitat never became large or permanent was that they were not bound by geographic barriers. And, as a result, they could adapt by expanding into virgin environments that permitted them to maintain an extensive subsistence base rather than selecting an intensive one. Carneiro argues that sociocultural evolution frequently occurs when a population is boxed into a restricted habitat and is forced to intensify subsistence through a variety of social and economic mechanisms. In Amazonia the narrow *várzea* floodplain represents such a circumscribed habitat, whereas the vast open *terra firme* habitat (occupying approximately 97% of Amazonia) is relatively uncircumscribed. In integrating Carneiro's resource concentration model (1970a) and circumscription model (1970b), we find that larger and more complex settlements arose in the *várzea* because it contained concentrated aquatic resources and was circumscribed by the *terra firme* habitat, which was relatively impoverished of protein resources. *Terra firme* societies, on the other hand, remained small because they could adapt to resource depletion by simple geographic expansion (Hames, this volume).

Lathrap (1968) uses an ecological historical approach to model Amazonian adaptations. He argues that hunter–gatherers initially entered the Amazon Basin along its main river systems. In these fluvial habitats aquatic resources supplied dietary protein and fat, and silt-rich *várzea* soils were adequate for near-permanent cultivation, which permitted the development of large and complex societies. As the population grew, segments were forced into fluvial or *terra firme* habitats and subsequently devolved because they were forced into unprofitable hunting and extensive shifting cultivation to supply their subsistence needs. Lathrap's argument, particularly his emphasis on differential protein richness between *várzea* and *terra firme*, stimulated a number of researchers to test his generalizations about protein resource availability (Gross 1975, 1982), population density differences in *várzea* and *terra firme* habitats (Denevan 1970, 1976), and to extend his model of warfare between *várzea* and *terra firme* groups over aquatic protein to warfare among *terra firme* groups over terrestrial protein (Durham 1976; Harris 1977; Morey 1970; Ross 1978).

In pursuing the protein hypothesis, Gross (1975) accepts Carneiro's (1961) argument that poor *terra firme* soils did not limit cultural evolution but, following Lathrap (1968) and Carneiro (1970a), a lack of animal protein did. To demonstrate his point, Gross collected quantitative data (see Gross 1982, for a new tabulation with a larger sample size) on a sample of eight neotropical native societies (three of which were outside the Amazon Basin) and concluded that although their protein intake was adequate, it was not impressively high. He suggested that if these populations were to have grown more dense, they soon would have become limited by a decline in protein consumption. As a result of this environmental fact, they instituted cultural mechanisms to control population growth and dispersion (for a list, see Meggers 1971:106). Importantly, for the first time, Gross noted that manioc and other staple crops are deficient as complete protein sources in their amino acid composition and populations who depend on such crops must heavily supplement their diet with protein of animal origin. (Following Spath [1971, 1981] he states that methionine may be the ultimate limiting amino acid). In contrast, the limits to population density are higher in *várzea* habitats due to a more plentiful supply of fish and game resources. Finally, the option of either raising protein-rich crops or domesticating wild animals to offset natural protein limitations were not available to all Amazonian populations. Gross's work is important in Amazonian human ecology because, instead of using impressionistic data and being theoretically vague, as in some past models, he supports his arguments with quantified data on plant and animal food values, fish and game biomasses, and native protein consumption, and he attempts to build a rigorous deductive model.

Roosevelt's (1980) archaeological monograph is the latest entry into the field of differential cultural evolution. After a painstaking review of the work of her predecessors, she develops a kind of synthesis of the soil and protein models. Applying the Mesoamerican model (Sanders and Price 1968), Roosevelt simply argues that the ability of *várzea* groups to grow maize and beans in alluvial areas was the primary factor that allowed the development of stratified societies. (Interestingly, Sanders and Price [1968:91–92] noted that higher population densities in parts of lowland South America were related to maize cultivation.) Roosevelt generally concurs with Carneiro, Lathrap, Gross, and Denevan that *terra firme* societies could not become large or socially stratified because of the dispersive tendencies of extensive shifting cultivation and terrestrial hunting. But, contrary to these theorists, Roosevelt argues that the *várzea* societies solved their protein needs not only by utilizing more bountiful aquatic protein resources, but rather (following Meggers's thinking to some extent), by using *várzea* soils that allowed the cultivation of maize to supplement protein needs; this led to the following chain of events: (1) increase in population density, (2) intensive agriculture, (3) the need for managerial authority to coordinate and organize production, (4) more efficient use of aquatic resources, and (5) the production of surpluses to support social stratification.

Environmental Models of Amazonian Adaptive Behavior

Most of the work mentioned previously focused almost entirely on broad aspects of differential cultural evolution. In contrast, publications by Harris (1977), Ross (1978), Gross (1979), and Zarur (1979) have attempted to explain variability in particular cultural traits (e.g., warfare, male dominance, game taboos, and combined systems of age-grading, moieties, and lineality in *terra firme* populations) as adaptations to scarce protein resources or other environmental constraints. Initially, Morey (1970) developed a resource competition model of warfare for the Colombian Guahibo with protein as a prime ingredient, and later (Morey and Marwitt 1978) attempted to evaluate it against Amazonian data from the Human Relations Area Files (HRAF). The results were inconclusive but the authors stuck to their model after some ad hoc modifications. Following a suggestion by J. Ross and E. Ross, Harris (1974:102) argued that Yąnomamö warfare in particular and Amazonian warfare in general was a form of competition over scarce game resources. Harris has developed his position over a number of subsequent publications (Divale and Harris 1976; Harris 1977, 1979a, 1979b, 1980), partially in response to critics and partially to make it more general. According to Harris (1980:220–221), increases in village size or regional density lead to a decrease in either hunting efficiency or protein consumption, which, if permitted to proceed unchecked, would jeopardize the survival of the population. Warfare counteracts these trends by promoting a need for male combatants, resulting in female infanticide (thus slowing the rate of population growth). It also stimulates the dispersion of settlements, which has the effect of creating game preserves between village hunting territories to maintain a stable equilibrium between humans and their game supply (Harris 1979b:91).

In the course of developing an elegant and general model of warfare as a mode of competition over limiting resources, Durham (1976) provisionally tests it with Mundurucú data. He identifies protein as the limiting resource and suggests that warfare functions to enhance individual reproductive fitness by reducing competition from neighboring villages and creating no-man's-lands between villages, which act as game preserves. Although he identifies protein as the limiting resource (see Chagnon and Hames 1980 for an extended critique of his use of Mundurucú data), his model is radically different from Harris's (1979b) or Ross's (1978, and following) because it is based on natural selection theory, it attempts to deal with the costs and benefits to individuals engaging in warfare in a sophisticated way, and it is much more parsimonious.

Ross's (1978) model of game taboos directly confronts a paradox in the protein limitation model: if game is so scarce for *terra firme* societies, why do they prohibit the hunting and/or consumption of many large game species? Although his reasoning is convoluted, Ross assumes that the goal of this hunting strategy is to maintain an equilibrium between human and game populations. In low-density environments, large game animals are not tabooed because they are abundant and are easy to hunt. As population density and game depletion increase, large game become more difficult and costly to obtain, so hunters increasingly focus their attention on smaller game species that have higher reproductive rates and, as a result, represent a more stable hunting strategy. Large game is tabooed to prevent its extinction and to enhance hunting efficiency; presumably the taboo is lifted when their numbers have recovered. In the course of developing the model, Ross additionally argues that warfare functions to disperse settlements and create no-man's-lands between village hunting territories.

The Central Brazilian Gê linguistic groups present a puzzle for ethnologists. They were classified as "marginals" by many contributors to the *Handbook of South American Indians*, but according to Steward and Faron (1959:362), "while their villages are independent sociopolitical units, each is divided into a number of kinship divisions, moieties, age grades, associations and sexual dichotomies unequaled in South America." Structural anthropologists have attempted to explain their unique social organization in strictly cultural and psychological terms. For example, Maybury-Lewis (1967) explains Shavante social organization in terms of the strain and ambiguity inherent in the male role and the institutionalization of aggression and sexual antagonism. And Bamburger (1968) claims that differences in Kayapo and Xingú environmental adaptations are a function of different world views (see Ross [1978] for a critique, and Werner [this volume] for an account of structuralist's explanations of trekking and his ecological resolution). In directly confronting the structuralists' position, Gross (1979:322, this volume) suggests that Gê sociocultural complexity functions as a "cultural means for effectively integrating semi-autonomous foraging units (and later refugee groups) into large unified village units." Gross (1979:330) argues that because resources in the *cerrado* (scrub forest) are unreliable and highly seasonal, Gê societies have developed mechanisms that permit easy dispersion and aggregation to exploit more effectively the environment. In addition, chronic warfare selects for large village size and the solidarity of large villages is enhanced by complex crosscutting social institutions. Zarur has also attempted to develop a cultural ecological model of Gê social organization by contrasting it to Xinguano social organization. He argues that Gê dependence on a single protein capture technique (i.e., hunting) leads to a complex division of labor based on age-grades to make hunting more efficient and that these age-grades provide the basis for an effective military organization (Zarur 1979:651). In contrast, Xinguano groups, have diverse protein resources available but choose fish, which leads to a simple division of labor but not to the development of effective military organizations. As a result of different protein capture techniques, Gê groups are larger, more mobile, and more warlike than Xinguanos.

From our account of ecological theories of Amazonian social organization and subsistence behavior, it would appear that ethnologists are of three minds in the identification of the major variable to which a population must adapt (1) soil quality for horticulture (Meggers 1971; Oberg 1973; Roosevelt 1980; Steward 1949); (2) game capture for meeting protein needs (Carneiro 1970a; Gross 1975; Lathrap 1968; Ross 1978); and (3) its own cultural institutions (Bamburger 1968; Maybury-Lewis 1979; Turner 1979). This three-fold typology oversimplifies, however. For example, Steward (1949), Lathrap (1974), and Meggers (1975) employ shopworn diffusionary arguments in their ecological models. Gross (this volume) suggests that although protein is crucial, it is not the only variable of interest (see also Carneiro 1970b; Ross 1978). Since the publication of most of the previously mentioned work, there has been a new wave of intense ethnographic research on native human ecology and a generalized reaction to established theories, which we shall examine.

One of the main problems faced by those who have developed ecologically based theories of social organization is the lack of high-quality ethnographic research on ecological problems of native Amazonians, and a similar lack of empirical data on natural ecology. As a result, theories cannot be evaluated unambiguously because they are developed by gleaning descriptions from explorers, missionaries, or ethnographers whose work is not ecologically inspired. A clear example of the unreliability of nonecological ethnographic research is Aspelin's (1975) critique of Levi-Strauss's and Oberg's work on the Mamaindê (Nambiquara). He convincingly shows that nearly all of their important observations concerning subsistence and environmental adaptation were inaccurate. Similarly, Lee (1968:46) used data from the HRAF to claims that the Yąnomamö (named in the HRAF as Shiriana, which is what they are called by the Ye'kwana) are hunter-gatherers, which they are not (Chagnon 1968a). Murdock's assessment of South American hunter-gatherers is also riddled with errors (1968:19). Other examples could be cited. One of the main purposes of this volume is to produce reliable, accurate, and sophisticated data on Amazonian adaptations which can be of use for those who hazard grand theory building.

Another problem in comparative research is intracultural variability. For example, Vickers (1976) found that the Siona–Secoya of the Cuyabeno River in Ecuador use fish and other aquatic animals as their major source of dietary protein, while people of the same culture living on the Aguarico River (only 40 km distant) derive most of their protein from terrestrial mammals. The observed differences in the patterns of protein acquisition are entirely attributable to differences in the terrestrial and aquatic habitats of the two areas. Hence the Siona–Secoya do not have a single rigid adaptation, but draw on those aspects of their technological repertoire that best fit a given locale. If a researcher were to work in but one of the communities, he might develop a skewed impression of the nature of Siona–Secoya cultural ecology. The point that we wish to make here is that other Amazonian societies likewise adapt their cultivation, hunting, fishing, and collecting practices to specific habitats, and the interactions between resource availabilities and exploitative activities exert a powerful influence on settlement pattern and sociopolitical organization. One fact that we must bear in mind is that the cultures that we study today do not represent the full range of aboriginal Amazonian societies. Today we are dealing with groups of survivors who are for the most part located in the peripheral regions of the Amazon Basin and, to a variable extent, have been affected by disease, economic exploitation, missionization, and introduced technologies. The most glaring omission is the absence of the once numerous societies of the middle and lower Amazon floodplain. Fortunately, these constraints do not totally preclude the study of these *várzea* societies because we can still learn much from their archaeology, from various historical accounts, and from the continuing scientific research on the ecology of the region. Likewise, the surviving groups whom we study provide essential information on human adaptation in Amazonia in the past as well as in the present, for most of them still employ traditional subsistence technologies as the basis of their economies. The need to trace and evaluate the degree to which the conquest and subsequent events affected aboriginal adaptations is simply another of the anthropologist's tasks.

One of the first major critiques of the protein limitation position was made by Beckerman (1979) in response to Gross's (1975) earlier article in the same journal. Beckerman suggests that the importance of wild and domestic sources of vegetable protein available to Amazonian populations has not been evaluated, that estimates of fish and game biomass have been underestimated, that pre-Colombian population density figures (e.g., Steward and Faron 1959:51–60; Denevan 1976) have been underestimated by ignoring the terrible impact of western diseases, and that there is a general lack of empirical research on protein capture. In a reply to Beckerman, Gross (1982) suggests that although vegetable and insect protein may be available, they may be of little use to a population from a cost–benefit perspective. He further suggests that Beckerman's own research (1980), and Lizot's (1977), confirm his position (as well as Lathrap's [1968]) concerning protein availability in *várzea* and *terra firme* habitats. Later, Lizot (1977, 1978a), and Chagnon and Hames (1979), attempted to supply needed data on protein consumption and its costs. Lizot documents a number of distortions and speculations that Harris (1974) made about Yąnomamö culture and supplies data on Yąnomamö protein consumption. Lizot then concludes, through a comparison of two Yąnomamö villages, that warfare, settlement pattern, and alleged high rates of female infanticide (see especially Lizot 1978a:80) have nothing to do with competition over scarce protein resources. Chagnon and Hames (1979) make similar points about the Yąnomamö, but more importantly supply comparative data on animal protein consumption in 11 Amazonian populations to demonstrate that Gross's earlier estimates (1975) were too low and were generally based on inferior data (see Gross 1982 for new data and a reply).

To a certain extent, the protein issue has led to a number of publications examining hunting efficiency as it varies among individuals, under different technologies, and among different societies. Employing Wynne-Edwards's (1962) group selection model in her study of the Shranahua, Siskind (1973) posited that culturally created scarcity of nubile females led men to compete among themselves for mates by trading naturally scarce game for sex, and this dynamic induced warfare and other mechanisms to disperse populations in relation to protein supplies. Vickers (1975) countered by quantitatively comparing 15 Siona–Secoya hunters to show that there was little difference in hunting abilities and no discernible pattern of meat–sex exchange in their society (see Symons 1979 and Hawkes et al. 1981 for support of Siskind's empirical position, but from an individual, sexual selection perspective; Aspelin [1979:325] concurs with Vickers's interpretation).

Because most native Amazonians now utilize shotguns, steel fishhooks, and other foreign technological aids, it is important to know how these aids have affected the efficiency of protein capture and what effect this increased efficiency has had on consumption, labor, and other activities. In a controlled comparison of shotgun-hunting Ye'kwana and bow-hunting Yąnomamö, Hames (1979) demonstrated that the shotgun was more than twice as efficient as the bow (in kg/hr of hunting) and even more effective when used in conjunction with headlamps for night hunting and outboard motors for travel to distant and undepleted hunting locations. Several tentative conclusions were drawn: (1) increased hunting efficiency has led to a decrease in time allocated to hunting; (2) as a result, the rate of game depletion has not increased; and (3) protein requirements appear to be relatively inelastic. Yost and Kelley (this volume) importantly provide the first quantitative data on the efficiency of the blowgun and spear, as well as comparative data on the shotgun for the recently contacted Waorani. Their results parallel Hames's but differ in some important ways (see subsequent material). But, for the hunting and gathering Aché, Hill and Hawkes (this volume) show that the increased hunting efficiency of the shotgun has led to increased hunting and probably higher levels of game depletion. Although it is clear from these studies that western technology increases the efficiency of protein capture, its effects in other areas of life are problematic.

Studies of time allocation, yield, and output–input ratios (kg/hr) show considerable variation within and among groups. Through a comparison between Ye'kwana shotgun-hunters and Yąnomamö bow-hunters, Hames (1979) demonstrated that hunting success (in kg/hr) was negatively correlated with time allocated to hunting. However, among Ye'kwana hunters there was no relationship between hunting efficiency and hunting time but there was a strong negative relationship among Yąnomamö hunters. Just the opposite conclusion was drawn by Werner *et al.* (1979) in a comparison of four different Central Brazilian societies exploiting different habitats; time allocated to hunting was positively associated with hunting success. Yost and Kelly (this volume) show that a switch to more efficient hunting technology has led to a slight diminution in time allocated to hunting, but this may be statistically insignificant. Stocks (this volume) correctly notes that these two strategies are of time minimization (efficiency depresses time allocation) and energy or protein maximization (efficiency increases time allocation), and he shows through a seasonal analysis of Cocamilla hunting and fishing activities that they are clearly at the time-minimization end of the continuum. But research by Hill and Hawkes (this volume; Hawkes *et al.* 1981) clearly demonstrate that the Aché are energy or protein maximizers. The issue of time minimization or energy maximization is emerging as an important issue in Amazonian ecology. The issue promises to be complex. For example, after a partial analysis of his data, Vickers believes that efficient hunters hunt more frequently than inefficient hunters; however, at the same time he has documented that hunting effort increases as its efficiency decreases (1980). In order for us to make progress in this area, researchers will have to specify under what conditions time-minimization or energy-maximization strategies occur and to what use protein is put. For example, if Amazonian hunters and fishermen are energy or protein maximizers, then one could conclude that protein consumption is either a crucial and limiting dimension of the environment, as argued by Werner *et al.* (1979) and Gross *et al.* (1979), or a means to acquire mates (Siskind 1973; Hawkes *et al.* 1981). But, if they are time minimizers, it could mean that after a certain level of protein intake is attained, time may be more profitably diverted toward other activities. Finally, although hunting time allocation is just one part of total time allocation, starting from this base may enable us to develop a general model of time allocation by using, for example, indifference curves and other techniques from microeconomic theory (Winterhalder 1982).

One of the main problems in developing ecological models of human subsistence behavior is the incompleteness of research in population ecology. For example, for lack of alternatives, Gross (1975) had to rely on the work of Fittkau and Klinge (1973) for estimates of huntable biomass. Yet, as Beckerman (1975) points out, their data are of little value because the area censused was located 64 km from Manaus and had been heavily hunted with modern firearms because it was accessible by highway. (Similar objections apply to Gross's use of Odum's and Pigeon's [1970] El Verde data, and Ross's [1978] use of the Barro Colorado Island data [Eisenberg and Thorington 1973].) Use of data from population ecologists does have its place in human ecology, but without high-quality data on human use of the environment, such data are almost useless since they can be easily twisted to suit nearly any position. The Amazon Basin is larger than the continental United States and there is certainly reason to believe that it is not a uniform environment. Therefore, data from an artificially created island outside the basin (Eisenberg and Thorington 1973) or from a hunted-out portion of the Amazon 60 km from a city of over 300,000 must be used with caution. For example, Stocks (this volume) shows that the Cocamilla catch fish at an annual rate of 17.6 metric tons/km^2,which is three times higher than Gross's (1975) estimate for sustained yields in the rich Xingú fishing grounds. Also, Hill and Moran (this volume; Moran 1981) and Vickers (this volume) cite recent research that indicates that Amazonian soils are extremely variable in their fertility, which counters the widespread belief (Meggers 1971; Roosevelt 1980) that *terra firme* soils are uniformly impoverished.

Concern with what the environment has to offer native Amazonians was first expressed in a rigorous way by Carneiro (1960) in his carrying capacity formula for shifting cultivation. He used this formula to calculate the maximum number of individuals that could be supported indefinitely within a particular area without modification of horticultural practices. Although Carneiro's limited use of the concept is instructive, it does not correspond to the biological ecological definition of the term and has led a number of cultural ecologists to assume that adaptation is the maintenance of equilibrium with the environment. Evolutionary ecologists define carrying capacity (or *K* in the classic Verhulst–Pearl logistic equation) as the "maximum number of individuals that the environment can support on a given area" (Emlen 1973:235). It is achieved when "a population in a given area exists in a state of stable equilibrium through time, that is when natality is equal to mortality and the population ceases to grow" (Pianka 1978:114). Thus, carrying capacity can only be guessed at when estimating the quantity of some limiting factor, but it can be known when a population ceases to grow. The basic modeling procedure in many of the environmental approaches we have reviewed is to assume some limiting factor and to assume further that the goal of a local population is to limit itself in order to harvest the limiting resource on a perpetually sustained basis. The quantity of the limiting resource is then crudely estimated and any behavioral process that alters a population's density, structure, or geographic distribution is called an adaptation. Although this approach is heuristic, it suffers from a number of problems if followed blindly.

In biology, ecologists use the term *single factor thinking* to characterize assumptions that a single factor limits a population (Pianka 1978:14, 125); in anthropology it is sometimes called *monocausal determinism.* A. Johnson (1974) and Vickers (1976, 1980) note that many factors, some of which interact, limit Amazonian populations, and that identification of limiting factors is a matter of empirical investigation. This is not to suggest that simple models are always inappropriate or that a single factor cannot be extremely important (see Gross, this volume), but rather that given our poor knowledge of Amazonian subsistence and demography, one is advised to proceed with caution. The second assumption, that a population adapts in order to maintain a balance between its demands on an environment and the environment's ability to meet those demands, is commonplace in cultural ecology theory (e.g., Vayda and Rappaport 1968). This assumption is consistent with, and to some extent inspired by, the systems ecology of E. Odum (1969) and H. Odum (1971) and the group selection model of Wynne-Edwards (1962). In this sense adaptation is defined as any process that maintains the demographic balance of the population or entire ecological community. These cybernetic approaches have been rather strongly challenged by the biologist Williams (1966), which has led to the strengthening of a neo-Darwinian individual model of selection in ecology (Pianka 1978). Although group selection is possible (Alexander and Borgia 1978), such a group would be designed by natural selection to compete and expand reproductively against other groups and not to maintain perpetual demographic equilibrium. But even more importantly, at the empirical level, Irons (1980) points out that Wynne-Edward's model has never been convincingly demonstrated for any human population (see also Bates and Lees 1979).

The effects of limiting resources can be studied only diachronically or through controlled comparisons. Since it is extremely difficult to measure how a resource limits a population (i.e., demonstration of a decline in the growth rate), studies must index the effects of hypothesized limiting resources in terms of either increased energy and time required to gain resources, or the decline in its rate of consumption, or both. Then comes the hard part: the demonstration of changes in subsistence behavior or demographic adjustment that are causally linked to resource depletion. This can be done only by studying a single village through time or by comparing a set of villages occupying similar habitats but that vary in size. (We assume that resource depletion tends to proceed more rapidly than resource renewal, and that humans respond rather quickly to depletion by altering their behavior.) Thus far, few studies of this nature have been made. For example, it is commonly asserted that *terra firme* groups respond to declining crop yields by clearing new areas of forest for gardens; or, when gardens become increasingly distant from the settlement, the settlement is moved in order to shorten walking distance to gardens, making horticulture more efficient (for examples, see Hames this volume, and Vickers this volume). Although this is undoubtedly true, we have but a few studies that only partially document this process (Hames in press a; Johnson this volume). The effects of game depletion has emerged as an important issue, but Vickers (1980; Hames and Vickers 1982) has made the only diachronic study of its effects directly, and Hames (1980) has done it indirectly through locational analysis (see also, Chodkiewicz 1980). In addition, we have but one study that links resource depletion to changes in subsistence technology, which was made by A. Johnson (1978) for the Machiguenga. Regional studies have been steadily emerging from the Hunter College research team headed by Gross (Flowers *et al.* in press; Gross *et al.* 1979; Werner *et al.* 1979). Strong environmental controls are lacking in these studies; nevertheless, they are giving us a much clearer idea of how populations respond to resource depletion, intrinsic environmental variability, and market incentives. Problem-oriented, empirical studies are beginning to tell us the conditions under which resource depletion leads to intensification of subsistence, demographic readjustment, and sociocultural change.

Some of the contributors to this volume use the protein debate as a springboard for the presentation of their research (Gross; Hill and Hawkes; Stocks; Yost and Kelly), but this work is not aimed primarily at devising ecological explanations of major aspects of social organization (for exceptions, see Hill and Moran this volume, and Werner this volume). Rather, most of the work in this volume can be characterized as microecological in scope (Pyke *et al.* 1977). Microecology seeks to understand the immediate problems individuals face in exploiting the environment. As a result, contributors attempt to specify carefully the local environment and historical context in their presentation of detailed quantitative data on time allocation, resource yields, returns to labor, and consumption. In taking this direction, researchers are forced to consider individuals, possible alternatives, goals, and the factors that influence subsistence decisions. One of the major reasons behind this shift is that, until recently, this sort of work has never been done in Amazonia. Also, major ecological theories of Amazonian cultural evolution and social organization have lacked the adequate empirical underpinnings that these studies can supply. For example, it was not until 1970, when LaPointe's doctoral thesis was written, that we had any relatively complete data on time allocation and yields for basic subsistence activities. In the early and middle 1970s this fact became apparent, so that today we have no less than 22 separate quantitative studies dealing with major aspects of Amazonian subsistence (Hames in preparation), with more in progress.

The strong emphasis on microecological processes has led to the development of better research methods. Beginning with Carneiro's (1957) pioneering work, it was customary for ethnographers interested in ecology to measure a few gardens, roughly estimate time allocation on a few "typical" adults, and qualitatively describe other subsistence activities. Recent research, some of it found in this volume, is employing sophisticated ethological sampling techniques such as instantaneous scan sampling (Altmann 1974) or spot checks (A. Johnson 1975). These methods allow the behavior of all members of a village population to be sampled (including infants and juveniles) in order to answer questions concerning individual behavioral variation and the division of labor (Gross *et al.* 1979; Hames 1978, 1979; A. Johnson 1975; Johnson this volume; Johnson and Johnson 1978; Werner *et al.* 1979). Daily diary and recall methods have been employed to measure particular behaviors of large samples of subjects when particularly fine-grained analysis is desired (Aspelin 1975; Beckerman 1975, 1980, this volume; Bergman 1980; Dufour this volume; Hawkes, Hill and O'Connell 1982; Hill and Hawkes, this volume; Lizot 1977; Vickers 1976, 1980; Yost and Kelly, this volume). Exacting input–output methods have been used to study patterns of energy flow, human development, and nutritional status (Beckerman 1975; Dufour, this volume; Flowers, this volume), with some researchers relying heavily on native assistants to monitor time allocation and consumption in order to gain broader and more diachronic perspectives (Berlin and Markell 1977; Yost and Kelley, this volume). Methodological precision has found its way into the study of demography, genealogy, and settlement patterns (Chagnon 1974) as well as subsistence (Vickers 1980) through return trips to the field. Finally, multidisciplinary projects involving the cooperation of geneticists, biologists, biomedical specialists, demographers, and ethnographers (Berlin and Berlin, this volume; Chagnon 1974; Larrick *et al.* 1979; Neel 1970), and team-ethnographic research on intracultural (Chagnon and Hames 1979) or intercultural (Gross *et al.* 1979) projects, are giving us stronger holistic and comparative perspectives. This use of refined field methods and multidisciplinary and team-oriented approaches indicates an increasing awareness that certain ethnological problems cannot be solved through comparative analysis of general ethnographies.

The study of human ecology of Amazonia is in an exciting stage because of the rapid increase in the amount of research being conducted, the increasingly sophisticated methods being used, and the higher quality of the data being generated. The chapters in this volume are evidence of these trends. Yet much work remains to be done by anthropologists and other scientists. The very nature of ecosystem requires an interdisciplinary perspective, and social scientists have become increasingly familiar with the research of botanists, zoologists, soil scientists, and other specialists and have borrowed some of their techniques for data collection and some of their analytical concepts. Excellent examples of this type of cross-fertilization are Dufour's and Flowers's chapters in this volume, which deal with biomedical and nutritional aspects of adaptation. Cooperative efforts are likewise on the increase and it is not at all unusual for anthropologists to collect soil samples, plant specimens, and even blood samples for analysis by specialists.

There are many gaps in our knowledge which future research should address. We definitely need to have more fine-grained work on soil characteristics and distributions (e.g., see Moran 1981:219–221), and this knowledge needs to be articulated with the adaptations of specific native communities. The same is true for other aspects of the physical environment, including the characteristics of microclimate, aquatic habitats, flora, and fauna. For example, we now know general characteristics of certain Amazonian animals, but there is a crying need for detailed censuses of animal populations from specific sites within Amazonia. We are now getting considerable data from anthropologists on hunting and fishing yields, but we need better data on the population bases from which these yields are being taken. This particular question is of tremendous significance to the continuing discussion over protein availability within Amazonia.

We also need more research that is problem-oriented, and in which variables have clear operational definitions and statements of the propositions that are to be tested. That is, ecological anthropologists working in Amazonia (and elsewhere) must improve their use of the scientific method and distance themselves from the tradition of soft description that has characterized the field in the past. The recent work in optimal foraging theory (represented by Beckerman, Hill and Hawkes, and Stocks in this volume) are showing advances in this regard. In order to accomplish better problem-oriented human ecological research, anthropologists will need better training and preparation for fieldwork, and will have to give more consideration to quantitative measurement and sampling techniques. All too often the debate on tropical forest human ecology has bogged down because various researchers compare "apples" and "oranges" because they have not collected comparable data, or do not use equivalent units of measure.

Another important point to be made is that we need more long-term studies of adaptive processes from Amazonia. Most research covers periods from several months to a year, yet there is wide agreement among ecologists that short observation periods are unlikely to yield representative data on environmental fluctuations and long-term cycles. The constraints (financial and otherwise) on long-term research are very real, yet anthropologists and other scientists should seek to expand their data bases as far as possible. Return trips to study areas are one means of accomplishing this (e.g., Chagnon 1974; Vickers 1980).

We also need better definitions of the specific home ranges or territories utilized by Amazonian communities. To date, very few studies of Amazonian peoples have presented extensive data on this important aspect of adaptation (Bergman 1980; Hames 1980: Smole 1976; Vickers this volume). Once the territory of a specific community has been described, the subsequent tasks are to inventory the resources within the area and the productive techniques that are employed to exploit the resources (e.g., Bodley and Benson 1979 on the distribution of palms). These steps, which are similar to suggestions made by Steward (1955:40–42), provide a solid empirical basis for the material and ecological analysis of a society and generate data for comparative analysis as well.

Archaeology and ethnohistory are also methods that can contribute much to our understanding of the human ecology of Amazonia. The archaeological investigation of the Amazon is in its infancy, and the analysis of the historical accounts of explorers and travellers have not been given much credibility to date. We feel that both of these approaches can and should be integrated into our data base. The work of Meggers and Evans (1957) and Lathrap (1968) demonstrates that such data are applicable to the issues of Amazonian adaptation. Historical data must be used with caution, of course, but previously discredited accounts such as Carvajal's description of dense populations along the Amazon (Heaton 1934) are being vindicated by modern research (e.g., see N. Smith 1980).

The development of new behavior sampling methods and rigorous output and consumption measures (just reviewed) now allows us to focus on individual variation. Since the costs and benefits of behavioral traits are sensitive to individual differences (e.g., age and sex) and vary with microhabitat and macrohabitat characteristics, we can begin to understand individual behavioral differences as well as intracultural and cross-cultural variation. We believe this bottom-up approach, which forces us to focus on individuals, bebavioral alternatives, and costs and benefits, complements traditional anthropological approaches to adaptation by allowing us to ask questions about some of the more proximate and situational causes of behavioral variation. In its focus on individuals, it allows us to examine how behavioral traits of ecological interest are transmitted intergenerationally or how innovations are spread through local or regional populations. For example, how children are trained in subsistence behavior is virtually ignored by ecologists, yet it is an important part of the adaptive process. Wilbert's article (1976) on canoe construction and geographic knowledge (1979) and Ruddle and Chesterfield's monograph (1977) on subsistence education among the Warao of the Orinoco delta are pioneering and, it is hoped, will lead to a more integrated human ecology. If individuals make choices about alternative patterns of behavior we must presuppose that they have some sort of cognitive scheme for ordering the environment and some means for evaluating the adaptive utility of alternative behaviors. The work of B. Berlin and E. Berlin (this volume and references therein) on Aguaruna faunal classification continues in the tradition of ethnoscience but breaks new ground by showing the relationship between cognitive schemes and patterns of faunal capture and consumption as it varies between three villages living in different habitats. Finally, the criteria by which individuals evaluate alternative methods of environmental exploitation are virtually ignored in ecological research. It is assumed that individuals either blindly follow cultural prescriptions or somehow automatically choose the most energetically efficient behaviors. Although these answers may be true, they simply beg a fundamental psychological question. By focusing on the proximate and ultimate causes of behavior ethnographers can arrive at a truly holistic human ecology.

The Problem of Adaptation

The concept of adaptation plays a central role in most human ecological analyses. But, in contrast to biology, anthropology has not settled on a consistent definition or measure of the concept. For most cultural ecologists, adaptation is the maintenance of an equilibrium between people and the totality of their environment or, in the words of Rappaport (1974:384),

> any process . . . which tends to keep the states of crucial variables . . . within ranges of viability or tends to return them to such ranges should they depart from them may be taken, other things being equal, to be adaptive.

This definition suggests that adaptation is a conservative process that keeps a population on an even keel in relation to its environment and bears close resemblance to organismic conceptions of society (Radcliffe-Brown 1940; White 1959). Most contributors to this volume give no explicit definition of adaptation, but many imply that it has to do with efficient or effective ways of utilizing resources or dealing with stress. However, there seems to be no unified assumption as to what the goal of efficiency is—survival, equilibrium, cultural maintenance, or maximization of reproductive success. This concern with efficiency has led a number of contributors to use optimal foraging theory derived from evolutionary ecology (Beckerman this volume; Hames and Vickers 1982; Hawkes, Hill and O'Connell 1982; Hill and Hawkes, this volume; Stocks, this volume) or linear programming models from microeconomics (Behrens 1981; A. Johnson 1980; Werner, this volume). For some of those who use evolutionary biological optimization approaches, adaptation is ultimately measurable in terms of reproductive success (Hames, this volume; Hawkes, Hill and O'Connell 1982; Hill and Hawkes, this volume), an orthodox evolutionary approach. For others (Beckerman, this volume; Stocks, this volume), this is not necessarily true. Since the reproductive consequences of alternative behaviors are impossible to measure with any precision, researchers are forced to employ currencies (Schoener 1971), such as the net rate of energetic return for an activity, which correlate highly with reproductive success. For those who employ microeconomic models, adaptation is seen in terms of efficient ways of achieving culturally defined goals (e.g., Behrens 1981; A. Johnson 1980). Thus, optimal foraging theory and linear programming models are remarkably similar in their use of a common mathematical technique (i.e., linear algebra) and a proximate measure or currency of adaptation (energetic efficiency), but the ultimate goal of the adaptive behavior differs radically (e.g., reproductive success versus achievement of culturally defined goals). In other words, the techniques used to generate testable hypotheses are all but identical, but the theory to be tested is not. These differences in theory lead to the generation of different hypotheses concerning adaptive behavior. This emerging debate, it is hoped, will lead to the refinement of models—which are often based on unrealistic assumptions—within each camp (e.g., Beckerman, this volume; Hawkes *et al.* 1981) and to attempts to reach some common grounds of agreement between camps (Gross, this volume). The disagreements, in part, are the result of differences in levels of analysis (A. Johnson 1982).

Overall, it is relatively easy to point out areas for future research in Amazonian human ecology, but the actual collection and analysis of data is much more difficult and time consuming. Yet the "mysteries" of Amazonia are fully amenable to understanding via the scientific method and researchers are making increasingly important progress toward this goal. Ultimately such research in Amazonia both contributes to and benefits from ongoing ecological research in tropical and other areas of the world. In this sense Amazonian human ecological research is not unique, but uses methods and generates propositions that are relevant to the general field of human adaptation studies. As scientists, we also recognize that tropical environments everywhere are now experiencing rapid change due to deforestation and the impact of development, just as temperate zones experienced in the past. This knowledge gives our research a sense of urgency and a concern that scientific findings be incorporated into the ongoing process of policy formulation and planning for these regions.

I
Cultivation

CHAPTER 2

Machiguenga Gardens

Allen Johnson

Introduction

The tropical rainforests of Amazonia offer the best remaining opportunities for us to observe the processes and consequences of shifting cultivation in indigenous systems of production. Although much has been done to correct the opinion of an earlier era that shifting agriculture is wasteful and inefficient, it is still common for economists, agronomists, and some geographers to maintain that "shifting cultivators . . . generally achieve low productivity at the expense of some damage to the land and other natural resources" (Webster and Wilson 1966:87). This conclusion is warranted for many modern systems of shifting cultivation, but only where population increase has disrupted the proper balance between cultivation and fallow that is the basis of successful shifting cultivation. In low population regions such as Amazonia, we may still observe conditions under which shifting cultivation is an appropriate, and perhaps even ideal, system of food production.

A slash-and-burn garden, as indeed the whole subsistence economy of shifting cultivators under low population density, is extraordinarily complex and fluid. In the evolution of a garden from first clearing to final abandonment, and in the evolution of a household's settlement from first gardens to the point where shifting to a new residence has become advantageous, we find a predictable progression of forms and strategies of production, and this I will attempt to describe here for a single case. But these regularities are almost lost amid the innumerable variations that arise from

ISBN 0-12-321250-2

differences in microenvironments and the personal histories and aims of individual decision-making cultivators.

These variations are the characteristic outcome of adaptation to an environment that presents many possibilities for exploitation—that is, an environment in which the press of population has not become so great that production strategies are perforce restricted to a small number of the most intensive methods. Hence, the household economy of extensive horticulturalists is typically a mixture of hunting, gathering, and fishing, combined with cultivation of a variety of crops that provide food, medicines, and materials for clothing, tools, storage containers, and other manufactures.

Despite the relative abundance of opportunities enjoyed by low-density shifting cultivators, however, they still confront scarcity at every turn (Which are the best soils? Where is a reliable source of pure water? How far must I go to encounter game? etc.) Each household continually faces decisions about how to balance competing demands, and as these decisions are made the household configuration of adaptive strategies changes. In the course of many years, an individual household will move in and out of close exchange relations with other households; will depend more on its gardens at one point, and less at another; will maintain two residences for periods of time; will produce more or less food depending on the number, ages, and sexes of offspring; and so forth. In short, there are many degrees of freedom in such systems, and our attempts at description necessarily catch the flow at a moment in time and petrify what must be seen alive to be appreciated.

The present description concerns shifting cultivation among the Machiguenga, Native American inhabitants of the tropical rainforest of the Upper Amazon; specifically, a community on the Kompiroshiato River, a tributary of the Urubamba River in the department of Cuzco, Peru. Their language is of the Arawakan family, and is closely allied to that of the Amuesha, Campa, and Piro Indians, who also inhabit the montaña of Southeastern Peru. As is typical of the montaña region, Machiguenga settlements vary from single families to small hamlets of related families, located on a stream that provides clean water for household needs, near a river suitable for fishing, and with abundant forest for hunting and from which gardens are cleared. Population density is 0.3 persons/km^2. The Machiguenga spend nearly as much time procuring wild foods as they do cultivating their gardens, but it is from their gardens that the vast bulk of their food derives, including a great overproduction of starchy tubers for food security under isolated and vulnerable living conditions.

The Machiguenga distinguish various stages of land use, from forest through cultivation to abandonment and regeneration. In Figure 2.1 we see that the highest level contrast is between garden, fallow, and forest. At the

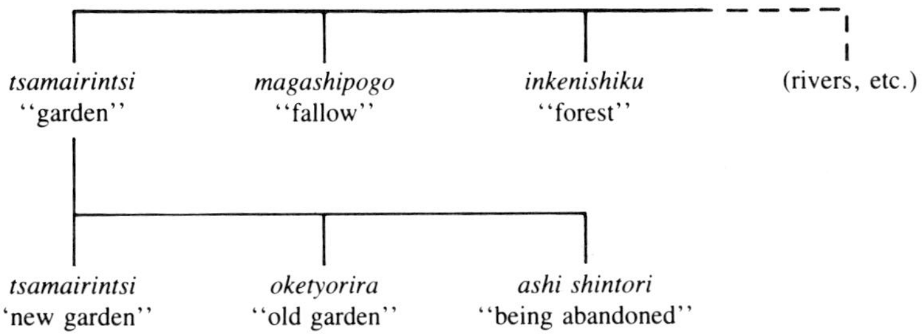

Figure 2.1 *Machiguenga garden stages.*

next level, gardens evolve from new to old to early abandonment (*ashi shintori,* literally, "it belongs to peccary").

In the normal course of events, when Machiguengas decide to move to a new settlement, they clear a large new garden and plant it. Meanwhile, they continue to reside at the old site, where they live from existing gardens. As the new garden nears production, a new house is built in the center of the garden. There follows a period of alternating residence in the old and new locations, but within a year residence is almost exclusively at the new site. There, a second garden will be cleared, and the garden around the house will begin a gradual transformation into a housegarden, a process whereby the garden tends to become smaller, to produce a smaller quantity of the staple crops, but to have a great qualitative diversity of cultigens that serve as condiments, specialty foods, teas, and medications. Other gardens, located farther from the house, tend to be abandoned as they age, although as long as the house is occupied abandoned gardens may be brought back into production.

The Machiguenga view the developmental cycle of a garden in terms of the tasks to be done. An informant lists these as follows

1. *nonkogakerora kametiri kipatsi*—"I select good land"
2. *nonkarasetakerora*—"I clear it"
3. *nompotakerora*—"I burn it"
4. *nompankitakerora*—"I plant it"
5. *nontsamaitakotakerora*—"I cultivate (weed) it"
6. *nogakerora shinki*—"I harvest maize"
7. *nompireatakerora nogakerora ovatsa nogakemparora*—"I cut and harvest manioc"
8. *nompankitairo pashini sekatsi*—"I replant manioc"
9. *tera nompankitairo aikyiro*—"I will not plant any more"

We shall proceed to examine each phase of this process more carefully. Then, we shall attempt to characterize the average horticultural behavior of a typical Machiguenga settlement.

Garden Making

SOILS

According to a Machiguenga myth, the earth was originally covered with water, until one day a man came and wondered, "Why aren't there gardens here; why is there no soil?" He proceeded to mound up piles of mud, and after many months succeeded in making dry land, although in this time there were neither hills nor mountains, trees nor weeds. The man said to the people who lived then, "I will go home; this earth is now ready for you to live better." Thereafter, until the Moon brought manioc, people lived by eating a particular soil known as "heart of the earth," and by fishing.

In the general classification of "things," earth or soil (*kipatsi*) is included among the nonliving (*tera ontime aniane*). Selection of the soil is regarded as the first step in gardening activities, and this is considered a purely male responsibility. Five primary kinds of soil are recognized (Table 2.1). As a rule, the first four types of soil are found on slopes, away from the river; the fifth type is found in small beaches that occasionally line the inside of a bend in the river.

The four hillside soils are clearly rank-ordered for suitability for gardens: *shimentya* soil is by far the most preferred, *potsitari* is also considered a good soil, *kiraari* is only adequate, and *kiteri* will not be used for gardens at all. The best soils are described as "plenty black, no big rocks, but soft and well drained." *Shimentyapatsari* is black, or very dark brown, and full of gravel of the consistency of crushed shale. This makes it a "soft soil" (*metsopatsari*) that is both easy to work and well drained. This soil is in short supply—at the time of the study it was only available across the river and those men who wanted a *shimentya* plot had to make a special raft trip across the river several times per week.

Imvanekipatsa is also a soft soil, and is a favorite for gardens because it is easy to work. The lack of rocks also means that tubers will grow big and

Table 2.1

Machiguenga Soil Classification

Term	Gloss
shimentyakipatsa	gravel soil
potsitapatsari	black soil
kiraapatsari	red soil
kitepatsari	yellow soil
imvanekipatsa	sandy soil

long. All of the sandy, riverside land near Shimaa in 1972 was either in cultivation or in short fallow following cultivation.

The other three hillside soils are generally considered to have "hard soil" (*kusopatsari*). Hard soil is difficult to work and slow to drain, and is given as the reason some gardens fail. "Black" soil is preferred as a "good" soil capable of growing all the Machiguenga repertoire of crops. "Red" soil raises some doubts—some men insist it is good for all crops, but several think it is a "bad" soil that would not grow maize, where "sometimes things grow, sometimes not." All agree, however, that "yellow" soil is no good for farming. Black soil is in short supply near Shimaa, but red soil is abundant. Yellow soil is uncommon, and may result primarily from soil leaching in older gardens.

Soil texture and color are the primary factors taken into account in selecting a new garden site. The Machiguenga do not believe the existing forest cover holds any clues to good soils; several men used the same phrase, "trees always grow in the forest" (i.e., whether the soil is good for crops or not). Steepness of slope is a concern, but not a major one, since nearly all gardens (except riverside plots) are on hillsides at a vertical angle of from 20 to 40°. Often, new gardens will be located adjacent to old ones, but if the soils are not suitable new gardens may be cleared at a 10 or 15 minutes walk, leaving a substantial block of uncut forest between gardens.

In general, Shimaa soils are typical tropical latosols, with relatively low amounts of soluble minerals, humus, and nitrogen. Four soil samples taken from various primary forest locations around Shimaa had an average nitrogen content of 0.32%, which is on the high side of the range for tropical rain forest (Wrigley 1969:40). Organic matter averaged 6.8%, which is also on the high side (Wrigley 1969:45), and the cation exchange capability of nearly 20 mg equivalents per 100 g is also moderately high for the tropics (Webster and Wilson 1966:52). Soils are on the average slightly acid, generally falling between pH 5.5 and 7.0, ideal for the kinds of crops the Machiguenga grow. Only one field showed a pH below 4.5, which is the point at which serious problems for many crops begin to occur (Wrigley 1969:52–53), and it was a field notorious for failing to grow food (although some men saw this as a result of spiritual forces). In texture, soil samples show a wide range from predominantly "sandy" to predominantly "clay" soils. Taken altogether, however, the samples show that Shimaa enjoys good soils in comparison with other tropical regions.

Soil analyses also reveal some of the basis on which the folk classification of soils rests (Table 2.2). It is apparent that the Machiguenga preference for *shimentya* soils is compatible with their unusually high content of nitrogen and organic matter. The differences between black and red soils are in the expected direction, but so small that it is not surprising that the Machiguenga disagree among themselves whether the differences are significant. The soft soils also show high content of sand and gravel, in contrast

Table 2.2

Analysis of Garden Soils, by Machiguenga Soil Category

Soil category	Organic matter (%)	Nitrogen (%)	Sand and gravel (%)	Silt (%)	Clay (%)
shimentya (gravel) (*n* = 6)	8.2	0.40	58	24	18
potsitari (black) (*n* = 5)	5.4	0.27	34	42	24
kiraari (red) (*n* = 9)	5.1	0.24	28	44	28
imvaneki (sandy) (*n* = 2)	1.8	0.08	72	20	8

to the hard soils that are high in silt and clay. The sandy soils of the river margin are, expectedly, highest of all in sand content, but unexpectedly low in nitrogen and organic matter. This probably reflects their sandiness, which does not encourage the accumulation of nutrients, but may also in part be due to their desirability and the consequence that all had been cropped with short fallows since 1968.

We can see that the Machiguenga environment offers a range of soils for the gardener to choose from. They differ in fertility, workability, slope, drainage, and other respects. The Machiguenga have words to express these differences, and clearly balance advantages against disadvantages in selecting where to clear a garden. These decisions are made a year or more before the fact, and one's intention to plant in a particular location is advertised so that disputes may be resolved before irreversible steps, such as building a house or giving up claims to alternate garden sites, are taken. Even though land is in general very abundant, there is still a scarcity of the best, most suitable, or most convenient lands, and this plays a role in settlement patterns and social relations between households.

CLEARING A GARDEN

The *montañna* experiences a rainy season around November through April, and a somewhat drier season around June through September. The period at the end of the rainy season, from April through July, finds most men occupied with the labor of clearing new gardens. Once this work has begun there is a tendency for it to become all-consuming: men will work steadily in their gardens 5–7 hours a day for days in succession until the work is complete. It is important that the garden have two to three low-rainfall months in which to dry before burning in September.

The labor of clearing (*nonkarasetakerora*) is viewed as having two distinct phases: *nonkarasetakero itiomiani inchato,* "cutting brush," and *nontogakero omarani inchato,* "felling trees." This means that a man goes through the whole garden once with a machete, cutting everything under 8 cm in diameter, then a second time with an axe, cutting everything else. For an example, we may follow "Carlos"[1] as he clears a new garden 0.44 ha in size in May, 1973.

Carlos has selected a location near the river, with a comparatively gentle slope (15°) and black soil. He says there was a garden here "long, long ago," but he has no idea who might have planted it. Now, at any rate, the larger trees are over 1 m in diameter and over 25 m tall. The underbrush is fairly dense, composed of small plants, brushy shrubs, and vines and creepers. Indeed, the most striking aspect of this phase of work is the ubiquity of tough vines that weave the forest together, acting as a kind of woof to the warp of the trees and shrubs.

Carlos leaves all saplings and trees over 8 cm diameter or 3 m high standing. The rest he cuts down as carefully as if he were weeding a garden. The small plants he pulls out by hand, and the larger, tough, or thorny plants he cuts with a machete, as close to the ground as possible. He further chops the brushy plants so that they lie in fragments close to the ground; the better to burn, he says. He also cuts the vines that climb the larger trees; here the danger of dislodging debris from high in a tree is particularly great, and Carlos is very careful.

He works at a relaxed but steady pace (about 40 machete strokes per minute). His movements appear haphazard—first to the left, then the right, forward a ways, then backtracking a bit and off in another direction. But after a while the outlines of a rectangular garden, with straight edges, takes shape. He pauses once to refresh his wad of coca, but otherwise works without interruption. This phase of the clearing takes about 64 hours in 8 days to complete.

As soon as the brush is cut, Carlos begins to fell trees. Actually, just before doing so he had planted a small patch of ground with manioc. These plants will sprout and grow even while the garden is drying for burning, and although the above-ground parts of the manioc will burn off in the fire, the now-formed roots will quickly regenerate and come to maturity many weeks in advance of the manioc that is planted after the garden is burned.

Carlos cuts completely through the smaller trees, but most do not fall because they are held aloft by the tangle of vines. In the larger trees, he makes a large-wedge-shaped cut facing downhill, then cuts through carefully from the uphill side, watching by the movement of the tree to see when it is ready to fall. When the tree is standing thus precariously, he leaves it for the next tree, working until perhaps 20 trees are ready. His wife and son are in the garden with him, gathering grubs from rotten logs and stumps, and collecting the hearts of felled palms. Then, Carlos fells completely one large tree that is linked by vines to the neighboring trees, and all tumble together in a spectacular roaring crash as Carlos hoots and whistles and everybody laughs. As the cloud of debris settles, Carlos moves on to another tree and begins the process over. Unlike the small brush, however, little effort is made to cut the tree branches up for better burning, probably because the branches are thick and the wood is tough and hard. This phase of the clearing takes Carlos about 31 hours.

Based on observations of a number of men, the average rate of both phases of clearing a garden is about 200 hr/ha. Carlos works a bit slower (219 hr/ha), and there is in general a good deal of variation in these figures. A number of factors influence the rate at which men work. First, the density of brush and thickness of the trees determine labor costs. Interestingly, there is an inverse relation between these: the bigger the trees, the lighter the underbrush. As a tree gets larger, it is geometrically more difficult to bring down: a 15-cm diameter tree is felled in a minute; a 30-cm tree in about 7 minutes; a 0.5-m tree in 18 minutes; a 60-cm tree in over 40 minutes, and a 1.1-m tree in an hour and a half. On the other hand, thick brush can slow a man down as well. For example, in very dense brush, Ricardo worked so slowly it would have taken him over 215 hours just to clear the brush from a hectare, whereas in very light brush Arturo cleared a hectare in only about 30 hours.

On these grounds I doubt that the frequent claim (cf. Vayda 1969:215), that primary forest is more difficult to clear than secondary forest, applies in the Machiguenga case. Each type of forest presents its own problems for the worker. For example, for Carlos, clearing brush took about two-thirds of the total labor in clearing his garden; but for Felipe, it was felling trees that took about two-thirds of his time; yet both men were able to complete both phases of clearing their gardens at a rate of about 200 hr/ha (Table 2.3).

In Felipe's garden the trees were somewhat bigger, and the brush somewhat lighter than in Carlos's. One might even want to argue from Table 2.3 that primary forest is actually easier to clear than secondary growth, but there is too much variation in the data for this to be warranted. For one thing, individual workers accomplish their tasks at different rates, even in the same garden. Carlos, for example, is notable for the care with which he

Table 2.3

Clearing Brush and Trees

Worker	Rate of cutting brush	Rate of felling trees	Total
Carlos	147 hr/ha	72 hr/ha	219 hr/ha
Felipe	67 hr/ha	119 hr/ha	186 hr/ha

does all phases of garden work; this makes him appear slower to the casual eye. But when Carlos clears brush, he clears out everything, cutting plants within 2–3 cm of the ground, and pulling up small plants, even those only a few centimeters tall. Felipe, on the other hand, leaves the small plants untouched and cuts some brush several centimeters from the ground (which is easier because it requires less accuracy and less stooping).

On the basis of my observations, it is not even certain that labor is saved by reclearing a garden that has been fallow for a year or two. To be sure, there are no large trees in it, but the weeds and brush are quite dense. As we have seen, direct measurements of Ricardo clearing dense brush give a rate of over 200 hr/ha, which is the same as it takes to clear primary forest.

Garden work occasions little cooperation, whether it involves men or women. Only men are involved in the labor of clearing, and it is usual for a man to work alone in his new garden. Men will sometimes accompany each other to a garden, although there is seldom any division of labor between them. For example, two brothers, Tito and Ricardo, worked together clearing a large garden. They worked side by side, first clearing brush, then felling trees, chatting occasionally or stopping to chew coca. But when they were done they were able to show how they had divided the garden between them, with the boundary between the two halves marked by rocks and fallen logs.

An important contrasting case involved the brothers Felipe and Mariano. They worked side by side in Felipe's garden in 1974, just as they had in Mariano's garden in 1972. They worked separately, but almost in unison. While cutting down large adjacent trees, for example, I timed them each at exactly 21 axe swings per minute over a period of several minutes, as the axe blows fell in steady syncopation. Characteristically, for the clearing phase, they worked long hours, day after day, with little rest or variation in activities. As the work neared completion, however, both appeared to become increasingly irritable until one day, after a bout of beer-drinking, they broke into a loud, angry shouting match in which Mariano accused his brother of exploiting his labor. The result of this fight was a split of the hamlet into separate households living at widely separated settlements for several years.

BURNING

Burning a garden is one of the chanciest, and most frustrating, tasks the Machiguenga undertake. Like clearing, it is exclusively a male activity. The basic principle is that strong sun and a good wind make a successful fire. But finding the right day on which to burn seems to be more a matter of luck than skill.

Most gardens are cleared in April and May, and burned in September and October. Thus they have 3–4 of the driest months in which to dry. But gardens can be observed to burn in any month, even in February, which is by far the rainiest month. Since it can, and does, rain in any month, however, Machiguenga gardens are never really dry, and never burn cleanly.

The rule of thumb is that 5 consecutive days of strong, hot sun make for a good burn. The problem for the individual gardener, however, is that it can rain at any time. Hence, after a few hot days one can sense the anxiety mounting as a man decides whether this would be a good time to burn. If he sees the smoke from someone else's fire, even if it seems too soon, he may rush out to burn his own garden, because it is believed that burning a garden brings rain.

A man waits for a day in which the temperature rises to about 30° C, and then for the afternoon wind, which is usually strong by 1:00 or 2:00 P.M. For example, on the day of September 5, 1972, several gardens within about a kilometer of our house went up in smoke at about 2:00. In the previous week, on August 31 and September 1, there had been a total of 13 mm of rain, followed by 4 days of almost no rainfall (total 4 mm). The daily high temperature over this period gradually rose from 22 to 31° C on the day of the burn. Of three gardens burned this day, the one that had been drying the longest burned well, and the other two burned badly. As it turned out, this day was followed by 4 more that were very hot and dry. On September 9, Ricardo burned his field and it burned very well. Had the others waited, theirs would have done better. But, of course, one never knows.

Burning a garden is hot, tiring, unpleasant work, usually involving several men and boys together. It is necessary to remain in the garden while it is burning because the fire catches, then fizzles, and must be restarted and encouraged. There is some danger of being caught by the fire, but the Machiguenga know which areas are going to burn hot and they avoid them; we know of no cases of people or their relatives who were seriously injured during a garden fire (although one myth tells of a man burning to ashes in his garden fire). As the fire catches, the men run from spot to spot, building a wall of fire downwind; from time to time they cup their hands and whistle to attract the wind. As the fire burns itself out in one part of the garden, they move to start it farther on, until the whole garden has had a chance to burn.

About half of the gardens do not burn well, and individuals express their disappointment and anger when this happens. A poorly burned garden is full of rubbish that must be mounted into piles for reburning, and all the weeds that have sprouted or regenerated since the clearing are now tough and well-entrenched and will require extra weeding. But even when a Machiguenga garden burns well, it is still cluttered with criss-crossing logs and areas of unburned brush. After lying several months in the dry season, the ground is still damp and logs are beginning to rot. Large logs are merely scarred by the fire, and one learns to navigate Machiguenga gardens by walking on top of the larger logs, stepping or leaping from one to another in a makeshift pathway through the garden.

It takes a man and his male kinsmen about 20 combined hours to burn a hectare of cleared garden. Depending on how well it burned, they will have more or less work to do in reburning, but all gardens require at least some reburning. And no matter how much care is invested in reburning, there will remain some areas of a garden so tangled and impenetrable that no cultivation will be attempted in them. With much individual variation, then, reburning takes an average of another 60 combined hr/ha.

The Machiguenga see the value of burning primarily in terms of clearing the garden and fertilizing the crops. They deny that it retards the growth of weeds: "As you can see, weeds begin to sprout again the day after you burn," they point out. But the better a garden burns, the less rubbish and the easier it is to work in. Also, they believe that the ash is good fertilizer for all garden crops.

Planting

New gardens are planted in September and October, immediately after burning is complete. Thereafter, as crops are harvested, some are replanted to keep the garden continuously productive. Through surveys of gardens, and in interviews with different men and women, I have collected the names of 80 cultigens, not counting varieties of each (Table 2.4). The majority of these are occasional cultigens tended by individuals in the vicinity of their homes. A smaller number are basic staples in the diet, and it is with these that we may begin.

MANIOC (*Manihot esculenta*)

In Shimaa only sweet manioc is grown. Sauer (1949:508) observed that whereas bitter manioc is frequently the primary staple food of native diets in Amazonia, "the sweet manioc is nowhere the staple." Although this may be true in the lower altitudes of the *montaña,* where bananas and plantains are basic stables, the Machiguenga are a striking exception to this

Table 2.4

Machiguenga Cultigens

Machiguenga name	Common name	Scientific name	Use
1. *ampee*	Cotton	*Gossypium barbadense*	Manufacture
2. *anona*	Soursop	*Annona* sp.	Edible fruit
3. *aroshi*	Rice	*Oryza sativa*	Experimental crop
4. *ashi merentsi*	Hibiscus	*Hibiscus* sp.	Medicinal herb
5. *cacao*	Cacao	*Theobroma cacao*	Edible fruit
6. *cahiki*	Coffee	*Coffee arabica*	Edible fruit, cash crop
7. *cebolla*	Onion	*Allium* sp	Condiment
8. *chakopi*	Arrow cane	*Gynerium sagittatum*	Manufacture
9. *choritoshi*	?	?	Dye (?)
10. *coca*	Coca	*Erythroxylon coca*	Anaesthetic
11. *iinge*	Peanut	*Arachis hypogaea*	Minor food
12. *imvogo* (*shanko*)	Sugar cane	*Saccharum officinarum*	Beverage, candy
13. *inchoviki*	?	?	Leaves for steaming fish
14. *intsipa*	Guaba	*Inga* sp.	Edible fruit in pods
15. *iratsipini*	?	?	Medicine protects males from female pollution
16. *irimoki*	Lemon	*Citrus limon*	Edible fruit
17. *iseka ataava*	?	?	Chicken feed
18. *ivenkiki*	Sedge (*piri-piri*)	*Dichromena ciliata* and *Cyperus* sp.	Medicinal herb
19. *ivinishi*	Spearmint	*Mentha spicata*	Medicinal herb
20. *kasankari*	Lemon grass	*Cymbopogon citrarus*	Medicinal herb
21. *kemi*	Squash	*Cucurbita* sp.	Minor food
22. *kepia*	?	?	Leaves used in manufacture
23. *kocama*	?	?	?
24. *kogi*	Barbasco	*Lonchocarpus* sp.	Fish poison
25. *kogi kiripeshianiri*	Barbasco	*Tephrosia toxicaria*	Fish poison
26. *komashiki*	Guava	*Psidium guajava*	Edible fruit
27. *kompiro*	(*Yarina*)	*Phytelephas microcarpa*	Manufacture (palm)
28. *korama*	?	?	Similar to *poi*
29. *korinti*	?	?	Decorative flower
30. *koriti*	Sweet potato	*Ipomoea batatas*	Minor root crop
31. *koviriki*	?	*Spilanthes ocymifolia*	Medicinal herb
32. *kuri*	(*Pijuayo*)	*Bactris gasipaes*	Edible palm heart, hardwood for manufactures
33. *kuro*	?	?	Magical herb protects garden (appearance similar to *onko*)

(*continued*)

Table 2.4 *Continued*

Machiguenga name	Common name	Scientific name	Use
34. *magona*	Yam	*Dioscorea* sp.	Minor root crop
35. *mananeroki*	(*Tomate del monte*)	*Solanum* sp.	Magical protection, children's toy
36. *manataro*	Palm	*Bactris chaetochlamys* (?)	Edible fruit, palm heart
37. *mandarina*	Tangerine	*Citrus reticulata*	Edible fruit
38. *manko*	Mango	*Mangifera indica*	Edible fruit
39. *maroro*	Kidney bean	*Phaseolus vulgaris*	Minor legume
40. *miminkoki* (*tsitsita*)	Pigeon pea	*Cajanus cajan*	Minor legume
41. *naranja*	Orange	*Citrus sinensus*	Edible fruit
42. *onko*	Cocoyam	*Xanthosoma nigra* (?)	Major root crop
43. *onko makato*	?	*Anthurium* sp.	Root crop, relative of arrowroot
44. *pamoko*	Tree gourd	*Crescentia* sp.	Serving bowls
45. *pankogiririra poreatsiri*	?	?	Purple flower for "sun to look at"
46. *pantiarigo*	?	?	Edible nut
47. *pao*	Tree gourd	*Crescentia* sp.	See *pamoko*
48. *parianti*	Plantain	*Musa paradisiaca*	Edible fruit
49. *pashiroki*	?	*Trema micrantha*	Typing fiber
50. *piarintsi*	Bottle gourd	*Lagenaria* sp.	Water jugs
51. *pocharoki*	Various	Various	Any small, sweet fruit with a seed
52. *poi*	Yam-bean?	*Pachyrhizus tuberosus* (?)	Edible seed and tuber
53. *porenki*	(*Canna*?)	?	Edible fruit, condiment
54. *poseiro*	Banana	*Musa* sp.	Edible fruit
55. *potogo*	(*Oje*)	*Ficus* sp.	Medicinal herb
56. *potsoti*	(Achiote)	*Bixa orellana*	Dye, face paint
57. *saaro*	Moonflower	*Datura* sp.	Narcotic, medicine
58. *sankonka*	Colander?	?	Medicinal herb for gripe
59. *santari*	Cedar	?	Medicinal herb for head cold
60. *sarioki*	?	?	Seeds for beads
61. *sekatsi*	Manioc	*Manihot esculenta*	Major root crop
62. *seri*	Tobacco	*Nicotiana tabacum*	Medicinal/magical herb
63. *sharovantareki*	Calabash	?	Containers; also, medicinal herb
64. *shikovana*	?	?	Medicinal herb for diarrhea, similar to *ivenkiki*

(*continued*)

Table 2.4 *Continued*

Machiguenga name	Common name	Scientific name	Use
65. *shimanteki*	?	?	Edible juice
66. *shimasheri*	?	*Cassia* sp.	Flowering tree for garden
67. *shinki*	Maize	*Zea mays*	Major food
68. *shinti*	?	?	Medicinal herb for pregnant women
69. *shirina*	?	?	Minor root crop, similar to *koriti*
70. *shovnaki*	(*Dali-dali*)	*Calathea allouia*	Minor root crop
71. *tamviapini*	?	?	*Cushma* ornament
72. *tinti*	Papaya	*Carica papaya*	Edible fruit
73. *tareko*	?	?	Dye (goes on green, dyes red)
74. *tivana*	?	*Bromelia* sp.	Fiber for nets
75. *tsigaro*	?	*Scheelea cephalotes*	Palm: edible heart; manufacture
76. *tsimoritoki*	Grenadilla	*Passiflora cuadrangularis*	Edible fruit, root
77. *tsirianti*	Pineapple	*Ananas comosus*	Edible fruit
78. *tsitikana*	Pepper	*Capsicum* sp.	Condiment
79. *tsivi*	Avocado	*Persea americana*	Edible fruit
80. *tsota*	Gourd	?	Serving bowls

rule. By any criterion, manioc is the central food in their diet. Manioc accounts for nearly two-thirds of the calories of food energy produced in Machiguenga gardens. Manioc is the first crop planted in a new garden, setting up the framework within which other crops will be interplanted. Even the word for manioc, *sekatsi,* from the root *-seka-* (eat), translates literally as "that which is eaten."

According to legend, *sekatsi* was brought by *Kashiri* (Moon) and given to the girl he wooed and later got with child (bearing the Sun as their first child). Before this, people ate *kipatsi ompotsavagete* (very black soil; also, *anigaki kipatsi*—"heart of the earth"). The myth translates here:

> *Kashiri* saw the man's daughter. Then she took out beer. Then he saw its blackness. *Kashiri* said, "What have you eaten?" She said, "Manioc." He said, "That is not manioc; that's soil." Then he gave her manioc, the kind he himself ate. Then she ate some. Then *Kashiri* went home (to the sky). Then her father came home and she gave him some of the manioc from *Kashiri*. He said to her, "Where did you get this; this isn't manioc?" "This I got over there; it is manioc—that other there isn't manioc, it's soil." Thereafter they ate manioc.

Although women are occasionally seen planting manioc, most often while replanting one they have harvested, manioc is primarily planted by men. One morning in October (1972), I found Javier in Julio's garden, gathering the manioc stalks (*ogati*) from which cuttings would be made for planting. Javier is fastidiously neat and orderly in his garden work, and now he is carefully selecting stalks from among the pile where Julio left them after harvesting his manioc. Javier selects from the pile of knobby, angular stalks only straight ones of 2 to 3 cm in diameter, and cuts them to a uniform length of about a meter. When he has about 50, he binds them into a neat cylindrical package and carries them tumpline style to his own garden 10 min away.

Some farmers prefer thicker stalks for planting, others thinner. If stalks are too thin they are thought to produce small roots, but if too thick they are hard to cut. Javier prepares the ground with a borrowed shovel. Facing uphill, he brings the shovel down straight in front of him, making a vertical cut in the hillside. He then loosens the soil with several quick strokes of the shovel. Men who lack a shovel do this with a short digging stick of palm wood, but the shovel is clearly more effective. The result is a flat "pad" of about 30 cm in diameter into which two pieces of cut stalk, about 15 cm in length, are firmly pressed. Javier prepares about a dozen pads at a time, locating them in a roughly equidistant distribution. He then cuts the stalks, holding one in his left hand while with a knife in his right he cuts pieces to the desired length with quick strokes. He plants these in the prepared pads, then picks up his shovel to begin preparing a new set.

Javier works more carefully than many men, so that whereas the average man requires about 50 sec to plant a single pad, Javier takes 65 sec. Javier also plants his garden more densely than usual; one informant explained that this is characteristic of men with smaller gardens: men with large gardens can plant their manioc plants more widely spaced and still have a sufficiently large crop. Data bear this out: of seven new gardens, the three that were larger than 1 ha averaged about 2500 manioc plants per ha, while the four that were less than 1 ha in size averaged 3500 plants/ha.

As an example of how small details can appear quite important to the Machiguenga, I may mention that the direction in which the cuttings are inserted into the ground, and the distance between cuttings in a single pad, are important. In planting my own garden I inserted several cuttings "backwards." Each knob on a stalk bears a tiny axillary bud pointing upward (as the plant grows); this should be facing the planter as it is inserted into the pad so that the plant, which will sprout at these buds, is sprouting upward toward sunlight.

Another example occurred when the Peruvian anthropologist Alejandro Camino visited me from the Machiguenga community of Monte Carmelo, a day's journey downstream from Shimaa. There they plant both cuttings side by side, touching each other, he told me; and he demonstrated the method for me in a few holes. The next day, Oscar, who was visiting my garden with me, drew up short before these holes and, looking startled, asked me, "Who planted those?" These are matters of style that do not seem to have large significance for the productivity of a garden, but reflect the detailed attention the Machiguenga pay their garden labors.

Manioc is the security plant of the Machiguenga. They plant much more of it than they are likely ever to consume. As such, it is the subject of certain magical beliefs and practices that protect and ensure its growth, but manioc is such a reliable crop that these practices are not nearly as stringent as those governing the more delicate maize. Certain nonfood plants (*kuro,* and a variety of *ivenkiki*) are cultivated to protect the manioc. When the manioc is young it is also liable to suffer if the gardener eats tabooed meats like howler monkey, but these bad effects are much less than those feared for maize, and as manioc is always being replanted when it is harvested, taboos regarding manioc are not taken too seriously. Indeed, it is believed that if the planter eats tapir meat when the manioc plants have grown to about a meter high, the roots will be robust like the tapir.

Machiguengas distinguish many varieties of manioc. I collected names for 15 different types. But the outside observer can, I believe, make more of this diversity than it warrants. Some of these names are common names for varieties appearing in every garden, but many others are unique names given by single farmers and are not recognized by others, even their neighbors. Common varieties include "red" (*kiraari sekatsi*), "yellow" (*kiteri sekatsi*), and "white" (*kutari sekatsi*), and several varieties named after particular plants or animals (*tsireriganire* [*tsireri* means "a palm tree"], *shimaganire* [*shima* is *boquichico* fish]). Other names are apparently individual constructions, such as *kiratonkishiari* (*kira* [red] + *tonki* [bone] + *shi* [leaf]), referring to the red-veined leaves of this particular variety. As with named varieties of other cultigens, such as maize and sedge, the Machiguenga are sometimes exploiting a shared typology and sometimes an idiosyncratic name based on the personal history of the plant or its immediate visual properties. Often, when I asked an informant to name the manioc varieties in another man's garden, he would say, "I don't know that one—it came from upstream [or downstream]."

There is a widely shared functional typology, however, into which most of the individual varieties fit. On one dimension, manioc is distinguished by whether its tubers have white or yellow flesh. White flesh is preferred for its delicate taste and fine texture. On another dimension short-rooted varieties are distinguished from long ones. Women prefer short-rooted varieties, they say, because they are less work to harvest. If these preferences were acted on one would expect to find mainly short-rooted, white varieties in the gardens, but this is not the case. Rather, it is the *variety* of manioc types that is striking, in any garden. When I asked men why they planted such a diversity, I got similar replies. One man said, "Because we want to." This is a better answer than it appears at first, because the Machiguenga do enjoy diversity for its own sake. Another man gave this answer: "I plant these varieties because I am afraid otherwise they will die out." Differences in taste are taken as part of the variety of life, and a Machiguenga would not disdain to eat any variety of manioc that was available.

MAIZE

After manioc, the most important crop to the Machiguenga is maize (*Zea mays*), *shinki*, which is also a crop grown by males. It is the predominant crop in new gardens, accounting for more than half of all crops planted there (manioc is another third of the plants, so manioc and maize together constitute about 85% of all plants in a first-year garden). Maize accounts for about a fifth of the total calories produced in gardens. It is a favorite food when roasted green, and when dried is ground up as an important ingredient in manioc beer. A ground maize beverage is given as an early food for infants and as a milk substitute during weaning.

Despite the universal belief that manioc should be planted first in a new garden, my notes record several instances in which maize was the first crop in the ground. This makes sense because maize grows faster than manioc and produces seed while manioc is still only a bush ½–1 m tall. But maize is believed to require so much from the soil that it can only be planted once in a garden; thereafter, "The soil isn't any good; no really good ears will grow." Hence, in a sense, maize *is* the new garden, and once it is harvested the garden takes on the familiar cropping pattern of an old garden.

Maize is planted in deep, open, conical holes in the ground made with the aid of a palm-wood digging stick about 1½ m long, 7 cm in diameter, and weighing about 5 kg. The stick is sharpened to a dull point at one end. Standing erect and holding the stick in his right hand, a man clears away surface litter from a small area with a few scrapes of his foot, then punches a hole in the ground, using about six strokes to make a hole 20 cm deep and well packed along the sides. He then bends over and with his free hand drops 5–8 seeds into the hole, leaving it uncovered; owing to the narrowness and depth of the hole, birds cannot get to the seed at the bottom.

On observation this labor appears casual and haphazard, but the labor is quite demanding. A man planting maize appears to zig-zag around the garden in a random manner, but practice shows that this is made necessary by the rubbish of logs and branches in the garden. The net result of his activity is maize planted in a nearly uniform pattern, with each plant at about 1 m distant from each other plant.

As in most garden labor, a man planting maize works steadily for 5–7 hours with virtually no rest, apart from two or three 5-min breaks to chew coca. Hence his casual appearance while working is really a form of pacing to get through the day. The speed with which his work is accomplished depends on the quality of the soil. Rocky soils are the hardest to work in, as the digging stick is frequently (and painfully) jammed into rocks. Sandy (*imvaneki*) soils are easiest to work; I once clocked Felipe in *imvaneki* soils planting 5 maize holes/minute. But the average for most hillside plots is close to 2 holes/minute. And Julio, a careful worker, took almost 50 sec/hole because he was weeding out small plants as he worked, as well as planting occasional cotton seeds.

Because maize is planted only in new gardens, and because most new gardens are planted in September and October, green maize is available as a food primarily from December through May. There are the expectable variations: we have been given green corn to eat in October and November and have seen maize planted in a new garden in January; one occasionally also sees small clusters of maize plants growing in housegardens, so that fresh corn is available for eating occasionally throughout the year.

Even more variable is the performance of the maize crop in the gardens. Many seeds are planted in each hole, but usually only two or three sprout and grow to maturity. Of these, all three may grow tall and yield many healthy ears, perhaps only one or two will yield, and sometimes none bear seed. Furthermore, whole areas of a garden may have maize plants that grow well enough, but produce no seed. The Machiguenga usually attribute these differences to differences in individual seeds; even in my garden, where I planted some hybrid corn of guaranteed uniformity, plants were as variable in size and robustness as in other gardens, and this was explained to me as owing to differences in seed.

Maize, however, is also considered to be particularly vulnerable to supernatural interventions. The behavior of the man planting the corn is of the utmost importance, not merely while planting the maize, but especially during the period from when the plants are about 30 cm high until the seeds begin to mature on the ear. For example, in a section of Pedro's large garden, where the maize plants were scattered quite far apart (up to several meters between plants), the plants bore ears but the seeds never matured. Pedro attributed this to the fact that his friend, Santiago, who helped plant this maize, had eaten *shito* monkey while the maize was in its vulnerable phase. *Shito* is among a number of animals, including howler monkey, peccary, tapir, and a kind of fish (*charava*), that can be harmful to women and children when they eat them, and to infants and certain crops when eaten by their fathers or those who planted them (i.e., their "progenitors"). Informants could not provide a mechanism for this magical effect.

During the vulnerable period of maize growth, it is customary to chew the leaves of a particular sedge (*shinkivenkiki*) and walk about the garden spitting here and there to protect the maize. It was when his maize was about 30 cm high that I wanted to measure Javier's garden and told him so in advance. But as I approached his field with my assistant, Javier stood in the trail, blocking our access, his bow and arrow firmly in hand, explaining that we might hurt his maize. I promised to be very careful and not step on any, but this was clearly not the point. When his corn was over 1 m tall and beginning to form ears, Javier very cordially permitted us to measure his garden. These and other examples are evidence of the extent to which magic is invoked to protect this unpredictable crop.

There is less variety and more agreement between informants about the varieties of maize than in the case of manioc. I recorded 10 named varieties of maize, of which 6 are widespread. These 6 are: *sanorira* (true), *kiraakiri* (red seed), *kitekiri* (yellow seed), *kovano* (Cuban), *tonkaigiri* (exploding seed, i.e., popcorn), and *tsoviroki* (?; also a popcorn). The other four names appeared to be idiosyncratic: *chakamishinkina* (*chakami* is "trumpeter," *Psophia leucoptera*), *tsivitaroki* (*tsivi* is "avocado"), *nenarometiki* (*nearometiki* is a common tree), and perishinko (?). Most farmers when planting a field will do so with mixed seeds, and it is common for men to exchange seeds with one another in order to increase the variety in their own fields.

"WOMEN'S CROPS"

Manioc and maize are clearly associated with men, and together provide the basic staples of the Machiguenga diet. The next most important crops, in order of frequency of planting, are primarily associated with women, and I give them the label "women's crops," although the Machiguenga do not use this label. When we ask either men or women, "What do women plant?" the standard answer is "*onko* and *magona*," just as "maize and manioc" is the standard answer for men. Women's crops are *onko* (cocoyam), *tsirianti* (pineapple), *ampee* (cotton), *tinti* (papaya), and *magona* (yam).

Women' style of planting contrasts with men's. Men plant one crop at a time, in large quantities. Women tend to plant many different crops at once, and to spend a relatively short time at any one planting. Usually, a woman goes out to plant accompanied by her young children. She carries a net bag full of various cuttings and seeds. In the garden she may nurse her infant or carry it on her back in a sling; if she has an older child, he or she will help with the care of younger siblings, or will carry the seedlings and hand them to her as she needs them.

Women work more slowly and with less apparent effort than the men (in fact, they expend considerably less energy in planting than men do). A woman's most common garden tool is a kitchen knife with a 15–20 cm blade. She loosens the dirt with the knife, spending as long as 1 minute if the soil is hard or if she is nursing an infant, but as little as 20 sec if conditions permit. She removes a cutting from her bag and drops it in the shallow hole, pressing loose dirt in around it. She then rises, looks about for a good place for another cutting, and repeats the process. Generally, within less than an hour, she will have exhausted the cuttings in her bag and will return home.

I have either observed or been told of men planting pineapple, cotton, sugar cane, papaya, and yam on occasion. But this will usually be in a housegarden, where many different household members will plant experimentally. Many other minor crops planted by women are rather unfamiliar to the men. When I asked men to identify them in their fields, they often had to say, "I don't know; I will have to ask my wife." On one occasion, when a man's wife was present while I was asking him the names of uncommon crops, he would answer me, only to be corrected in a matter-of-fact way by his wife in nearly half the cases. These crops include squash (*kemi*), sweet potato (*koriti*), bean (*maroro*), *dali-dali* (*shovnaki*), *porenki*, and *shirina*, among others.

Women's crops vary in taste and texture. They are often sweet or bitter, piquant, nutty, or otherwise unusual in taste. If it is true that no meal is complete without large portions of manioc (and maize, in season), it is equally true that it is not complete without small portions of one or more women's crops, which bring a special enjoyment to the meal. It would be a fair generalization to say that men spend the most effort producing food, and produce the most food energy, whereas women, in their more casual way, contribute diversity to the diet.

OTHER CROPS

According to crop frequencies based on counting the crops planted in a 2% random sample of each of 15 gardens, the 8 major crops of the Machiguenga are, in order of frequency: manioc, maize, cocoyam, pineapple, cotton, sugar cane, papaya, and yam. After these, the frequency of crops drops off sharply, and only another 18 crops appear at all in the random samples. The rest are so infrequently planted that they were missed altogether in the random samples, and appear in my records only because of detailed study of several housegardens. Of these remaining crops I will mention several that stand out for one reason or another.

Coffee and cacao are the most numerous of the nonfood crops. These have been introduced only recently and the first marketable harvest of coffee, amounting to several sacks of beans, was sent by the schoolteacher to market in 1973. Most farmers have a few trees of each planted and some, planning on raising trees for cash production, have planted large gardens of coffee trees. As a measure of the transitional significance of these crops (from subsistence to cash production), both the sweet red fruits encasing the coffee beans and the purplish cacao beans are eaten raw as snacks.

Plantains and bananas are major food crops for the Machiguengas living at lower altitudes (about a 500-m altitude seems to mark the transition), but in Shimaa these are minor foods. Most farmers have perhaps 20 trees under cultivation at any one time, but do not have fruits to harvest more than every 2 months or so. In my judgment, the fruits could not compare in size or taste to those grown at lower altitudes. Other minor foods that were at least numerous enough to be mentioned several times in the random samples include pigeon pea (*miminkoki*), peppers (*tsitikana*), bean (*maroro*), avocado (*tsivi*), "guaba" (*intsipa*), squash (*kemi*), orange (*naranja*), *kocama, shirina, pantiarigo,* and *pocharoki.*

A number of Machiguenga cultigens serve as medicinal or magical plants. By far the most important of these are the varieties of *ivenkiki,* sedges that appear to fall in at least two genera (*Dichromera* and *Cyperus*). *Ivenkiki* is planted around every house and in many gardens as well. This plant clearly demonstrates the tendency of the Machiguenga to see plants as individuals as well as members of classes or categories. Each *ivenkiki* plant has a special function. Many are used magically to protect the household or specific crops from disease, predation, or spiritual danger. Passing through a garden with an informant, one learns that this individual plant protects maize, that one manioc, a third is for headache, a fourth for stomachache, and so on. These plants do indeed look different: some are short, others tall; some leaves are short and stubby, others long and slender. What all have in common is the appearance of tough blades of grass topped off by umbelliform flowers nested in several narrow, pointed leaves.

When a household moves to a new site, both men and women bring along cuttings of their individual plants for replanting at the new site. As they settle in, they will obtain additional varieties through exchange, according to their needs. In each case, the individual plant has a known lineage as a descendant of an ancestor that possessed the same properties. Hence, the distinguishing character of an individual *ivenkiki* plant is not, in a botanical sense, its physical similarities and differences with other kinds of *ivenkiki,* but is a more personal or spiritual quality. We have seen that something like this is also true of many varieties of manioc, and this is a general characteristic of Machiguenga classification.

Many other cultivated plants are used as medicines, including tobacco, *Datura, Hibiscus,* and spearmint. These may be ingested but are also often used as the basis for ointments and poultices. Other nonfood cultigens in the gardens include materials for manufacture, such as balsa, *tivana* fiber, a number of palms, and other useful items, such as gourds and arrow cane. Housegardens also contain flowering plants, many of them native to the area, that are simply enjoyed for their beauty. Often, these cultigens are of wild plants that may not be available in the forest if another household has harvested them. The Machiguenga believe that many of their domesticates will survive in the wild. Pineapples, bananas, oranges, avocados, and other cultigens are often discovered to be growing in the forest and are assumed to be wild. These will be spared and nurtured if the land is cleared for a garden.

Weeding

Weeding (*notsamaitakotakerora*) is continuous work for the Machiguenga men. When I asked one how often he weeded, he replied, "We are always weeding—weeds don't wait." Weeding for men has that quality that spinning cotton thread has for women: though all other tasks be done, there are still weeds to pull.

GENERAL DESCRIPTION

Ideally, weeding is done every 6 weeks. The Machiguenga say that after 6 weeks weeds enter a growth spurt that is so rapid that it is difficult to catch up. Furthermore, many fast-maturing weeds go to seed after this period and make future weeding even more difficult. One day, I came upon a man weeding in a garden that appeared to me to be rather free of weeds. Since his field was no more weedy than my own garden, I asked him when he thought I should weed mine and he said, "Now."

Accordingly, I went to my field and began weeding, working primarily with the machete until coming to the conclusion that weeding by hand was far more effective. I later observed that the Machiguenga rarely use the machete when weeding, except for tough vines and plants with nettles, thorns, or sharp edges. What struck me most was how easy it is to damage the tender young crops (another disadvantage of the machete). In order to work quickly, the main skill is to make instantaneous discriminations between cultigens and weeds. Working mainly by hand, it is possible to pull the weeds out by the roots, while not damaging the crops at all; many men even take a few moments over a bent or damaged cultigen to prop it up or repair it.

I worked alongside Mariano and Carlos for 3 hours, weeding Mariano's garden. They had chewed coca and during this period they took no breaks at all. When I left, around noon, they paused to renew their coca wads, but then worked more or less continuously throughout the afternoon. They work straight up the slope of the garden. I quickly fell into this pattern because it puts your body in the best position for weeding: legs and knees downhill, body leaning towards weeds that are close at hand on a steep slope, balance natural and easy to maintain. Each man works independently, clearing a swath about 3–6 m wide. The width is often determined by the positioning of unburned logs (most of which have been felled downhill and hence lie parallel to one another). Men tend to work between parallel logs, using them as borders. They may work 30–40 m uphill this way, before returning to start a new swath off to one side.

At one time or another I observed nearly all the men of Shimaa weeding, and they were invariably meticulous about it. Ideally, each weed is pulled out by its roots. This is easier in soft, sandy soils than in clay. When a root came up covered with dirt, the dirt was brushed away to prevent the weed from rerooting in the damp soil. Weeds that had to be cut were cut very close to the ground (within a centimeter).

In smaller gardens, it is usual to remove the weeds and other small rubbish to the edges of the garden. It is not uncommon to come across a Machiguenga garden that is completely weed-free, like one of our own formal gardens. This is especially true of first-year gardens, which are much better weeded than older gardens. But I know of a few gardens 4 or 5 years old that are continuously maintained free of weeds. Julio, for example, passes through an old garden near his house frequently, on his way to do other tasks; and as he does so he stops here and there to pull whatever new weeds have recently sprung up. Even when this kind of care is not taken, most gardens are rather clean because weeds and other rubbish can hide snakes. Hence gardens, like major trails and the clearings around houses, are kept free of debris.

LABOR COSTS OF WEEDING

The main factors influencing the amount of effort expended in weeding are the amount of time that has elapsed since the previous weeding, and the kind of weeds the predominate in a garden. This latter, in turn, appears to depend on a combination of the kind of soil and the number of years the field has been in cultivation. At the extremes, my measurements show a variability of a whole order of magnitude, from less than 1 m^2 of land weeded/minute to nearly 9 m^2/minute. For the great majority, however, the average rate of weeding seems to fall between 1 and 2 m^2/minute.

The weeding is easiest when weeds are young. As time goes by, weeds become denser and so firmly entrenched they cannot be pulled but must be cut. Within just a few days, the cut weeds have sent up new shoots and will be even more stubborn at the next weeding. For one period of 3 to 4 months Mariano neglected his garden. Many men commented that he was falling behind and would be unable to catch up. When he did resume weeding, with the help of his male relatives, weeds throughout the garden were moderate to heavy, and work proceeded at a low average rate (2 m^2/minute). His garden was over 1 ha in size, and would have required about 100 hours of labor to weed completely. Mariano handled this by weeding in the center of his garden, allowing weeds to encroach at the borders. He explained to me that he planned to weed the rest of the garden later, but he never did. Two years later this garden was completely abandoned, having produced for 3 years (about average). Mariano blamed his problem with weeds on the poor burning that had taken place. But another informant discounted this reason in Mariano's case: "There are always weeds. The day after you burn your garden, the weeds are beginning to sprout—the fire doesn't kill them. Weeds must be controlled. Mariano let the weeds get out of hand, and you see the trouble he has."

The fastest weeding I observed was in a new garden where the weeds were small (about 15–45 cm high) and the soil was soft, gravelly *shimentya*. Here weeds were scattered and could be pulled virtually without effort. At the opposite extreme, the toughest weeding is in older gardens with more clay in the soil. These gardens, after long use, tend to be taken over by grasses. By comparison with bushy weeds like the ubiquitous *tsiregiro* (*Bidens pilosa*), grasses are tough and impossible to pull by hand. They must be cut by machete in an arduous sort of "hand mowing." Such was the case in my first garden, on land that was given to me because no one else wanted it, following about 5 years of continuous cultivation and less than a year of fallow. The worst grass of all was on the airstrip, where periodically the schoolteacher browbeat the men into forming a line of swinging machetes to keep the strip in shape for planes. Hence, my experience in Shimaa would tend to support Boserup's (1965:24, 29–30) argument that the hoe is a technological response to the predominance of grasses among weeds as a consequence of the intensification of agriculture. Under normal conditions in Shimaa, weeds never get to the condition where a steel tool is crucial to the task.

WEEDING AND HARVESTING

Weeding in the new garden is frequent. But after the harvest of corn, when the garden is about 6 months old, the frequency of weeding tends to decline. The maize harvest is also a major weeding time, and the weeding goes slowly because of the frequent stops to harvest the maize in the area just weeded. Thereafter, whenever any crop is harvested, the area around it is thoroughly weeded and all weeds and rubbish are removed to the side of the garden. Then the crop (except maize) is replanted. In between harvests, while the replanted crops are maturing, weeding is much less frequent than before (see Table 2.5). Along with other evidence to be re-

Table 2.5

Density of Weeds According to Age of Field

	Density of weeds (number of gardens)	
Age of field	None to lightly moderate	Moderately heavy to abandoned
First year	8	1
Older	2	8

viewed later, it will be clear that the "return to fallow" begins even before the end of a garden's first year, as the demands of starting a new garden increase the neglect of the aging one.

Harvest

Harvest (*nagakerora*) is, like weeding, an activity that goes on continuously throughout the year. With the exception of maize, Machiguenga crops are harvested only as they are needed or, as in the case of pineapple or papaya, as individual fruits reach the desired state of ripeness. Harvest is also the garden activity most likely to include women and youngsters.

GENERAL DESCRIPTION

The harvest of maize may be said to begin the transition from a new garden to an old one. We have already seen that maize is planted only once in a garden, and its harvest is the occasion of a major weeding. Actually, maize is harvested green in small quantities during October and November. The Machiguenga treat this corn-on-the-cob as a special dish, not as a staple. But the main harvest of maize begins in December, when the kernels have ripened to hardness and begun to dry. Owing to the labor invested in weeding at this time, a man might work all day and harvest only about 25 kg of maize. Therefore, the harvest of maize in a field may take 10 or more visits to the garden over a period of 2 months. Since individuals have planted their gardens at different times, maize will be harvested throughout the period from October to May.

Commonly, a man and his wife will cooperate to harvest maize. The man will weed a certain area (perhaps 100 m^2) and then harvest all the maize within, cutting the stalks close to the ground and piling them near his wife. She then removes the ears and partially or completely husks them. Most men will, in the meantime, remove all weeds and other litter to the edge of the garden. This struck me at first as a waste of green manure, but there is less danger of snakes if the garden is continuously swept clean of debris, and it seems the soils are fertile and abundant enough to make manuring and other labor-intensive techniques unnecessary.

Maize is the first crop harvested from a new garden. Indeed, most of the root crops will not be harvested during the first 12 months, and the 20% or so that are will tend to be harvested toward the end of the first year. By this time the garden will have entered the less orderly manner of cultivation characteristic of old gardens.

The harvest of manioc is typical of the way older gardens are maintained. Either men or women may harvest manioc. Men use a machete, women a kitchen knife; for this reason, women prefer to harvest less mature plants or varieties known to have small roots. Women also work much more slowly than the men. Apart from these differences, however, the following description of Felipe harvesting manioc in May 1973 gives a general idea of the process. Starting at about nine in the morning, he spent about 10 minutes cutting the leaves and stems to use for cuttings in replanting, discarding the smaller stems (*gara oshivoi,* "it won't grow"). He then took 12.5 minutes to sweep the area carefully, removing all the weeds and other loose rubbish to the edge of the garden.

Felipe then began to dig up the tubers, using a slender, pointed stick of palm wood to dig carefully around the tubers. This was a long-tubered variety (*ogatsantsapairi*) and one of the tubers was almost 1.5 m long; Felipe cleared the dirt around the whole length of the tuber before pulling it out of the ground. The manioc tubers grow out from the stalk like the legs of a starfish, and during the harvest it is common to clear away the dirt from all the tubers, then pull them all out at once while they are still attached to the main stem. Felipe then replanted new cuttings in the same holes he had just harvested, where the dirt had been softened in the course of the harvest. Altogether this took another 35.5 minutes. Thus, in 58 minutes Felipe harvested and replanted seven mature manioc plants, yielding exactly 30 kg of tubers.

In a few days it would be necessary for someone from Felipe's extended family to harvest more manioc. Each time a man goes to harvest manioc—this is less true when women do the harvesting—he takes the opportunity to do some weeding in addition to that involved in the harvest itself. Indeed, any time a man passes through a garden, he takes a few minutes to pull out weeds. In this way the more careful farmers manage to maintain old gardens continuously weed-free.

Most other crops are harvested on a casual basis, by women, during short visits to a garden. The women's crops, such as cocoyam, yam, *dali-dali,* cotton, and so on are scattered in low density throughout gardens. A woman typically will harvest only one plant, or perhaps two, at a time. She will bring home no more than a kilo or two of one of the food crops, which will serve as a side dish to the main meal of manioc and whatever wild foods are available. In this and other cases, women's work appears relaxed, almost desultory, compared to the men's, and we see that women provide quality and diversity in food just as men provide quantity.

YIELDS

The mix of foods yielded by a garden depends on its stage of development. Maize is grown exclusively in new gardens, for example, whereas manioc and other root crops are harvestable only after the maize harvest is

Table 2.6

Garden Yields by Type of Garden for an Average Downstream Household, 1972–1973[a]

Crop	New hill (0.56 ha)	Old hill (0.36 ha)	River margin (0.06 ha)	Total (0.98 ha)
Manioc	1149	4687	526	6362
Pineapple	443	780	0	1223
Maize	731	0	0	731
Plantain, banana	207	380	0	587
Papaya	35	274	0	309
Cocoyam	21	210	11	242
Sugar cane[b]	17	201	0	218
Yam	12	26	0	38
Cacao	0	22	0	22
Pigeon pea	2	9	0	11
Guava	0	0	11	11
Beans	0	6	0	6
Coffee[c]	0	0	0	0

[a]In kg edible.
[b]In liters of juice.
[c]Immature.

over; hence the large increase in production of many crops in the old gardens as compared with new gardens (see Table 2.6). This mainly reflects the coming to maturity of many crops already planted in the new garden. However, it is also true that a greater variety of crops appears in old gardens as a result of their evolution, as discussed in the next section.

The figures reported in Table 2.6 are conservative; that is, actual production is probably higher than reported here. Figures have been revised downward to account for loss due to pests and shrinkage of gardens due to weed encroachment. The figures are based on two kinds of data. First, the frequency of plants in gardens was estimated by counting frequencies of crops in random garden samples and taking the mean for each type of garden. Second, the average yield in edible portions of each crop was estimated from direct observation of harvests and weighing of the yield. Note that Table 2.6 reports only the most frequently planted crops. Other cultigens listed earlier were too rare to appear in the random samples.

The Life Cycle of a Garden

As we saw briefly in the opening pages of this chapter, the Machiguenga begin their move to a new location by planting a garden. As it matures, they gradually shift their residence to the new garden. Each year, a household or group of cooperating households will plant a new garden to keep the cycle of gardens going, and in particular to ensure a supply of maize for the coming year.

The new garden is called by the general term, *tsamairintsi* (garden); after the harvest of maize, it is referred to most commonly as *oketyorira,* "old garden." In old gardens, the frequency of all crops except maize is greater per hectare than in new gardens. Furthermore, there is a greater variety of cultigens in an old garden than in a new garden. On the other hand, old gardens tend to be about a third smaller than new gardens because weeds encroach at the borders, shrinking the garden, and, as we have seen, older gardens are less well weeded, on the whole.

After the garden is about 3 years old, it is likely to enter the first stage of abandonment, known as "*ashi shintori*" or "*itsonkatakero shintori*" ("it belongs to peccary" or "peccary will finish it"). In this stage a garden is no longer being weeded, but still has many crops left that may or may not be harvested, depending on need. A garden in this stage may still be weeded and returned to production. Felipe's garden 11-2 (see Table 2.7) was *ashi shintori* in 1972, but was cleared and reburned in the summer of 1973 and planted in maize, like a new garden. In 1975 it was still in production as *oketyorira.* An *ashi shintori* garden can also serve as a lure for peccary. Felipe's field 11-3, which was *oketyorira* in 1972, was *ashi shintori* by 1975. In it he had constructed a *caña brava* hunting blind where he often spent nights lying in wait for peccary. If the *ashi shintori* garden is left untended, the food crops are gradually smothered by the burgeoning secondary growth. For example, in my two gardens, which were abandoned in September 1973, virtually all cultigens had been obliterated by December of 1974. This land is now referred to as *magashipogo,* which we could translate either as "fallow" or as "secondary growth," depending on whether the land will be brought into production again.

As far as I have been able to discover, the Machiguenga do not believe the land loses its fertility through continuous use. Whenever I asked them why certain crops had not done well, they always blamed the seed, in keeping with the individualistic view of plants discussed previously. If pressed, they specifically denied that soils had anything to do with it, even in land that had been in continuous production for 5 years (and had extremely low values of nitrogen and organic matter in the soil analyses). In any case, I have seen a number of cases in which land has been recleared for planting after only 1 or 2 years of fallow.

Usually, the garden closest to the house will be kept in continuous production. This may be referred to as *novankireshi* (house garden—literally, "my plants"; the term may refer to all of a man's plants, in addition to the housegarden). For example, Julio's 12-2 garden was cleared in 1971 and was still in production in 1975 with no visible changes (except that it was 50% larger in 1975). House gardens have by far the greatest diversity of

Table 2.7

Machiguenga Gardens (Downstream Cluster) 1972–1975

			1972 (August)		1975 (August)	
House	Garden	Year planted	Size (ha)	Description	Size (ha)	Description
10	1	1972	0.29	New	—	Second growth
	2	1971	0.47	Old	[a]	Old (housegarden)
	3	1970	0.25	Old	—	Second growth
	4	1972	0.51	New	0.82	Old (housegarden)
	5	1975	—	—	0.68	New (drying for burning)
11	1	1972	0.30	Riverside (new)	—	Abandoned
	2	1970	0.40	Abandoned	0.40	Old (recleared 1973)
	3	1971	0.65	Old	—	Abandoned
	4	1975	—	—	0.85	New (drying for burning)
	5	1975	—	—	0.30	New (riverside, drying for burning)
12	1	1972	0.89	New	0.42	Old
	2	1971	0.36	Housegarden (old)	0.63	Housegarden (old)
	3	1970	—	(River, lost in flood)	—	—
	4	1975	—	—	0.38	New (burned 8/75)
13	1	1972	0.05	Old (cleared)	—	Second growth
	2	1973	—	—	0.44	Old
	3	1975	—	—	0.76	New (drying for burning)
14	1	1972	1.12	New	—	Abandoned
	2	1974	—	—	0.34	Old (river margin)

[a]In 1975, Fields 10–2 and 10–4 were one continuous garden.

crops. A new garden will have about 7 different cultigens in it; an *oketyorira* garden about 40% more (10 different cultigens). But a house garden has closer to 30 distinct cultigens (this is not counting all the named varieties of single species such as sedge or manioc). These will include many medicinal plants, represented by only 1 or 2 plants for each variety. And, new or experimental crops are tried out in the house garden. For example, Aradino had planted rice in his house garden as an experiment in 1972; by 1975 he was planting it in substantial amounts in his new garden.

As old gardens go out of production, new ones are added. In any given year a single household may or may not have a full range of gardens, depending on its relations with other households. Sometimes, a man will devote a year to working in the garden of his brother or brother-in-law; he will then have access to the crops in that field. But ownership will remain in the hands of the other, and if relations between the households cool, as they generally do after a period of time, the man is likely to move away and plant his own garden.

Between 1972 and 1975 the downstream cluster underwent remarkable changes (cf. O. Johnson 1978:74–77). In 1972 the five households (10–14) lived together in a tight cluster a few hundred meters in diameter, surrounded by their gardens. Household 10, however, had already set up dual residence with a house near gardens 10-2 and 10-4, about 4 km from any other houses. In early 1973, members of household 10 considered the new location their primary residence and allowed their old house and gardens (including the newly planted 10-1) to revert to forest.

The members of household 13 had been planning a move for some time, and cleared garden 13-2 in 1973. By 1974 they had set up new residence in the center of that garden, which was about a kilometer away from the original downstream cluster. The brothers in households 11 and 14, meanwhile, had a serious argument and household 14 moved about 2 km downstream where they subsisted from a new river-margin garden after 1974.

Hence, in the space of about 2 years the original cluster had virtually disintegrated. This centrifugal movement is a characteristic phase of Machiguenga settlement, which moves cyclically between aggregation into clusters and dispersion into isolated households. The five households of the downstream cluster had come together and split apart several times before 1972. The reasons for this are too complex to be addressed here, but they are primarily ecological. The Machiguenga enjoy the economic security of large cooperative kin clusters, but become irritable and yearn for isolation when local competition for resources, including the best soils, fishing areas, hunting and gathering areas, and clean water, become intense through protracted settlement (A. Johnson 1982).

Between 1972 and 1975 the population of the five households changed only slightly with the birth of two girls (in household 14). Of course, all the children grew 3 years older, so dietary needs of the households probably increased significantly, but not dramatically. We see in Table 2.8 that the amount of land in new hillside gardens in 1975 was virtually identical to that in 1972. The major change in this period was that the area in cultivation in old hillsides in 1975 had increased by half the 1972 amount. This

Table 2.8

Summary of Size of Downstream Gardens, 1972–1975[a]

	New hill	Old hill	River margin	Total
1972				
Cluster total	2.81	1.78	0.30	4.89
Mean per house	0.56	0.36	0.06	0.98
1975				
Cluster total	2.66	2.71	0.64	6.01
Mean per house	0.53	0.54	0.13	1.20

[a]In hectares.

major difference has more to do with modern change than traditional economics. In 1972 members of the downstream group had no cash crops under cultivation, but were watching with interest experiments in coffee, cacao, and rice undertaken by the men associated with the schoolteacher of Shimaa. By 1975, men in the downstream group had planted large numbers of coffee trees, and were continuing to care for them in old gardens that formerly would have been returned to fallow.

Input–Output

In 1972 the downstream cluster included 37 people in five households. The composition of an average downstream household appears in Table 2.9.

The field observations reported earlier in this chapter, in combination with time allocation estimates based on random sampling (A. Johnson 1975), make it possible to estimate the hours of labor contributed to these gardens each year (Table 2.10). Males contribute over 75% of all garden labor and are involved in the widest range of tasks. Women and girls are primarily involved in the harvest.[2]

Using energy expenditure estimates on adults (Montgomery and Johnson 1977), and adjusting for the average body weight in each category in Table 2.9, the time estimates can be converted into energy expenditures (Table 2.11). Because men work at higher rates of energy expenditure than women, they account for fully 85% of garden labor, measured as calories expended. Tables 2.10 and 2.11 show that weeding is the major category of labor input, followed by harvest. Together, these account for 75% of labor expended in gardens.

For purposes of comparing output to input, Table 2.12 provides a breakdown of the nutrient content of the foods listed in Table 2.6. Total garden production for an average Machiguenga household provides an abundance of nutrients, without even counting the rich contributions made by fishing, hunting, and gathering.[3] Tables 2.6, 2.11, and 2.12 may be used to develop measures of efficiency of garden production useful for comparisons with other studies. For example, dividing the total calories produced (from Table 2.12) by the total calories of work expended (Table 2.11), we find that the energetic efficiency of Machiguenga gardens is close to 20:1. I would note, however, that this input figure represents only labor expended in the garden itself. Tool manufacture, food processing, and other costs of production should also be taken into account, but that is too extensive a task to be undertaken in this chapter.[4]

Table 2.9

Mean Number of People and Body Weight in Average Downstream House

	Males		Females		
	Mean number per house	Mean body weight[a] (kg)	Mean number per house	Mean body weight[a] (kg)	Total number per house
Household head	1.0	51.8	1.4	40.5	2.4
Unmarried adult	0.2	51.8	0.8	41.8	1.0
Youth (age 6–13)	0.6	31.8	1.6	24.6	2.2
Toddler (age 2–5)	0.4	10.9	0.2	[b]	0.6
Infant (age 0–1)	0.8	8.2	0.4	4.1	1.2
Total	3.0		4.4		7.4

[a]Used to calculate energy expenditure in Table 2.11
[b]No data.

Table 2.10

Garden Inputs: Labor Time in an Average Downstream Household

Household member	Task (hours/year)							
	Clear	Burn	Plant	Weed	Harvest	Travel[a]	Other[b]	Total
Males								
Head	115	17	217	681	268	40	21	1359
Unmarried adult	—	4	14	86	40	—	—	144
Youth	—	26	—	307	143	—	—	476
Toddler	—	—	—	—	9	—	—	9
Infant	—	—	—	—	—	—	—	—
Subtotal								1988
Females								
Head	—	—	—	67	277	40	—	384
Unmarried adult	—	—	—	—	138	35	—	173
Youth	—	—	—	—	76	—	—	76
Toddler	—	—	—	—	—	—	—	—
Infant	—	—	—	—	—	—	—	—
Subtotal								633
Total	115	47	231	1141	951	105	21	2621

[a]"Travel": all trips to garden, including bringing in the harvest.
[b]"Other": primarily seed selection and preparation of disposable tools (e.g., digging sticks, tumplines).

Table 2.11

Garden Inputs: Energy Expenditure in an Average Downstream Household[a]

Household member	Task (kcal/year)							
	Clear	Burn	Plant	Weed	Harvest	Travel	Other	Total
Males								
Head	49,700	8,800	61,200	241,100	75,600	13,900	6,800	457,100
Unmarried adult	—	2,100	3,900	30,400	11,300	—	—	47,700
Youth	—	8,300	—	68,200	24,900	—	—	101,400
Toddler	—	—	—	—	500	—	—	500
Infant	—	—	—	—	—	—	—	—
Subtotal								606,700
Females								
Head	—	—	—	9,200	44,900	7,000	—	61,100
Unmarried adult	—	—	—	—	23,200	6,300	—	29,500
Youth	—	—	—	—	7,800	—	—	7,800
Toddler	—	—	—	—	—	—	—	—
Infant	—	—	—	—	—	—	—	—
Subtotal								98,400
Total	49,700	19,200	65,100	348,900	188,200	27,200	6,800	705,100

[a]Figures rounded to nearest 100.

Conclusion

I have summarized aspects of Machiguenga horticulture, touching both on ethnoecological descriptions of how the environment and system of production are conceptualized, and on observer-oriented descriptions of behavior. These provide strong evidence that the Machiguenga are thoughtful and knowledgeable horticulturalists who successfully meet their subsistence needs without causing irreversible degradation of their environment.

The productivity and stability of their horticultural system is undoubtedly a reflection of the low population density at which they now live. In a broader sense, their low density permits eclectic, opportunistic procurement strategies such that members of each household make individual assessments of the needs and the relevant scarcities confronting the household. These vary with time and place: at one moment labor may be especially scarce; at another, good garden land; at still another, drinking water or fishing spots.

The Machiguenga are remarkably free from social constraints in developing procurement strategies. There is no warfare in their region, no villages

Table 2.12

Annual Production of Nutrients from the Gardens of an Average Downstream Household[a]

Crop	Calories (kcal)	Protein (g)	Calcium (mg)	Phosphorus (mg)	Iron (mg)	Vitamin A (mcg)	Thiamine (mg)	Riboflavin (mg)	Niacin (mg)	Vitamin C (mg)
	Nutrients									
Manoic	8,397,800	63,600	2,544,800	2,163,000	89,100	0	3,200	2,600	38,200	1,208,800
Maize	2,638,900	68,700	65,800	2,119,900	18,300	511,700	3,100	700	13,900	0
Plantain, banana	716,100	5,900	47,000	199,600	4,700	1,027,300	400	200	3,500	117,400
Pineapple	636,000	4,900	220,200	97,800	6,100	183,400	1,000	500	2,400	746,000
Cocoyam	222,700	3,900	232,300	213,000	2,900	12,100	200	100	1,700	16,900
Sugarcane	178,800	700	28,300	26,200	1,500	0	50	20	200	4,400
Cacao	100,300	2,600	23,300	118,100	800	3,300	0	30	400	700
Papaya	98,800	1,500	61,800	40,200	1,200	339,900	100	100	900	142,100
Yam	44,100	500	11,800	14,100	400	11,400	50	20	300	11,800
Pigeon pea	13,000	800	3,900	13,600	200	2,200	0	20	300	3,300
Beans	9,000	600	3,500	12,800	200	600	0	0	100	400
Guava	7,600	100	2,400	2,900	100	8,800	0	0	100	24,000
Total	13,063,100	153,800	3,245,100	5,021,200	125,500	2,100,700	8,100	4,300	62,000	2,275,800

[a]Estimates of the nutritional content of foods are from INCAP (Institute de Nitrición de Centro América y Panamá) (1961). Figures are rounded to nearest 100.

or superordinate political structures, no lineages or other named social groupings beyond the household, and a very loose "kindred." At any time, a married couple and their offspring can move and set up a new residence, if present circumstances are sufficiently irksome. From a comparative perspective, the Machiguenga enjoy freedom at the expense of a more efficient use of their land. Their horticulture is wasteful in the narrow (and, I believe, misleading) sense that a great deal of food energy is produced that is never consumed. An average household produced 13 million calories of food energy in the 1972–1973 season, but probably consumed less than 5 million calories. Even allowing that this year was possibly one of greater-than-ordinary production (people the previous year complained there was "no manioc," and increased their production to compensate), overproduction of calories is clearly a basic feature of the system. What people meant in 1972 by "there is no manioc" was that their own gardens were producing not quite enough for a period of time and they had to request manioc *occasionally* from other households. There was plenty of manioc in the area and never any possibility that anyone would go hungry. But the Machiguenga prefer to be completely self-reliant in their production of staple foods. Hence, where other communities might reduce their need for a security margin of food by establishing extensive networks of mutual obligation, the Machiguenga maximize autonomy through overproduction. In a sense, their abundant land per capita and a regional absence of warfare give them the luxury to do so. Under modern change, which encourages population growth and aggregation into densely settled villages, we can expect to see this pattern disappear and to be replaced by the characteristic economic and social ills of densely settled horticulturalists in tropical regions.

Notes

[1]The Machiguenga rarely use personal names. Many men and most women do not have personal names; they use kin terms. They have been given Spanish names in recent years, by missionaries and schoolteachers, and I use those here for the reader's convenience.

[2]Note that Table 2.11, being based on random samples of behavior, probably misses some small labor inputs due to sampling error. For example, women plant women's crops and probably require about 7 hr to do so, even though in Table 2.11 this input is overlooked.

[3]See Johnson (1980). Note that the data reported here may differ from data reported in earlier publications because the population and some features of the model used to calculate production data have been revised for greater accuracy and authenticity.

[4]If the Machiguenga planted bitter manioc, its heavy processing costs would have to be included before any comparisons could be drawn. Time and energy spent in manufacture and food processing by the Machiguenga may be estimated from data in Johnson (1975) and Montgomery and Johnson (1977).

CHAPTER 3

The Cultivation of Manioc among the Kuikuru of the Upper Xingú

Robert L. Carneiro

Introduction

The cultivation of manioc is the principal subsistence activity of many Indian tribes of Amazonia. Probably none, though, grows this plant more extensively than do the Kuikuru of central Brazil. In the pages that follow I describe in detail the full cycle of manioc cultivation among this group, from the clearing of garden plots in the forest to the consumption of the processed root.

The Kuikuru are one of three Carib-speaking villages located at the headwaters of the Xingú River. I first studied them in 1953–1954, when they lived near the shore of Lake Lamakuka, south of where they are now located. By 1975, when I visited them again, they had moved north in order to be included within the boundaries of the Xingú National Park (see Carneiro 1977:215n.). Most of my data on Kuikuru manioc cultivation were obtained during this second visit. However, in the following description, I shall also draw on information from my first trip. Kuikuru agricultural practices seemed not to have changed in the intervening 21 years.

The Habitat

The habitat occupied by the Indian villages of the Upper Xingú basin, which lies between 12 and 13° S, is largely one of tropical rain forest.

ISBN 0-12-321250-2

However, because the Indians repeatedly burn the grass and scrub that come up in fallow fields, large areas of savanna have been created in the inhabited parts of the basin. These savannas are not suitable for cultivation.

The Kuikuru distinguish several types of forest. First and foremost is *itsuni*, primary forest. *Itsuni* is the climax vegetation of most of the upper Xingú, and still covers the greater part of the region. The Upper Xingú lies close to the southern limit of tropical rain forest in this part of Brazil, and indeed, what we see on a vegetation map of the region is a tongue of forest, representing the Upper Xingú, extending south and flanked on either side by savanna. It seems likely that the presence of forest here is associated with conditions of relatively high moisture attributable to the many rivers and lakes of this circumscribed region.

The forest near the present-day Kuikuru village of Ipatse is not as tall as that near their former village of Lafatuá, some 15 miles (24 km) to the southeast. Since I saw only one really large tree in the *itsuni* surrounding Ipatse, it is clear that the primary forest here is relatively young. No doubt, at one time or another, all of this area was cut over for cultivation by its earlier inhabitants. The forest that has grown back since then is now primary forest in terms of the tree species that make it up. But it is still, I would judge, a century or more away from being climax.

The vegetation that invades an abandoned garden is called *tafuga* by the Kuikuru. *Tafuga* is, essentially, secondary forest. The trees most typical of it have light green foliage, soft wood, grow rapidly, and need ample sunlight. However, under the name *tafuga*, the Kuikuru include saplings and seedlings of primary forest trees, many of which have come up as coppice shoots from stumps left over after the felling and burning of *itsuni*. The Kuikuru never clear *tafuga* for planting a garden, but wait until it has evolved into a more advanced form of forest.

Vegetation that has advanced beyond the stage of *tafuga* but is not yet *itsuni* is called by several different names. What name the Kuikuru give to it depends on which of several species of trees predominates in such a forest. Thus, *agïpe* is forest in which the *agï* tree is common, *agafagïpe* is forest in which *agafagï* is common, and *kejitepe* is forest where *kejite* is common. (The suffix *-pe* means "thing" in Kuikuru.)

These forests are not successive stages of a single cycle of forest succession. Rather, they are local variants of the same general type—forest derived from *tafuga* and on its way to becoming *itsuni*.

Agïpe-, *agafagïpe-*, and *kejitepe*-type forests seem to develop slowly, primarily because the scrubby undergrowth beneath them is intentionally burned every dry season. Indeed, it seems likely that the reason *agï*, *agafagï*, and *kejite* are especially common in these forests is that they are fire-resistant species that are able to gain ground at the expense of other species less tolerant of fire.

Although burning the undergrowth under these intermediate-type forests retards the return of *itsuni* to the area, fire seems to enrich the soil. In fact, *agïpe, agafagïpe,* and *kejitepe* are all suitable for clearing and planting manioc. Analyses of soil samples I made in garden plots of newly cleared and freshly burned *itsuni* and *kejitepe* showed that soil under *kejitepe* was slightly higher in mineral content (see Table 3.1).[1] Had I sampled the two plots after clearing, but before burning, the results would have been a more accurate gauge of their relative fertility. Still, it seems safe to say that *kejitepe* (and *agïpe* and *agafagïpe* as well) are no less fertile than *itsuni.*[2]

Another type of forest recognized by the Kuikuru is *egepe.* This type, though, is not named after a particular tree. That is, there is no "*ege*" tree. The defining feature of *egepe* seems to be the type of soil underlying it. Most *itsuni* grows in red, sandy soil called *ŋoŋo,* but underlying *egepe* is black earth, called *tumbutïñï.* (For a recent article proposing an anthropogenic origin for the black earth in Amazonia see N. Smith [1980].) *Itsuni* and *egepe* soils differ markedly in their productivity. The Kuikuru say that black *egepe* soil will produce much bigger manioc tubers than the red earth of *itsuni,* sometimes as thick as a man's thigh. And corn will grow poorly in red earth, so it is always planted in *egepe.* Although the Kuikuru would prefer to plant their manioc in *egepe* too, if they could, there is not enough of it around. *Itsuni,* though, is plentiful and nearby, and manioc, which is not very demanding of soil nutrients, yields more than adequately in cleared *itsuni.*

The Kuikuru distinguish one more type of forest, which they call *indagïpe.* As in the case of *egepe,* there is no "*indagï*" tree. *Indagïpe* is simply forest growing along the margins of rivers and lakes. Thus the term is essentially equivalent to our "gallery forest." The species of trees that make up *indagïpe* seem pretty much the same as those comprising *itsuni.*

Table 3.1

Comparison of the Mineral Content of Two Soil Samples from Areas Near the Kuikuru Village of Ipatse[a]

				Pounds per acre (2,000,000 pounds of soil)										
	Organic matter	pH	Ex H (me/100 g)	P	K	Mg	Ca	Mn	Fe	Al	NO_3 N	NH_3 N	Zn	Soluble salts K × 10^5
Itsuni (Plot A)	3.5	3.8	13	3	140	80	150	8	53	175	210	70	1	50
Kejitepe (Plot B)	3.6	4.7	11	5	155	85	150	19	26	165	5	65	1	10

[a]Plot A represents an area of *itsuni* forest recently cleared, burned, and hoed. Plot B represents an area of *kejitepe* forest also recently cleared, burned, and hoed.

Clearing the Forest

SITE SELECTION

A man who is ready to clear a new manioc plot has two choices as to its site. He can clear a tract of forest just beyond his current plot, or he can clear a plot in some other part of the forest. Less work is required if a new plot is cleared adjacent to an old one. For one thing, manioc cuttings for planting do not have to be carried so far. For another, there will be less fencing to do. But there is also an advantage to having one's garden plots widely separated. By doing so, one provides a certain degree of insurance against crop losses to marauding troops of peccaries (*Tayassu pecari*) The latter advantage seems to outweigh the former, for in more than three-quarters of the cases, a man clears his second manioc plot in a different location from his first.

In selecting the site for a new garden, a man is free to pick any part of the forest. The Kuikuru concept of ownership does not extend to standing forest. Only land that has been cleared and planted can be claimed as private property. And property rights to this land cease when the plot is abandoned. Thus, anyone can cut *sapé* (*Imperata* sp.) grass for thatching from another man's abandoned garden without any feeling that he has to ask permission. The only qualification here is that if a man has planted *piquí* trees (*Caryocar brasiliensis*) in his garden before abandoning it, he and his heirs after him continue to own the trees and their fruit.

Nevertheless, more as a courtesy than as an obligation, a man who wants to clear a tract of forest next to someone else's garden will ask his permission before doing so. This is done in case the latter has plans to clear that tract himself. If so, he is thought to have prior rights. All this is handled informally and amiably, and disputes over clearing rights are all but unheard of.

Looking at Kuikuru manioc gardens from the air, one can see that some occur alone in the forest, some exist in pairs, while others form part of a large composite plot comprising perhaps a dozen adjacent gardens. (These gardens, although sometimes cleared collectively, are always owned individually.) Over time, a composite plot "advances" through the forest, as older parts of it are abandoned and new areas of forest cleared ahead of it.

CLEARING THE UNDERBRUSH

Once a man has selected the site for a new garden, the next step is to clear the undergrowth. All swidden cultivators clear the low vegetation first, in order to make the trees more accessible for felling. The Kuikuru call the undergrowth *ofiña* and the trees *i*. Clearing the undergrowth is *ofiñakenïgï* and felling the trees is *íkenïgï*. Today, every Kuikuru has a machete for clearing the underbrush, but until the Upper Xingú basin was discovered by Europeans in 1884, and probably for a couple of decades thereafter, the Kuikuru had to rely on their aboriginal tools for clearing and felling.

For felling trees, the Kuikuru formerly used a stone axe, called *ï*. For cutting the small vegetation between the trees (up to an inch (2.5 cm) or so in diameter) they used half the mandible of a large, yellow-faced piranha, *kuakueŋgï,* with the teeth left in place. With this implement they sawed through woody vegetation too slender to offer enough resistance to be cut readily with the stone axe. Smaller saplings and seedlings were either pulled out or broken and bent over by hand about two feet (60 cm) above the ground.

Now let us see how the undergrowth is cleared today. If a garden is to be made just beyond an existing one, the low vegetation is first cut around the perimeter of the proposed new plot, to mark its extent. After this, a man simply clears the area, beginning at the end of his old plot and working his way to the outer edges of the new one. If the plot to be cleared is not adjacent to an old one, but is off by itself in the forest, no measuring is done beforehand to determine its size. A man just picks a spot and starts clearing out around it. When I asked an informant if the Kuikuru paced off a new plot to determine how large it would be, he replied, in Portuguese, 'só cabeça'—"only the head"—meaning he gauges it only by eye.

Traditionally, the Kuikuru make circular gardens. After clearing as much of the low vegetation as he thinks adequate, a man looks over the plot. If some part of the perimeter looks straight to him, he cuts down more of the undergrowth to round it out. However, when several plots are cleared next to one another, then, of course, they cannot all be circular, and are either rectangular or subrectangular.

Although the Kuikuru say that in certain parts of the forest there are lots of *etseke* (evil spirits) whom they fear, they do not allow this feeling to interfere with clearing the forest.[3] Asked about this, one informant said that when a man cuts the trees, the spirits take off.[4]

There is some variation as to when a man begins clearing a new manioc plot. Most men fell the trees in June, nearly 2 months after the rains have stopped, but Afukaká, the young Kuikuru cbief, did not do so until the second week of August. In the Upper Xingú the dry season is completely dry, so that even if the trees are left on the ground for only a month and a half after felling, they will burn well enough.

Nowadays, the low vegetation is cut with a machete. In doing so, a man leans over at the waist. He cuts the saplings with a diagonal stroke about a foot or a foot-and-a-half (30–45 cm) above the ground. Even after the plot is burned, sapling stubs protrude above the ground and, being charred, are difficult to see. The unaccustomed visitor, threading his way through a newly burned plot, is forever tripping over them.

When clearing the undergrowth, a Kuikuru generally begins cutting a certain area and then moves elsewhere to cut another area, rather than clearing systematically out from a single spot. Nor are the Kuikuru very thorough in clearing the low vegetation. A good deal of it is left uncut, in fact, to be consumed by the fire.

FELLING THE TREES

Once they have cleared the undergrowth to their satisfaction, the Kuikuru proceed to fell the trees. As I have said, before they adopted steel axes early in the century, they cut down the forest with stone axes. The Kuikuru did not make the axheads themselves but got them in trade from the Trumai, who in turn obtained them from the Suyá, who actually made them (Steinen 1940:249). However, the Kuikuru hafted the axheads, using *kų́ǫ́* wood for the handle. The axes were of the "imbedded celt" type (Carneiro 1974:107, 120), and looked like the specimen shown in Figure 3.1. When they got dull, axes were sharpened on stones—probably slabs of abrasive sandstone found in riverbeds. In all likelihood, when engaged in felling a tract of forest, the Kuikuru probably had to sharpen their stone axes every day (see Carneiro 1979a:41).

No living Kuikuru remembered the days of the stone ax, so I cannot offer an eyewitness account of how it was used. But there is no question that felling a tree with such an ax was arduous and time consuming. When Karl von den Steinen entered the Upper Xingú basin in the 1880s and introduced steel axes to the Bakairí and other Xinguanos, the Indians were very impressed. Indeed, Steinen left us a graphic account of how one Bakairí dramatized the difference in efficiency between the two implements:

> "How hard a Bakairí must work to fell a tree. Early in the morning, when the sun, *tschischi*, rises in the east, he starts work with his stone axe. *Tschischi* climbs the sky and the man keeps working hard, *tsök, tsök, tsök*. His arms get tired." (The mimic's arms slacken and hang down, he acts winded and looks exhausted.) "He keeps chopping, no longer vigorously, *tsök, tsök*, but groaning. The sun reaches zenith. His belly is empty." (The actor rubs his stomach with his hand and looks miserable.) "How hungry the Bakairí is. Finally, as *tschischi* is setting, the tree falls." (He says *tokále*, and indicates "one" with his little finger.) "But you, Caraíba, white man," (suddenly the mimic is full of life and vigor), "you take your steel axe, raise it high, and bring it down with force, *tsök, tsök*, and boom: the tree is down. One firm kick and it's on the ground. All around trees are falling. And now we say to you, Caraíba: give us your steel axes [Steinen 1897:74]."

Today, every Kuikuru owns a steel axe, or can readily borrow one, and it is with this implement that the forest is cut down.

In felling the trees in his plot, a man may work alone or be helped by a close kinsman or friend. Usually, he attacks the biggest trees first, beginning with those near the center of the plot. These he tries to fell in such a way that most of their length will fall outside the plot. The idea is to reduce as much as possible the area of the plot covered by felled vegetation.

Figure 3.1 *Stone ax of the type used by the tribes of the Upper Xingú basin in the nineteenth century. (After Steinen 1897:197.)*

Another reason for felling the large trees toward the outside is to knock down some of the trees just beyond the perimeter of the plot, which would otherwise remain standing. This helps to open up the plot a bit, allowing more sunlight to reach the growing crops when the sun is low in the sky.

In attacking large trees, careful attention is paid to the direction in which they are felled. The Kuikuru prefer to have trees fall parallel to each other, rather than having one land at right angles on top of another, since this makes it harder to get around in the plot. But the way a tree falls cannot always be precisely controlled, and a certain amount of criss-crossing of fallen trunks is inevitable.

If a tree is growing straight, a Kuikuru can usually make it fall much as he wants it to. This is done by notching a tree on the side in which he would like it to fall, and then cutting it through on the opposite side. A tree with a decided lean, though, must be felled in the direction in which it is leaning. Trying to make it fall in some other direction is risky because it might land on top of the man cutting it. If vines link the crown of a tree to others, this may well cause it to fall in a direction different from that intended.

The Kuikuru, who now live at Ipatse, clear their gardens in young primary forest where there are no huge trees with flaring plank buttresses. Thus, they have no need to build scaffoldings around the buttresses in order to cut above them. But they are familiar with the technique, and used it when they cleared gardens in the larger forest around their older village site of Lafatuá.

THE DRIVING TREE FALL

In clearing the forest, the Kuikuru make use of the driving tree fall. Smaller trees growing in the path of a large one are partially cut through so that in falling, the large one will easily topple the small ones in its path. A falling tree may also pull down surrounding ones if its crown is attached to theirs by vines. The driving tree fall, then, saves the Kuikuru the considerable labor of having to cut down each tree individually.

A tree left bent over and under tension after being hit by another one is cut through without delay. Otherwise it may snap at any time and hit someone. Even for an experienced Kuikuru, tree felling is not without risk. A tree may easily fall in an unexpected direction, making the man felling it scurry out of the way.

NOTCHING

Notching is the first step in felling a tree. After chopping into it at a certain point, a man shifts his position and chops a little to the left or right of the first cut. When he is finished, he leaves a notch which forms an obtuse V. The part of the tree cut through in making the notch may represent as much as 40% of its cross-section area. Having finished the notch, a man moves to the opposite side of the tree and begins chopping. The Kuikuru chop hard, fast (almost a stroke a second), and with little or no interruption until the tree falls.[5]

Nowadays, with the steel ax universal among them, the Kuikuru fell trees of any size. Occasionally, though, a very large one may be girdled and left to die. When dead, it will shed its leaves and thus not shade the growing crops. Before steel axes, though, a large tree presented more of a problem. Sometimes it was killed by girdling, but sometimes it was burned down. Firewood, including stacks of arrow cane, were piled around the base of the tree and ignited. Every few hours someone would go out to the tree and throw more wood on the fire, or rekindle it if it had gone out. Two or three days later, the tree would fall. As part of felling a tree with fire, the Kuikuru may have hacked out the charred portion of the trunk and then rekindled the fire, as other tribes are reported to have done (Carneiro 1974:114, 117).

After the trees have been felled, a Kuikuru plot has the usual messy appearance of any newly cut swidden, with trees lying every which way on the ground. An isolated tree may even have been left standing, to be felled later. In this condition, the plot is left to dry for 2 or 3 months.

BURNING THE PLOT

The next step in preparing a manioc garden is burning the slash. This process is called *ifotelï* by the Kuikuru. Most men burn their plots during the first two weeks of August. They know the time for *ifotelï* has arrived when turtles start laying their eggs on the sandy beaches of the Kuluene River, and the constellation *ofoŋo,* the duck, is seen in the eastern sky just before sunrise. Because the first rains do not fall until the fourth week of September, burning occurs more than a month before the start of the rainy season. The purpose of early burning is to permit early planting. A Kuikuru wants to have as much manioc planted as possible when the first rains come, so the cuttings can take full advantage of the fresh moisutre and grow rapidly.

On the day a man picks to burn his garden plot, he waits until about noon to do so. Were he to try early in the morning, the plot would burn poorly because it would still be too wet from the dew. Before starting a garden fire, a man carefully observes the direction of the wind. Once he has determined this, he begins setting fire to his plot near the downwind edge. To set the fires he may break off some large pieces of bark from a fallen tree and use these as torches. After setting several fires near the perimeter of the plot, he moves toward the center, setting additional ones as he goes. In burning a new garden, Afukaká started 14 separate fires, in a zigzag pattern, moving across the garden.[6]

Because not a drop of rain has fallen for 2 or 3 months when a man burns his garden, the slash is extremely dry. When ignited, it blazes up rapidly and burns furiously. Each little fire set in the plot quickly expands and soon all of them coalesce into a roaring inferno.

Afukaká took just 20 minutes to set all the fires in his plot. Then, after watching it burn for about 10 minutes, he left for home.

Because it burns rapidly, in an hour or two a garden fire has pretty well subsided. Here and there, though, charred logs and stumps will continue to smolder for days. The day after a man named Ulïti burned his garden, a dozen or more logs in it were still burning at the ends. One fallen tree, a *fala,* was still burning vigorously along its entire length. Much smoke accompanies a garden fire, and a blue haze hangs over a plot for a day or two after it is burned.

No fire breaks are cleared along the edge of a plot, and it is very common for a garden fire to escape into the surrounding forest. Because it is very dry at this time of year, the forest supports the flames that lick their way into it. As I watched the burning of Afukaká's plot, the wind carried the flames toward some of the surrounding trees. Even though green, the foliage of one such tree became so superheated that it suddenly burst into flame with a tremendous "woosh!"

Fed by the dry leaves lying on the grounds, here and there tongues of fire penetrate 200 feet (60 m) or more into the neighboring forest. They scorch the tree trunks in their path, causing their leaves to shrivel and eventually fall. Sometimes even the trees themselves catch fire. Rather than bursting into flame, though, they usually smolder until finally their trunks burn through enough for them to fall. For example, the fire made by

Ulïti in his garden plot on August 7 escaped and burned its way hundreds of feet into the nearby forest. Feeding mostly on dry leaves and dead logs, the fire nevertheless attacked an occasional standing tree and started it burning. Incredible as it may sound, on September 20, fully 44 days later, this same fire was still smoldering. During the 1½ hours I spent in the forest that day, collecting tree leaves for identification, I heard three trees fall, their trunks having been burned through by the fire.

The Kuikuru showed no apprehension about escaped fires. In this respect they are like their fellow Caribs of the Maroni River in Surinam, about whom Peter Kloos (1971:31) wrote: "When the surrounding forest begins to burn . . . it is regarded without much concern."

PILING AND REBURNING

After a garden fire has burned itself out, the plot is a mass of blackened stumps, trunks, and branches. The Kuikuru tidy up their plots a bit by a practice they call *itsufutelï*, known in Portuguese as 'encoivarar.' This consists of piling up and reburning the smaller logs and branches not consumed by the first burn. A few days after first setting fire to his *tufiñafo* (as a burned but unplanted manioc garden is called) a man returns to it and begins the reburn. Smaller logs that he wants to take home for firewood are piled up just outside the plot.[7] If a charred log is too big to carry, he chops it in half and throws both halves off to the side. When he finishes working for the day, he shoulders two or three of these logs and heads back to the village.

Those logs and branches not wanted for firewood are cut with an ax or broken by hand and piled here and there throughout the plot, some 30 to 50 feet (9 –15 m) apart. These piles are usually made around a stump, but also around a termite nest or a natural hummock in the ground. Charred vines are chopped out at ground level and tossed on the heap. The branches of any small tree that failed to burn well are broken off and thrown on the pile too. After a man has finished piling up unburned vegetation, he goes through his plot and sets fire to each pile.

The reason the charred slash is usually piled on top of a stump is to help burn it away. Stumps that catch fire may continue to smolder a long time. Eventually, some of them burn deep into the roots, leaving a sizable hole in the ground. Burned-out termite nests, being hollow inside, also leave holes when they collapse from the heat. Altogether, 'encoivarar' takes some 9 or 10 hours of work for a plot of average size.

After the initial burning of a plot, some 15–20% of its surface is covered with logs, stumps, and branches. But by the time the Kuikuru finish *itsufutelï*, the proportion of the plot so covered is reduced to about 5% if done carefully, or 7–10% if not.

Plot Characteristics

The color of the soil in a plot that has just been burned is predominantly black. This is due to the large amounts of charred wood fragments left on the surface. Here and there, where a large branch has been reduced to ashes, the soil is gray. Only where a charred log has been removed, or a termite nest has collapsed, does one see the natural red color of the soil.

Burning the slash in a plot raises its fertility considerably, as will be shown by figures presented later. Ashes increase the fertility immediately, but the mechanical and chemical breakdown of charcoal into a form usable by plants takes months. Thus, charcoal provides a reservoir of nutrients for the growing crops that is tapped more slowly than are ashes, and consequently can be drawn on for a longer time.

SIZE OF PLOTS

The size of gardens varies considerably. During my field work with the Kuikuru in 1975, I measured nine manioc plots, and the results are shown in Table 3.2. The range in plot size for this sample was between 0.98 and 2.72 acres (0.40 ha to 1.10ha), the average being 1.61 acres (0.65 ha). Differences in plot size did not seem to correlate closely with differences in the family size of the men making the gardens. There was, perhaps, more of a correlation between the size of a man's newest plot and that of his previous one. Thus, Mugí, who had a large producing garden of 2.12 acres (0.86 ha), made his next plot only half that size.

Table 3.2

Sizes of Nine Kuikuru Manioc Gardens, with Names of the Owners

	Size of plot	
Owner	Acres	Hectares
Mafukakumá	1.40	0.57
Fugatú	1.40	0.57
Ulïti	2.72	1.10
Jakalu	1.17	0.47
Tauagaku	1.23	0.50
Mafá	1.94	0.79
Kamáŋgagï	1.55	0.63
Mugí (No. 1)	2.12	0.86
Mugí (No. 2)	0.98	0.40
Mean:	1.61	0.65

An informant once told me that Kuikuru gardens are smaller today than in the past. To account for this he said the Kuikuru do not make them as big as they used to because today there are too many peccaries around the village. Later, though, he attributed the supposed smaller size of present-day gardens to the Kuikuru's having fewer manioc cuttings available than they once did. Neither explanation struck me as convincing.

Since the Kuikuru do not measure their plots, it would be difficult for them to say accurately if plot size had, in fact, changed over the years. In 1953, I measured the garden of a man named Kïlafi because it seemed to be of average size. Kïlafi's plot turned out to be about 1.625 acres (0.66 ha), virtually the same size as the average plot in 1975. My feeling is that there has really been no essential change in plot size in 20 years, and probably for a good deal longer. Thus, the assertion that plots used to be bigger may simply reflect the familiar attitude that "there were giants in those days."

Whether plots were smaller a hundred years ago, when the Kuikuru had to rely on stone axes to fell trees, is another question. Even the oldest informants were not born until after the Kuikuru had adopted the steel ax, so their opinions on this question, which differed anyway, were not those of eyewitnesses. Because the consumption of manioc does not appear to have increased, my guess is that in stone-ax times, manioc plots were probably about the same size, since they had to yield about the same amount of tubers as today.

Planting

PREPARING THE SOIL

After reburning his *tufiñafo,* a Kuikuru prepares the soil for planting.[8] This is done by hoeing up low mounds, *ifuŋu,* into which manioc cuttings are later inserted. Aboriginally, these mounds were made with a wooden spade called a *ku̯ǫ́,* after the tree whose wood was used to make it. This spade was about 2 feet (60 cm) long, and 5 inches (13 cm) wide at the widest part. In using it, a man drew the soil toward him, passing it back between his legs. The *ku̯ǫ́* has long since been replaced by the iron hoe. The Kuikuru received most of their iron hoe blades from the Indian Service, but hafted them themselves with a wooden handle about 5 or 6 feet long (1.5 m to 1.8 m).

In making a mound, a man first scrapes away any charred pieces of wood with his foot. As he starts to hoe, he is facing east and, continuing to face in that direction, he draws the soil toward him, building up a low mound. He then moves to the opposite side of the mound and again hoes the earth toward him, increasing the size of the mound, and leaving a shallow depression next to it. Altogether, he takes about 15 hoe strokes from the west side of the mound and about 30 strokes from the east. Most strokes are aimed at widening the mound rather than at deepening the depression. It takes about a minute or two to hoe up a mound, and about a minute to move from one mound to the next.

A finished mound is about 3 feet (.9 m) across and some 6 to 10 inches (15 to 25 cm) high.[9] On the west side of the mound a shallow depression is left about 9 to 12 inches deep. Forest soil is very friable to begin with, and hoeing loosens it even more, leaving it very easy to plant.

In the process of hoeing, a man comes across many roots. Since they would interfere with planting and, if still living, might compete with manioc for nutrients, they are chopped out and thrown aside. To keep a sharp edge on his hoe blade for this purpose, a man sharpens it from time to time with a steel file. Roots too large to chop out with a hoe are cut with a machete or an ax. Before steel tools were available to them, the Kuikuru used a piranha mandible for this.

Most hoe strokes draw soil to the top of the mound and thus collect and concentrate the top soil. As a mound is hoed, charcoal and ashes get mixed with the earth, but this seems incidental to loosening the soil and making the mound. No special effort is made to work into the mound the ashes and charcoal that lie beyond its immediate vicinity. The Kuikuru appear to be unaware of the effect of ashes on soil fertility.

While the predominant color of the surface of a *tufiñafo,* just after burning, is black from the charcoal strewn all over it, a change occurs when a mound is hoed. As the thin layer of blackened topsoil gets mixed with the subsoil, the surface color of the mound becomes red. That mixing in the ashes from the burned slash increases the fertility of the soil is well-known to students of tropical agriculture. Analyses of the soil samples I took from Kuikuri plots further demonstrates this (see Table 3.3). The two plots compared in Table 3.3 were only a few hundred yards apart in what seemed to me indistinguishable areas of primary forest. The only difference was that

Table 3.3

Comparison of the Mineral Content of Two Soil Samples from Areas Near the Kuikuru Village of Ipatse[a]

				Pounds per acre (2,000,000 pounds of soil)										
	Organic matter	pH	Ex H (me/100 g)	P	K	Mg	Ca	Mn	Fe	Al	NO_3 N	NH_3 N	Zn	Soluble salts $K \times 10^5$
Plot C	2.7	4.0	10	1	40	25	100	4	45	250	105	32	1	23
Plot D	3.5	3.8	13	3	140	80	150	8	53	175	210	70	1	50

[a]Plot C represents an area of *itsuni* (primary forest) sampled while the trees were still standing. Plot D represents a nearby area of the same type of forest sampled shortly after felling, burning, and hoeing had taken place.

Plot C was sampled while it was still under standing forest, while Plot D was sampled some 4 months after the forest had been felled, 1 month after the slash had been burned, and about 10 days after the rains had begun. (It had rained 3 or 4 times during that interval.) Moreover, the soil cores taken from Plot D were drawn from freshly hoed mounds in which ashes from the burned slash had been mixed with the soil. As Table 3.3 clearly shows, in everything but pH the soil in the plot that had been burned and hoed had much more favorable characteristics for cultivation than that still under standing forest.[10]

PREPARING TO PLANT

Kuikuru men not only clear the forest, burn the slash, and hoe the mounds, they plant the manioc as well.[11] This may strike some readers as anomalous, since among so many tropical forest tribes it is the women who do most of the gardening. But in my opinion, the Kuikuru are not anomalous. In an informal comparative survey I have made of the division of labor in subsistence in Amazonia, I have found that wherever warfare and hunting are absent or unimportant, it is the men who will do most of the gardening. This is precisely the case among the Kuikuru.

Most men prefer to hoe up all the mounds in their *tufiñafo* before planting any of them. Once in a while, though, a man hoes up a section of his garden, plants this, hoes up another section, plants that, and so on.

The distance between mounds in a manioc garden (or, to be more precise, the distance between the centers of the mounds) varies from 6.4 feet to 8.4 feet, (1.95 m to 2.56 m), the average being about 7 feet (2.13 m). How far apart mounds are made is, to some extent, a matter of individual choice. But it also reflects the number and distribution of stumps and logs remaining in a plot. The ideal distance between the edges of adjacent mounds is an *aseni*. This is a Kuikuru armspan—the distance between the finger tips of the outstretched arms. Mounds planted closer than this are called *iŋufiŋï*, the term applied to a hammock twined more closely than usual.

The Kuikuru give two reasons for making their manioc mounds an *aseni* apart. First, if mounds are spaced too tightly, the branches of the manioc plants cannot spread and develop the way they should. Second and more important, if manioc plants grow too closely together, it is difficult to move between them for weeding.[12]

When the average distance between the centers of mounds is 7 feet (2.1 m), the average area of a mound and its surroundings is about 50 square feet (4.65 sq. m).[13] This means that in an acre there will be some 870 mounds, and in a plot of average size (1.61 acres) (0.65 ha), about 1400 mounds. Since, on the average, about 10 manioc cuttings are planted per mound, planting a new garden involves setting out some 14,000 cuttings.

This is a sizable number of cuttings. Where does a man get them all? As a rule, entirely from his own gardens. Some come from plants that are being harvested, while others come from plants that are to be left in the ground to grow for another year. The latter are pruned, and the stems derived from them are used for planting.[14] (More will be said about pruning in a subsequent section.)

A man with two manioc gardens should have no trouble producing all the cuttings he needs to plant a new one. A man with only a single plot can generally do so too, especially if his plot is an old one and he is harvesting it for the last time. In this case, all the stems coming from it can go to plant his new garden, since none will be required to replant his old one.

But even a man with but a single garden can usually still manage to come up with enough cuttings to replant that garden and plant a new one as well. One of the virtues of the manioc plant is that it yields more stems than are needed to replant it. From measurements I made in a new garden being planted by Mugí, I found that he used some 112 inches (284 cm) of cuttings per mound. However, from the plants he was pruning in an adjacent garden he was getting an average of 313 inches (795 cm) of plantable stems per mound. Thus, a manioc plant yields 2 to 3 times the length of stems required to replant it. Thanks to this, one can readily increase his area of planted manioc. The area of manioc a man has under cultivation, then, is limited more by his willingness to work than by his supply of planting stock.

When cut long, manioc stems are called *kuigife.* After the stems are cut from a harvested or growing plant, they are usually piled up next to a fence or under a nearby log. Large leaves are often placed over them to protect them from the sun. Thus shaded, they may last several weeks, until needed for replanting.

If a man's new garden is next to his old one, stems for planting will be ready to hand, but if his old garden is some distance away he will have to bring stems from it. Every day during the planting season some 60 or more stems, each about 4 or 5 feet (1.2 to 1.5 m) long, may be tied together in a bundle weighing 75 pounds (34 k) or more and carried to the new garden site.

If, despite everything, a man finds that he does not have enough stems to plant a new garden completely, he may get more, without payment, from a more affluent relative or a close friend. But if he has to go to someone else for cuttings, he must usually pay for them. A few bundles may be purchased for an arrow, some urucú (*Bixa orellana*), or a cotton waistband. For a large number, a man may have to give a machete or something of comparable value.

THE PLANTING OF MANIOC

Before he can begin planting, a man must first reduce his stacks of *kuigife* stems to cuttings. This he does either by breaking them by hand or cutting them with a machete. The usual length of cuttings is 12 to 16 inches. Once the cuttings are ready, a man loosens the soil on the side of a mound next to the shallow depression, using only his hands. Any roots still left in the ground are removed. Then the cuttings are inserted into the mound.

Generally, a man plants alone, but often a close kinsman or friend may help. Occasionally, a man who has cleared a new plot will invite all the men in the village to plant it for him. In return, he rewards those who worked with liberal amounts of food and drink which his wife prepares especially for the occasion. Not only the planting of a new plot, but the clearing of it as well, may be arranged and paid for in this manner.

The number of cuttings planted per mound varies. In the hundreds of mounds I saw being planted, the range was from 5 to 14, with the average around 9 or 10. I was told, though, that if cuttings are very thick (2 or 3 inches [5 to 7.5 cm] in diameter), as few as 3 may be planted in a mound. Because thick cuttings produce large tubers, a man plants fewer of them in order to leave ample room for the roots to grow. Thinner cuttings, on the other hand, produce smaller roots, so more of them can be planted in a mound.

Manioc stems less than half an inch (1.25 cm) or so thick are considered too thin to plant and are generally discarded. However, a man with a meager supply of stems may resort to planting some he would otherwise disdain.

Cuttings are planted close together in a mound, parallel to each other, and are inserted about a third of their length into the ground. All cuttings are slanted to the west at an angle of about 45°. A bit of earth is placed around the spot where they enter the ground and then tamped down slightly with the hands. The reason the Kuikuru give for slanting the cuttings toward the west is that as the young plants begin to grow, they will be leaning away from the prevailing easterly winds. During the early part of the rainy season, when the new plants are just coming up, these winds blow with great force[15] and could more easily damage the plants if they were facing into it. Soon, though, as they grow taller, the plants assume an upright position.

Instead of inserting them into a mound parallel to each other, one man I watched splayed his cuttings at their lower ends so the tubers developing from them would have more room to grow. Occasionally, a man plants manioc cuttings vertically instead of at an angle. This is called *kuakuagï*, after an *etseke*, or spirit, of that name. The story goes that this spirit, who always plants his manioc this way, was once observed doing so by a Kuikuru, who decided to copy him. One man, Mugí, had planted a few mounds of a variety called *fefegiku* by placing 2 large cuttings horizontally on the ground and covering them completely with earth. This method of planting he had copied from Brazilians.

When a man finishes planting a mound, he often takes a bunch of small rootlets he has previously pulled out of the ground and places them over the cuttings. This protects them from the sun. Some men prefer to use pieces of charred bark for this. One man planted his cuttings so that the protruding stems criss-crossed one another instead of running parallel. While the sun might burn the top 3 or 4 cuttings, he said, the lower ones would be protected.

Planting is done rapidly. It takes only about a minute to insert a dozen or so cuttings into a mound. If we include the time it takes to break long stems into cuttings, and to move from one mound to the next, the total time required in planting a mound is about 2½ minutes. To plant an entire garden of about 1400 mounds, then, it takes some 58 hours. This figure, however, excludes the time spent cutting the stems from the parent plants and carrying them to the new garden.

A hard-working man may finish planting a new garden by the time the rains begin in late September. Usually, though, men will still be planting into October. And sometimes, having been away from the village during part of the planting season, or simply out of laziness, a man may still be planting in November.

Once it is entirely planted, a manioc plot is no longer a *tufiñafo*. Now it is called a *kuigíanta*.

MANIOC AND MAIZE

The Kuikuru cultivate many varieties of manioc. Indeed, I was able to elicit the names of nearly 50 of them. However, only about 4 or 5 varieties are ordinarily planted in a single garden. The name for manioc in general is *kuigi*, and the variety most commonly planted is called *kuigífekugu* ("true manioc"). *kuigífekugu* yields heavily, and usually more mounds of it are planted in a garden than of all other varieties combined.

Manioc is usually segregated within a garden so that only one variety is planted in a single mound. Also, cuttings of the same variety are commonly planted in contiguous mounds, so each variety is clustered in a particular part of a garden. (Further details about varietal differences will be presented later in the chapter.)

Occasionally, crops other than manioc are planted in a *kuigíanta*. This is most often true of *afati* (urucú),[16] whose bushes sometimes serve as boundary markers between plots. A few sweet potatoes may be grown in a manioc plot, too. However, maize (*aná*) is never planted in the same garden with manioc. Instead, it is planted in a separate garden called an *anáito*. Maize requires more fertile soil than manioc, and so an *anáito* is always cleared in an area of *egepe* forest, where the soil is black and rich. Unlike manioc, maize is not planted by every man in the village. In fact, only 4 or 5 men plant cornfields, as a rule, sharing their harvest with everyone else.

ABSENCE OF FERTILIZERS

The Kuikuru appear not to have a clear grasp of the concept of soil fertilization. They seem unaware of the importance of wood ash in increasing the fertility of their gardens and, as I have said, make no special effort to work it into the soil. Not surprisingly, then, they make no use at all of animal manure. Should a Kuikuru find some tapir dung in his garden, for instance, he will probably spit in disgust. Indeed, one man told me he would actually go to the trouble of burning it. So strong is the Kuikuru's aversion to animal droppings that, on one occasion, the mere fact that a tapir had defecated near a certain lagoon was enough to deter them from fishing there, claiming that the fishing had thereby been ruined.

People never defecate in their gardens. If nature calls while a man is working in his field, he will go into the nearby forest to answer the call. Should a child defecate in a garden, his father carefully picks up the feces and throws them into the woods.

THE MANIOC PLANT

A man who has planted a *kuigíanta* does not have to wait long to begin seeing the fruits of his labor. Within a couple of weeks after planting, fresh green leaflets can be seen sprouting from a cutting. The new leaflets do not grow from the ends of the cuttings, nor from the leaf scars, but from tiny ridges just above and below the leaf scars. And these leaflets begin to emerge even though no rain has fallen for 3 months. When the rains finally arrive, they naturally give a big boost to plant growth. To allow his cuttings to grow unhindered, a man now removes the clumps of rootlets or slabs of bark that he placed over them when he planted them.

If some manioc cuttings in a mound have died, or were "born ugly," they may be pulled out and new ones planted in their place. Even so, not all planted cuttings in a mound live and grow. However, some that look dead may nevertheless send up fresh stems from below ground.

When the stem of a growing manioc plant reaches 2 or 3 feet (.6 to .9 m) in length, it divides into three branches, which splay out from their common point of origin. When each of these new stems has in turn grown another 2 or 3 feet, they again divide into three branches. Some manioc plants even rebranch a third time before leafing out in full.

The same little ridges on a manioc stem that give rise to leaflets above ground, produce tiny rootlets below ground. These rootlets continue to grow until eventually their ends begin to enlarge and form tubers. Or it may happen that the lower ends of 4 or 5 cuttings may coalesce underground. From the point where they fuse, a single root, called *intsï* by the Kuikuru, begins to develop. This root grows straight down into the soil, reaching a length of a foot or more and a thickness of up to two inches. The surface of this root is brown and scaly, much like that of the manioc tuber itself. One informant said that this root could itself be planted and would produce tubers. From the sides of this thick root thin rootlets begin to grow, fanning out laterally in all directions. Anywhere from 1 inch to 7 or 8 inches from the main root, these rootlets begin to engorge with starch, forming the manioc tubers, *ñuŋu*. The tubers continue to enlarge as long as the plant is allowed to grow. (The number and sizes of the tubers produced in a mound will be discussed later in the section on harvesting.)

Crop Care

PREDATION

As manioc plants develop, the Kuikuru are not the only ones who cast covetous eyes on them. Manioc, in fact, has several predators. Peccaries, agoutis, and deer are especially fond of either the tubers or the leaves, and it is not uncommon to see their tracks and droppings in or around a garden.[17]

The greatest danger to a manioc garden by far comes from peccaries. While both collared peccaries (*Tayassu tajacu*), *feukueŋgï*, and white-lipped peccaries (*T. pecari*), *feu*, live in the forests around the village, it is the latter, traveling in troops of up to 80, which take the heaviest toll of the crop.[18] Zoologists consider peccaries to be diurnal, but the Kuikuru assure me that those which inhabit their forests are nocturnal. These peccaries, it seems, have, over the years, changed their life habits so they can prey on the manioc gardens at night while the Kuikuru are asleep.

A garden not protected by a fence falls easy prey to the peccaries. In an unfenced garden owned by a man named Tabata, I saw 10 or 12 adjoining mounds of manioc that had been completely denuded of tubers by peccaries. Worse yet was what happened to the garden of Tefuku, a young apprentice shaman. Tefuku had been in shamanistic seclusion for several months and was thus unable to fence his garden. As a result, it had been completely devastated by the peccaries, with perhaps not a single mound left untouched.[19]

FENCING

The perimeter of a manioc garden bordering the forest is always protected against predators by fencing. However, that part of a plot abutting another is never fenced. To tell where his plot ends and his neighbor's begins, a man often plants a single row of a variety of manioc called *maisafalu,* which is unique and distinctive in having a silvery stem. His neighbors may do the same, so it is common to see, here and there in a composite manioc field, two adjacent parallel rows of *maisafalu* wherever two gardens come together. Until planting occurs, some natural feature, such as a prominent stump or a big log, may be used to tell where one plot bounds another.

As mentioned earlier, a row of urucú bushes may serve at times instead of *maisafalu* to mark the boundary of a plot. Sometimes, though, no special divider is planted at all, and even so, a man knows exactly how far his garden extends.

In November, when a man has finished planting his new garden and has done some preliminary weeding, he usually begins the task of building a protective fence. This is not an easy job. If a garden is of average size and stands by itself in the forest, its owner faces the prospect of having to put up about 1000 feet (300m) of fence. And this fence must be sturdy since peccaries are ready to throw their full weight against it. (One can readily see the advantage in clearing a new garden next to one or two existing ones: the length of fence one has to erect is thereby substantially reduced.)

The Kuikuru build two types of fences, one called *tafiti* and the other *kagafuku.* A *tafiti* fence is made by cutting and splitting logs from the new garden into lengths of about 4 or 5 feet (1.2 to 1.5 m). A shallow trench about 8 or 9 inches (20 to 23 cm) deep is then dug along the outside of the garden, and into it the ends of the split posts are inserted vertically, next to one another. The trench is then filled and the earth tamped down, leaving a solid row of upright posts. Some men slant their *tafiti* fences outward a bit to make it harder for peccaries to scramble over them. The ends of the fence posts are left jagged and irregular to make it more difficult for the peccaries to scale them.

The other type of fence *kagafuku,* is made by taking longer, unsplit logs and piling them, one on top of another, between pairs of vertical posts which have been hammered into the ground. The tops of these paired posts are then lashed together with vines to hold the horizontal logs securely in place.

Tafiti fences are more common than *kagafuku,* probably because they are harder for peccaries to climb. But peccaries are not easily deterred by any kind of fence. As one informant put it, "peccaries are very tough." If a peccary cannot climb a fence, he will try to burrow under it, or even chew his way through it.[20]

A Kuikuru, then, may have to try other means to protect his garden from these predators. At a low or weak place in a fence, unplanted stems of manioc may be piled high to keep the peccaries out. One man, Jakalu, planted a row of pineapple (*Ananas* sp.) plants across the middle of his garden, hoping the peccaries would be deterred by the spiny leaves from passing through them.

Another stratagem is to make a small opening in a fence, hoping a peccary will find it, and then to dig a deep pit just inside the opening and cover it with leaves. Any peccary falling into the pit would find it too deep to climb out of, and when a man next visited his garden, he would kill it.

The most time consuming but probably the most effective way of dealing with an infestation of peccaries is by means of a *jati*, a peccary trench. A *jati* is built only on rare occasions, and only by an energetic and determined man whose manioc crop is in serious jeopardy. Just inside the entire length of his garden fence, he digs a trench about 4 feet (1.2 m) deep. Then, at regular intervals, he inserts sharpened stakes about two feet (.6 m) long into the bottom of the trench. Any peccary vaulting the fence will land in the trench and impale itself on the stakes.

In spite of all precautions, though, peccaries do considerable damage to the manioc crop. Some Kuikuru get angry over this, while others get sad, depending on their temperaments. One man is said to have cried openly over the loss of manioc to the peccaries. So frustrated had the Kuikuru grown over these losses that in 1975 they were talking seriously about overcoming their aversion to peccary meat and beginning to eat it. In this way they not only would make good some of their losses, but avenge themselves on the peccaries in the bargain.

To try to minimize his losses to the peccaries, a man usually builds his new manioc garden in a different part of the forest from his old one. This is the best way the Kuikuru have of spreading the risk. And, of course, in making their manioc gardens as large as they do, the Kuikuru are no doubt making allowance, in a rough way, for expected depradations by the peccaries.

Deer do not pose the same dire threat to manioc plants that peccaries do, but they still are an annoyance. Two kinds of deer inhabit the area, the smaller forest deer, *oŋiŋo*, (*Mazama americana*) and the larger savanna deer (*Ozotocerus bezoarticus*) *asá*. Deer have no interest in manioc tubers, but are very fond of the leaves, especially if they are young and tender. Fences are less effective against deer than against peccaries because deer can jump over them more readily. If deer are getting into a plot and eating too many leaves, a man may drive a series of 3 foot-long (.9 m) sharpened stakes, called *asáfotsitofo*, into the ground some 6 or 7 feet (1.8 to 2.1 m) inside low places in the fence. These stakes are slanted outward, so when a deer jumps over the fence at a protected point, it is liable to impale itself.

Since coming into contact with whites, the Kuikuru have hit upon a new device for warding off deer. It is called an *asáeŋetofo*, or "scare deer," and consists of an iron hoe blade and a metal plate hung from a pole or nearby tree, which clang together in the breeze.

Third on the list of manioc predators is the agouti (*Dasyprocta azareae*). Agoutis burrow under a fence and then into a mound, where they nibble at the roots. While harvesting, the Kuikuru would sometimes point out to me tubers whose ends had been nibbled in this fashion. Wherever it is chewed, a manioc tuber begins to rot. If the damage is extensive, the Kuikuru throw the tuber away. If not, they lop off the rotten end and save the rest.

The 'jacú' bird (*Penelope* sp.) likes to eat manioc leaves, but the harm it does to the standing crop is negligible. There is some insect predation of the manioc gardens, but not much. A large caterpillar called *tukutiñá* eats the leaves, but there are few of them around, and the damage they do is very minor.

Leaf-cutter ants (*Atta* sp.) pose a serious threat to manioc plants in certain parts of Amazonia, especially the Guianas. Thus, Richard Schomburgk (1922:I:185), who traveled widely in British Guiana, wrote of the leaf-cutters there:

> A cassava . . . cultivation in which they are nested soon resembles our timber areas after the caterpillars have devastated them. In a short while the ants eat away the whole of the leaves and drag them into their subterranean dwellings. If a field is once visited and robbed by them, the destruction of the whole is to be feared.

And for the Maroni River Caribs of Surinam, Kloos (1971:36) reports much the same thing: "The insect cuts off the leaves of the manioc . . . to bring the fragments to its nest. A whole garden can be destroyed in a few nights."

In the upper Xingú, though, the leaf-cutter ant hardly seems to bother with manioc leaves.[21] What these ants do covet, however, are the granules of processed manioc flour. At the Kuikuru village of Lafatuá, there was a large leaf-cutter ant nest under the very center of the village plaza. And every day one could see columns of ants marching from the nest to one or another of the houses, where each ant would shoulder a granule of flour, often larger than itself, and head back to the nest.

WEEDING

It does not take long for weeds to make their presence known in a new Kuikuru garden. Indeed, within a month after a plot is burned, even before it is planted, weeds begin to grow. This occurs although it is still the depth of the dry season and rain has not fallen for months. If a Kuikuru is to harvest a good crop, it behooves him to keep the weeds under control. In an unweeded garden, such as one belonging to a man named Ñukau, the weeds were so tall and thick that at first I was unaware that manioc was growing there at all. No doubt, this infestation of weeds had adversely affected the size of the tubers.

The Kuikuru use the word *tafuga* to refer collectively to the weeds growing in a garden. As I noted earlier, *tafuga* may be made up of various types of vegetation. Many *tafuga* plants are seedlings or saplings of secondary forest trees that have grown from seeds blown into the garden from some nearby source. A good many *tafuga* plants, though, have grown from burnt stumps that were part of the *itsuni* vegetation before it was felled and burned. Despite the intense heat of the burn, many *itsuni* trees are not entirely killed, and later send up sprouts from the roots (not the tops) of their charred stumps. In its early stages, all *tafuga* vegetation is green and succulent, and only a practiced eye can distinguish which weeds represent primary forest trees and which secondary forest.[22]

Weeding, called *ñoŋkilï* or *uatafugakiŋu,* is done at any time of the year, but probably more frequently during the rainy season, when weed growth is heaviest and men are less occupied with planting and harvesting. Weeding generally begins in earnest in November or December, after fencing has been completed, or at least is well under way.

Weeding is generally men's work, but women occasionally help out. Sometimes a man weeds alone in the garden, but more often he is accompanied there by his wife. Often she will pull tubers from the manioc plants he has just weeded.

Men vary greatly in the diligence with which they weed. Some pull everything out by the roots, and always keep their gardens pretty well clear of weeds. Others, less fastidious, let the weeds get a head start, and then, finding it too difficult to pull them out, cut them with a machete instead (see Johnson this volume). This, of course, leaves the roots in the ground to sprout again.

Like garden work in general, weeding is usually done early in the morning, from around 5:30 to 9:00 A.M. A serious weeder, though, may work past 9:00 A.M. and a really dedicated gardener may return to his plot in the afternoon to continue weeding.

The weeds pulled or cut from a garden are thrown on a pile, a new pile being made every 10 or 15 feet. Small stems pruned from growing manioc plants, unsuitable for planting, are also thrown on the weed piles. In a day or two, when the weeds are dry, they are burned.

Weeds are often piled on top of dead stumps in the garden so that each time they are burned the stumps will burn a little too. In this way, the stumps in a plot gradually disappear. Since over the course of a garden's life the logs lying in it are cut up for fencing or for firewood, after 3 or 4 years of cultivation a plot may be almost entirely free of any trace of the trees that once covered it. Eliminating stumps and logs makes room for additional mounds when second and third plantings of manioc are made. And, in a limited way, burning the stumps and weeds restores some nutrients to the soil.

A well-weeded garden may last up to 5 years, but a poorly weeded one lasts 2 or 3 at most. Once weeds have been allowed to become really thick, it is simply too much trouble to clear them. If a man decides he is not going to replant a garden, he tends not to weed it so thoroughly. In fact, he may actually end up weeding only those mounds from which his wife is about to extract the last batch of tubers.

SECOND PLANTING

All manioc planted during the dry season is planted from relatively short cuttings, averaging about 14 inches (36 cm) long. The planting of short cuttings is called *ifelï*. However, if manioc is planted during the rainy season, it is done with much longer cuttings, about 2 or 3 feet (0.6–0.9 m) long. The planting of long cuttings is called *egipokegi*. The long stems planted during the rains are inserted into new mounds hoed up between already existing ones in a growing garden. The upper ends of the long stems planted by *egipokegi* may die, but the lower part of the stems will sprout.

I am not sure I understand the reason for *egipokegi*. Two informants said it was done because short cuttings would die if planted during the rainy sason. Two other informants, though, said long cuttings were planted during the rains so they "would be born quicker." Either or both reasons may be true. Or some other reason may be involved.

REPLANTING

During the first 3 or 4 months of the rainy season there is little harvesting of manioc. A family lives off the stores of processed flour it put away during the previous dry season. (More will be said about these stores when harvesting and processing are described.)

By the time the rains cease in April, though, most families have run out of stored flour. They must then go to the fields every 2 or 3 days to dig up more tubers. By this time, manioc planted the previous year has been growing for 8 or 9 months and its roots are of edible size. Still, unless they have no other producing plants, a family refrains from eating manioc that young. They prefer to leave these tubers in the ground to grow bigger. Instead, they will harvest manioc that has been growing since August or September of 2 years before, which is now 20–22 months old. The roots of these plants are not only bigger, but have a higher proportion of edible starch (Brand 1943:94; Tracy 1903:19).

As the mounds bearing these older plants are harvested, they are immediately replanted. If this is the third crop of roots pulled from these mounds, though, the mounds are abandoned. Replanting continues throughout the dry season, the plants being harvested providing the stems for it. Indeed, the cuttings derived from the harvested plants are inserted, after a little hoeing, in the very mounds from which they came. The soil is no doubt more fertile a couple of feet (0.6 m) away, but either the Kuikuru do not know this, or simply disregard it.

Manioc is replanted in much the same way as it is planted in the first place. This time, though, masses of rootlets and pieces of bark may no longer be at hand, so to protect newly planted cuttings from the sun they may be covered with large leaves picked from trees in the forest nearby.

LEAF LOSS

As I have said, the dry season in the Upper Xingú is bone dry. Not a drop of rain falls for 3 months or more. As an adaptation to this condition of drought, the manioc plant sheds its leaves during July and August. In one plot I examined late in the dry season, the few leaves growing on the plants were tiny recent sprouts, the older leaves having fallen off earlier. Whether the growth of tubers underground is stopped, or even slowed, by this leaf loss I do not know. It seems likely that it is. At any rate, the plant is merely quiescent at this stage, and not dead as it may appear.

PRUNING TO OBTAIN CUTTINGS

A man who is planting a new garden, and not merely replanting an old one, may need more stems than those provided by the plants he is currently harvesting. These extra stems are obtained by pruning the year-old manioc plants he put into the ground the previous August or September and left undisturbed. This practice is called *feguketagï*. First, the leaves and thin outer branches of the plant are cut off and discarded, leaving bare stems, 4 or 5 feet (1.2 to 1.5 m) long, growing out of each mound. These stems are then cut off and stacked by the fence, ready to take to the new garden.

A man may prune all the plants in his garden or only some of them. While pruning, he cuts the stems low, leaving only some 8–12 inches (20–30 cm) protruding from the ground. Attached to these stems beneath the soil, the tubers are left undisturbed. By early September, the stems begin to sprout, and when the rains start again later in the month, they quickly leaf out. If the pruning of their stems caused the tubers to stop growing, they now resume their growth.

Besides providing stems for planting, pruning serves two other purposes. For one thing, it permits easier weeding. Year-old manioc may be 6 or 7 feet (1.8–2.1 m) tall and its branches may have spread out so much that they intertwine with adjacent plants. After this foliage has been lopped off, one can weed around the base of the plants quite readily.

The other advantage of pruning is that the roots of the plants that are pruned are left in the ground to grow for another year. Thus, when finally picked, they will have been growing for 20–22 months and will be correspondingly bigger. A man I was once watching prune his year-old manioc plants partially uncovered a growing tuber which was 2½ inches (6 cm) in diameter and told me that by the time it was harvested next dry season, it would be twice as thick.

Manioc left in the ground for a second year is called *ufigi*. The Kuikuru emphasized that *ufigi* is not a particular variety of manioc but, rather, a stage of growth that any variety might be allowed to attain.

Overall Cycle of Cultivation

Let me now present an overview of Kuikuru manioc cultivation before proceeding to discuss the harvesting of tubers in detail. Two cycles can be distinguished: one involves the successive phases in the life history of a single plot, while the other involves the succession of plots planted and abandoned over the years.

First of all, there is much flexibility in the planting, harvesting, and replanting of a single garden. And this flexibility is due largely to the unique characteristics of the manioc plant itself. Most crop plants ripen at a definite time and must be harvested as soon as they ripen. Not so with manioc. Seven or 8 months after manioc is planted, the tubers are big enough to eat, but there is no need to pick them yet. They will not spoil if left in the ground. Indeed, they will continue to grow indefinitely. A mound of planted manioc may be likened to a savings bank. Not only is the "principal" secure, it continues to accumulate "interest" until withdrawn. And a manioc tuber can continue to grow for 24 months or more and still remain edible.[23]

This fact gives the owner of a manioc plot considerable leeway as to when to harvest. And not all tubers growing in a garden need be treated the same way. Some may be picked now, and others left till later. Couple this variation in the time of harvesting of a particular crop with the fact that a garden is replanted 2 or 3 times during its lifetime, and it is clear that there is often much staggering of the plants in a garden. Several stages of development in the growing tubers may be represented in a single field.

But there is a limit to this flexibility. If a man has only a young manioc field to draw from, he and his wife may have to harvest it before it is a year old. If he has two mature gardens, though, he can afford the luxury of picking only the older and bigger roots, leaving the small ones in the ground to grow large. Thus, a man's decision when to harvest a certain crop of tubers depends in part on how many gardens he has. So obviously it is an advantage to have at least two.

The Kuikuru say that a really diligent man will clear and plant a new manioc garden every year. And to help inculcate this ideal they tell of the semilegendary figure, Fágima, a man of the now-extinct Jïgamï tribe. Fágima was said to be such a hard worker that every year he cleared and planted, not one but *two* plots of manioc. Fágima's exploits are on a par with those of Paul Bunyan. No Kuikuru regularly plants two gardens a year. In fact, very few even plant one garden a year. We can see this by looking at the number of gardens each Kuikuru had under cultivation in 1954 and again in 1975. Because gardens last 3 years or longer, if a man were to plant a new one every year, he would have at least three in existence at any one time. Yet only a few Kuikuru had that many (see Table 3.4).

Out of the total of 77 men, only 6 (8%) had 3 gardens. Another 30 (39%) had 2 gardens, while 37 men (48%) had only one. And there were 4 men who had no gardens at all.[24]

Actually, the replanting of a manioc garden as it is harvested is in some ways equivalent to planting a new one. But there is a difference. Replanting can be done only so many times. After 2 or 3 replantings, the yield of a manioc garden falls low enough so that the return from it is no longer worth the effort. It is cheaper, in terms of hours of work, to clear and plant a new garden. Thus, if a man thinks that next year's harvest from a given plot will be his last, he generally clears a new garden this year in anticipation of this. Then, when his old garden gives out, he will have a new one with which to replace it.

In order to understand the relationship between the frequency of plant-

Table 3.4

Number of Manioc Gardens under Cultivation among the Kuikuru in Lafatuá and Ipatse

Lafatuá, 1954	
Men with no manioc plots	3
Men with 1 manioc plot	22
Men with 2 manioc plots	14
Men with 3 manioc plots	3
	Total 42 men, 59 plots
Ipatse, 1975	
Men with no manioc plots	1
Men with 1 manioc plot	15
Men with 2 manioc plots	16
Men with 3 manioc plots	3
	Total 35 men, 56 plots

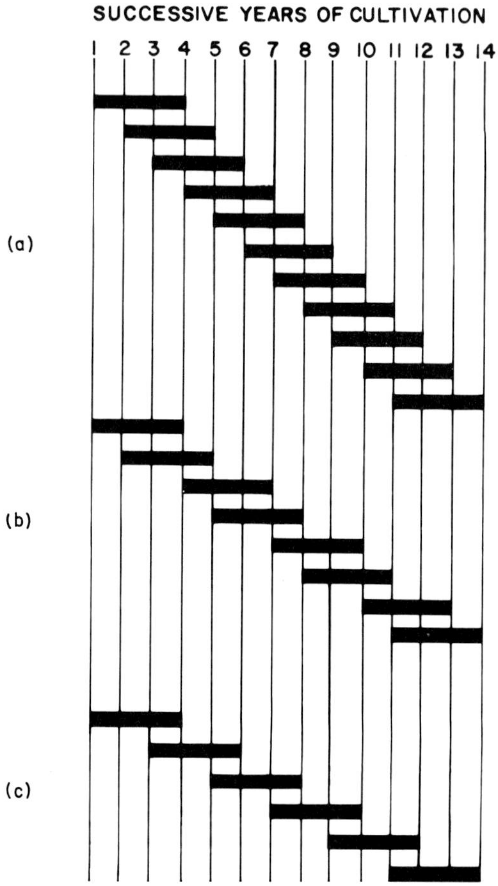

Figure 3.2 *Diagram showing three possible regimens for the successive clearing and planting of manioc plots by the Kuikuru over an extended period of years.*

ing a new garden and the number of gardens one has at any one time, let us look at Figure 3.2. This diagram assumes that each manioc plot lasts 3 years, a conservative estimate. If a man plants a new plot every year (Figure 3.2a), once the cycle of cultivation is well under way he will always have three manioc gardens. If a man plants a new garden one year, skips planting one the next, plants one the third year, skips the fourth, and so on (Figure 3.2b), he will always have two manioc gardens growing at once. And if a man plants a new garden only every second year, (Figure 3.2c) he will have two plots growing some years, but only one during others. Most men—67 out of 77, according to Table 3.4—were represented by categories 3.2b and 3.2c in the diagram.

Harvesting

Let us now return to the techniques of cultivation and describe the harvesting of tubers. Although manioc tubers are pulled from the ground throughout much of the year, extraction of the roots occurs most intensively during the dry season. Women are anxious to pick and process as many roots as they can at that time in order to lay in large supplies of flour before the rains begin. This way they will not have to go to the gardens every 2 or 3 days during the rainy season, thus avoiding the annoyance of getting drenched if caught by a sudden storm. Moreover, not having to go to the fields during the first months of the rainy season allows them more time to gather 'piquí' fruit, which ripen in abundance in November and December. And the extra free time from subsistence also lets them twine hammocks and do other needed household chores.

So heavily engaged in picking and processing manioc does a woman become during the dry season that if the garden is more than 3 miles (5 km) or so from the village, her husband may build a small house very close to it. The family lives from July through September in this garden house called a *fifitsiŋofo*.

In the present village of Ipatse, where the Kuikuru were living in 1975, almost all garden plots were near enough the village so no family thought it worthwhile to build itself a *fifitsiŋofo*. However, one man whose three gardens were about 4 miles (6.5 km) away told me that next year he would build a garden house, and that he and his wife would remain there, working up manioc, during much of the dry season.

At the site of Lafatuá, where the Kuikuru were living in 1953, the gardens were considerably farther from the village than they are at Ipatse. Consequently, several families had built garden houses near their plots. They stayed there most of the dry season, but returned to the village occasionally for feasts and ceremonies. When the rains begin at the end of September, though, garden houses are abandoned and families move back to the village, bringing with them the many pounds of manioc flour they have processed there.

When a man decides to build a garden house, he starts work on it as soon as he finishes fencing his new plot. Working on it intermittently through the rainy season, he usually completes it by April or May. A garden house is said to last 3–5 years before it has to be rebuilt.[25]

Let us look now at the harvesting process itself. Early in the morning, usually by 5:30, husband, wife, and children leave the village and head for a producing garden. Quite often, a second woman—an older daughter, the sister of either spouse, etc.—goes along to help with the work.

The tubers, which are the only part of the manioc plant eaten by the

Kuikuru,[26] are lodged firmly in the ground at the base of the growing plant. If the garden has not been weeded recently, the surrounding weeds must be cleared before harvesting can begin. After a man has pulled out these weeds, he cuts the stems of the manioc plants close to the ground and tosses the stems and leaves off to the side. If he is going to replant cuttings in the same mound, he cuts the stems short, but if the stems are to be accumulated for later planting elsewhere, they are cut into longer 2- or 3-foot [0.6–0.9 m] sections.

Once a man has cleared a mound of weeds and cut the manioc stems, his wife goes to work. The process varies a little from person to person, but this is how it is usually done.[27] First, the woman clears off the area around the plant with her hand and then, using a wooden digging stick called a *tuŋufi,* she begins digging into the mound. Nowadays a machete often replaces a digging stick for this. However, the two are sometimes used in tandem, a machete being employed to loosen the soil in the mound, and the digging stick to dig close to the roots. As the woman uncovers tubers, she pulls them out of the ground with her hands. In this way she removes most of the shallow tubers as well as those growing a little deeper but more or less horizontally. Then she grabs hold of the manioc stems protruding from the ground, which her husband has just cut short, and gives them a pull. Out comes the root system of the manioc plant, including, at its lower end, perhaps a dozen tubers.

After the stems are pulled up and the attached tubers removed, the woman again picks up her digging stick and begins to scratch the soil for any remaining tubers. Often a manioc tuber breaks in half as it is being dug out, in which case the lower half is simply left in the ground. Likewise, a woman will not bother to pull out or pick up very small tubers. Manioc tubers grow to such size and in such abundance that there is no need to recover every last one of them. If part of the woody root or stem is still attached to a tuber, it is broken off and thrown away. Then the tuber is tossed into a rectangular woman's carrying basket called a *tátofoŋo,* lying on the ground nearby.

When a mound is dug up, a lot of tubers come out of the ground. Not all of the 12 or 14 cuttings that may have been planted in a mound grow but, as if to compensate for this, each live cutting produces more than one tuber. In fact, from the lower end of a living stem, 2 or 3 tubers may develop. In a sample of 35 mounds I watched being dug up, the number of tubers pulled out per mound ranged from 5 to 53. And in size they varied even more. The smallest tuber I saw pulled (which the woman who unearthed it did not bother to retrieve) weighed only ⅛ of an ounce (3.5 gm) and the largest weighed 4.9 pounds (2.2 kg). However, two tubers I once saw in back of a house, both of the variety known as *talakiñï,* weighed 8 and 9 pounds (3.6–4.1 kg). In length, these tubers varied from barely more than an inch to 32 inches.

The average weight of tubers taken from a mound varies also. The aggregate weight depends primarily on how long the tubers have been growing and whether they represent the first, second, or third crop planted in that mound. A first crop of manioc, harvested after 1 year of growth, will yield an average of about 14.5 pounds (6.5 kg) of roots per mound. A second crop, also harvested after a year's growth, will yield about 13 pounds (5.9 kg) of tubers. And a third crop yields around 8.5 pounds (3.9 kg).[28]

Now, the average plot is about 1.6 acres (0.65 ha) in size and has some 1400 mounds of manioc growing in it. This means that an average-size garden of year-old plants will yield about 10.2 tons (9275 kg) of tubers if they are first crop. If they are second crop they will yield 9.1 tons (8275 kg), and if third crop, 6 tons (5450 kg). Per acre, this works out to 6.2 tons (5635 kg), 5.6 tons (5090 kg), and 3.7 tons (3365 kg), respectively.[29]

This yield of manioc tubers, moreover, can be counted on pretty reliably from year to year. The Kuikuru have no tradition of "bad years." To be sure, a particular individual may have a rough time of it if he is lazy and fails to fence his plot or weed it properly. But that is his own fault. With reasonable care, which most Kuikuru give their gardens, no one need ever fear of running short on manioc.[30]

I was unable to weigh a crop of *ufigi,* that is, manioc that has grown for 2 years. From all accounts, though, such a crop would weigh almost twice as much as a one-year crop. Of course, it takes twice as long to produce a crop of *ufigi,* and thus the per-acre yield should perhaps be divided by two. Still, the only work involved during the second year of growth for *ufigi* is pruning and weeding, and so the yield in calories returned per calorie invested, is very likely higher than for a one-year crop.

Returning to the details of harvesting, once a woman has picked a good number of tubers, she sits on the ground and begins processing them. If two women work together, one may continue pulling tubers while the other scrapes them. Using a mussel shell scraper called an *efete,* a woman first cuts off the far end of the tuber and then scrapes off all the heavy brown rind.[31] When scraped clean of its cortex, the tuber is either white or cream colored, depending on the variety. Scraped tubers are then thrown into an empty *tátofoŋo* (woman's carrying basket).

A woman prefers to scrape the smaller tubers first, leaving the largest to be brought back to the village unscraped. Some unscraped tubers are put on top of the scraped ones on a *tátofoŋo,* but the biggest of all are generally handed to a youngster to carry home.

Work proceeds in the garden until the women who are scraping tubers have filled their baskets, at which point they hoist them on their heads and start back to the village. A woman's husband accompanies her home unless he decides to stay longer and continue planting or weeding. If he does go home with her, he may carry a few tubers himself, but never in a woman's carrying basket. Instead, he uses a man's carrying basket, called a *tsaŋgo,* which he carries on his back by means of a tumpline around the forehead.

It usually takes about 1½ hours of work for women to fill their carrying baskets with peeled tubers. Thus, by 7:00 A.M. or so, the trek home begins. The weight of tubers carried by a woman is not inconsiderable. Indeed, a *tátofoŋo* piled high with manioc roots may easily weigh 75 pounds (34 kg) or more. Because 2 or 3 women, or a woman and one or two girls, often harvest together, the total weight of roots brought home by a family after a morning of harvesting is usually between 100 and 200 pounds (45 kg to 90 kg). As they carry home their *tátofoŋos,* filled to overflowing, the women walk double time, their hips undulating rhythmically as they keep their loads in perfect balance.

During an average morning's work, some 15 mounds will be picked clean of tubers. Since the family's aim is usually to pull out all the roots in a garden, and since the average garden has some 1400 mounds, this means 3 or 4 months of work. And indeed, between June and September, a family goes to the garden almost every morning. Even on those days that a man goes fishing, his wife may still go to the gardens by herself to harvest tubers.[32]

The Processing of Manioc

Although she may have pulled 200 pounds (90 kg) of tubers before the sun was well in the sky, a woman's work for the day has just begun. Once out of the ground, manioc roots spoil quickly, so by 7:30 A.M. she is usually hard at work reducing them to flour.

During the rainy season a woman processes manioc indoors, sitting near the small doorway that faces the plaza. This way she is protected from any rains that might fall, and at the same time has a fair amount of light to work by. During the dry season, though, manioc is processed outdoors. To make it easier for his wife to do this, a man builds a large drying rack, called an *ogo,* in front or in back of the house, depending on where she can best catch the morning sun. On top of four corner posts of palm logs he lays a platform about 6 feet (1.8 m) above the ground. The top of the rack not only provides a surface on which to dry manioc, but also a roof under which a woman can work, shielded from the rays of the sun.

Reducing manioc tubers to flour is a long and complicated process, and I will present only the highlights of it here.[33] In preparing manioc a woman is frequently assisted by a housemate, each woman performing one of the successive steps as if she were doing it alone, which she often does.

Sitting on a mat under the *ogo,* the woman takes from her *tátofoŋo* the tubers that remain unscraped and proceeds to scrape them with a mussel shell, just as she did in the garden. Manioc tubers are scraped vigorously so that both the thin, scaly brown skin covering the tuber, as well as the thicker, tougher, pinkish-white phelloderm underlying it, are removed in the same operation. As she finishes scraping each tuber, she puts it into a medium-sized, flat-bottomed pot, called an *afukugúŋgugu,* with some water.

When she feels she has scraped enough tubers, she washes them in the pot and transfers them to another pot. She then takes an *afukugu* (a larger flat-bottomed pot) and places it on the ground against one of the posts supporting the *ogo.* On top of this pot she puts a grating board (*iñagi*). The grater is about 30 inches (76 cm) long and 8 inches (20 cm) wide in the middle. Into an area about 7 inches (18 cm) square in the midsection of the grater are set some 2000 fine sharp bits of palm wood to serve as teeth.

Sitting on a mat on the ground, the woman either straddles the big pot with her legs or else stretches one leg along one side of it and doubles the other leg back under her. She braces the grater by placing the far end against the corner post of the drying rack and the near end against her belly. This keeps it steady during the vigorous motions of grating.

Holding a scraped tuber vertically with both hands, the woman rubs it back and forth against the grater, whose sharp teeth break it down into a watery pulp. Since the grater has a narrow waist, not much pulp accumulates on top of it, most of it falling off to either side into the pot below. On the average, it takes about 20 seconds to grate a tuber.

As soon as the tubers have been grated, the woman goes on to the next operation. Taking another *afukugu,* she places 3 or 4 wooden slats, called *apo,* on top of it, and over these extends a flexible mat strainer, known as a *tuafi,*[34] With a gourd ladle, she scoops up a good-sized mass of grated pulp and empties it on top of the *tuafi.* With the same ladle, she then pours some water over the mash, most of which runs down through the strainer into the pot below. She then begins kneading and squeezing the manioc mash with her hands, and pressing it against the strainer to express the liquid from it. Every so often she rolls up the strainer with the moist pulp inside and twists the two ends in opposite directions, squeezing more liquid out, which also runs into the pot.

What remains on top of the strainer after the pulp has been squeezed is the coarse flour, called *timbuku.* Despite the squeezing, *timbuku,* at this stage, is still slightly moist. The milky juice, called *kuigíikugu* ("manioc sap"), that has gone through the *tuafi* into the pot below contains in suspension the fine starch flour, which the Kuikuru call *kuigiñu.* We know it as tapioca.[35] The juice also contains deadly prussic acid, HCN, and must be processed further before being rendered safe for human consumption.

The coarse *timbuku* flour which accumulates on top of the strainer after each batch of pulp is squeezed is then molded into a loaf, (*gipafa*). Each loaf is about 9 inches (23 cm) long and 4 inches (10 cm) in diameter and bears the impression of the *tuafi,* much as tamales bear the impression of the corn husks in which they are baked. The loaves are placed on the overturned grater, which sits on a nearby pot. Every so often, the woman takes a bunch of *gipafa* loaves and puts them on top of the rack to dry. She proceeds in this way until all the grated tubers have been squeezed.

The pot containing the poisonous juice is allowed to stand for several hours. During this time, the *kuigiñu* gradually settles to the bottom, and the woman then slowly pours off the poisonous liquid, leaving a layer of tapioca 2–3 inches (5–7.5 cm) deep in the bottom of the pot. Using a few handfuls of old *timbuku* (coarse flour), she absorbs as much as she can of the juice remaining on the surface of the tapioca, and then throws the coarse flour away.[36] With a wooden spatula called a *kutigu,* the woman next cuts this layer of *kuigiñu* into five pieces, always in the same shape. These pieces are then carefully removed from the pot and placed on a mat to dry.

Both the coarse *timbuku* and the fine *kuigiñu* flour still contain a residue of prussic acid, and are left several days in the sun to help volatilize some of the remaining poison. To accelerate this process, both *timbuku* loaves and *kuigiñu* slabs are later broken up by hand and spread out in the plaza on a long drying mat (*fugutaŋa*). Even so, not until the flour has been thoroughly cooked will the final traces of HCN be driven off.

Despite its high toxicity, the liquid decanted after the tapioca has settled is not thrown away. With further treatment, it can be made into an appetizing drink. This treatment merely involves putting the liquid over the fire and boiling it vigorously until all the prussic acid has been driven off. The resulting liquid, a trifle thickened as a result of the boiling, is now a potable, slightly sweet drink called *kuigiku.*

The coarse and fine flour also undergo further processing after drying in the sun. The chunks of coarse flour are broken up by hand and pounded in a mortar (*kofi*) and pestle (*afutofo*). The pounded flour is then sifted through a sieve called an *aŋagi,* which is much like a *tuafi* strainer except that the weft elements are twined further apart. The purpose of this sifting is to remove the woody fibers remaining in the coarse flour. It may be well to explain the source of this fibrous material.

Running through the center of the manioc tuber for its entire length is a taproot about as thick as a match stick, and very woody. Grating the tuber reduces this taproot to shreds which make up much of the woody fiber found in the grated pulp. In the straining process, most of these fragments remain on top of the *tuafi* with the coarse flour. Later, after pounding, when the coarse flour is sifted, the starch flour falls into the pot below leaving on top of the sifter a dry, woody, fibrous, inedible residue called *jakatïpe* which is eventually thrown away.

The fine flour, having already passed through the *tuafi* as wet pulp, does not have much fibrous matter left in it. Thus, it does not need to be pounded in the mortar and pestle. But it is sifted through the *aŋagi* to remove whatever tiny bits of fiber may remain.

By methods I need not describe here (see Dole 1978:218–232), two different kinds of nonalcoholic manioc drink, *tilisiñï* and *tukuniñe*, are prepared. Varying proportions of fine and coarse flour go into these drinks.

Cassava cakes, called 'beijú' in Brazil and *ikine* by the Kuikuru, are made on a large circular clay griddle called an *alato*. Both kinds of flour normally go into beijú, but the proportions will vary depending on the occasion. For everyday fare, beijú is made largely with coarse *timbuku* flour. For special occasions, though, a very high quality beijú is made using only fine flour. In general, the higher the proportion of fine flour in a cassava cake, the better it is.

Varieties of Manioc

I have spoken of manioc so far almost as if it were all the same. But this is not the case. While manioc belongs to a single species, *Manihot esculenta*, it has many varieties. And I was able to elicit from the Kuikuru the names of no fewer than 46 of them which they raise in their gardens.[37] These varieties vary in such things as the height of the plant, the color of the stem, leaves, and petioles, the distance apart of the leaf scars, and—most important of all from the Kuikuru's point of view—the amount and quality of the two kinds of flour their tubers yield. The Kuikuru are thoroughly familiar with every feature of their manioc varieties, and have definite preferences among them.

It should be emphasized that all varieties of manioc planted by the Kuikuru are "bitter." that is, poisonous.[38] And it should be pointed out as well that the toxicity of a variety of manioc and its sweetness—that is, the amount of sugar it contains—appear to be unrelated. Indeed, a Kuikuru who had once spent a few months in Rio de Janeiro and eaten a lot of "sweet" (nonpoisonous) manioc there, assured me that *maisafalu*, the Kuikuru's sweetest (but still poisonous) variety, was distinctly sweeter than any he had tasted in the city.

The 46 varieties of manioc grown by the Kuikuru are listed in alphabetical order in Table 3.5. To be sure, these varieties are not all of equal importance. The 6 varieties most grown by the Kuikuru, which together comprise over 95% of all the manioc they plant, are, in order of importance, *kuigífekugu*, *fekïgi*, *tifagi*, *jalífutisï*, *maisafalu*, and *kïake*.

The leading variety by far is *kuigífekugu* ("real" or "true" manioc). It is the king of the crop. Perhaps three-quarters of all manioc planted is of this

Table 3.5

Names of the 46 Varieties of Manioc Elicited from the Kuikuru as Being Planted in Their Gardens

afátufi	*kuatagi*
afefítufi	*kugitsaka*
agafïtáasïtisï (also called *ikobesene*)	*kuigífekugu*
agaka	*kumigi*
agapisútufi (=*apisútufi* ?)	*kuoŋgu*
agï	*maisafalu*
akulá	*naasi*
apisútufi (=*agapisútufi* ?)	*ŋikogótufi* (=*kajápitufi*)
asuti	*otigi*
egiŋgógafetisï	*takatiñï* (=*mukuga* ?)
ekege	*talakiñï*
ekïgi	*tañétufi* (= *kaŋátufi* ?)
fefegiku (=*tatïfekú*)	*tïfefegiliñï*
fekïgi	*tifagi*
fopijoŋo	*tiñakafiŋo*
ifïpe	*tïgafefe* (=*tïfesesekiño*)
iŋe	*tïiséi*
igapisï	*tugumáitufi*
jalífutisï (=*ikobesene*)	*tupaga*
isutisï	*uagitífekugisï*
juapunaku	*uagitítufi*
kagafesekekiñï	*utigi*
kaŋátufi (=*tañétufi* ?)	
kïake	

variety. It does not grow exceptionally tall, but it yields many roots. Moreover, the tubers, which are very white after peeling, give lots of high-grade *kuigiñu*. And the *kuigiku* drink made from it is thick and tasty.

The second most planted variety is *fekïgi,* which is quite similar to *kuigífekugu*. It is said to produce the largest roots, and also yields lots of *kuigiñu*.

Tifagi ("sting ray") is a variety that grows tall, with long stems. Its freshly peeled tubers are light yellowish-cream in color, and they too produce much *kuigiñu*.

Jalífutisï (*jalí* means "tapir") grows very tall—up to 8 or 9 feet (2.4–2.7 m) as compared to 6 or 7 feet (1.8–2.1 m) for most varieties. It gives less *kuigiñu* than the three previous varieties, but it is relatively sweet.

The sweetest *kuigiñu* of all, though, is given by *maisafalu*. As already noted, this variety is distinct in appearance from all others in having a silvery stem. And its petioles are purple and some of its leaves purplish-green. Its tubers have a lighter and thinner skin than most varieties. Since the *kuigiñu* of *maisafalu* is so sweet, it is favored in making cassava cakes (*ikine*), as well as manioc drink (*kuigiku*). But its coarse *timbuku* flour is not highly regarded and is usually thrown out once the *kuigiñu* has been separated from it.

Finally, there is *kïake* ("leaf-cutter ant"), which, like *maisafalu*, has reddish petioles. The stem of this variety is reddish-maroon, like the ant after which it is named. It does not produce much *kuigiñu*, but what it does yield is sweet, and its *kuigiku* is said to be very sweet, something the Kuikuru value highly. As with *maisafalu*, the coarse flour yielded by *kïake* is not thought much of and is usually discarded.

Since the many varieties of tubers have different properties, they are often processed separately. To facilitate this, on a given day a woman may pick only the roots of a single variety.

Storage

All the manioc processed during the dry season and not needed for immediate consumption is stored inside the house. *Timbuku* and *kuigiñu* flour are stored separately. In the form of *gipafa* loaves, *timbuku* is stored in a large version of the men's carrying basket, called an *asaŋguati*. *Kuigiñu* is stored as loose flour in a special silo-like basket (*kuigiñúofefijo*). Both types of basket are lined with the large leaves of the *tsuitsï* tree to keep flour from coming through the loose mesh of the baskets.

By the time the dry season ends, the amount of manioc flour stored in a Kuikuru communal house is prodigious. My figures for this are based on observations made in two houses. The first observation was made in the house of Agatsipá around the middle of the manioc processing season. The other was made in the house of Funupa just after the end of the processing season.

On August 18, I made a careful inventory of the manioc stored in Agatsipá's house with the results shown in Table 3.6.[39]

To the total I obtained of 2333 pounds (1060 kg) of stored flour should be added the flour put away by the women of Agatsipá's household between August 18 and the end of the dry season, a period of some 35 days.

Table 3.6

Weight of Manioc Flour Stored in Baskets in the House of Agatsipá, August 18, 1975

Type of manioc	lbs	Kg
Timbuku stored in *kuigiñúofefijo*	233	106
Kuigiñu stored in a large *asaŋguati*	1091	496
Kuigiñu stored in a medium-sized *asaŋguati*	398	181
Balls of *kujao* (a finer form of *timbuku*) stored in a small *asaŋguati*	167	76
Balls of *kujao* stored in a smaller *asaŋguati*	77	35
Weight of 550 dry *timbuku* loaves still on drying rack	367	167
	2333	1060

Assuming that manioc was harvested during 25 of these 35 days, that the average weight of tubers picked per day was 150 pounds (68 kg), and that, reduced to flour, this weight would diminish by 60%, it would mean that an additional 1500 pounds (682 kg) of manioc flour were stored in Agatsipá's house during the remainder of the dry season. Thus, by the time the rains began, this household of 11 persons had some 3850 pounds (1750 kg) of stored manioc flour to draw from.

On October 4, two weeks after the rainy season had begun, I inventoried the house belonging to Funupa, in which 14 people lived. Here I counted a total of 15 baskets of manioc flour, 9 of which were filled with coarse flour and 6 with fine flour. Several of these baskets, which were tied closed when first filled, had now been opened, showing that people were already drawing on this stored manioc for their daily meals. Since an *asaŋguati* full of manioc flour weighs between 300 and 400 pounds (136 and 182 kg), the total weight of manioc stored in Funupa's house was between 4,500 and 6,000 pounds (2045 and 2725 kg).

This manioc flour, incidentally, is not collectively owned by the entire household. It is the property of each family that processed it.

How long such stores of manioc will last depends on how much each family has put away, how many people there are in the family, and the rate at which they consume it. An industrious family with a lot of manioc in storage is said to be able to go until April—virtually the end of the rainy season—before it must resume picking fresh roots. A less industrious one, on the other hand, may run out of stored flour as early as December or January.

In the aggregate, though, there is no question that Kuikuru manioc cultivation yields bountifully. So abundant and reliable are these yields that the village is virtually guaranteed a secure and rewarding subsistence, year after year. Whatever worries the Kuikuru as a group may have, running short of manioc is not one of them.

Manioc and the Supernatural

One manifestation of a secure subsistence is that little supernaturalism surrounds it. And this is pretty much the case with the Kuikuru. In the interest of touching all bases in my description of manioc cultivation, though, let us see what supernaturalism there is in this aspect of Kuikuru subsistence.

The Kuikuru have a myth that manioc first grew at the bottom of a river. Then, through the agency of a woman named Kefesugu, it was obtained by the Kuikuru for planting on land. The 'trairão' (*Hoplias* sp.) fish is said to be the "owner" of manioc, except for the variety *maisafalu*, which is "owned" by the 'pirarara' (*Phractocephalus* sp.), a kind of catfish. The Kuikuru explain why *maisafalu* is planted in rows along the boundary of a garden plot by saying that it is the "river of manioc" (*kuigínakagagï*), where certain *etseke*, or spirits, come at night to bathe.

Although the trairão and pirarara are the owners of manioc, *isufo* is the *etseke* of it[40] and, according to some informants, the owner of manioc stems. The *isufo* (there are, apparently, lots of them) live underground in the manioc gardens. In fact, "*isufu*" is the name given to rounded pellets of earth found at the bottom of some manioc mounds, as if these pellets somehow represented the *isufu* spirits. The *isufu* act somewhat as guardians of manioc, taking care of the plant so it won't become ill. Should it sicken anyway, people may ask the *isufu* to cure it.[41]

Manioc is a hardy plant, not susceptible to many ailments, so when plants do not grow right, the Kuikuru are quick to attribute the trouble to witchcraft. It may be the fault of a *tijakitse* bird, which, on its own, without human intervention, may bewitch a plant, and make its leaves sick. More likely, though, if manioc is not growing well, a human sorcerer has put a magical object of some kind in the garden to harm the plants. Young plants are especially vulnerable to this, it is said, and if bewitched, their stems may remain thin and their tubers small. Should that happen, a shaman may be asked to locate and remove the offending object.[42]

Predation of the manioc fields may also be attributed to sorcery. Thus, when a man's plot has been ravaged by peccaries, he is likely to think it is due to sorcery by someone in the village. In that case, the sorcerer (who usually goes unnamed) is said to have made a *feúfutofo*—a likeness of a peccary—and placed it secretly in the victim's garden. The effect of this *feúfutofo* is to draw peccaries to the garden to eat the roots. Thus, one reason given for making one's second garden in a different area from his first, is that if the two were growing side by side, a sorcerer could more easily ruin them both.

The Kuikuru have a ceremony (*ailofo*) for manioc which they call *ago*. In this ceremony, which is performed during the rainy season, songs are sung about manioc and the myth of its origin is recounted.

Abandonment and Reforestation

Two things act to shorten the life of a manioc garden. One is the loss of soil fertility, and the other is the infestation of weeds. The playing out of the soil is reflected in the thinness of the stems. And when a man sees the stems of his plants coming up thin, he knows he can expect small tubers.

Small tubers are, of course, the indisputable sign that the time has come to abandon a plot.

An abundance of weeds may affect manioc cultivation even more than does impoverished soil. Every year, more and more weeds invade a garden, and it becomes progressively harder to keep them under control. Finally, when the weeds are just too thick and too tall to cope with, a Kuikuru simply gives up the effort. A last crop of roots may be dug up, and then the plot is abandoned. An abandoned plot is called a *kuigítsufigïpe* ("old manioc thing").

How long does a plot last? The garden of an indifferent cultivator may last only 2 years, but usually a Kuikuru can count on 3 years of crops from it. Sometimes a garden is planted for 4 years, and occasionally even for 5. By and large, the longevity of a plot reflects the care taken by a man in fencing and weeding it.

As soon as a man has pulled the last tuber from the ground, his ownership of the land ceases.[43] Now, anyone can use it. Ordinarily, though, no one wants to. An abandoned plot is of interest to someone only if it has grown up to *iñe,* a tall grass called 'sapé' in Brazil (*Imperata* sp.). *Iñe* is used to thatch houses, and an old manioc plot may be a good source of it.

Once a plot is abandoned, the slow process of plant succession begins. If not drastically interfered with, this process will eventually see the return of *itsuni,* primary forest. What the first stage of this succession will be varies from plot to plot. Usually it is *tafuga,* the weedy second growth that may have forced the abandonment of the plot in the first place. However, if a man has weeded his plot thoroughly, extirpating every tree root growing in it, the first invader may be *iñe* grass. With little or no *tafuga* to compete with, grass may quickly sprout and establish itself. If so, a thick, chest-high stand of *iñe* may completely cover an old garden for the first 5 years. Only then will trees begin to grow up through the grass and eventually shade it out[44]

The rate at which *itsuni* returns to a plot depends on how it is treated after abandonment. If the grass or scrub growth has been repeatedly burned, *itsuni* will return slowly. Burning, however, may favor the development of fire-resistant scrub forest, such as *agïpe* or *agafagïpe,* which will dominate the area for a time.

How long it takes for *itsuni* to reestablish itself is difficult to estimate. One informant, who had a better grasp of numbers than most, guessed that, starting with *tafuga,* it might take 30 years for *agïpe* to establish itself, and then another 40 years for *itsuni* to take over. These figures, though, probably vary a good deal.

Since the undergrowth in *agïpe* and *agafagïpe* (scrub forest) are burned every dry season, the soil under these forests probably regains its fertility fairly soon. At least it becomes fertile enough to support a new manioc garden. Thus, as far as Kuikuru cultivation is concerned, it is not essential for *itsuni* to reestablish itself. But the Kuikuru do prefer *itsuni* for planting. Even if the soil underlying it is no better than that under *agïpe* or *agafagïpe,* weeds are likely to be fewer.

Best of all, the Kuikuru would like to plant virgin, or at least climax, rain forest. Climax *itsuni* grew within 3 miles (5 km) of their old village of Lafatuá, but in their present location it is lacking. My guess is that the *itsuni* they clear and plant nowadays is 100 years or more from climax.

But although the soil underlying their *itsuni* is not of high fertility, it is still adequate for the purpose. It is well able to sustain the growth of an ample crop of manioc, thus providing the Kuikuru with a bountiful and secure subsistence.

Overview and Interpretation

Practiced in the productive way they do it, manioc cultivation has profound consequences for the Kuikuru. I have discussed some of these consequences in earlier publications (Carneiro 1957:164–196; 1960; 1961:48–49), and do not intend to cover the same ground here. But in concluding this chapter I would like to look more closely at the magnitude of Kuikuru manioc production.

The massive scale on which the Kuikuru plant, harvest, and store manioc is most impressive. One gets the feeling of bountiful, almost lavish accumulation. But yet, not of ostentation. There is no invidious display of one's stored flour, no competition as to who can produce the most. And certainly no tubers are ever left to rot as a show of affluence. What we find is a controlled supply of manioc reflecting a comfortable abundance.

On the basis of my field work in 1953–1954, I estimated that the Kuikuru produced about 2½ times as much manioc as they actually consumed (Carneiro 1957:159). Based on a fuller study in 1975, I would like now to reestimate the ratio of manioc planted to manioc consumed.

We saw earlier that the average Kuikuru garden contains 1400 mounds of manioc, and that the average yield of year-old tubers per mound is 14.5 pounds (6.6 kg). This means that the total weight of roots obtainable from a completely harvested plot is 1400 × 14.5 = 20,300 pounds (9227 kg). Since there were 41 plots with growing tubers as of September, 1975, the total harvestable weight of manioc was then on the order of 20,300 (9227 kg) × 41 = 832,300 pounds (378,318 kg).

What is the caloric equivalent of 832,300 pounds of manioc? First of all, of the weight of a manioc tuber only some 25% is edible starch, the rest being water, fiber, etc. (Barrett 1928:373). Thus, the 832,300 pounds of roots represents 832,300 × 0.25 = 208,075 pounds (94,383 kg) of edible starch.

For convenience in calculating the caloric value of this amount of starch let us translate this figure from pounds to grams. Because there are 453.6 grams in a pound, 208,075 pounds of manioc is equivalent to 208,075 × 453.6 = 94,382,820 grams. Pure digestible carbohydrate, whether suger or starch, yields very close to four Calories per gram. Assuming that manioc flour, as the Kuikuru process it, is 90% digestible, a gram of it would yield 4 × 0.90 = 3.6 Calories. Thus, the total caloric value of the Kuikuru's harvestable manioc is 94,382,820 × 3.6 = 339,770,000 Calories.

Having calculated the Kuikuru's potential manioc production, we next need to determine their actual consumption. The caloric intake of the average Kuikuru (age and sex differences being taken into account) is about 3000 Calories per day. Since manioc constitutes about 80% of the diet, it can be said to supply 3000 × 0.80 = 2400 Calories of their daily intake. If a person consumes 2400 Calories of manioc a day, his yearly consumption of it will be 2400 × 365 = 876,000 Calories. Then, because there were 120 Kuikuru at the time of the study, the intake of manioc for the entire village for one year would be 876,000 × 120 = 105,120,000 Calories.

Let us now compare the two figures we have arrived at:

Potential maximum production of manioc per year	339,770,000 Calories
Estimated actual consumption of manoic per year	105,120,000 Calories

By simple calculation we can see that the amount of harvestable manioc growing in Kuikuru gardens at a given time, if fully harvested and processed, would yield about 3.2 times the amount actually consumed in a year. This is a large discrepancy. What happens to the difference between potential production and actual consumption?

To begin with, we should note that of the estimated 832,300 pounds (378,318 kg) of manioc tubers growing at a given time, not all are intended to be consumed during the ensuing year. A certain amount of it (perhaps a third) is meant to be left in the ground to grow for a second year. So discounting this fraction let us look at those tubers that are intended to be harvested and eaten during the current year. What happens to them in their journey from mound to mouth?

As I have said, a substantial number of tubers are lost to peccaries before they are ever pulled from the ground. I cannot give an estimate of the aggregate loss involved. It varies greatly from plot to plot. In most cases it is relatively small, but in a few cases it is large, and occasionally even total.

To the losses sustained to peccaries must be added those smaller ones resulting from the timid nibblings of agoutis and deer.

There is, of course, a large and inevitable reduction in the "live weight" of manioc tubers as their indigestible components are removed through processing. As I have said, only about 25% of a manioc root is digestible matter. The other 75% is water, skin, phelloderm, tap root, woody fibers, and so on. The removal of these constituents accounts for the single greatest decrease in the weight of manioc once the roots have left the ground.

In addition to this, there is the fact that the *timbuku* flour of some varieties of manioc is not prized and, though edible, is thrown away rather than eaten.

Another trifling loss of manioc occurs when stored flour, especially *kuigiñu,* is attacked by a bright orange mold called *mufiso.* This mold attacks only the surface of the flour and is simply scraped off, but a little of the underlying flour goes with it.

Then, besides the necessary removal of indigestible matter in preparing the manioc root, there is the loss of edible flour through carelessness. So ample is the Kuikuru's supply of manioc that they do not concern themselves with saving every half-tuber or husbanding every half-loaf of *timbuku.*

In the old village of Lafatuá there was an unobtrusive but steady loss of manioc flour to leaf-cutter ants, which, in the aggregate, was probably considerable. I never estimated the size of this loss, but we can form some notion of how large it might have been from a remark of the naturalist Henry Walter Bates. Bates (1864:14–15) reported that one night he almost lost the contents of two bushel baskets of manioc flour to leaf-cutter ants. While the Kuikuru might grumble about the leaf-cutters, they never made a serious attempt to get rid of them.

Less common, more sudden, and of greater immediate magnitude is the loss of manioc in a house fire. No house fire had occurred at Ipatse since the Kuikuru had been living there, but 2 years before, such a fire had burned more than half the houses in the village of Afaŋítafagï, where the Kuikuru had lived immediately prior to moving to Ipatse. Moreover, house fires sometimes destroy a whole village, as the wind whips the fire entirely around the village circle. House fires had happened at least 4 other times in Kuikuru memory, including twice in 1953 while I lived with them at Lafatuá.

In summary, then, at every stage there is a whittling away of the gross weight of manioc that once grew in the fields. This reduction eventually sees the diminution of 832,300 pounds (378,318 kg) of tubers down to about 60,000 pounds (27,272 kg) of edible flour.

We cannot say that the Kuikuru "overproduce" manioc. There is never an "excess" of it to be disposed of in some way. It is simply that when the Kuikuru plant, they make allowance for the losses they are liable to sustain. They build a cushion into their manioc cultivation which permits them to surmount not only incidental, routine, and trivial losses of manioc, but also sudden, large, and, one might think, catastrophic ones.

No matter how great a loss might be, the Kuikuru can always draw on growing manioc tubers to tide them over, tubers which otherwise would not be drawn on for months or even a year. And if no major loss occurs, then this cushion can be carried forward, to be used when its time comes. Here again we see the effect, not only of the size of the manioc crop, but also of its flexibility. No manioc ever spoils because it stood in readiness to cover some emergency that never came. Either as raw tubers growing in the ground or as processed flour stored in the houses, manioc waits imperishably and imperturbably until it is needed. Indeed, if allowed to grow undisturbed, today's emergency ration will become tomorrow's dividend, as the tubers grow larger and larger with each passing month.

Manioc cultivation, then, provides the Kuikuru with a broad and solid subsistence base. On this base they have developed a culture that is relatively rich by Tropical Forest standards. For example, there are many ceremonies in their ritual cycle, and plenty of time to engage in them. Moreover, I am convinced that prior to European contact, this culture was even richer. The Kuikuru and their neighbors probably had bigger villages, stronger chiefs, mobilized labor on a larger scale, and perhaps even had social classes. If manioc cultivation did not create this culture, it at least provided the economic foundation on which it could be reared.

Notes

[1]These and other soil analyses presented in this paper were very kindly done for me by Prof. Matthew Drosdoff of the Department of Agronomy, New York State College of Agriculture and Life Sciences, Cornell University, Ithaca, New York.

[2]The Kuikuru miss no opportunity to burn the savanna, as well as the brushy undergrowth in relatively open forest areas such as those of *agïpe, agafagïpe,* and *kejitepe.* Indeed, I would call them compulsive pyromaniacs! When one asks a Kuikuru why he burns the grass, the stock answer is "*taloki,*" "for nothing." And I don't believe they were witholding their real motives from me. Could it be that we have here another instance of a practice carried out thoughtlessly which neverless results in a beneficial ecological outcome?

[3]There are three trees that the Maroni River Caribs will not cut down when clearing the forest because "they house important or powerful and dangerous spirits and no ordinary man dares to damage one of them" (Kloos 1971:28). They can, however, ask a shaman to cut them down.

[4]The Kuikuru attributed a mysterious fever I once came down with to the fact that I had spent an hour and a half in the forest the previous day, thus subjecting myself to the baleful influences of the *etseke* who dwell there.

[5]They chop a little slower than the Yąnomamö, but hit harder and chop longer without stopping. The Yąnomamö chop in great bursts and are more accurate in laying their axes than the Kuikuru (see Carneiro 1979b:47).

[6]As he walked through his plot setting fires, Afukaká called out, "*kuá, kuá, kuá,*" the song of the *kuakuagu* bird, as the Kuikuru are said to do when burning a plot. The informant I asked about it could not give me a reason for doing so.

[7]The Kuikuru will not take home just any log, but only those deemed to make good firewood because they split easily, burn well, have a pleasant scent, etc. Their favorite firewood is 'pindaíba' (*Xylopia* sp.), which also yields inner bark for lashing.

[8]A Kuikuru does not use the position of the stars or the fruiting or flowering of trees to tell him when to plant. Nor does he wait for the rains to begin. As soon as he has reburned his plot, he is ready. In the case of maize, though, the Kuikuru do wait for the first rains before planting.

[9]One informant said that in the days of the *kų́ǫ́* spade, mounds were made smaller than they are today. This I am ready to believe.

[10]Using a soil tube, a total of 15 cores were taken from Plot C, which was ⅙ of an acre (0.07 ha) in size. The soil from these 15 cores was then thoroughly mixed, and a sample drawn from that. Plot D, which was 1.4 acres (0.57 ha) in size, was sampled in the same way except that 20 cores were taken from it instead of 15.

[11]The one exception to this in the memory of the Kuikuru was a woman named Agufuká, who in 1973 or 1974 took over a *tufiñafo* cleared by a man who died shortly thereafter, and planted it to manioc herself.

[12]It is often the case, though, even in the plot of a careful cultivator, that the branches of plants in adjacent mounds intertwine, making it difficult, if not impossible, to walk between them. Thus, in order to weed his garden, Mugí had to stoop or squat around the base of his plants.

[13]Because the distance between them is as much as 7 feet (2.1 m), a man can situate manioc mounds between the stumps and logs in his garden in such a way as not to lose too much planting area to these encumbrances.

[14]It is one of the great advantages of manioc over a grain crop like maize that none of its edible part need be sacrificed to provide planting stock.

[15]So hard, in fact, that while I was living with the Kuikuru in 1975, a wind storm blew the roof off a large house, nearly killing the residents.

[16]This is the name given to the growing plant. The pigment derived from it is called *umïŋi.*

[17]The tapir, *jalí,* does not eat any part of the manioc plant and so cannot be considered a predator of it. But it does damage the plant with its big bulk as it moves through a garden. In 1975 at least one man had pineapple plants (acquired from Brazilians) growing alongside his garden, hoping the sharp spines at the tips of the leaves would discourage tapirs from barging through his field.

[18]The Kuikuru are convinced that the great number of peccaries in their habitat is attributable to sorcery. A sorcerer, they said, had made a beeswax doll representing a peccary and placed it in a manioc garden to draw peccaries to it.

[19]The tropical ecologist Pierre Gourou (1953:28) once made the flat statement about bitter manioc that "animals do not eat the . . . roots of this plant." But the fact is that they do. And the reason they can do so and not be poisoned is that the growing tuber has no free HCN, but only a cyanogenetic glucoside, which liberates HCN only after the root has been picked and begun to wilt, being then acted on by an enzyme (see Watt 1908:767). Thus, as long as it is still growing underground, peccaries can feast on it with impunity.

[20]The same sorcerer who ostensibly attracted all the peccaries to the vicinity of the Kuikuru village is also said to have equipped them with wings to help them vault the manioc fences!

[21]One reason suggested for this difference is that the leaf-cutter ant attacks only sweet manioc. But this theory seems not to hold, since the Maroni River Caribs plant almost nothing but bitter manioc (Kloos 1971:32).

[22]For an account of the Kuikuru's detailed and precise knowledge of forest trees and their characteristics see Carneiro (1978).

[23]As a matter of fact, the Kuikuru spoke of manioc roots being left in the ground for 3 years on occasion. The skin becomes hard, too hard to scrape with a mussel shell, so a knife must be used. And the inside of the root becomes hard too. But while harder, the tuber is also sweeter,

and the Kuikuru will not hesitate to reduce it to flour and eat it, just as they would a younger one.

[24]Such a person must get his tubers from a man who has an ample supply growing in his gardens. This is usually a close kinsman or housemate. In exchange, a man assists his benefactor by helping him weed or fence his plot.

[25]When a man dies, his son may occupy his garden house. However, I was told that a proper Kuikuru would feel sad about doing this, and would burn his deceased father's garden house instead.

[26]The Kuikuru do not eat the leaves of the manioc plant, but told me that the Kamayurá do. A man named Iŋgu, whose father was Kamayurá, said they boil it for a couple of hours and then eat it like a cooked vegetable. Eating manioc leaves is rare among Amazonian Indians, despite the fact that it is very high in protein. Another tribe that eats manioc leaves is the Bora of the northwest Amazon (Whiffen 1915:133).

[27]Men sometimes pull up tubers too. While in the sexual division of labor harvesting manioc is assigned to women, practical considerations sometimes override this norm. Once, for example, when a man named Kamáŋgagï went to the garden with his wife but no other accompanying female, he pulled all the roots while she scraped them.

[28]These are not very firm averages because the sample of mounds and plots used to calculate them was small. But they are probably not far from the mark.

[29]By way of comparison, Irving Goldman (1963:85) writes "Among the *Bahúkiwa* [a sib of the Cubeo] I estimated the acreage yield [of manioc] to be between five and six tons, a productivity that reflects considerable agronomical skill and effort."

[30]In 1953, the house belonging to a man named Manïfá caught fire and burned to the ground, destroying hundreds, if not thousands of pounds of manioc flour. But Manïfá did not ask anyone for help in replenishing his larder. He and his wife simply went back to their gardens and began pulling more tubers.

[31]Nowadays mussel shell scrapers are being replaced by coffee can tops. Formerly, after prolonged use, mussel shells were resharpened on an abrasive stone.

[32]During the month of August, though, the entire family may take a few days off from their gardening chores and go to the Kuluene River to gather turtle eggs on the gleaming white sandy beaches exposed by the receding waters. And late in the same month, most families take a few days off again to attend the feast of the dead (*kuarup*) at one of the other Upper Xingú villages.

[33]For a fuller account of Kuikuru manioc processing, along with some comparative data from other Amazonian tribes, see Dole (1978).

[34]The *tuafi* looks very much like a common Japanese place mat. Its warp elements are palm leaf midribs, and these are twined together at intervals of an inch (2.5 cm) or so with wild pineapple, palm, or cotton cord. In manioc processing as it occurs in the Upper Xingú, this mat strainer takes the place of the more familiar and widespread tubular press of Amazonia known as a *tipití*.

[35]Actually, we ordinarily see tapioca in this country, not as fine flour, but as pearl tapioca. The starch granules, with a little water added and subjected to heat, form tiny spherical beads, or "pearls."

[36]Actually, what is used to soak up the remaining poisonous juice is not *timbuku* proper but *jakatïpe*, the even coarser and otherwise useless residue left after *timbuku* has been pounded in a mortar and pestle and sifted through a sieve.

[37]By way of comparison, Kenneth Kensinger *et al.* (1975:44) record 22 varieties of manioc among the Cashinahua, and James Boster (n.d.:13) reports the Aguaruna and Huambisa each have more than 100 varieties.

[38]The one exception is a variety the Kuikuru have recently obtained from Brazilian settlers on the Rio Sete de Setembro, called *tïiséi*, which is nonpoisonous. However, in 1975, scarcely any of it was being planted.

[39]My procedure was to measure the volume of each basket of flour, and weigh samples of known volume of different types of flour. Then, from these figures, the computation was made. Allowance was made for the compaction of manioc flour in a basket under the weight of the overlying mass.

[40]The Tupí-speaking Kamayurá of the Upper Xingú recognize 3 spirits, or *mama'é*, of the manioc plant. For a description of the supernaturalism associated with Kamayurá manioc cultivation, which is more elaborate than that of the Kuikuru, see Oberg (1953:20).

[41]However, like all *etseke, isufu* are dangerous. They can attack a woman in a manioc field and give her a severe nosebleed, or even kill her. They can attack a man, too. Accordingly, when a man asks an *isufu* to cure his manioc, he does not expect a reply, or even want one. A person who hears an *etseke* is said to cry out in terror.

[42]However, the same informant who told me this said, on another occasion, that Kuikuru shamans do not cure manioc plants, only Mehinaku shamans do.

[43]The one exception occurs if a man has planted piquí in his garden, in which case the trees and their fruit continue to be his property.

[44]At the old village site of Lafatuá I saw more abandoned plots with sapé grass growing in them than I did at Ipatse.

CHAPTER 4

Adaptive Strategies of Wakuénai People to the Oligotrophic Rain Forest of the Rio Negro Basin

Jonathan Hill and Emilio F. Moran

Introduction

Discussions of native Amazonians have often been characterized by debates over the relative importance of protein availability, soil fertility, or tribal politics in shaping overall adaptive strategies. What these debates have often obscured are the significant differences in habitat productivity and consequent human organizational adjustments to variable production factors. One distinction that is becoming increasingly accepted in the ecological literature on rain forest ecosystems is the difference between oligotrophic (nutrient-poor) and eutrophic (nutrient-rich) rain forests (Herrera *et al.* 1978). This chapter will focus on the adaptive strategies of two Wakuénai villages of the lower Guianía River (upper Rio Negro) in the Venezuelan Amazon.[1] The area currently occupied by these populations is oligotrophic to a degree rare even in Amazonia, and thus provides an opportunity to study adaptation to an extremely fragile Amazonian ecosystem. Adaptive strategies can be expected to be quite different in eutrophic Amazon ecosystems (see Stocks, this volume).

As this chapter will show, even in the apparently isolated areas of Amazonia, native adaptive strategies can be studied in at least two planes or levels: one horizontal, the other vertical. Ecological studies in anthropology are becoming increasingly aware of the limitations of treating communities as if they were relatively closed to the impact of external influences coming not only from neighboring communities, that is, at the same hierarchical level, but also those coming from higher levels of organization—

ADAPTIVE RESPONSES
OF NATIVE AMAZONIANS

ISBN 0-12-321250-2

such as regional and national systems (Davis 1977; Gudeman 1978; Moran 1981, 1982; C. Smith 1974). Whereas past studies have benefitted a great deal from treating native villages as being almost closed systems for analytical purposes, such an assumption addresses only a limited set of interactions relevant to the study of adaptations (Moran 1981; Vickers 1981a). Human behavior includes responses to internal and external influences but they may not be separable as observable phenomena. Adaptive behavior is the product of complex evaluations by individuals of both internal and external constraints. Thus, to appreciate fully human–habitat interactions, it is necessary to explore how individuals organize themselves to achieve satisfactory ends. The discussion that follows will, therefore, discuss adaptive strategies as a complex system of horizontal and vertical relations. This chapter is a preliminary assessment of the social reorganization of two populations. The bulk of the soil analyses and production data remains unavailable at this time. The chapter begins with a discussion of the oligotrophic Rio Negro basin, proceeds to a discussion of Wakuenai social organization for production, and is completed by a discussion of the relationship between each native community and state-level institutions.

The Rio Negro Basin

The Rio Negro has its headwaters in Colombia and runs through Venezuela and Brazil for 1700 km until it runs into the main channel of the Amazon. The Rio Negro is known as the Guáinía in its upper reaches (Figure 4.1). The Casiquiare Canal connects the Orinoco River system with that of the Rio Negro near the Venezuelan town of San Carlos de Rio Negro. From this point on the river is known by the more familiar name of Rio Negro which it keeps for 900 km until it joins the main channel of the Amazon near the Brazilian city of Manaus. All significant settlements can be found along its margins because the river is the chief and surest means of transport.

A cluster of Northern Arawakan (Key 1979), or Maipuran (Lathrap 1970), language speakers today centers around the upper Rio Negro, the Casiquiare, and several rivers draining into the Orinoco above the Ature rapids. At the time of earliest contact with Europeans, this family of languages extended from the Manao Indians in the central Amazon floodplains near the mouth of the Rio Negro all the way to the Achagua and Maipure peoples along the Guaviare and middle Orinoco rivers. On the basis of linguistic and archaeological data, Lathrap (1970:72–76) hypothesizes that Maipuran-speaking Arawakan groups migrated out of ancestral lands on the central Amazon floodplain between 1000 and 500 B.C.

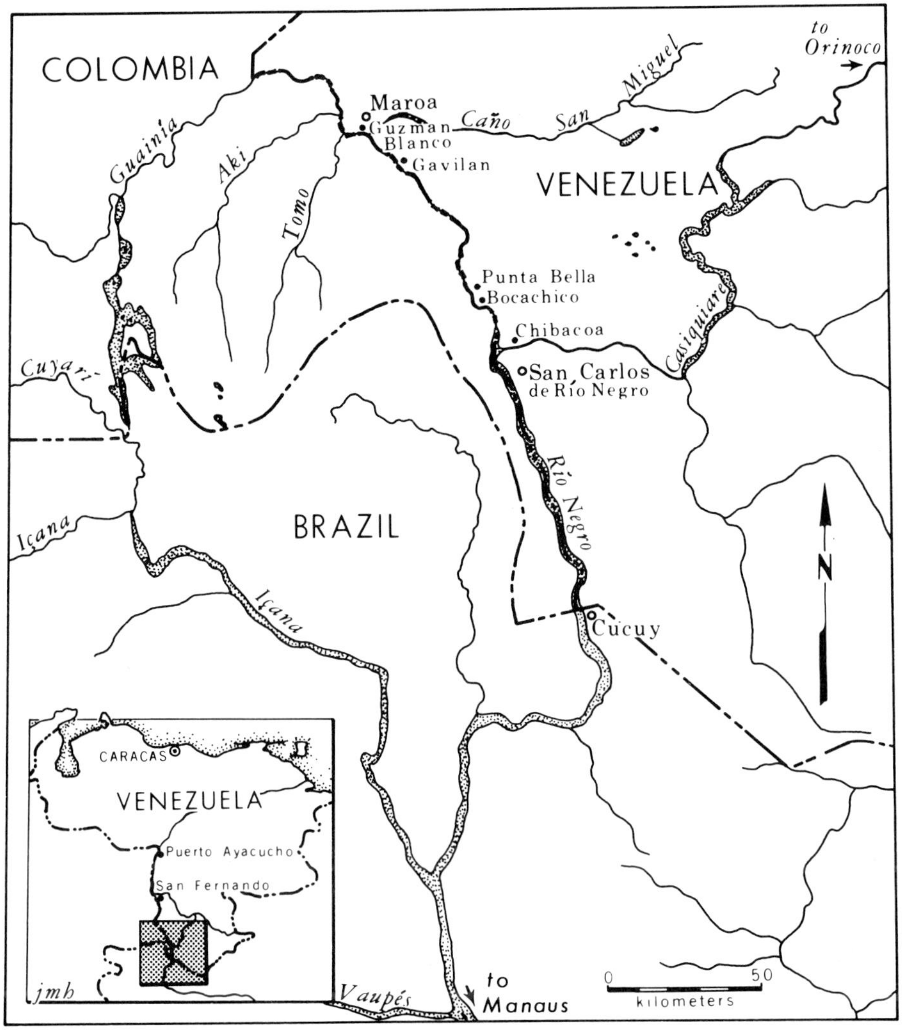

Figure 4.1 *Venezuelan Rio Negro Basin.*

in search of agricultural lands. They continued on through the Casiquiare to settle on more fertile lands along the Orinoco. According to this view, the Arawakan groups of the upper Rio Negro region today are descended from this ancient migratory wave of Maipuran-speakers who settled in areas of restricted alluvial lands located between more fertile floodplains to the north and to the southeast.

The Wakuenai and other Arawakan groups of the upper Rio Negro have developed close contacts with neighboring Tukanoan groups along the Vaupés river and its tributaries. Together, the Arawakan and Tukanoan groups form parts of the widespread riverine cultures of the northwest Amazon (Steward and Faron 1959). The northwest's populations are characterized in the literature by segmentary patrilineal clans that regularly fission to form satellite communities (Goldman 1963). The consequence of such lineage segmentation has been to maintain the number of persons who form the working domestic units of a given territory at a very low level. However, the area has served as a refuge for many Amazonian aboriginal peoples fleeing enslavement, warfare, and concentration in missionary villages.

The Rio Negro rain forest is nutrient-poor and is drained by black-water rivers. The area receives about 3600 mm of rainfall annually, with none of the months receiving less than 100 mm. Of the 6.6 million km^2 of tropical rain forest in Amazonia, it is estimated that 3.3% are composed of Podzols associated with black-water ecosystems (Nicholaides *et al.* 1982). Although these areas have never been of major economic significance to native populations or national systems, they have become a refuge for many Amerindians fleeing from the economic expansion into Amazonia by all seven countries with such territories (Scazzocchio 1980).

The Venezuelan Amazon, which is the site for the study, covers 300,000 km^2 and is drained by both black-water and white-water rivers. Within that larger area, the study includes a number of native settlements along the Guainía. In this study area one can find three principal forest types. One type of forest, Amazon *caatinga*, or heath forest, is seasonally flooded and is found growing on sandy Spodosols (podzolic soils) distinct from the more loamy soils found elsewhere on *terra firme*. Typically these soils are characterized by an upper layer of dark, partly decomposed, organic matter and a lower layer much poorer in organic matter and having a grayish to whitish color. Below these two layers is one characterized by a classic eluvial or spodic horizon. Table 4.1 illustrates the nutrient status of these Spodosols, which experience regular waterlogging and flooding. The most conspicuous tree species in Rio Negro *caatinga* is *Eperua leucantha* (Herrera 1979) because of its resistance to reduced oxygen presence in the soil and its ability to resist flooding lasting several days. The heath forests of Borneo and Sarawak are similar to the *caatingas* of the Rio Negro and grow on the same type of sandy podzolized soils drained by black-water rivers.

The second type of forest is found on slightly higher elevations (above 20 m) and is located mostly on Oxisols with occasional patches of Ultisols. It is richer in species diversity and includes taller trees (25–30 m). These soils are rich in iron compounds and low in organic matter. These sites are never flooded and are the soils most often used by the native inhabitants for agriculture. Table 4.2 illustrates a typical soil profile, although readers should note that there is considerable variation in color and exchangeable bases among the Oxisols in response to climatic and vegetational history.

A third type of forest, with a wide range of flood exposure is *yévaro*

Table 4.1

Chemical Analysis of Soil under Amazon Caatinga[a]

Horizon	Depth (cm)	pH	Organic C (%)	Total N (%)	P (ppm)	Exchangeable cations (mEq/100 g) Ca^{++}	Mg^{++}	K^{+}	Na^{+}	H^{+} + Al^{+++}
O_2	0–8	3.80	8.21	0.116	45	2.40	0.32	0.57	0.14	46.1
A_1	8–23	4.05	4.62	0.157	35	0.57	0.18	0.12	0.097	10.0
A_{21}	23–36	4.28	2.84	0.081	14	0.18	0.02	0.06	0.098	1.80
A_{22}	36–55	4.29	0.61	trace	8	0.06	0.01	0.05	0.010	1.20
A_{23}	55–75	5.16	0.18	trace	5	0.05	0.006	0.003	0.001	0.06
A_3	75–105	5.16	0.16	trace	3	trace	trace	trace	trace	0.62
B_1	105–113	4.97	0.44	nd[b]	nd[b]	0.01	0.02	trace	trace	1.10
$B_{2(h)}$	114–125	4.11	14.56	0.475	36	0.93	0.22	0.03	0.0003	28.30

[a]Source: Herrera (1979:72).
[b]Not determined.

forest. It is thus composed of a varied range of plants—from the flood-resistant *Eperua leucantha* to the well-drained Oxisols supporting a more typical high-diversity rain forest with no dominants. More common on this intermediate gradient is *Eperua purpurea,* which is found on the better-drained but still periodically flooded mounds between the two forest types. Table 4.3 illustrates the soil nutrients in these sites. They are poor like the others but with less organic matter and lower pH.

Research on nutrient cycling in this region has identified very efficient nutrient capturing and cycling mechanisms (Jordan 1979). Forests in these oligotrophic environments act like gigantic filters, a role played predominantly by a thick, aboveground surface root mat. This root mat is capable of

Table 4.2

Chemical Analysis of Soil under Mixed Forest (La–site)[a]

Horizon	Depth (cm)	pH	Organic C (%)	Total N (%)	P (ppm)	Exchangeable cations (mEq/100 g) Ca^{++}	Mg^{++}	K^{+}	Na^{+}	H^{+} + Al^{+++}
O_2	3–0	3.75	5.65	0.717	268	0.23	0.057	0.081	0.086	18.4
A_1	0–30	4.52	3.41	0.252	378	0.178	0.014	0.027	0.082	14.4
A_2	30–80	5.61	0.64	nd[b]	21	0.13	0.031	0.022	0.033	5.2
B_1	80–140	5.90	0.36	nd[b]	21	0.10	0.023	0.020	0.041	3.2
B_2	140–200	5.90	0.49	nd[b]	13	0.10	0.025	0.024	0.045	4.0

[a]Source: Herrera (1979:73).
[b]Not determined.

Table 4.3

Chemical Analysis of Soil under Eperua purpurea *Forest*[a]

					Exchangeable cations (mEq/100 g)					
Horizon	Depth (cm)	pH	Organic C (%)	Total N (%)	P (ppm)	Ca^{++}	Mg^{++}	K^{+}	Na^{+}	H^{+} + $A1^{+++}$
O_2	0–5	4.62	3.51	0.21	58	0.44	0.089	0.20	0.11	nd[a]
A_{11}	5–30	4.45	1.95	0.14	35	0.19	0.014	0.012	0.075	12.6
A_{12}	30–55	4.75	1.01	0.081	41	0.22	0.010	0.008	0.095	0.61
A_2	55–80	5.00	0.47	trace	49	0.19	0.006	0.003	0.078	nd[b]
A_3	80–100	5.40	1.07	0.067	93	0.19	0.006	0.003	0.075	nd[d]
B_2	100–120	4.75	4.28	0.13	236	0.19	0.006	0.003	0.090	3.0
11C	120–140	5.05	1.36	0.074	158	0.18	0.006	0.003	0.079	nd[b]

[a]Source: Herrera (1979:74).
[b]Not determined.

absorbing nutrients in the water and the litterfall before such materials ever reach the mineral soil (Herrera *et al.* 1978). The root mat varies in thickness from 5 to 40 cm. The presence of most of the feeder roots above the soil surface, even when the soils are well drained, speaks to the oligotrophic nature of the environment (Uhl 1980). Water draining from these sites is clear and very poor in nutrients.

Besides the root mat, plants appear to (1) reduce herbivore pressure through accumulation of alkaloids, polyphenols, and other toxic chemicals; (2) return some nutrients to the stem before leaf shedding; (3) have sclerophylly; and (4) have symbiotic associations with epiphytic organisms (Herrera *et al.* 1978). These adaptations of plants in oligotrophic ecosystems permit them to approximate the biomass of more nutrient-rich areas, but the whole system is based upon a nearly closed nutrient cycle. Indeed, the forest feeds on itself in what must be one of the most remarkable nutrient cycles in the biosphere. Clearing of forests in these areas breaks the cycle, exposing a forest with very low inherent productivity to leaching and rapid degradation (Herrera *et al.* 1981).[2]

The nutrient-poor environment is reflected in the height, above-ground biomass, and mean basal area of trees. The mean basal area in three sites was 27.8 m^2/ha for trees more than 10 cm in diameter at breast height. This is considerably lower than the figures reported for a forest near Manaus (33.2 m^2) or for eutrophic forests of Costa Rica (40–50 m^2). Aboveground forest biomass is also low with 391 TM/ha (Jordan and Uhl 1978). Nearly 20% of the total basal area is made up of trees in the 1–5 cm diameter class. Forests are under 30 m in height, with a preponderance of small trees. The diversity of forest species is also lower for comparable forests at that humidity and temperature. It is more similar in diversity than one might expect to Holdridge's lower montane wet forest than to the moist rain forest (Jordan 1979). Uhl (1980) suggests that the small stature of the Rio Negro forests is a reflection of the oligotrophic conditions, an evolutionary adaptation of plant species that thereby make lower demands upon the restricted nutrient stocks of the soil.

The Rio Negro Population

It is difficult to consider the population of the Rio Negro as isolated or traditional in any exact sense of these terms. The indigenous societies of the Rio Negro are at various stages of integration into a regional subculture that dates to when the colonial government of New Spain sent a military expedition to drive out Portuguese slave traders and missionaries who had been operating as far north as San Fernando de Atabapo during the seventeenth and early eighteenth centuries (Wright 1981). Spanish missionaries and armies sought to pacify the area by concentrating indigenous peoples in regional centers called *reducciones*. The process of cultural change that began in the seventeenth century with Portuguese and Dutch penetration continues unabated to this day. Many Arawakan groups that survived the decimation of colonial times have become part of the regional Venezuelan society (Aguirre 1941), an ongoing process that can be observed today among the Baré, Baniwa, and other groups whose languages are on the verge of disappearing (Gonzalez 1980).

Throughout this long period of contact, the indigenous populations have been recruited by Colombians, Brazilians, and Venezuelans for extractive activities—especially the collection of rubber (*Hevea* spp.), *chiquechique* (*Leopoldina piassaba*), *chicle* (*Manilkara bidentata*), and *seje* (*Jessenia bataua*)—and as producers of manioc flour and flat cakes for local consumption and trade. *Regatões*, or river traders, plied the waters and bartered manufactured items in exchange for indigenous food and forest products.

Most of the indigenous Arawakan populations currently inhabiting the Rio Negro basin of Venezuela have migrated to the lower Guainía over the past 20 years. The Wakuenai are expanding from their ancestral lands along the Brazilian Isana River and the upper Guainía in Colombia to the unoccupied territories in the Venezuelan territory. Figure 4.2 illustrates the migration routes followed by the two phratries discussed in this chapter.

The Wakuénai are an Arawakan-speaking indigenous population of the northwest Amazon region. They include speakers of five distinct, yet mutually intelligible, dialects of the language called Wáku: Curripaco, Curricarro, Oh-hon, Eh-hen, and Ñame. All five dialects are spoken today along the Guainía River of Venezuela, but the Curripaco dialect is the most

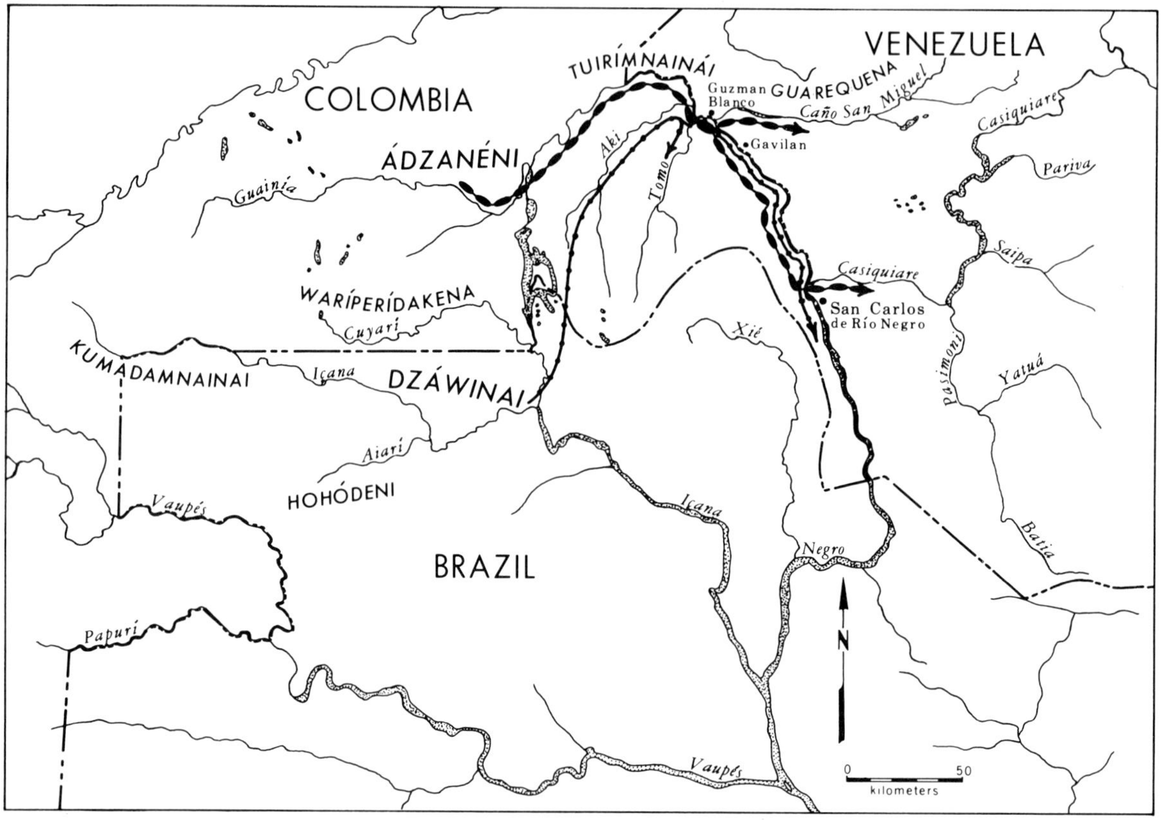

Figure 4.2 *Territories and migratory routes of the Wakuénai phratries.*

common in the region. Partly for this reason, all Wakuénai villages of the region are referred to by non-Wakuenai as Curripaco.

Wakuénai living across the border in Brazil along the Rio Isana and its tributaries have come to be known as the Baniwa. To the Wakuénai of Venezuela, the Baniwa are neighbors who speak an unintelligible language and they deny the existence of Baniwa kinsmen in Brazil. Baniwa, it turns out, is the term in *lingua geral*[3] for Wakuénai and explains the widespread and indiscriminate use of the term Baniwa for many Arawakan peoples of the Rio Negro. Wright (1981) notes that for over two centuries the Isana has been called the "river of the Baniwas" and in Brazil it would appear that the Wakuénai living there have long accepted being called Baniwas (Galvão 1959; Galvão and Oliveira 1973; Nimuendajú 1950; Oliveira 1975). Galvão and Oliveira (1973) noted that the names of Wakuénai clans or sibs had been changed from Arawakan to *lingua geral*.

The Wakuénai were originally subdivided into 20 or more exogamous patrilineal groups (*náiki*), each one named after an animal that represents the group's collective ancestral spirit (*liwarúna*). Iñápirrikúli, the mythical creator and transformer of the world, is believed to have created all the *náiki* by pulling the totemic ancestors out of a hole in the earth near Jipana, a village on the Rio Aiari in Brazil. The descendants of these groups of totemic ancestors have become phratries of four or five named patrilineal groups which are hierarchically ranked within the phratry (Galvão and Oliveira 1973). The phratries are the highest level at which the rule of exogamy is enforced. Thus, the hierarchical ranking of patrilineal groups within a phratry is not based on wife-giver/wife-taker inequalities but ritual and mythical principles. The phratries have no overall political leaders, but each local patrilineal group has a number of ritual specialists who guide the group's members in ceremonial and everyday social contexts. The Wakuénai have four types of ritual specialists who promote both horizontal integration among phratries through ceremonial activities and vertical hierarchical integration among local patrilineal groups within each phratry through sacred ritual activities (J. Hill n.d.). The phratric name is important to Wakuénai groups living along the lower Guainía as a term of self-reference and as a convenient means for indicating the social distance required for intermarriage.

The principle of territorial control by the phratry was particularly important as an adaptive strategy and has been severely weakened among the dislocated Wakuénai in the lower Guainía of Venezuela. Segments of patrilineal groups affiliated with the Ádzaneni, Dzáwinai, and Waríperídakena phratries have migrated to the lower Guainía over the past 20 years, and the traditional association between a phratry and a continuous territory has given way to a new pattern in which members of all three phratries are interspersed in a single territory. The Wakuénai share this territory with the few remaining Baniwa (see Figure 4.3). This dislocation has reduced the familiarity of the Wakuenai with their traditional territories and has placed them in an inferior physical environment. The phratry is the highest level at which the principle of local territoriality operated, but specific uses of microenvironments for agriculture and ancestor worship were negotiated at the level of individual patrilineal groups. Each patrilineal group identified specific areas within its phratry's territory for use as agricultural lands and as sacred lands where no one was permitted to hunt. The Kadawpurirri, for example, lived near Tonowi and cleared gardens from a large forest of *mori* trees (unidentified) that stood on *terra firme* not far from their village.[4] They called their garden lands *Káaliwári,* the rib cage of *Káali,* a younger brother of the mythical creator Iñápirrikúli, and owner of all cultivated plants.

Territorial delineation of agricultural and sacred lands may have been untenable except for the affinal alliances among phratries created by the traditional Wakuénai custom of bride service. Years ago, young men had to

Figure 4.3 *Wakuénai phratries along the lower Guainía.*

spend a period of 2 years living in the villages of their wives' kinsmen before they could take up residence in the villages of their own patrilineal kinsmen. During this period of uxorilocal residence, the young men had to cut, burn, plant, and weed manioc swiddens for their fathers-in-law (see, e.g., similar findings in Kracke 1978:37ff.). The harvest formed a bridewealth payment to the wives' kinsmen and helped to establish lasting alliances among phratries. "Women are acquired by the quality of the swidden" (Matos Arvelo 1912:44).[5] The Ádzaneni are particularly insistent about the importance that the young men's performance of agricultural activities had in determining the attitudes of their wives' senior kinsmen. A father-in-law had the authority to send young men away empty-handed if he detected any evidence of incompetence or idleness in their performance of agricultural activities. Conversely, the wives' kinsmen could force their daughters to leave their natal villages with young men whose parents had properly arranged for their marriages and who had proven themselves to be capable gardeners. The conflicts engendered by this control are manifested in Wakuénai myths.[6]

Men of the Ádzaneni and Dzáwinai phratries report that the traditional custom of bride service was prohibited by missionaries of the New Tribes Mission. Details of the process of missionization among the Wakuénai may be found in J. Hill (n.d.) and Wright (1981). In addition, the missionaries encouraged interethnic marriage with *lingua geral* speakers, a process of *mestizaje* that is accelerating along the lower Rio Isana in Brazil (Galvão and Oliveira 1973:29) and in the lower Guainía and Casiquiare in Venezuela.

The end of the bride service and arranged marriages between patrilateral cross-cousins undermined the traditional authority of senior men of the patrilineal group over their daughters' husbands. However, some village headmen have compensated for the loss of authority traditionally vested in the custom of bride service by persuading their daughters' husbands to live and work permanently in the wives' villages after marrying. Kracke reports on the same process of group formation around the father-in-law among the Tupí-speaking Kagwahiv of the Rio Madeira in Brazil. "If the father-in-law is the headman and at all persuasive, however, the result tends to be permanent uxorilocal residence" (1978:14). The end of bride service also meant that fathers were no longer under the traditional obligation to force their daughters to leave the village upon marriage and live with arranged husbands. Thus, young women's personal preferences to continue living in their fathers' villages after marriage can also lead to uxorilocal residence. Kracke points to the father–daughter relationship as a factor that reinforces father-in-law authority in uxorilocal residence among the Kagwahiv. "A foundation stone of the father-in-law's hold is his daughter's emotional attachment to him" (1978:37). Among the Wakuenai of the lower Guainía, both the persuasiveness of village headmen as fathers-in-law and the personal preferences of headmen's daughters to remain in their fathers' villages have contributed to the emergence of uxorilocal residence.

Of particular interest in this chapter are two Wakuénai villages along the lower Guainía: Punta Bella and Gavilán.[7] The Ádzaneni of Punta Bella (for location, see Figure 4.1), a small village of six families (with a population of 28), reflect the process of *mestizaje* due to interethnic marriages and the emergence of uxorilocal residence as a strategy for reorganizing the productive labor of women in the absence of the traditional mechanism of bride service and patrilocality. As in the traditional social organization, the father-in-law/son-in-law relationship is fraught with tension—except that now the tension is life-long rather than lasting only 2 years. To the extent that the ideology of village unity has managed to contain the dynamic potential for conflict among affinal relatives in Wakuénai culture within a transformation of the group's social organization, the village of Punta Bella has adjusted its local organization in ways consistent with its traditions and with the new emerging relationships with outsiders as manioc surplus producers (see following discussion).

The Dzawinai village of Gavilán is made up of seven families (with a population of 36) and presents an interesting contrast to the Ádzaneni of Punta Bella. Although uxorilocal residence exists in Gavilán, social organization contrasts in the two villages. In Gavilán uxorilocality forms a peripheral ring of affinally related men around a central core of patrilocal Dzáwinai men and uxorilocal Dzáwinai women, instead of a structural core of Ádzaneni women as in Punta Bella. The village was founded about 8-years ago after a split between evangelical Dzáwinai and traditional Dzáwinai shamans.

Uxorilocal Residence: Father-in-Law Leadership in Economic Production

The father-in-law/son-in-law relationship continues to play a pivotal role in the social organization of Punta Bella due to a high rate of uxorilocality. But the crucial relationship is no longer that between uncle and nephew because the preference for patrilateral cross-cousin marriage has almost disappeared. Four of the six families in the village conform to the rule of uxorilocal residence (see Figure 4.4).

The high rate of uxorilocality in Punta Bella may be seen as a transformation of the traditional custom of bride service from a 2-year process to a lifetime status. In the absence of bride service, the current tendency toward lifetime uxorilocality helps to buttress the traditional authority of fathers-in-law over affinal relations and agricultural activity in the local group. Uxorilocality has increased the value of cooperative labor among groups of closely related kinswomen as the Wakuénai entered into trade relations with the Whites, *criollos,* and other indigenous groups of the region, all of whom are willing to pay cash for manioc products (cf. Murphy and Murphy, 1974).[8]

The sale of surplus manioc products forms the basis of socioeconomic relations with outsiders and calls for a high degree of cooperation among workers within the village. Although each family has its own swiddens for manioc cultivation, the six male household heads collectively slash-and-burn new swiddens each year, and all adult men and women collectively participate in the planting of new swiddens. Afterward, the married cou-

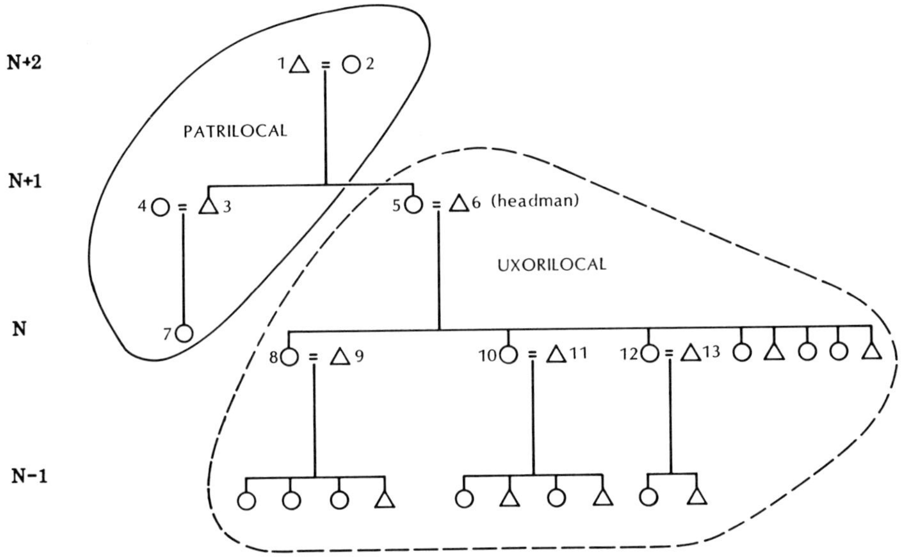

Figure 4.4 *Kinship and residence in Punta Bella (N = generation).*

ples who own the swiddens are responsible for weeding, harvesting, replanting, and other activities.

The processing of raw manioc tubers into a marketable product is associated with an even higher degree of cooperation among households. Uxorilocality in Punta Bella has facilitated the formation of two teams of women able to produce surpluses that are bought by communities lacking the production advantages of uxorilocality. Each team is led by a pair of adult women and is supported by girls over 3 years of age, by which time they have developed skills in peeling manioc tubers with large knives, keeping fires lit, and carrying flat breads to the racks for drying. By the time they reach puberty, girls participate in all phases of manioc processing and provide a significant labor input. The men normally enter into the process only long enough to run the manioc grating machine (*cigueña*) before carefully oiling and cleaning it for the next user. A harvest by one production team takes 2–4 days to process in the village, depending on the availability of sunlight for drying the flat cakes (*casabe*). The yield is about 65 flat cakes and five *latas* (one *lata* equals about 12 kg or about 19 liters) of manioc flour (*mañoco*) in addition to what is locally consumed.

Every 2 weeks several men and women travel from Punta Bella to San Carlos de Rio Negro to sell large quantities of *casabe* and *mañoco*. On occasion they sell smaller amounts to visitors from neighboring villages. Unlike most Wakuénai and Yeral communities of the region, Punta Bella is heavily dependent on sales of manioc products for cash to buy steel tools, clothing, fishhooks, soap, and other basic necessities. They did not participate in any Venezuelan government programs because they lived on the Colombian side of the Guainía River until 1979. Unlike many other Waukénai villages, none of the Punta Bella village members receives a salary, and there is no generator to produce electricity. The manioc grating machine is the only major piece of government-donated equipment that they have received so far. They are hoping to receive a small outboard motor from the government to help them transport manioc products to market. Their average monthly production for commercial sales is estimated to be over 400 flat cakes and 25 *latas* of manioc flour. Estimated total value of these goods sold is around Bs (*bolivares*) 2000 (US$470) or about Bs70 per capita each month.

The Adzaneni of Punta Bella have managed to retain a high level of ethnoagronomic knowledge that is part and parcel of their traditional ethnic identity as agriculturalists, but which has not been lost in their new role as commercial manioc producers. They are quite specific about the variable potential of soils in the region. They have ethnoagronomic criteria for choosing appropriate soils for cultivation and rejecting other ones based on color, depth, and taste of soils. The best soils of the region are said to be black or brownish black (5 and 7.5 YR. 2/2 in Oyama and Takehara 1967).[9] Once one of these soils has been located, they dig holes to determine its depth. If it is more than the length of a man's forearm throughout the area to be cleared, then it is considered to be of sufficient depth to withstand cultivation. They also taste a handful of the black topsoils to see if it is bitter or sweet. Only sweet or semisweet soils are considered suitable for cultivation. Even better are what they term "yellow" soils (i.e., clay loams). The Wakuenai noted that these soils are not found in Venezuela but only on the Colombian side, and then only in small patches along the lower Guainía. White sandy soils are considered not to be good for cultivation of favored crops such as bananas, manioc, and sweet potatoes. Also considered inappropriate for cultivation are flooded forest (*rebalsa*) and soils infested with leaf-cutter ants (*Atta* sp.). Only pineapple (*Ananas* spp.) and *mapuey* (*Dioscorea* spp.) are sometimes planted in white sand Spodosol areas (see Table 4.1). The accuracy of their criteria is shown in Figure 4.5 in which the size of the hexagon illustrates the relative fertility of the soils. Figure 4.5a represents a typical soil under *Eperua purpurea* (see Table 4.3), whereas Figure 4.5b illustrates a well-drained Oxisol. As both hexagons clearly show, both of these soils have a very acidic pH, very low calcium (Ca) and magnesium (Mg) content, and extremely high aluminum (Al) content. These three factors serve as limiting factors within a year after a garden is burned. The Spodosols are even poorer, but the indigenous population does not attempt to burn them. They also weed and clean swiddens after planting to improve productivity by reducing competition and allowing tubers to grow larger. The most impressively cultivated swidden in Punta Bella had manioc plants that averaged 2.5 m in height and were ready for harvest after 8 months—unlike other areas where up to 2 years are required to grow comparable-sized tubers.

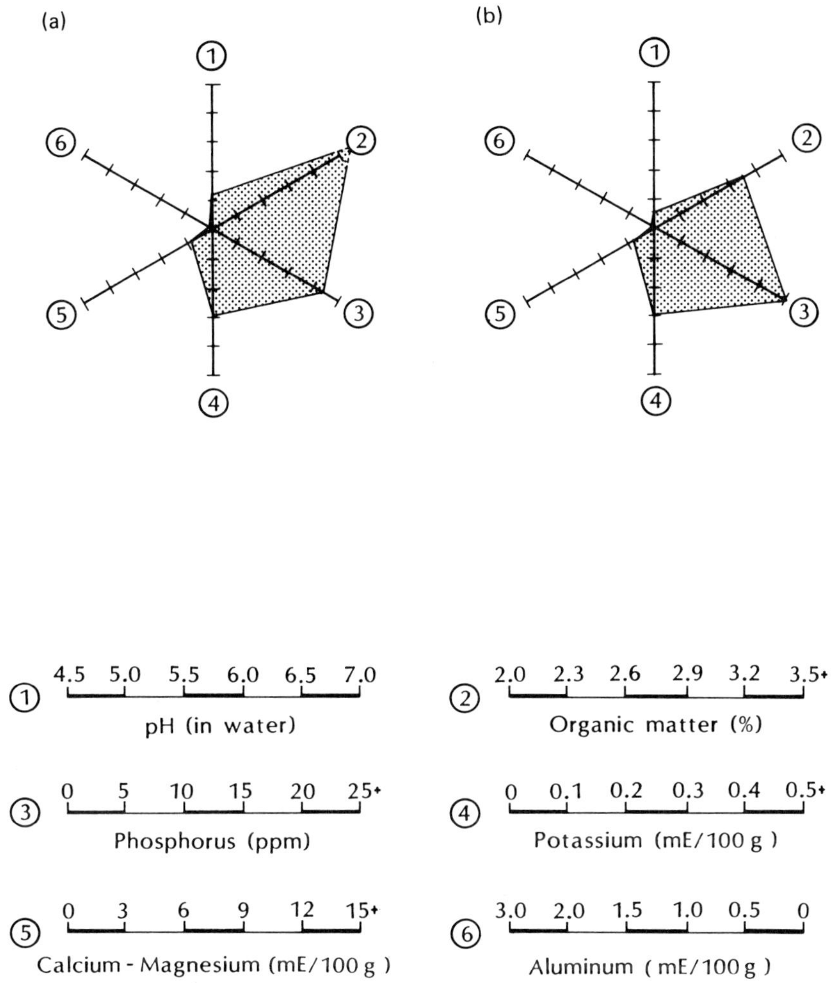

Figure 4.5 *Soil under (a)* Eperua purpurea *and (b) Oxisol.*

Patrilocal/Uxorilocal Residence: Father-in-Law Leadership in Horizontal/Vertical Integration

The Dzáwinai village of Gavilán presents an interesting contrast to the Ádzaneni village at Punta Bella. Unlike the latter, the former's social organization is based on a patrilocal–patrilineal cluster of four Dzáwinai men, one of which is headman (see Figure 4.6 for an illustration of village kinship relations). As was the case with the children of uxorilocal marriages in

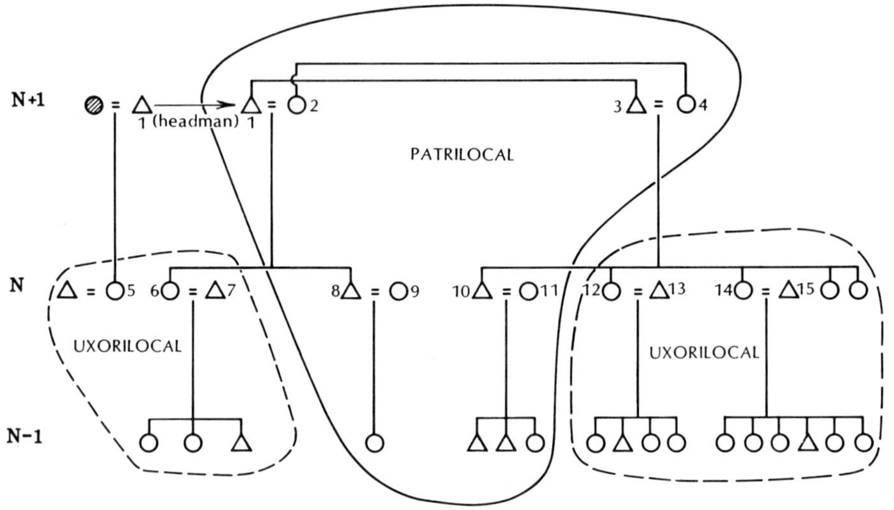

Figure 4.6 *Kinship and residence in Gavilán (N = generation).*

Punta Bella, the transmission of patrilineal names passes through the mother from the maternal (great) grandfather and the children learn to speak the Curricarro dialect of their Dzáwinai mothers. In both Punta Bella and Gavilán, uxorilocal residence provides a measure of social and linguistic unity in villages. On the other hand, despite the presence of some uxorilocal households in Gavilán, uroxilocality is not a basic principle of labor organization for agricultural production as it is in Punta Bella. Instead, the Gavilán population maintains a closer link to its past through the patrilocal residence in four out of seven households. This traditionalism has its basis in the leadership role of the Dzáwinai men, and their claim to vertical and horizontal integration through ritual in traditional villages has helped the transition to participation in vertical (i.e., state-level) relations. As Kracke (1978) has noted, the role of the headman can make a significant difference in the organizational strategies of small Amazonian communities.

Although the Gavilán village shares the same ethnoagronomic criteria for soil selection discussed earlier, their agricultural performance is less market-oriented and they have picked sites for agriculture with inferior soils. The organization of agricultural activities in the village expresses the unity of three of the four Dzáwinai families who form the hub of village relations and, with one notable exception, the peripheral situation of the families that reside uxorilocally. With reference to Figure 4.6, cutting, burning, and planting of new swiddens are collectively carried out by three of the four patrilocal families and the family of the headman's daughter (6) who is married to a non-Wakuénai man (7). A fourth patrilocal family head

(10) has teamed up with his Waríperídakena brother-in-law (13) in cutting, burning, and planting manioc swiddens. The Hohódeni man (15) and his family carry out these agricultural activities alone, a reflection of his marginal status in the village. Finally, a daughter (5) of the headman by a previous marriage lives across the river in a single family settlement with an Ádzaneni husband, and they provide a variety of cultivated produce in exchange for meat and manioc products. Lands on the Colombian side of the river near Gavilán are more fertile, resembling the yellowish soils that the Dzáwinai used to cultivate in their former village along the Rio Isana. They appear to be more fertile than the sandier soils that characterize the soils in manioc swiddens on the Venezuelan side of the Guainía and are rich enough to support such crops as corn (*Zea mays*), beans (*Phaseolus* sp.), watermelon (*Citrullus vulgaris*), squash (*Curcurbita* sp.), sugar cane (*Saccharum officinarum*), and plantains (*Musa* spp.). On the Venezuelan side, bitter manioc (*Manihot esculenta*), pineapple (*Ananas* sp.), cashews (*Anacardium occidentale*), and hot peppers (*Capsicum* spp.) are among the few staple crops that can grow well.

The labor of processing manioc in the village follows the same pattern as the organization of agricultural activities in manioc swiddens. The group of three patrilocal Dzáwinai families and the headman's daughter's family share a single manioc oven in a kitchen near the headman's house. A government-donated manioc grating machine, identical to the one in Punta Bella, is housed in the same building. This is the only group in the village in which women of more than one household cooperate in the processing of manioc. The group includes a pair of sisters (2 and 4) in Figure 4.6 and a mother and daughter (2 and 6). The other three women of the village have their own kitchens and separately process manioc with help from their unmarried daughters. There is no production of surplus manioc for sale in neighboring villages or in Maroa, the nearest town. The emergence of uxorilocal residence in Gavilán is not directly connected to the increased value of women's labor as manioc harvesters and processors in the way that the two are related in Punta Bella. There is no evidence that the headman of Gavilán and his brother have encouraged their married daughters to reside in the same village with them and their sons. Uxorilocality appears rather to have emerged due to personal and cultural preferences of individual women of the N generation whose marriages took place after the demise of traditional customs of bride service and arranged marriages between patrilateral cross-cousins (e.g., see Kracke 1978, on father–daughter relations).

The village of Gavilán was founded about 8 years ago after a series of fights along the Isana with Dzáwinai kinsmen who had been converted by New Tribes missionaries. The two senior men of Gavilán are both ritual specialists, and the missionaries had trained their converts to alert the Brazilian military whenever they heard that traditional ritual specialists had been active. The Brazilian military supported the missionaries and their converts and forced the Dzáwinai of Gavilán out of their native territory. Other Dzáwinai ritual specialists settled with their families in Colombia along the Rio Tomo after their expulsion from the Isana, while the founders of Gavilán immigrated to the lower Guainía in Venezuela (see Figure 4.2). They are still persecuted by evangelical Wakuénai for continuing their traditional religious beliefs and practices. However, Colombian and Venezuelan officials have refused to act against the Dzáwinai of Gavilán on the basis of complaints made by the evangelical Wakuenai. In fact, Venezuelan officials in Maroa, many of whom are of Baniwa descent, have patronized the Dzáwinai ritual specialists on more than one occasion for cures and divinations.

The Dzáwinai of Gavilán arrived in Venezuela in 1973, the year before the *Conquista del Sur* (CODESUR) program came to an end with the changing of national administrations. The men were too busy building their houses and preparing the first manioc swiddens to seek employment in any of the CODESUR projects. During the Acción Democrática administration of Carlos Andrés Perez (1974–1979), the village received economic aid in the form of a kerosene-powered electric generator, electrical wiring and lighting installations, classroom materials for a village school, and two 6-hp outboard motors. An adopted son of the village headman attended the Catholic mission school in Esmeralda on the upper Orinoco and became the village school-teacher. His government salary of Bs1800 (about US$430) per month is the only source of cash income that regularly enters the village since the current administration put an end to almost all economic aid programs to indigenous villages of the region.[10]

Patrilocal residence as a principle of village organization provides a relatively stable basis for collective decisions. The bonds of kinship and affinity that hold together the various households of the Dzáwinai village of Gavilán are less fragile than interfamilial relations in the predominantly uxorilocal, Ádzaneni village of Punta Bella. However, the potential for conflict and a breakdown of social relations is quite high whenever outside institutions intervene in the local economic or social life of the village. A serious conflict erupted in 1977 when the *Instituto Agropecuário Nacional* (IAN) cattle-raising project failed, due to corruption and mismanagement. When the IAN officials first approached Gavilán as a potential location for pastures, the younger men of the village voted against participating on the grounds that it would take away time from the subsistence activities that they needed to feed their families. Besides, they had no guarantee that they would ever receive any benefits. The headman, his brother, and his Ye'Kwana son-in-law were in favor of the IAN project as a way of earning extra cash. After a series of negotiations with the headman acting as a mediator to reassure the more reluctant men, they all agreed to clear an area of about 9 ha for use as a pasture. For this work they received Bs8000

(US$2040) in two payments. The IAN officials promised to pay the Dzawinai men a proportionately larger sum of money to clear a larger, 25-ha area during the next year. At the urging of the village headman, the men cleared about 20 ha. Later, they found out that the IAN project had been terminated and that they would not receive any of the money that they had been promised. The blame fell on the headman, his brother, and son-in-law; the three men who had favored participation in the project since its inception. The negative impact of the IAN project on the subsistence economy of Gavilán is apparent in the poor quality of manioc swiddens from 1977 and 1978. Instead of searching for pockets of more fertile soils in the vicinity of their village, the men simply planted within the area that had been indiscriminately cut and burned for use as pasture. In this way they managed to salvage a fraction of the time and energy that they had spent on the IAN project.

Conclusions

Traditionally, the Wakuénai adapted to the rain-forest environment of the northwest Amazon region by organizing themselves into mythical associations, or phratries, of several patrilocal–patrilineal groups that collectively controlled aquatic and terrestrial resources along the Isana and Guainía rivers and their tributaries. Within the territory controlled by the phratry, each patrilocal group defined specific areas as its own sacred and agricultural lands. The development of a complex ancestral cult might have played an important role in the formation of exogamous phratries through the creation of a more stable relationship between local patrilineal groups and specific microenvironments than that characterizing most other lowland South American Indian groups. The populations of northwestern Amazonia appear more territorially fixed than other lowland populations and their myths are remarkably detailed in their ecological data.

The ability of Wakuénai groups to reconstitute themselves in the same location even after severe depopulation and dislocation has been documented for Hohódeni and Waríperídakena groups along the Isana and Aiarí rivers in Brazil (Wright 1981:182ff.). However, the current migrations of Wakuénai groups from their native territories along the Isana and upper Guainía to the lower Guainía in Venezuela are the result of permanent, irreversible changes in traditional social organization. The Wakuénai were in a period of reconstitution after dislocations caused by the rubber boom of the 1940s when Protestant missionaries of the New Tribes Mission arrived and began to eradicate traditional social customs governing marriage and ceremonial rituals of exchange between affinally related patrilocal groups. More recently, the national governments of Brazil, Columbia, and Venezuela have all acted to increase their political presence along the Isana and Guainía rivers: in Brazil through an increase in military installations,[11] in Colombia through private enterprise, and in Venezuela through government economic programs such as CODESUR.

The Wakuénai groups immigrating into the lower Guainía and its tributaries in Venezuela have fled from situations of social breakdown that are unintelligible within the framework of their traditional culture and have moved into an area that has been largely abandoned through rural-to-urban migration of its indigenous Baré and Baniwa inhabitants (Primov 1979). Many of the Wakuenai groups migrating into this demographic vacuum chose to settle in Venezuela because of better economic conditions and less direct outside intervention, even though they know that they are occupying a region of inferior soils for their traditional subsistence agriculture. The Ádzaneni and Dzáwinai villages compared in this study both fall into the category of Wakuenai groups seeking better economic opportunities in Venezuela: one is in the process of developing strategies more oriented at vertical than horizontal adaptations, the other is attempting to maintain a measure of economic independence as it adjusts to *criollo* contact (see Figure 4.7). They differ from the evangelical Wakuenai groups who have settled in remote villages on the Caño San Miguel and the Casiquiare rivers and who seek isolation from all outside contacts. Both the Ádzaneni of Punta Bella and the Dzáwinai of Gavilán have chosen to increase rather than decrease the amount of contacts they have with nonindigenous outsiders. Both groups have transformed the traditional rule of patrilineal descent to allow phratric names and language dialects to pass through either male or female lines to younger generations. This social adjustment, accompanied by a strong tendency toward uxorilocal residence, has the effect of approximating the social unity that had once been maintained by the localization of patrilineal groups.

The Ádzaneni of Punta Bella have opted for an economic strategy of temporary local autonomy based on manioc cultivation and maintenance of traditional ethnoagronomic practices. The emergence of uxorilocal residence as a principle of village organization that maintains a core kinship structure of closely related Ádzaneni women is the social foundation of this strategy. The nutrient-poor soils of the lands along the lower Guainía in Venezuela are suitable at best for small-scale commercial manioc production, and their allocation of earnings into production is easing their way into contact with Venezuelan society. Their organization facilitates production from their cottage industry.

The Dzáwinai of Gavilán have deemphasized the role of traditional agriculture in their village and have elected to participate in economic assistance programs sponsored by the Venezuelan government. Although uxorilocal residence plays a role in Gavilán, it is not tied to an economic strategy and does not form a central core group of closely related Dzawinai

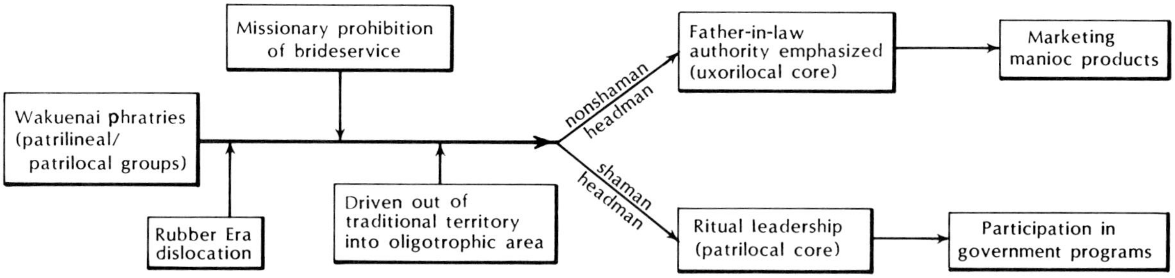

Figure 4.7 *Schematic summary of transformations in Wakuénai local organization.*

women. Instead, uxorilocal residence unites a core group consisting of both male and female Dzáwinai that dominates socioeconomic relations in the village. Also, an emphasis on sacred rituals that strengthens the cultural identity of individuals as members of the local Dzawinai group has greatly enhanced the stability of interfamilial relations in the village. This stability has allowed Gavilán to survive the impact of poorly managed development programs that could have undermined the traditional leadership of village elders and to profit from more carefully managed programs of technological and educational aid. The economic inequality caused by the Bs 1800 salary paid to the young man who had become the village schoolteacher is partially offset by sharing according to traditional rules of reciprocity. The schoolteacher's monthly salary is nearly the equivalent of an entire month's earnings from manioc production by the six Ádzaneni families of Punta Bella. Such economic power in the hands of a young, unmarried man has a tremendous potential for undermining the prestige and authority of senior men as local group leaders, but its redistribution is assured by the headman. This study has shown some evidence that Wakuénai groups along the Guainía in Venezuela are in the process of redefining traditional rules of marriage and descent and revitalizing traditional rituals, a process which could lead to a greater unity among local groups in the future. (See Figure 4.7 for a summary of the transformation.)

Contemporary Wakuénai adaptive strategies in the Rio Negro ecosystem reflect varying degrees of awareness of the limited potential for production in a nutrient-poor region and a range of transformations in social organization geared at different levels of participation in the Venezuelan economy. Vertical integration into a nonindigenous socioeconomic system is causing local leaders to play a greater role in group formation (e.g., see Kracke 1978:54). Wakuenai village headmen have responded by augmenting the two basic forms of integration available in traditional social organization: the authority of the father-in-law over his sons-in-law, and ritual specialization. Punta Bella has managed to achieve a considerable degree of economic self-sufficiency through a strategy of manioc production that emphasizes the authority of the father-in-law over his sons-in-law. Due to its small scale and lack of competition from other producers, this strategy is succeeding despite the nutrient-poor soils along the lower Guainía. However, in the long run, such soils may prove to be a limiting factor to economic self-sufficiency, especially if village population increases or if a number of other local groups choose to follow the same strategy. The second village, Gavilán, has expanded the traditional tendencies toward horizontal and vertical integration in Wakuénai ritual specialization and has given expression to the strong emotional attachments of daughters to their fathers. Their response to having migrated into a poorer environment has been to deemphasize the role of agricultural production as a medium for vertical economic integration into the Venezuelan economy and to seek a more direct participation through government assistance programs. In the long run, this response may turn out to be a more viable adaptation to the nutrient-poor, oligotrophic environment of the lower Guainía because it acknowledges the fact that the tendency toward vertical integration of indigenous groups in the Venezuelan Amazon through government-supported programs is likely to continue and that such integration will further restrict their freedom to exploit better patches of soil along the international frontiers. The Rio Negro ecosystem can only support much smaller populations than any of the state-level societies are willing to accept in connection with their geopolitical stance. The concentration of people in large communities cannot be sustained by local systems of production given the oligotrophic conditions. The rational response is to link up with state subsidies to maintain a concentration of sedentary populations otherwise unsustainable with current knowledge and production costs.

Acknowledgments

The coauthors gratefully acknowledge the support provided by NSF grant DEB77–14981–A01 to the Institute of Ecology at the University of Georgia; funds from UNESCO/MAB; and assistance from colleagues at the *Instituto Venezolano de Investigaciones Científicas* (IVIC), in particular Ernesto Medina, Rafael Herrera, Nelly Arvelo-Jimenez, Howard and Kate Clark, Carl Jordan, Haydee Seijas, Becky Holmes and Les Sponsel. Without their support this research would not have been possible. They should not, however, be held responsible for the views expressed herein.

Notes

[1]The terms Baniwa and Curripaco have been used to refer to the Wakuénai. Nimuendajú registered his suspicion that *Baniwa* was an outside term of reference for the Wakuénai rather than an indigenous term of self-reference when he wrote of the "So-called Baniwa of the

[Isana] River" (*Os chamados Baniwa deste ultimo rio*) (1950:164). The Wakuénai of the Isana and its tributaries in Brazil could accept the name *Baniwa* without causing any great confusion because the real Baniwa lived across two international borders in Venezuela. In the lower Guainía region today, Wakuénai and Baniwa live in neighboring villages, and the distinction between them is made for the social reason that the Wakuénai are immigrants whereas the Baniwa are not.

[2]The major factor appears not so much to be in cultivation but the rate of decomposition of the root mat. This is due to the function of the mat in nutrient absorption and cycling.

[3]*Lingua geral* became the *lingua franca* of the Amazon region during the colonial period. It is a mixture of a Tupian lexicon and Portuguese syntax.

[4]Soil samples of all swiddens currently in use in Gavilán and Punta Bella, including two samples from what Wakuénai claim are more fertile swiddens on the other side of the river, are still awaiting analysis in the Centro de Ecología at IVIC at the time of this writing.

[5]"*La mujer es adquirida por la fuerza del conuco*" (Matos Arvelo 1912:44).

[6]The relationship of father's sister's husband (FZH) and wife's brother's son (WBS) is central to the myth on the origin of Iñápirrikúli, the mythical creator of the world. In the myth, FZH unsuccessfully tries to kill WBS (Iñápirrikúli) by setting fire to the entire perimeter of the garden. Conflict between Iñápirrikúli and his FZH and, more generally, between Iñápirrikúli's kinsmen and their affinal relations is important in several other important myths, all of which deal with everyday social rules and the conflicts that they tend to generate.

[7]Names of the villages have been changed throughout this chapter.

[8]Murphy and Murphy (1974) noted a similar process among the Mundurucú of the Tapajós river in Brazil.

[9]Some of these soils may well be anthrosols, that is, soils resulting from previous human occupation and organic debris accumulation (N. Smith 1980).

[10]The impact of high rates of government subsidy in the form of cash has had a deleterious effect on indigenous agricultural systems of Venezuela. An extended discussion of the topic is found in Moran and Hill (in press) and more broadly in Salisbury (in press).

[11]The Brazilians have not yet constructed the planned extension of the North Perimeter Highway that, if built, will cross directly through native Wakuenai territory along the Isana River. However, they started work in 1977 on an air and naval base along the Aiari River at Wapui, or Jipana, the Hohodeni village near where Wakuénai groups are believed to have emerged in mythical times (Robin Wright, personal communication).

II
Hunting and Fishing

CHAPTER 5

Neotropical Hunting among the Aché of Eastern Paraguay

Kim Hill and Kristen Hawkes

Introduction

Hunting has long been considered a behavior of primary importance in the physical and social evolution of humans (e.g., Dart 1953; Washburn 1960; Washburn and Lancaster 1968). The importance of hunting has been stressed in theories concerning bipedalism, cranial expansion, tool manufacture, sharing, territoriality, warfare, development of the nuclear family, sex roles, and a wide variety of other topics. Recent debates about hunting have centered on the importance of meat in the diet of hunter–gatherers (Hawkes et al 1982; Lee and Devore 1968; Lee 1979; Teleki and Harding 1981), the relative efficiencies of various hunting technologies (Hames 1979), local game depletions in South America (Hames 1980; Vickers 1980), and, especially in lowland South America, the amount and adequacy of protein in the diet supplied by hunting (Chagnon and Hames 1979; Gross 1975; Harris 1974, 1977; Meggers 1971; Ross 1978).

Despite the long-standing interest in human hunting, surprisingly few precise and quantitative data have appeared in the anthropological literature (however, see K. Hill [in press] and R. Hames [n.d.]). In this chapter we present data on the hunting behavior of the Aché of eastern Paraguay that should be useful in the comparative analysis of lowland South American economies in which hunting plays a key role, and more generally in developing models of both the evolution of hominids and the temporal and spatial variations in their behavior.

Aché hunters are quite efficient predators. Hunting supplies over 1 kg of

ISBN 0-12-321250-2

meat (live weight) and over 150 g of animal protein per person per day, accounting for 80% of the calories in the diet (Hawkes *et al.* 1982). This contradicts currently held ideas about the importance of meat in the diet of nonarctic hunter–gatherers (Hayden 1981; Lee and DeVore 1968) and the abundance of game in tropical South America (e.g., Gross 1975; Meggers 1971). We present a tabulation of time utilization and hunting returns for adult hunters. Differences in returns from different technologies are compared, and the effect of hunting pressure in a heavily hunted area is reported. We use models derived from optimal foraging theory to account for the total set of species hunted, using different technologies. Hunting returns at the group level are calculated, and then a more detailed analysis of variance in returns between individuals and across age categories is presented. Finally, we explore the problem of optimal group size based on maximizing returns and reducing risk. The implications of these results are discussed with respect to hypotheses concerning lowland South American economies, and the evolution of human hunting behavior.

Environmental and Historical Background

An area of about 5000 km² between 54–56° W and 24–25° S in the forests of eastern Paraguay is the traditional range of the northern Aché (see Figure 5.1). The area is drained to the east by the Paraná river and to the west by the Paraguay. It is characterized by gentle rolling hills composed of soft sedimentary rock and weathered basalt, interspersed with low flat valleys. The hills are covered with tropical broadleaf evergreen forest, while the valleys are filled with tall broadblade grasses. The Aché prefer to inhabit and hunt in the forest which comprises many microhabitats including wet lowlands; floodlands adjacent to large rivers; forest thick with several species of small or large bamboo; forests of smaller trees, dense vines and undergrowth; and forests with large towering trees and little ground level plant growth. The Aché themselves divide all forest areas into two categories: nice (clean) forest and ugly (tangled) forest, the latter being defined by difficulty of mobility, thorns, spiny plants, and thick, tangled understory.

The eastern extreme of the Aché territory is more tropical than the western part. This correlates with a rainfall gradient ranging from approximately 1500 to 1700 mm annually, as well as to an increase in altitude from about 250 to 350 m and to a 1°C decrease in the mean temperature as one moves eastward. The dominant characteristic of the weather pattern is extreme unpredictability from month to month and year to year. Although March–September is a somewhat drier time of the year, with the usual

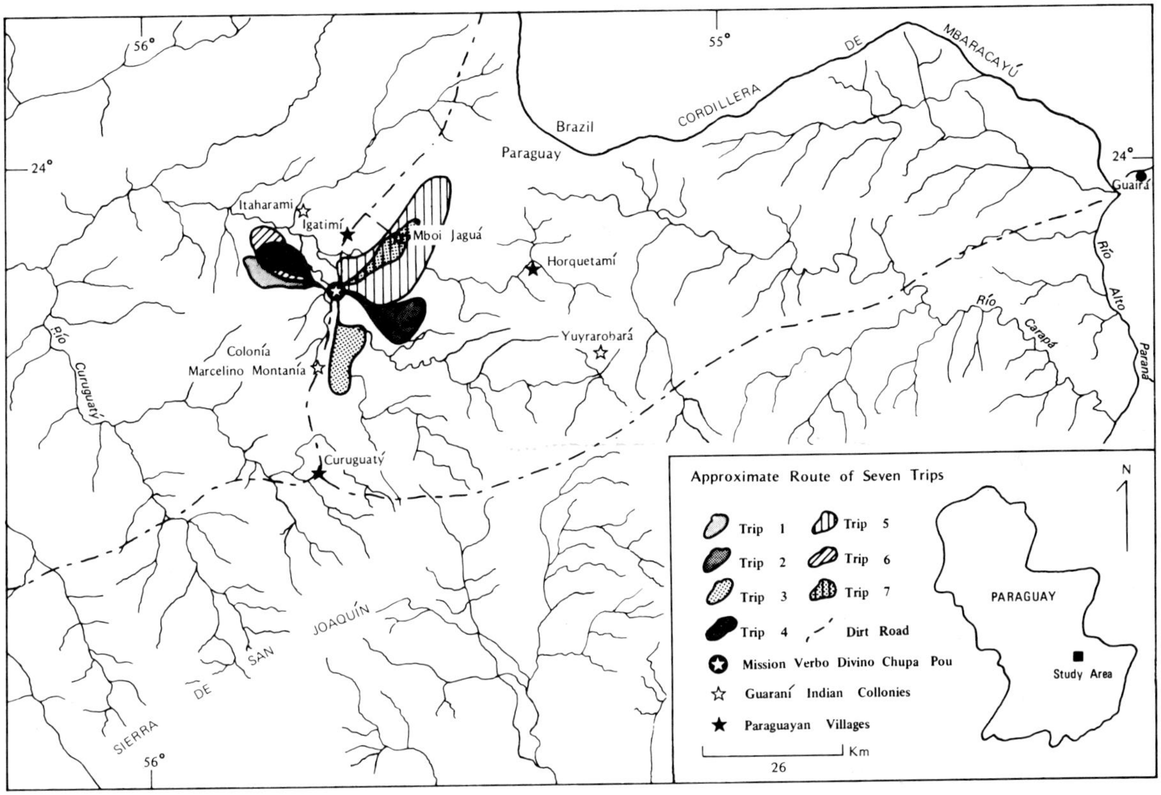

Figure 5.1 *Study area and route of foraging trips. The route followed by the women and children on each foraging trip is that described by the* circumference *of each shaded area. Hunters spread out in all directions from this route, and the shaded areas do not represent hunted areas, but are only used to illustrate clearly the routes.*

"rainy season" falling between October and February, in any single year almost any month may be the wettest or driest. The fluctuations in temperature are far more predictable, and it is temperature change rather than precipitation that marks seasonality. The average daily temperature for the region is 17°C in July and 27°C in January. The extremes are approximately 41°C and −3°C with several days of frost every year. The Aché break down the year by referring to "hot time," "cold time," or the time when a particular resource is in season.

The flora and fauna in this region, as is the case in most tropical areas, are very poorly described. We have identified 33 mammals hunted by the Aché, and they have named several others which we have not yet seen or identified. In addition, the Aché eat at least 10 species of reptiles and amphibians, more than 15 species of fish and a seemingly endless number of birds. We have thus far identified only a few of these. Edible products of more than 40 plants and trees are exploited in most years. The major species, used for both food and a variety of tools, is the palm *Arecastrum romanzofianum* which is very abundant throughout eastern Paraguay. Cerambycid larva, which feed on rotting palms, and at least 10 other larval types are eaten frequently. The adult forms of at least 5 insects are eaten, and at least 14 different kinds of honey are taken, the most common being that of *Apis melifera*.

Due to the ample rainfall and numerous streams, water is rarely in short supply. On occasion, when camping far from a water source, the Aché extract water from hollow bamboo, bottle trees, or the center of large vines. In addition to hundreds of small streams and rivers, there are a handful of large rivers that crosscut this region and present temporary barriers to the Aché in times of heavy rainfall.

The Aché (also called Guayaki) divide themselves into three categories based on differences of culture, dialect, and geographic range. All speak a language of the Guaraní family. They have been foragers in the jungles of eastern Paraguay, surrounded by hostile sedentary farmers, since before the arrival of the Spanish.

There are some references to the Aché in historical accounts before the 1960s (see Metraux and Baldus 1963) but the first extensive ethnographic reports come from Pierre Clastres (1968, 1972) who studied two of the three Aché groups. This research deals with the third, the northern Aché who have only come into unarmed contact with outsiders within the past decade (see K. Hill n.d.).

Ecology and Area of Study

The area covered by this study is located near two drainage systems, the Jejui Guasu and the Jejui Mi rivers near Igatimí, Alto Paraná (see Figure 5.1). This represents the extreme northern portion of the traditional Aché foraging range, and the forest of this area appears to be slightly less tropical than in the majority of the traditional home-range forests. Nevertheless, the animal species found in both areas seem to be identical, even though their densities may differ slightly. Most of the economically important plant species are found in both areas, although, once again, densities may differ somewhat.

The area covered by this study is approximately 600 km^2, or about 12% of the total traditional home range for the northern Aché. During the study period, approximately 155 Aché were resident in this area, for a density of 1 person/4 km^2. This is slightly higher than the precontact density of approximately 1 person/5–10 km^2 and in addition there is considerable population of Paraguayans, Chiripa, and Mbya Indians (about 2000 total) who also live in the area.[1] Summing up these conditions, it is our impression that the area is more heavily hunted than was the case for traditional hunting ranges before contact, even though the Aché currently spend a good deal of time at a mission settlement where they are provisioned.[2]

The ecology of the area has not been well studied. Five primary ecotypes which are relevant to hunting can be distinguished. Four of these are similar to those described by Hames (1980)[3] for the Ye'kwana and Yąnomamö in southern Venezuela. (One of the ecotypes described by Hames, that produced by human gardening activity, is not hunted by the Aché.)

PRIMARY FOREST

The vast majority of eastern Paraguay is covered by this tall dense forest which develops on high ground with adequate drainage. Direct sunlight rarely reaches the ground level and, thus, ground cover is medium to sparse. But the density of vines hanging from trees is a strong impediment to travel in many areas and large stands of bamboo also make traveling difficult at times. Important prey items commonly encountered in this ecotype include tapir, deer, monkeys, peccaries, and coatis.

SEASONAL FLOODLAND FOREST

Found behind natural levees of rivers, floodland forest floods seasonally, or once every few years. Areas which may have been near major river courses within the past 10 years, but are now situated some distance from the current water flow, may also be covered with this forest. Trees are usually small-to-medium size, and several species of bamboo are found. All forest areas immediately adjacent to large (greater than 10 m across) streams or rivers have characteristics of this ecotype. Important prey items of this zone are paca and armadillo.

RIVERINE LAGOONS

Riverine lagoons are connected to the river systems during the wettest times of the year, but throughout the driest seasons they slowly shrink, sometimes becoming mere swampland. If not too deep, they are filled with tall broadblade grasses. Important prey items of this zone are capybara, otter, fish, and caiman.

SWAMP FOREST

Wet areas are found in low undrained depressions or at water sources anywhere within the primary forest, where the water table comes to the surface over a large area. This ecotype is often marked by an abundance of palm trees, which are of primary importance in the Aché diet, and is frequented by feeding tapirs and peccaries. The Aché prefer to camp within reasonable walking distance of this ecotype (but never in it, due to the water on the ground) in order to exploit the palms found there.

MEADOW GRASSLAND

Meadows are characterized by tall grasses of various species and by a complete lack of trees. It is frequently, though not always, somewhat swampy. These grasslands are rarely greater than ½ km in diameter and are seldom used by the Aché except to cross from one forest area to another. The large Paraguayan antlered deer and the rhea (a large flightless bird) frequented these meadows in times past, but both now seem to be locally extinct. This zone has been of some importance historically because many Paraguayan horses and cows found in meadows near the fringes of white settlements were killed and eaten by Aché hunters, leading to a recent period of intensified hostilities between the Aché and their neighbors (see K. Hill n.d.)

Prey Species and Hunting Techniques

Although the overall foraging pattern of the Aché, and the return rates for different resource items have been described (Hawkes *et al.* 1982) the precise details of hunting should be elaborated further. Table 5.1 lists the animal species that are known to be eaten by the Aché. In addition to their use as food, the young of most mammals are captured live and kept as pets, especially monkeys, coatis, and peccaries. While the mammalian list presented is nearly complete, there are many more species of birds occasionally included in the diet that have not been listed. This list should potentially include all animal species in eastern Paraguay, as the Aché seem to have no generally tabooed food items.[4]

The Aché are bow hunters, and their techniques seem to be similar to those of many other lowland South American archers. The bow is made of palm wood and is approximately 2 m in length. It is nearly identical to the bow of the Siriono, and is slightly larger than the Y̨anomamö bow. Arrow shafts are made from reeds about 1.5 cm in diameter, and the head is

Table 5.1

Aché Prey Species

Common name	Scientific name	Aché name	Weapon used[a]
MAMMALS		Bue	
(Rodentia)			
Agouti	*Dasyprocta aguti*	Kuchi	H, B
Capybara	*Hydrochoerus hydrochaeris*	Pywa	B, C
Paca	*Cuniculus paca*	Bywa	H
? Rat	?	Buja	H
(Edentata)			
Giant anteater	*Myrmecophaga tridactyla*	Kware	C
Tamandua	*Tamandua tetradactyla*	Kwaremini	B, C, H
Nine-banded armadillo	*Dasypus novemcintus*	Tatu	H, S
Giant armadillo	*Priodontes giganteus*	Krypura	H, S, C
? Armadillo	?	Kry'y	H, S
(Artiodactyla			
Red brocket deer	*Mazama americana*	Wachu	B
Pampas deer	*Ozotoceros bezoarticus*	Wachupuku	B
White-lipped peccary	*Tayassu pecari*	We'e (chachu)	B
Collared peccary	*Tayassu tajacu*	Kanje	B, C, S
(Perissodactyla)			
Tapir	*Tapirus terrestrus*	Brevi	B
(Primates)			
Black howler monkey	*Alouatta caraja*	Kraja	B
Brown capuchin monkey	*Cebus apella*	Pwa'a	B
(Carnivora)			
Crab-eating racoon	*Procyon carnivorus*	Bejywa	H, B
Coati	*Nasua nasua*	Kane	H, B
Bush dog	*Speothus venaticus*	Juape	H, B
Crab-eating fox	*Cerdothyon thous*	Krachowa	B
Jaguarundi	*Felis jaguarondi*	Betapa	H, B
Margay	*Felis wiedi*	Kajamini	H, B
Ocelot	*Felis pardalis*	Kaja	B
Puma	*Puma concolor*	Juku	B
Jaguar	*Panthera onca*	Jamo	B
Tayra	*Eira barbara*	Aira	H, B
Otter	*Petonura brasilensis*	Chery	H, B, C
Giant otter	?	Cherywachu	B, C
Prehensile-tailed porcupine	*Coendu prehensilis*	Gui'i	
(Lagomorpha)			
Forest rabbit	*Sylvilagus brasilensis*	Ata	B, H
(Marsupialia)			
? Opossum	?	Kyre	H, B
(Not identified)			
?	?	Bechi	H
?	?	Bywa tapi	H
?	?	Beche kruma	H, B

(*continued*)

Table 5.1 (*Continued*)

Common name	Scientific name	Aché name	Weapon used[a]
REPTILES and AMPHIBIANS			
Iguana (various species)	?	Cheju	H, C, S
? Turtle	?	Krombe	H
Caiman	?	Jamongi	B, S, H
? Toad	?	Wypy	H
? Frog	?	Pepa	H
(Snakes)			
Anaconda	*Eunectes murinus*	Memboruchu	C
? Boa	?	Membopura	C
? Rattlesnake	?	Chinipura	C
Jararaca (fer-de-lance)	*Bothrops atrox*	Membo	C
?	?	Kimina	C
?	?	Chei	C
BIRDS[b]		Wyra	
? Vulture	?	Tayja	B
? Vulture	?	Chivura	B
? Buzzard	?	Briku	B
? Curassow	?	Jaku	B
?	?	Nambu	B
?	?	Kuachi	B
?	?	Uru	B
? Toucan	?	Toka	B
Macaw	*Ara macao*	Wyraraka	B
? Parrot	?	Akwa	B
? Parrot	?	Bero	B
Rhea	?	Bewyra	B
? Eagle	?	Kane	B
?	?	Doko	B

[a]B refers to bow; H, hand; C, club; S, bow or arrow used as spear.
[b]Not all bird species listed.

carved from hardwood, with barbs along one edge. Occasionally, arrow heads are more flattened and sharpened on both edges to produce a double-edged knifelike point useful for killing very large game. Bird arrows with blunt striking points are also employed. The overall length of the arrow varies considerably, but an average seems to be about 1.8 m. Clastres (1972) has described Aché technology in some detail.

Several real and potential differences between Aché hunting and other hunting patterns described in the literature should be mentioned before elaborating on the specifics. First, the Aché do not hunt with dogs. Recently this has begun to change, but all information in this chapter concerns human hunters using only their skills. Aché hunters are of course, very good trackers, and we witnessed a group of four men track a single

Aché adult, who had become lost, for an entire day through thick jungle without losing the trail. Tracking animals through thick forest with much organic litter on the ground is very difficult, yet the Aché almost never lose the track of a single animal unless it begins to travel a heavily run trail, or it passes through a swampy area. If the animal is not wounded badly, the hunters will not track it, as it is likely to cross a swamp (especially tapir); however, if it is well shot, and does not move further than ½–1 km they can almost always find it. Thus, the estimation of the initial wound is very important when deciding whether to track. Second, the Aché frequently kill animals by hand; the bow is used as a supplementary weapon only for certain animals, rather than as a kill-all tool. This is discussed in a later section, as weaponless hunting may provide important information concerning early hominid hunting behavior. Third, the Aché employ no poisons of any kind for hunting or fishing and thus differ from most other South American groups. Fourth, the Aché use all of the generalized capabilities of humans in order to pursue game. This includes digging, climbing, running, entering deep water, using fire, and inventing a wide variety of on-the-spot tools in order to reach less accessible prey. They also frequently use touch and smell to supplement other sensory input. Only animals which are much swifter on the ground and run long distances rather than seeking shelter (i.e., climbing trees or entering burrows) are safe from continued pursuit. Finally, it should be noted that although hunting is "men's work," due to the daily movement pattern of the band, women are frequently present at or near the kill site and help in the hunt when necessary as spotters. Women make occasional kills (never with arrows, however). The young of both sexes, if they are present (but especially males), are used to climb trees in order to fetch arrows or killed game, or to flush out arboreal game for continued pursuit. Descriptions of several commonly used techniques (see Table 5.2) for various game animals follow.

PACA (*Cuniculus paca*)

The paca is a large rodent that makes its burrow near a permanent water source. The burrow has one primary entrance and several escape exits that are difficult to see but which can be found by noting small disturbances in the leaves that cover the exits. When the Aché hunt in an area with signs of paca, they stay quite close together and, when a burrow with fresh signs (as determined visually, or by feeling for warmth, or by smelling) is located, the finder whistles to the other searchers. The men who are nearby respond with a similar whistle, and the finder then whistles twice, which gives an exact fix on his location and means "come here." The additional hunters approach slowly in order not to disturb the sleeping paca. They locate all escape exits, and assign one man to cover each. Positioning is

Table 5.2

Aché Kill Verbs[a]

Verb	Technique
Biri	To choke, strangle, or suffocate, by applying pressure with foot or knee on animal's chest or neck.
Bio'o	To break its neck by twisting its head with both hands.
Bopo	To stun with bird arrow. Small animals and birds suffer broken ribs and neck.
Bro'o	To puncture in the gut region with a projectile (arrow or bullet).
Bwapi	To slam to death by grabbing tail or limbs and whacking head against tree or on ground.
Ity	To knock it down or make it fall into a hole (trap).
Juka	(Generalized kill verb).
Jywo	To shoot with an arrow (anywhere on body).
Krapi	(Literally means "to gut it.")
Pacho	To beat it to death with a club.
Pycho	To stab or puncture it with a hand-held object (usually knife or spear).
Ruka	To split it open with a machete or the sharp edge of a club.

[a]These verbs would all translate as "kill" in an Aché narrative.

done without the use of language because the process is well understood by all. If there are more exits than the number of men, some exits are plugged with dead wood. When all outlets are attended, a log is introduced into the main entrance and thumped vigorously. This noise causes the sleeping rodent to dash out one of its escape exits where an awaiting hunter grabs it and suffocates (*biri*) it.

CAPYBARA (*Hydrochoerus hydrochaeris*)

The capybara is the world's largest rodent and represents the second largest game item in Paraguay, weighing up to 60 kg. Large grassy lagoons which are their primary feeding grounds are inspected closely. If fresh tracks are discovered on the bank, the lagoon is surrounded by hunters. Several men enter the lagoon and try to drive the animals to one end where they will attempt to flee onto land. Waiting hunters shoot the escaping prey with arros (*jywo*), club them to death (*pacho*), or split them open with a machete (*ruka*). In the rainy times of the year, capybaras are not usually hunted due to the depth of the lagoons and their ability to remain submerged for long periods.

GIANT ANTEATER (*Myrmecophaga tridactyla*)

The giant anteater is the largest of the two anteaters found in Paraguay. They are slow but have long sharp claws which are supposedly able to kill both man and jaguar. To kill anteaters, the Aché use a long club in order to stay out of range of the claws. The club is often fashioned on the spot.

ARMADILLOS

There are at least four species of armadillo in Paraguay ranging in size from about 4 kg to over 50 kg. When encountered aboveground foraging, armadillos are run down and their necks are broken by hand (*bio'o*). Frequently they manage to reach their burrows, and the pursuing hunter immediately thrusts in a foot in order to prevent the animal from escaping deeper or digging. A companion will unstring the hunter's bow and it is thrust into the hole to spear the prey (unstrung bows are almost straight). If the armadillo manages to escape deep into its burrow, or if a fresh burrow is discovered and determined to contain a sleeping animal (the smell and feel of the entrance are the major clues) it is dug out using the hands or an unstrung bow as a digging tool. Success rate with armadillos in their burrows is virtually 100%. If a burrow is discovered late in the day it is often plugged up with a log in order to be dug out the following day.

DEER (*MAZAMA AMERICANA*)

The brocket deer found in eastern Paraguay weighs about 30–35 kg. They are usually discovered sleeping during the day, and the single hunter who happens upon the animal will approach quietly to within 15 m and shoot an arrow at the heart. If the animal is shot well, the hunter will track it, especially noting blood loss. Sometimes others are called in order to track more rapidly. If blood loss is minimal after about 150 m of tracking, the chase is abandoned. Deer call imitations are also used in order to lure feeding deer within bow range.

TAPIR (*TAPIRUS TERRESTRIUS*)

Tapirs are the largest game animals in South America, weighing up to 150 kg. The primary hunting technique for this prey is essentially identical to that used for deer. Hunters often imitate tapir whistles while walking, in hopes of receiving an answer. The second method of hunting tapir, which we have never witnessed but have heard retold in many narratives, is the pit trap. A large hole is dug (about 3 × 3 × 3 m) on a tapir trail and covered over with branches, leaves and dirt. When a tapir (or jaguar) falls in (*ity*), it is clubbed to death or shot with arrows.

COATI (*NASUA NASUA*)

The coati is a raccoonlike animal that is often found in groups and can weigh up to 7 kg. Coatis seem to be much more common in the Aché diet than has been reported for many other South American groups. Although sometimes shot with arrows, they are most frequently killed by hand. Hunters are alert to signs of very recent coati foraging and will track down the animals if the signs appear less than an hour old. Once the animal, or more usually, animals, are found, the hunter stands at the base of the tree in which they take refuge and begins to repeat a low sound ("ooh, ooh") which seems to prevent the coatis from jumping immediately to the ground to escape. This effect is mysterious, but may be due to the similarity of the sound to the grunting of a jaguar (*Felis onca*). Meanwhile, the hunter calls for help and continues chanting at the base of the tree until a sufficient number of companions arrive. If there is much undergrowth around the tree it is quietly cleared out of the way so that the hunters are not impeded when they chase the animals. When all the undergrowth is cleared, and there are a sufficient number of hunters in position, one person will grab a nearby sapling and begin bending it back and forth and make a swishing noise with the mouth. This frightens the coatis sufficiently that they begin to leap from the tree (often from a height of 20–30 m or more) and as they hit the ground they are grabbed by the tail or hind limbs and slammed against the ground or a tree (*bwapi*). This can be quite dangerous, and bites to the arms are common. Although traditionally the Aché sometimes wrapped their forearms with a cord to protect themselves from bites, only once during the study period did we observe this technology used.

WHITE-LIPPED PECCARY (*Tayassu pecari*)

The white-lipped peccary would by all accounts seem to be the primary target prey of Aché hunting. As these peccaries live in large hordes, a well-executed hunt can bring several animals, each weighing in excess of 30 kg. Signs of white-lipped peccaries that are less than ½ day old are usually followed. Many hunters track the horde together. Due to their numbers and foraging behavior, these animals leave a trail that is hard to miss (for an Aché). Men will often trot for 2-hour stretches trying to catch up with the animals. Once they are within hearing distance of the feeding horde (which makes a great deal of noise) the lead hunters slow down and allow all others to catch up. Group hunting is most important, and an attempt by a single hunter to kill one of the animals is severely chastized regardless of success. If a lone hunter comes upon a horde of white-lipped peccaries, he is expected to run and find other hunters, if he is aware of their location, in order that a group effort may be made. When all hunters have surrounded the peccaries, they begin to close in and shoot. No signal is given, but all approach the prey at about the same time and the same speed. Once the hunters begin to shoot, the peccaries will either run to escape or to charge. The latter behavior is preferred by the hunter as it allows the hunter to jump into a nearby tree and continue shooting. It is also quite dangerous, as peccaries can bite viciously. If the animals begin to stampede they are not pursued further because they can easily outdistance the hunters. The Aché seem to spend a good deal of their hunting time in what could be considered as pursuit of these animals (i.e., they are following fresh tracks). However, encounter with several other prey items seems just as likely even when peccary tracks are being followed. The pursuit of this animal can be time-shared with the search for others (see the section on Methods and Discussion).

COLLARED PECCARY (*Tayassu tajacu*)

Collared peccaries are frequently encountered in small groups, or individually. They can weigh up to about 25 kg. Unlike white-lipped peccaries, they have a permanent burrow in the ground or in a hollow tree, to which they run if disturbed. If the hunter can pursue them to this burrow, or find one that is occupied, a kill is ensured. Usually, the main entrance is blocked off and then small holes are cut into the burrow to locate the prey. We have witnessed the Aché spear the animal through one of these holes, or smoke it out and club it to death, or, ingeniously, introduce a looped bowstring into the hole and around the neck of the peccary to strangle it. Imitation calls occasionally are used to lure it. The size of these animals along with the high probability of a kill upon locating the burrow, seems to make them especially vulnerable to predation. Colonists using shotguns and dogs are extremely efficient at wiping out local populations in a very short time.

CAPUCHIN MONKEY (*Cebus apella*)

Capuchin monkeys are extremely abundant in the forests of eastern Paraguay, and the largest adults weigh about 4.5 kg. They are killed almost every day of the year by Aché hunters. Capuchins can flee quite quickly through the trees, but hunters on the ground can always keep up with a single monkey. Pursuit lasts quite some time after the prey is located because capuchins are skilled at hiding in the branches of tall trees and hunters sometimes have to climb the tree to flush them. They move rapidly and erratically and lay low to the branches, making a difficult target. They are never shot on the move. The hunter waits until they sit still in a spot which is not blocked by branches. Considerable maneuvering is required in order to get one good shot. Upon encountering monkeys, many hunters will be called in order to pursue more efficiently the troop without losing the majority of the members. If fewer than 10 hunters are present, some members of a monkey troop almost always escape. Hunters occasionally climb trees with bow and arrows in order to shoot monkeys in high branches. Imitation calls are often successful at both locating a troop, and inducing it to come close. Monkeys kept as pets also elicit the calls of nearby troops.

HOWLER MONKEY (*Alouatta caraja*)

Large howler monkeys can weigh more than 10 kg. They live high in the treetops in single-family groups. Once their feces are discovered at the base of a tree (they are extremely odorous) or their calls are heard, they have a high probability of being killed. This is because they are very territorial, and if the hunter fails to kill them on one day, he can usually find the same family again nearby on a subsequent day. Frequently, the hunter must climb the tree in order to flush them originally, but aside from that they are hunted with the same methods as for capuchin monkeys.

OTHER MAMMALS

Other mammals are not major prey items but are all taken upon encounter. Small mammals are usually killed by hand and larger ones are shot with arrows.

SNAKES

There are many species of snakes in Paraguay, but most of those eaten weigh between 1 and 4 kg. Poisonous snakes are killed as a defensive response, but they are also eaten. When a snake is encountered, a long pole or unstrung bow is used to club it do death without getting too near. The heads of poisonous snakes are immediately cut off and buried.

CAIMAN

The Paraguayan caiman that we have seen are rarely larger than 1.5 m in length. Although the caiman is hunted with bow and arrow from the banks of lagoons, the Aché also hunt them by entering shallow lagoons and feeling along in the mud with their feet, hoping to locate the body of a sleeping reptile. Once the body is located, an arrow is used to spear the prey and bring it to the surface. Surprisingly, we have never heard of anyone being bitten.

BIRDS

Most birds are hunted with bow and arrow. If the bird is large, a pointed arrow will be used. Blunt bird-arrows are used for smaller birds. Hunters will sometimes plan to return to a rotting carcass that has been found, in hopes of killing the vultures likely to be there. A variety of bird calls are used to locate or lure particular species. Sticks are sometimes thrown in boomerang style at small birds; however, most small birds are ignored by adult hunters.

Daily Hunting Routine

An Aché hunter's day begins with the first predawn light. Occasionally, calls from capuchin monkeys will be heard at this time and one or more men will quickly take up their weapons and start off in the direction of the prey. Other men in camp imitate the monkey calls, trying to keep the prey continually voicing its location. As the sun rises, men begin eating whatever meat is left over from the previous night and sometimes they spend a few minutes extracting palm fiber from nearby trees. More often, however, women do the palm extraction while men tell stories and jokes and begin to sharpen their arrows. Men spend between about 10 and 30 minutes restringing bows, sharpening arrows, readjusting the feathers on their arrows, and making other minor repairs. Any man who is in the process of manufacturing a new bow or new arrows will be continually working on that project throughout the morning.

There is always some discussion as to what direction the hunters should set off, and what will be the approximate camping destination for that day. Both men and women participate in this discussion, voicing their desires and the information that they may have concerning resource sightings that have been made in the recent past, or even sometimes in the distant past. At some point, a consensus is apparently reached although we have not been able to pinpoint the decision nor the most important factors in reaching it. Jokes and conversation continue for about 2 hours after sunrise and then suddenly, with little warning, the men take up their weapons and leave the camp, walking in single file. The lead man often breaks a trail, either with a machete or with his bow. Women back in camp begin packing up their things and follow behind in the men's trail. Aché camps move almost every day.

During the day, a small number of men will always continue in a specific direction with the women following behind them, or just out of sight, tracking them. The trailbreakers are frequently old men or older boys. The group of trailbreakers, women, and children travel in a reasonably straight line in single file. Changes in direction are minor and seem predictable to the Aché (but not to us). Feedback from the hunters concerning a successful kill usually turns the group in the direction of the kill site. The majority of the adult hunters begin spreading out perpendicular to the direction of travel after about ½ hour, and continue spreading and coming back together throughout the day. Their direction of travel is determined by game encounters and animal tracks, and may double back considerably. The direction of the women's movement is kept in mind as a general reference for orientation of travel during search for game. The men seem to have a good idea of where the women are throughout the day, although we have not been able to determine how this is possible (it is probably due to years of experience with the overall foraging pattern) or how accurate their knowledge is at any point in time. Men search widely and probably cover an average of three times the distance covered by the women's group. Because the camps move about 5 km/day (measured with a pedometer), men are probably covering 15 km on an average hunting day. This, of course, varies considerably, and it should be kept in mind that the jungle is quite dense and thus the effort is considerably greater than would be the case in open country.

Adult hunters usually break up into smaller groups of two to four men and space themselves while searching. Although the men often space themselves so as to be out of sight, but within earshot, this is by no means always the case. Small groups of men frequently walk together in single file and men who spread out often totally loose auditory contact with each other. The importance of this spacing will be discussed further in another section.

Several game items are taken by single hunters upon encounter. The most important of these are deer, collared peccary, tapir, armadillos, birds, and other solitary animals. Social animals, as well as those for which cooperation can yield significantly higher results, are usually preyed upon by multiple hunters. When a single hunter encounters such prey he will yell or whistle to nearby hunters. They usually reply and he signals once again his exact location. All hunters within earshot respond by converging on the spot. The range of these signals depends greatly on the topography and precise location of the sender and the receiver. Based on walking times to received signals we estimate the range between 0.5 and 1.5 km. Single hunters who have wounded large solitary animals will sometimes call for help to track that animal.

The hunter's day, then, consists of spreading out to search for game, a common period of recounting the hunt after a successful pursuit (the length of which seems to correlate with the size of the kill), occasional short periods of gathering and eating vegetable foods, and a continuance of the search–pursuit cycle. We rarely recorded cases of men spending more than 3 hours without auditory or visual contact with other hunters.

Near the end of the day the hunters begin to find their way back to the women and children, or call the women to come to their location if a large animal has been killed. Men who become widely separated from the group are called by group members as dusk approaches. Several times men were guided back to the camp in complete darkness by yells from camp members. Although we have not witnessed such a case, Aché narratives do mention that men occasionally wander too far and are forced to spend the night alone in the forest. This is especially arduous in the winter because men do not carry fire and have no method for producing it.

At the end of the day the meat is butchered by both males and females and is cooked at several fires. The hunter who kills an animal sometimes butchers it, but rarely cooks it. For many animals there are taboos against the hunter's consumption of his own kill and these taboos extend to his siblings and parents as well. These taboos were not broken in any case in which we monitored closely the precise consumption of a game item, although we do not have enough data to feel confident that they are never broken. The day's total take is widely shared among band members.

Data Collection

Hill has spent about 450 days living with the Aché, of which approximately 165 days were spent in the jungle hunting and gathering, and the remainder at a mission settlement. The longest period spent in the jungle before returning to the mission camp was 28 days. All qualitative data presented are extracted from interviews, taped narratives, or firsthand observation during hunting trips and periods of jungle residence. All data from informant accounts will be identified as such, and all other data presented come from direct observation.

Quantified data concerning subsistence were collected on 66 observation days between March and July of 1980 (see Table 5.3). Data were collected on seven separate trips out from a mission base camp, ranging in length from 4 to 15 days. On one trip (4 days) men used only shotguns, on one trip (12 days) some men used shotguns and some used bows, and on five trips (50 days) men used only bows, at our request. The study produced measurements on rate of hunting returns per hour and per day. Because six of the observation days were spent primarily walking on dirt roads rather than hunting in the jungle, they have been eliminated from the data set. Two half-days of hunting are eliminated from the daily returns data, but the hours spent actually foraging on those days are included in the calculations of hourly return rates. The remaining 58 days are the basis of all calculations of average hunting activities or returns per day. They can be divided into 53 days of active hunting, 1 day of fishing, and 4 days of heavy rain (see Table 5.3). Time and returns for fishing have been excluded from this analysis because they have been discussed previously (Hawkes *et al.* 1982).

Table 5.3

Days of Data Collection, 1980

March	31[a]						
April		1[a]	2[a]	3[a,b]	4	5[a]	6[a]
	7[a]	8[a]	9[a]	10[a]	11[a]	12[a]	13[a]
	14[a]	15[a]	16[a,b]	17	18	19	20
	21	22	23	24	25[a]	26[a]	27[a,c]
	28[a]	29[a]	30[a]				
May				1[a,d]	2[a]	3[a,e]	4
	5	6	7	8	9	10	11
	12	13	14	15[a]	16[a]	17[a]	18[a]
	19[a]	20[a,b,d]	21	22	23	24	25
	26	27	28	29	30[a,b]	31[a]	
June							1[a,d]
	2[a]	3[a]	4[a]	5[a]	6[a]	7[a]	8[a]
	9[a]	10[a]	11[a]	12[a]	13[a,b]	14	15
	16	17	18	19	20	21[a]	22[a]
	23[a]	24[a]	25[a,d]	26[a]	27[a]	28[a]	29[a,e]
	30						
July		1	2	3	4	5[a]	6[a]
	7[a]	8[a]	9[a,d]	10[a]	11[a]	12[a]	13[a]
	14[a]	15[a,b]	16	17	18	19	20

[a]= Days foraging in jungle for which quantitative data were collected.
[b]= Days excluded from data set.
[c]= Days spent fishing.
[d]= Days of heavy rain.
[e]= Incomplete foraging days.

We accompanied the Aché on most foraging expeditions between March and July of 1980. Hawkes stayed with the women and recorded all activities pertinent to time allocation and returns from women's subsistence efforts. In addition, she clocked all adult men arriving and departing from the women each day, and weighed all game killed during this period. This was possible because animals killed early in the day were brought to the women to carry and all game was brought back to the camp before consumption. Hill accompanied one or more hunters each day and recorded the time breakdown of the men's activities for that day at the end of the day. The name of the hunter who killed each game item was obtained by questioning the person who brought the item to the women or into camp (not necessarily the hunter himself), if not observed directly. The men who were monitored were not chosen randomly but were picked because of Hill's proximity to them at the time that the hunters began to split up. This tends to weight the sample in favor of those men who carried on the most amicable relations with Hill; however, nearly every adult hunter was monitored for part or all of one day using this procedure.

Time hunting is defined as the number of hours between taking up weapons and starting to walk in the morning, and the arrival of hunters at the location that is to be the camping spot for that night. In all cases, this includes time spent eating gathered items along the way, time spent in pursuit of those items, time spent in the company of other hunters resting and talking (especially after a successful hunting episode), and time spent carrying the kill to the campsite. Estimates will be presented of time spent in these activities, considered as part of the gross hunting time, in order to allow comparison with other hunting studies that also include these activities. Periods of rest or eating with women and children present are subtracted from men's hunting time. These events were rare.

The optimality approach that we use in this analysis requires measurements of encounter rates, pursuit times, and processing times for all resources. In order to collect these data, Hill timed the pursuit of all game hunted by the men whom he accompanied. These measured pursuits for any given species were then used to calculate the expected pursuit time for each kilogram of that species killed, regardless of whether or not each kill event was actually observed. Pursuit time starts with the first behavioral indication that a specific prey target was being hunted. This was usually visual or auditory contact with the prey item followed by a rapid reaction on the part of the hunter. In some cases, pursuit time was defined by intensive tracking of a single prey item (or items) to the exclusion of total search alertness. This, however, is somewhat ambiguous. Pursuing a fresh track did not always exclude the simultaneous search for other items. These "warm pursuits" were sometimes temporarily halted in order to engage in the "hot pursuit" (i.e., with visual and auditory contact) of other prey items. Further problems with precise measurement of pursuit time were encountered due to the continual arrival of hunters at a prey site over a period of time when hunting cooperatively. For these cases the number of man hours of pursuit is an approximation rather than a precise measurement. It is unlikely, however, that the pursuit times given in this paper are more than 20% off the actual value if it were precisely measurable. Such a margin of error would not change any of the conclusions presented in this paper. A revision of the defining moment that pursuit begins might have strong consequences for the rank order of resources (see Table 5.7) but once again would not alter the conclusions presented.

Pursuit time

Search time in this paper is considered to be the total of all time hunting, minus the total of all pursuit time necessary to kill the total game take. This means that all time spent during the day that is not recorded as pursuit time, is considered search time. Once again, there are some difficulties with this due to the activities that take place throughout the day that are not actually active search for prey items. Our search time estimates are therefore rough but, again, a 20% error would not change the conclusions presented in this chapter. Refinements in concepts and methods are necessary. We look for further improvement and hope to collect more accurate measurements of search and pursuit times for further analyses.

Search time = all hunting time less pursuit time.

Processing time for resources in this paper is equivalent to butchering time. Observations were made on the required butchering time for each species, and these are extrapolated to give the total processing times for any resource item listed. All animals were weighed with hanging scales (10, 20, or 50 kg scales) and are presented as live weights. Several measurements have indicated that the Aché consume between 65 and 85% of the kill by weight, depending upon the particular species. Many internal organs are eaten, as well as blood, skin (with the hair that remains after singeing), brain, bone marrow, and, occasionally, intestinal contents. We use 65% of live weight as the edible portion in order to calculate caloric value; however, we consider this to be a minimum estimate. Many South American game animals are quite fat (e.g., peccaries, coatis, armadillos, tapirs, caviomorph rodents) and we estimate their caloric value at 3000 Cal/kg (as taken for medium-fat whole pork). In addition, we estimate deer at 1250 Cal/kg, monkeys at 2000 Cal/kg, birds at 1900 Cal/kg, reptiles at 1500 Cal/kg and fish at 1370 Cal/kg. These estimates are derived from USDA Agricultural Handbook No. 456, food composition table for use in Latin America, or Meehan (1977), and are extrapolated where no equivalent items are listed, taking into account primarily the proportion of fat for each animal. These estimates should be considered rough and subject to revision when more precise measurements are made.

As the flora and fauna of this area are poorly studied and described, we have provisionally identified animal species ourselves. This should be taken as tentative. A detailed study of the area by a qualified ecologist would be especially appropriate and welcome.

The study population from which data were recorded totalled 35 adult men, 6 old men (over 50 years), and 13 young men (between 10 and 17 years), although all hunters were not present at any one time. For most of the analyses presented, only adult men are considered because the time spent and the hunting returns for the other two age groups were quite erratic and were not accurately recorded. One old man and one young man are included with the adult hunters, as their time allocation and returns were comparable to the adult group.

Time Allocation for Subsistence

Hunting is the major activity of adult men in Aché society. During foraging trips, all adult men hunted or fished every day except the 4 days of heavy rain.[5] Young men (under 17) and old men (over 50) hunt somewhat less frequently, although we do not have precise data on these groups. Activities directly related to subsistence (i.e., food procurement and pro-

Table 5.4

Time Allocation for Adult Hunters[a]

Activity	Man-hours	Percentage
Outside normal foraging hours[b]		
Gathering	56.4	30.0
Butchering[c]	131.9	70.0
	188.3[d]	100.0
Normal foraging hours		
Gathering vegetable	21.4	0.5
Extracting honey	37.8	0.9
Fishing	85.0	2.1
Active pursuit of game	1005.1	25.2
Search	2839.3	71.2
	3988.6[e]	99.9
Total subsistence work	4176.9 hr/604 man-days[f]	

[a]Includes all whole foraging days.
[b]Refers to activities that take place in the morning before hunters take up their weapons, or in the evening, after returning to camp.
[c]Some of this work may have been done by males in the older or younger age groups, or women.
[d]Equals 5% of total subsistence work.
[e]Equals 95% of total subsistence work.
[f]Approximately equal to 6.9 hr/man-day.

cessing) were recorded for all men, and the time breakdown in man hours is shown in Table 5.4. It can be seen that 95% of an adult hunter's subsistence effort is hunting, and over 70% of this time is spent searching for game. Spot checks on the time recorded as search time allow us to estimate that about 10% of this time is spent resting, talking or eating, and about 5% of the time is spent transporting the kill to the camping spot (although some search does continue during transport). It is notable that the time spent gathering by men was less than 2% of their total work time, and most of this occurred on rainy days or outside normal hunting hours. During the study period there were approximately 11 hours of daylight each day. An average of almost 7 hours daily (including rainy days in the average) were devoted to the procurement and processing of food by adult men. The remaining 4 hours of daylight were spent in manufacturing and maintaining weapons, playing with children, joking, eating, and various other activities.

7/11 of total time spent on subsistence.

Around these averages there is marked variation in the length of the hunting day, from a maximum of 10 hours average among hunters on a specific day, to a minimum of less than 2 hours (excluding days of no hunting). Daily return rates vary from a maximum of 14.3 kg/hunter to a minimum of 0.0 kg/hunter. Figure 5.2a shows the correlation between total hours hunted on a given day and the total weight of game for that day. Grouping the data emphasizes this relationship. On the 22 hunting days

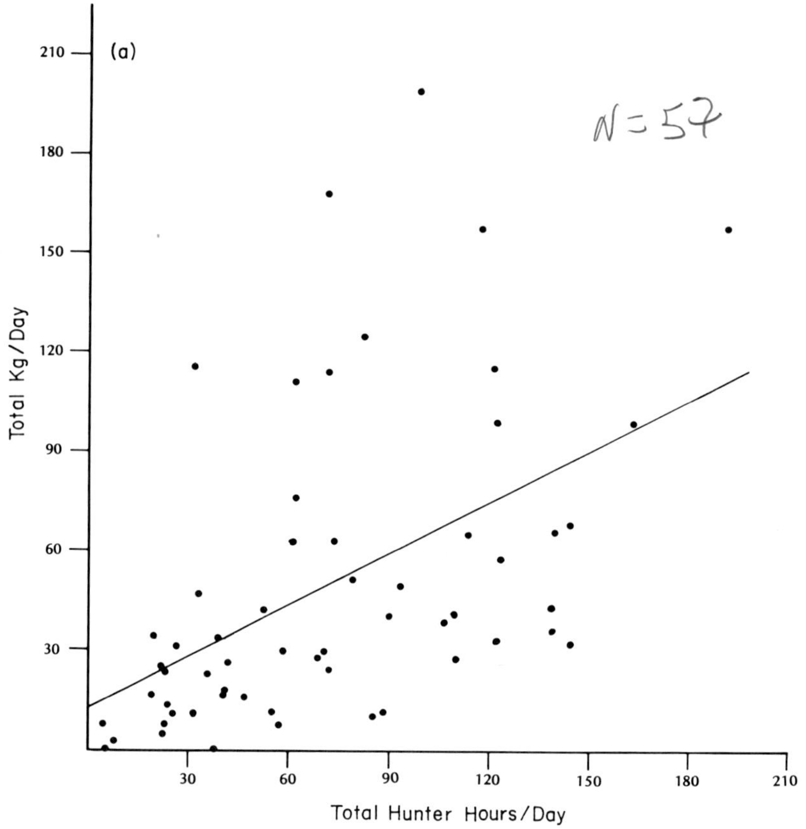

Figure 5.2 *Hunting returns per time spent foraging. (a) Total number of hours hunted by men in the foraging band, and total kg of game killed on that day. Each point represents one day* ($r = 0.50$, $p > .01$)*. (b) Average returns/hunter, and the average number of hours hunted/hunter on the day following those returns. Each point represents one day. There is no correlation* ($r = -0.04$)*. (c) Relationship of average hunting return rate (kg/hr) during sample period to average number of hours hunted per day, for individual hunters. Each point represents one* hunter. *Sample includes all men for whom seven or more days of bow hunting returns were recorded. The correlation is significant* ($r = .64$, $p > .01$).

on which the hunters brought more than 2 kg of meat per consumer, the average number of hours hunted was 7.3 hours per hunter-day. On 31 days on which they brought back less than 2 kg per consumer, the average was 6.0 hours per hunter-day. Hunting success does correlate with the number of hours hunted for a group of men over a period of 53 days. In other words, if the group desires more meat, one strategy that will be successful is simply to hunt more hours per day. In this sense, hunting as an occupation is not a gamble, and average daily rewards are a function of labor input.

Table 5.5 shows the relationship between hours hunted and days of high

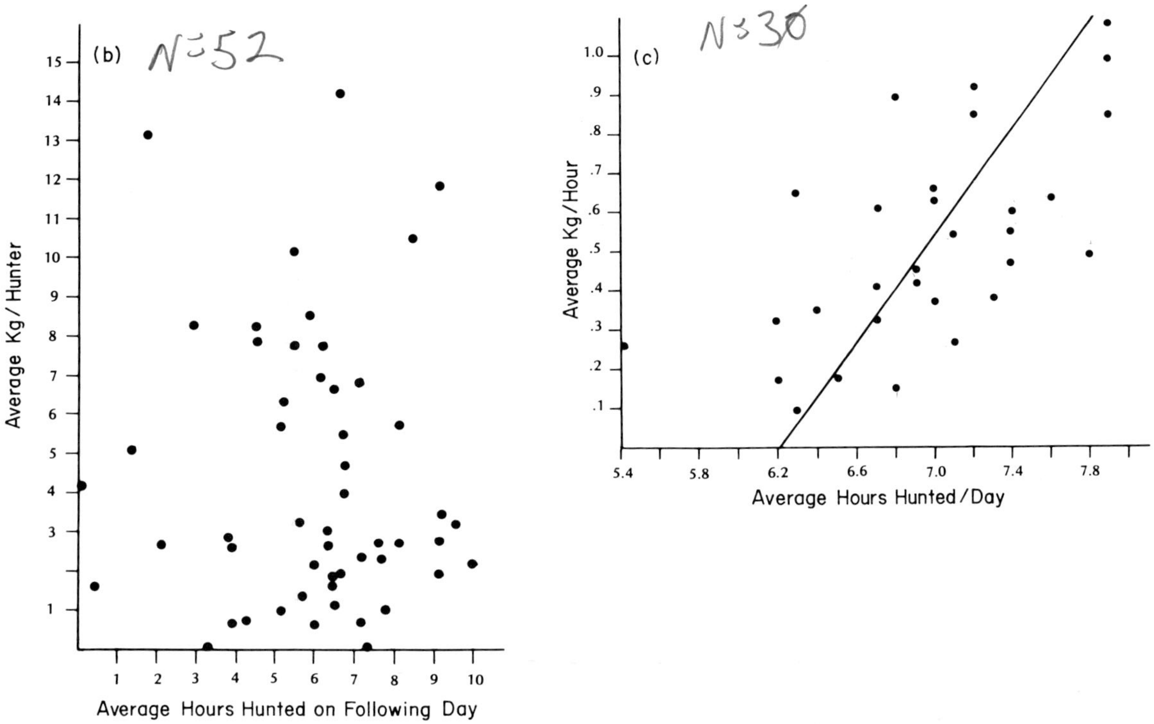

Table 5.5

Hunting Returns and Time Spent Foraging

	More	Same	Fewer	Significance (χ^2)
Individuals[a]				
Day-of-return rate > 1.5 kg/hr	25	00	16	not significant
Day after total returns > 9 kg/day	13	01	26	$p < .05$
Groups[b]				
Day-of-return rate > 1 kg/ man hour	05	—	08	not significant
Day after total return > 2.5 kg/consumer	04	—	12	$p < .05$

[a]Each individual was scored daily as to whether he hunted more, the same, or fewer hours than his own average throughout the study period. Numbers are in man-days (604 man-days sampled).

[b]Each group was scored as to whether it hunted more or less than the group average over the study period (6.9 hr/man-day). Numbers are in group days (53 group days sampled).

return rates, and the hours spent hunting on the day *following* high total meat returns for both individuals and groups. The data show a tendency for individuals to hunt more hours than average on days that they get a high hourly return rate (kg/hr); however, this is not significant. Groups do not hunt more or fewer than normal hours per man per day on the days when the return rate is high. On the other hand, total meat returns for individuals or groups on any single day *do* influence the number of hours hunted by those individuals or groups on the *following* day. Both individuals and groups hunt fewer man-hours per day on the day following high returns ($p < .05$ for both). However, as Figure 5.2b shows, there is no linear correlation between the average returns/hunter on a given day and the average number of hours hunted on the following day.

Despite the lack of *linear* correlation between the same variables, the significant X^2 correlation for the grouped data results because long average hunting days rarely follow high returns, whereas short average hunting days are not related to the preceding day's success rate. In other words, hunters as a group are unlikely to spend long hunting days subsequent to an unusually successful day (the relatively empty upper right quadrant in Figure 5.2b), but they are as likely to quit early on a day following poor hunting as on a day following high returns.

Adult men spend an average of 6.9 hours/day foraging on days of good weather. The averages for individual men ranged from 5.4 to 7.9 hours per day for the subset of 30 adult men that we observed for 7 or more complete foraging days. Figure 5.2c shows the relationship between individual hunting returns in kg/hour over the study period, and the average number of hours hunted per day for those individuals. There is a significant correlation ($r = 0.64$) between a man's average return rate, and his average number of hours hunting per day. (Hames [1979:241ff.] demonstrates that this correlation is not found for Yąnomamö or Ye'kwana hunters.) It is important to note that this is a correlation between return *rate* and time hunted, and thus the two are not interdependent (as would be the case if overall returns were used instead of rate). Multiplying the hourly return rate by the average number of hours hunted gives an even greater discrepancy between the average daily returns for the best and worst hunters.

One final point concerning time allocation is the consistency that hunters display throughout the sample period in the average number of hours that they hunt daily. This study samples a period of about 100 days from the end of the warm season to the middle of the coldest part of the year. Dividing the sample period into thirds allows us to determine if there are trends to hunt more or fewer hours as days become somewhat shorter and significantly colder (morning temperatures dropped from about 24 to 2°C from March to July). The average number of hours hunted per man-day was 7.1 during the final third of the study period, and 6.7 during the first third. The difference is not significant.

Hunting Returns and Comparison of Technologies

About one-eighth of the men in the study population have acquired a shotgun, by trading with local colonists or mission personnel. Although most of the guns are quite old, they are cleaned and oiled daily while hunting and are maintained in good condition. Most gun owners possess fewer than five brass cartridges which are reloaded in the morning or evening with black powder, newspaper wadding, 00 (double-ought)-sized lead fishing weights, and are sealed with beeswax. Occasionally, the lead weights are melted down in a can and poured into a hole made in the ground with the finger in order to produce a slug which is then carved or pounded to the correct gauge (12, 16, or 20). Slugs are used for tapir and jaguar, while all other game are killed with the large shot.

The techniques employed are similar to those described by Hames (1979) for the Ye'kwana, except that only very large shot is used; thus, the scatter pattern is not as wide and small arboreal animals are difficult to hit. Hunters refrained from using shotguns, at our request, for the final 50 foraging days, and thus we were able to make a comparison between bow and shotgun hunting.

It is our impression that Aché hunters are not more accurate with a shotgun than they are with a bow, nor do they seem to use it to increase shooting range. However, the effective knockdown and kill rate at any distance is much higher than with a bow. In addition, the hunter does not need to position himself as carefully because leaves and branches do not deflect buckshot to the same extent as they do an arrow. Nevertheless, most game taken with shotgun is approached to within 15 m or less before firing. The Aché rarely use shotguns to kill monkeys or birds. This will be discussed in the following section.

The key to bow-hunting success is persistence rather than high accuracy. Bow hunters pursuing arboreal game will shoot again and again until the game is mortally wounded. Men sometimes organize target shoots and take turns shooting and laughing. In two target shoots organized by the Aché during our study period, 60 arrows out of 288 shot (21%) stuck in a target 15 cm in diameter placed 25 m away at an angle 45° from horizontal. It was our impression that the men were shooting rather poorly compared to other target shoots we have witnessed, and the hunters definitely shoot more accurately as the angle to the target approaches 90°.[6] Arboreal game is usually shot at an angle of between 75 and 85° in order that the calculation of the trajectory arc not be too difficult. Aché bow hunters rarely shoot horizontally more than 15m.

The results from the 66 days of observations on hunting returns for 37 adult hunters, compiling a total of 3984.6 hours of hunting, show that

2622.3 kg of game (live weight) were killed, or 0.66 kg/hour. The effects of technology can be shown clearly by comparing the returns for bow hunting to those for shotgun hunting. Shotgun hunters killed 744.1 kg of game in 464.1 hours, for an average return of 1.60 kg/hour. The same men bow hunting in the same areas killed 1878.2 kg of game in 3520.5 hours, for an average return of only 0.53 kg/hour. Hames (1979) has also demonstrated that shotguns are more effective for Yąnomamö and Ye'kwana hunters.

Table 5.6 demonstrates the difference in the composition of animal species taken by hunters using either shotgun or bow. Most notable is the fact that large game makes up about 87% of the take with shotgun and only 24% of the take with a bow. Examining the returns for each species in kg/hour of hunting, we see that the returns for the two peccary species and for deer are 5–30-fold higher with shotgun than with bow. For smaller game, there is no increase in the return rate with shotguns, and in fact there is a significant decrease in the kilograms taken/hunting hour for both monkey and paca. These results will be examined in more detail in the next section.

Table 5.6

Game Species Killed by Bow and Shotgun Hunters

Game species	kg	kg killed/ man-hour hunting	% of total	
Shotgun hunting (464.1 man-hours)				
Collared peccary	163.8	0.35	22.0	
White-lipped peccary	212.1	0.45	28.5	86.8
Deer	270.0	0.58	36.3	
Paca	12.4	0.03	1.7	
Coati	36.8	0.08	4.9	
Armadillo	31.5	0.07	4.2	
Monkey (capuchin)[a]	7.8	0.02	1.0	13.1
Birds	6.2	0.01	0.8	
Snakes	3.5	0.01	0.5	
Total	744.1	1.60	100.0	
Bow hunting (3520.5 man-hours)[b]				
Collared peccary	77.0	0.02	4.1	
White-lipped peccary	307.1	0.09	16.4	24.3
Deer	71.5	0.02	3.8	
Paca	315.5	0.09	16.8	
Coati	242.2	0.07	12.9	
Armadillo	348.3	0.10	18.5	
Monkey (capuchin)	484.2	0.14	25.8	75.8
Birds	18.4	0.01	1.0	
Snakes	4.8	0.00	0.3	
Miscellaneous	9.2	0.00	0.5	
Total	1878.2	54.00	100.0	

[a]In these cases shotgun hunters called for a bow and took the game with an arrow.
[b]The following items were also captured live: 14 monkeys, 3 coatis, 1 deer, 1 tamandua, and 1 collared peccary.

A second and extremely interesting comparison of technology can be made by considering the take in kg/hour for bow hunters using only their hands. Since all armadillos and pacas were taken by hand, and almost all the coatis and miscellaneous small animals, we can calculate that Aché foragers are able to kill a minimum of 0.27 kg of game/hour hunting using only their hands! This represents over 50% of the total game taken by bow hunters. It is likely that they could raise the return rate even higher if bows were not available and if all foraging time were devoted only to those animals which can be taken by hand or with a club. It is also noteworthy that all fishing done by the Aché during this period used only the hands and the take was greater than 10% (by weight) of the total game take listed above for nonfish items. These results should be of some interest in the calculation of possible hunting efficiencies of early hominids and other prebow hunters.

Several authors (e.g., Hames 1980; Vickers 1980) have recently demonstrated quantitatively that hunting returns diminish as a function of the time that an area has been hunted frequently. Mission Verbo Divino, Chupa Pou, where the Aché are now settled, was established in August 1978 and has over 150 Aché residents. In order to test for diminishing hunting returns as a function of hunting pressure, we calculated the average return rate per hour on the first day of each foraging trip and compared it to the average return rate for all days. The first day usually represents a distance of not greater than 8 km from the mission. The average returns for first days of each trip, including all shotgun and bow hunting, was only 0.28 kg/hour (N = 686.9 man-hours). This is less than half the average returns for all days combined for all hunters, and demonstrates that a population this size can have considerable impact on the nearby game populations in less than 2 years.

The Optimal Diet Model and Hunting Strategy

As we have noted in the previous section, the composition of game species taken by bow or shotgun hunters differs dramatically. Not surprisingly, large game is killed more effectively with a shotgun. The rate of returns from coatis and armadillos, which are found throughout the primary forest and are usually killed by hand, is approximately the same for bow or shotgun hunters. Pacas are found in much greater concentrations near riverbanks in seasonal floodland forest. Because all three of the large game animals taken occur mainly in the primary forest, and because shotgun hunters do quite well hunting large game, it is not surprising that pacas are somewhat underrepresented in the game taken by shotgun hunters. Shotgun hunters spend larger portions of time in the primary forest and thus do not encounter pacas as frequently as do bow hunters. Monkeys, however, do inhabit the primary forest and are quite frequently encountered there. Nevertheless, Aché men killed only three monkeys while shotgun hunting, whereas 218 monkeys were killed by bow hunters. Our hypothesis is that the returns from shotgun hunting are just enough higher than bow hunting that they eliminate monkeys from the optimal diet of shotgun hunters, even though all other game items are still in the optimal diet.

We have previously used optimal foraging models to account for the overall mix of resources exploited by an entire Aché foraging band (Hawkes *et al.* 1982). The underlying assumption of this theory is a simple one: people will continue to use or adopt foods and techniques which increase the ratio of returns to cost (measured conventionally as calories per unit time) and they will stop using or fail to copy the use of foods and techniques which decrease their ratio of returns to cost. Although seemingly simple and logical, some nonintuitive propositions flow from models built on this assumption.

The optimal diet model (Charnov 1976; Charnov and Orians 1973; Emlen 1966; MacArthur 1972; MacArthur and Pianka 1966; Pulliam 1974; Pyke *et al.* 1978; Schoener 1971) can be used to predict whether or not a given resource will be taken by foragers upon encounter with that resource. Food items are ranked according to the ratio of returns they provide (Cal) to the cost of acquiring and processing the resources (handling time) once they have been encountered. This model shows that returns will be maximized if foragers take only those resources for which this ratio is equal to or higher than the average returns they get from foraging in general, *including search time.* They should ignore all potential resources for which this ratio is lower than their average returns will be if they keep searching for some other resource. Thus, according to this model, whether or not a potential resource is in the optimal diet *does not depend upon its encounter rate.*[7]

The resource rankings of this model say nothing about the quantitative importance of a resource to optimal foragers. High-ranked items may be so rarely encountered that they contribute only a very small proportion of the diet. Low-ranked items, which are nevertheless in the optimal set, may be encountered with sufficient frequency to contribute the bulk. The ranking shows instead which resources are likely to enter or eave the diet, and in what order. This knowledge, combined with density estimates (encounter rate) for resources at any given time, can be used to predict the actual diet. If the encounter rate with high-ranked resources fluctuates widely, the optimal diet will also fluctuate widely and only the very highest ranked resources will never be excluded.

Table 5.7 shows the caloric returns to handling time for all the game hunted, and thus gives the rank of each item according to returns upon

Table 5.7

Caloric Returns to Handling Time for All Game Hunted

Item	(total kg taken)			(number of measured pursuits)			($\bar{X}$ hr pursuit/kg)		($\bar{X}$ hr butchering/kg)	(Cal/kg live weight [assume 65% edible])		($\bar{X}$ Cal/hr upon encounter [total cal/total handling time])		(rank of item)		(Cal/hr if item is taken upon encounter)	
	Shotgun	Bow	Shotgun		Bow	Shotgun		Bow			Shotgun		Bow	Shotgun	Bow	Shotgun	Bow
Collared peccary	163.8	77.0		est.		0.01		0.06	0.02	1950	65000		24375	1	1	1290	58
Deer	270.0	71.5		est.		0.01		0.06	0.02	819	27300		10237	2	2	1580	81
Paca	12.4	315.5		33			0.24		0.04	1950		6964		3	3	1648	309
Coati	36.8	242.2		20			0.22		0.06	1950		6964		3	3	1803	474
Armadillo	31.5	348.3		20			0.27		0.06	1950		5909		4	4	1949	693
Snake	3.5	4.8		03			0.01		0.16	1000		5882		5	5	1956	695
Birds ($\bar{X}$ = 1.4 kg)	6.2	18.4		est.			0.10		0.25	1240		3542		7	6	2373	701
White-lipped peccary	212.1	307.1	8[a]		13[a]	0.51		0.55	0.02	1950	3679		3421	6	7	2372	858
Capuchin monkey	7.8	484.2		37			0.97		0.10	1300		1215		8	8	2362	910

[a]Some estimates.

Handling time = pursuit time upon encounter + handling time after kill.

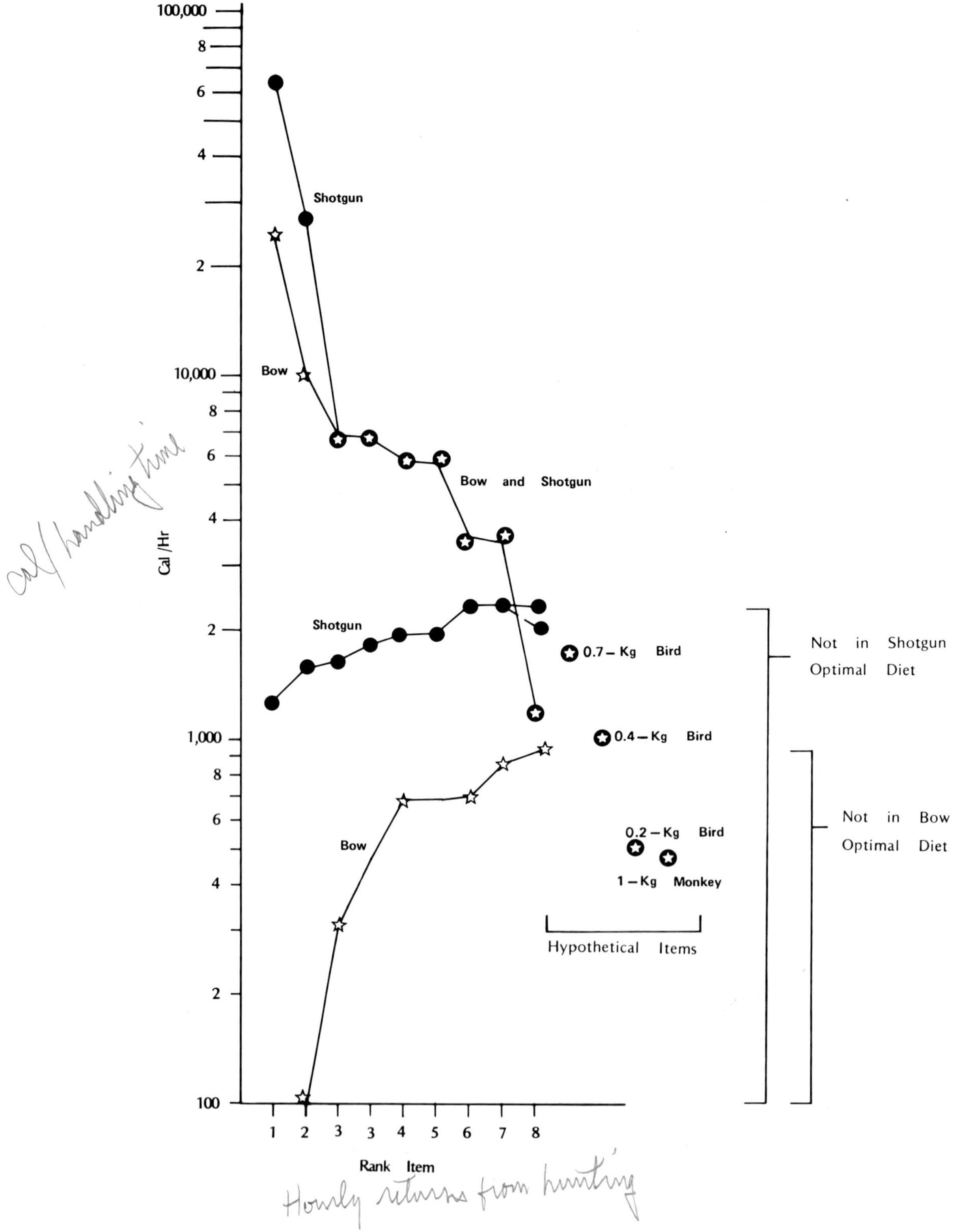
100,000
8
6
4
2
10,000
8
6
4
2
1,000
8
6
4
2
100
Cal /Hr
cal/handling time
Shotgun
Bow
Bow and Shotgun
Shotgun
Bow
0.7 – Kg Bird
0.4 – Kg Bird
0.2 – Kg Bird
1 – Kg Monkey
Hypothetical Items
Not in Shotgun Optimal Diet
Not in Bow Optimal Diet
1 2 3 3 4 5 6 7 8
Rank Item
Hourly returns from hunting

encounter. Handling time for a specific item is the sum of the pursuit time for that item upon encounter (including unsuccessful pursuits) and the necessary processing time before consumption of the item. It also shows the number of Calories per hour that a hunter can expect to get foraging by adding each ranked resource to the set of items that he will take upon encounter, starting from the highest ranked item. Shotgun hunters spent 464.1 hours hunting and 26.3 hours butchering their take, for an average foraging return of 2362 Calories/hunter hour. Bow hunters spent 3520.5 hours hunting and 111.1 hours butchering their take, for an average foraging return of 910 Calories/hunter hour (note that both of these return rates include all search time).

Figure 5.3 shows the ratio of Calories returned to handling time for each of the game items ordered by rank (the circular points and stars descending from left to right) and the average returns for hunting that will result from the addition of each animal to those that will be taken upon encounter (circular points ascending from left to right). The latter numbers are derived as follows: Search time for the entire set is calculated by subtracting the pursuit time of all items from the total number of hours hunted. If hunters took only the number 1-ranked item, their returns would be the total Calories of the number 1 item taken per total search time, plus the total pursuit time for the number 1 item, plus the total processing time for the number 1 item. In order to calculate the returns provided by adding in subsequently lower ranked items, one need simply add in their caloric contribution to the numerator and their pursuit and processing times to the denominator. The optimal diet model predicts that all items which increase the overall rate of returns for foraging should be taken upon encounter, whereas any item which upon addition to the diet decreases the overall rate of returns for foraging will not be taken.

The difference in returns for shotgun and bow hunting, and the effect upon foraging returns of taking capuchin monkeys with either weapon, strongly supports our hypothesis. Monkeys should not be taken by shot-

---

Figure 5.3 *Rank of prey items according to returns upon encounter (total Cal/total handling time) shown descending left to right, and hourly returns from hunting, as each ranked item is added to the diet shown ascending left to right. The optimal diet model predicts that ranked items should not be included in the diet when returns upon encounter with an item fall below the overall average foraging returns excluding it (the point at which the line descending from the left crosses the line ascending from the left). Thus, rank item 8 (monkey) should be ignored by shotgun hunters, but included in the diet of bow hunters. The dotted line shows the expected decrease in overall foraging returns from shotgun hunting if monkeys were taken at the same rate (kg/hr hunting) as they were taken by bow hunters. Additional items show the expected returns from birds of 0.7 kg* (kuachi)*, 0.4 kg* (uru)*, and 0.2 kg (> 50 species), as well as returns from small capuchin monkeys (1 kg). The first two birds were taken by bow but not shotgun hunters. The small birds were ignored by all hunters. Small monkeys are discussed in text.*

gun hunters because adding them to the list of items taken upon encounter would decrease rather than increase the returns for a shotgun hunter. Conversely, bow hunters will increase their foraging returns by taking monkeys, even though they are the lowest ranked game item.

Although the optimal diet approach may explain why shotgun hunters spent only 1.6% of their foraging time pursuing monkeys, whereas bow hunters spent 13.3% of their foraging time pursuing monkeys, the size distribution of the monkeys taken presents an interesting complication. Monkeys are one of the few game items for which hunters can get a reasonably good estimate of size before deciding whether or not to pursue. Assuming that pursuit time is the same for any monkey regardless of size, and using the measured butchering time, it can be calculated that any monkey under 1.8 kg will decrease the average foraging returns for bow hunting and thus should be ignored.[8] The hunter may have some difficulty estimating the precise size of a potential prey high in a tree, but prey of one-half this size should be easily recognizable as too small. Nevertheless, 10 monkeys of less than 1 kg each were taken by bow hunters (5.3% of all monkeys). These items are clearly not in the optimal diet (see Figure 5.3). Possibly smaller monkeys require less pursuit time; however, this cannot be determined from our data and will have to be investigated further.

The cutoff point for the size of birds that should be hunted can be determined from these data as well. Because the pursuit and processing time is nearly equivalent for all birds of 1.5 kg or less, and the returns for birds averaging 1.4 kg is 3542 Cal/hr upon encounter (see Table 5.7), one can calculate that shotgun hunters should not take birds smaller than 0.9 kg if they take only the items predicted from the optimal diet model. The smallest bird taken by shotgun hunters weighed 1.4 kg. Bow hunters, on the other hand, can take birds as small as 0.4 kg before decreasing their foraging returns (see Figure 5.3). The smallest bird taken by bow hunters weighed exactly 0.4 kg.[9] Many species of birds smaller than 0.4 kg were present in this area, but were not pursued by bow hunters.

It is well known that coatis can and do bite, and the use of hands rather than bow and arrow to kill them seems to be a peculiar and dangerous method of hunting. Some strong motivation must be behind this strategy, or the wise hunter would quickly abandon it. Shooting coatis with arrows rather than catching them by hand would be similar to shooting capuchin monkeys (they both are found in troops high in the treetops and try to stay hidden there). Using the figures in Table 5.7 and assuming the pursuit time per item to be the same as for capuchin monkeys, we calculate that hunting coatis with arrows would give only 2600 Cal/hr upon encounter whereas hunting them by hand gives 6964 Cal/hr upon encounter. This is because the method of hand hunting forces the animals to jump quickly to the ground rather than moving through the tree tops.

Variance in Hunting Returns between Individuals and Age Groups

Figure 5.4 demonstrates the difference in average hunting returns for individual men throughout the study period. There is approximately a 10-fold difference in kg meat/hour returns between the bow hunter with highest returns and the bow hunter with the lowest return ratio. This difference along with the trend for hunters with higher returns to hunt more hours per day (see Figure 5.2) produces a daily difference of 14-fold in the kilograms of meat brought back by the most successful and least successful bow hunter during this period. The variance between individual shotgun hunters appears lower but is based on very limited data.

Table 5.8 shows the average daily kill of all adult men who were sampled for 7 or more days (kilograms calculated by multiplying the horizontal by the vertical axis for all points in Figure 5.2b). Although the return ratio for the hunter with the highest return rate is 14-fold higher than for the hunter with the lowest return rate, the overall spread for the group is somewhat less than we had expected. Indeed, it would appear that the hunter with the lowest returns is unrepresentative of the lower return hunters as a subset because his daily returns were less than half those of the hunter with the next lowest returns. During this study period, we calculate that 32% of the men brought back 50% of the game or, alternatively, 57% of the hunters brought back 80% of the game.

Figure 5.5 shows the spread for individual hunters in the percentage of days that each got 0-kg meat returns for an entire foraging day. There is

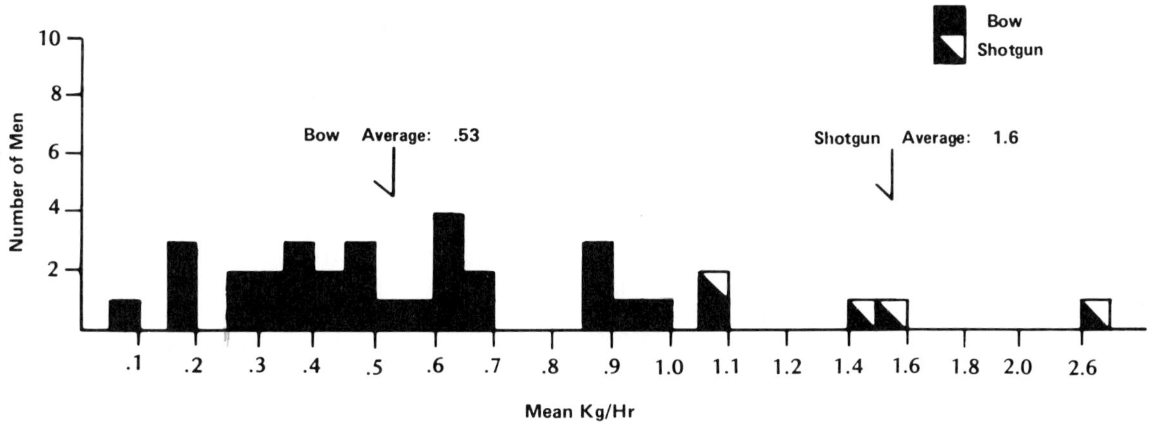

Figure 5.4 *Mean hourly return rate (kg/hr) for all adult hunters whose returns were measured for 7 or more days.*

Table 5.8

Average Yield for Individual Hunters

Hunters (ranked from lowest to highest)	Average yield (kg/day)		
1	0.6		
2	1.3		
3	1.3		
4	1.4		
5	1.4		
6	2.1		
7	2.2		
8	2.2		
9	2.2		
10	2.5		
11	2.7		
12	2.8		
13	3.5		
14	3.7		
15	3.7		
16	3.9		
17	4.0		
18	4.0		
19	4.1		
20	4.1		80% meat
21	4.5		57% men
22	4.5		
23	4.9		
24	5.8	50% meat	
25	6.0	32% men	
26	6.6		
27	7.4		
28	8.5		

(Brace: hunters 19–28 = 50% meat, 32% men; hunters 10–28 = 80% meat, 57% men.)

[a]Calculated by multiplying horizontal × vertical axis for each point on Figure 5.2b.

approximately a sixfold difference between the hunter with the lowest percentage of 0-kg return days and the hunter with the highest percentage of 0-kg return days. Thus, there is a greater difference in the quantity of game killed per day by individual hunters, than in the probability of getting some game item on any single day. In other words, the discrepancy between the number of days that each hunter is a contributor to the camp's meat supply is not as great as the discrepancy in the quantitative amounts of meat contributed by each hunter.

Overall returns are a function of both consistency and larger game kills, thus neither can be independently tested for significant correlation. Percentage of days with no kill is, however, not necessarily related to the number of larger animals killed, and thus we can ask whether the same

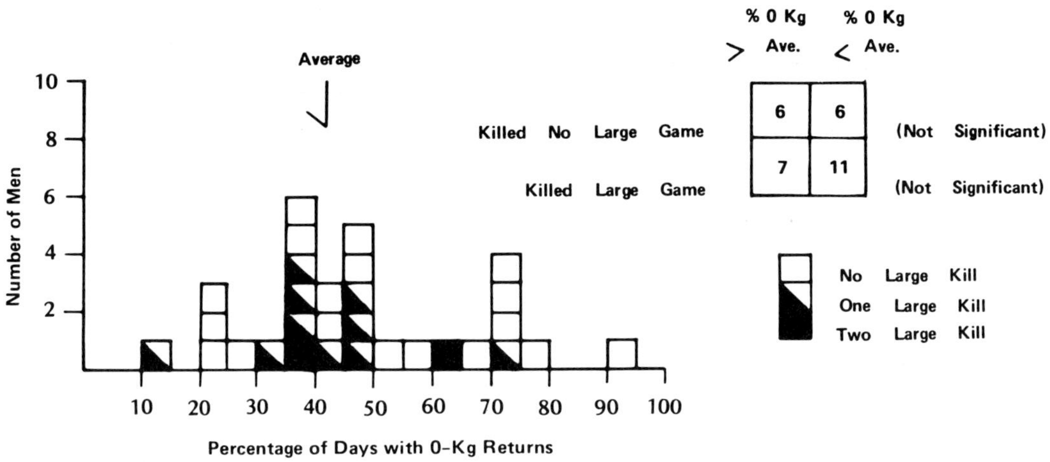

Figure 5.5 *Daily consistency of kills (percentage of days with zero kg returns) plotted for each individual hunter versus successful kill of large animals. There is no significant correlation between the two.*

subset of men have many large kills and a low probability of 0-kg returns per day (two independent measures of hunting skill). The subset of men who killed large game ($\geq$ 20 kg) during this period is indicated in Figure 5.5. Killing a large game item does not correlate significantly with a low probability of getting 0-kg returns on any foraging day. We interpret this to mean that there are two subsets of men who could be called the best hunters by the criteria of either consistency (low percentage of days with 0-kg returns), or of number of large game items killed. Either of these two strategies will produce high overall hunting returns.

Although there is no significant correlation between estimated age and hunting returns for the set of men that we call "adult hunters," there are some striking differences between these hunters and the other two age groups of hunters. Six old men (over 55 years) and 13 young men (between 10 and 17) accompanied us on some of the hunting trips. One young man who hunted every day and did well (0.35kg/hour) and one old man who also hunted every day with results comparable to adult hunters (0.63 kg/hour) have been included in the adult hunter data set. The remainder of these two age groups hunted much less frequently and less successfully. Although our data concerning time expenditures for these groups are only approximate, we do have exact measures of their kills. The five old men who have not been included in the data presented thus far spent 41 total man-days on foraging trips and contributed 10.4 kg meat in total. We estimate that they spent an average of about 4 hours per day hunting, for a return rate of about 0.06 kg/hour. Their contribution per man-day is therefore only about 7% of the average adult hunter. The 12 young men who have been excluded from the data presented thus far spent a total of 102 man-days on foraging trips and contributed 45.9 kg meat in total. We estimate that they also spent an average of about 4 hours per day hunting, for an average hourly return rate of 0.11 kg/hour. Their contribution per man-day is about 12% of that of the average adult hunter. Both of these age groups spent considerable portions of their time gathering vegetable items, trailbreaking, and butchering the game killed each day.

Band Size and Hunting Returns

The efficiency of subsistence strategies has long been considered to be a function of, and thus a limiting factor in, the size of hunter–gatherer bands (e.g., Lee and DeVore 1968; Steward 1938). Figure 5.6 demonstrates the effect of band size on two parameters of hunting returns, the returns per man-hour, and the percentage of the men who get 0-kg returns on any single day. The number of points is not sufficient to allow us to construct a

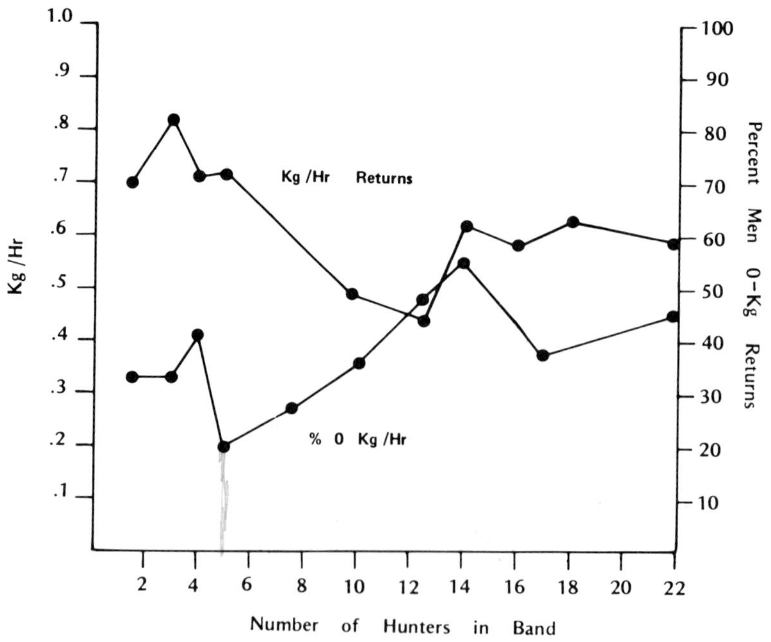

Figure 5.6 *Bow hunting returns as a function of foraging band size. Mean return rate (kg/man-hour) and percentage of hunters getting 0-kg returns on any single foraging day, are plotted against the number of adult hunters in the foraging band.*

curve in which we are totally confident; however, two trends are apparent. First, the returns per man-hour are higher in very small bands, and the percentage of men who make no kill on any single day is lower. Second, it appears that the drop in the return ratio, and the increase in the percentage of men who get 0-kg returns, levels off as the band size becomes larger. We tentatively conclude that bands of between three and five men get higher hunting returns per man-hour than larger groups, but it is not clear from our data why very large hunting bands are not as common as medium-sized bands. Figure 5.7 shows the actual frequency of different sizes of hunting bands we observed during the study period. It also shows the frequency of randomly selected band sizes before contact, elicited from informants by using what we consider to be an event independent of band size (e.g., "the day your father was bitten by a snake") as a reference point for naming all members in a foraging band on a specific day. The data show that bands rarely get larger than 25 men (80–100 persons), although,

optimal size for a man is probably 5 men. * if time is crucial.

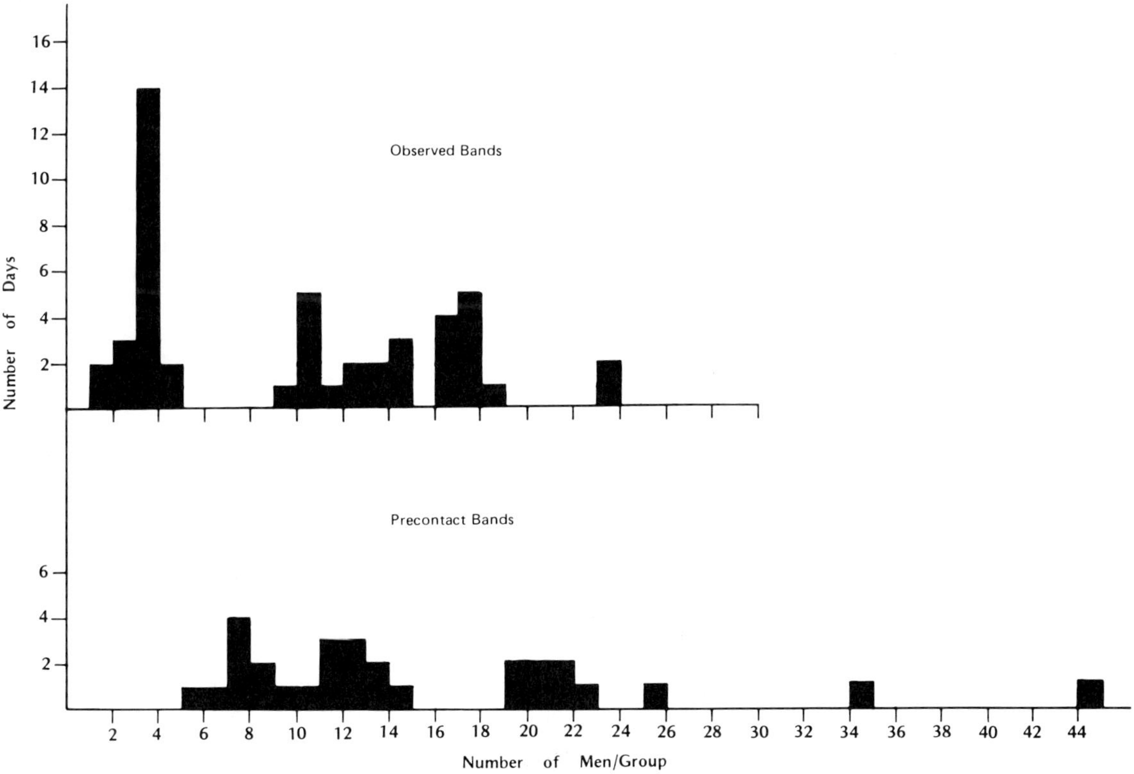

Figure 5.7 *Frequency of various-sized foraging bands as measured during the study period for bow hunting trips, and as recalled by informants. Events that we judged to be independent of band size were used to elicit informant recall of band size on a particular day prior to contact.*

as we mentioned, it does not appear that decrease in hunting returns can explain this phenomenon.

In order to explore this further, we can calculate the possible effect of having 40 men in the hunting band (120–160 persons). We assume that the major decrease in hunting returns for larger groups will be due to the crowding effect, or an overlap in the area searched. Forty hunters starting off from camp in the morning would take just 15 minutes to be spaced 50 m apart (they walk about 4 km/hour in the jungle), and only 30 minutes to have 100 m of spacing between each of them. The same of course would be true at the end of the day, as they were all coming together to the camping spot. This means that about 7% of the average hunting day the hunters would be less than 50 m apart, and about 14% of the day they would be less

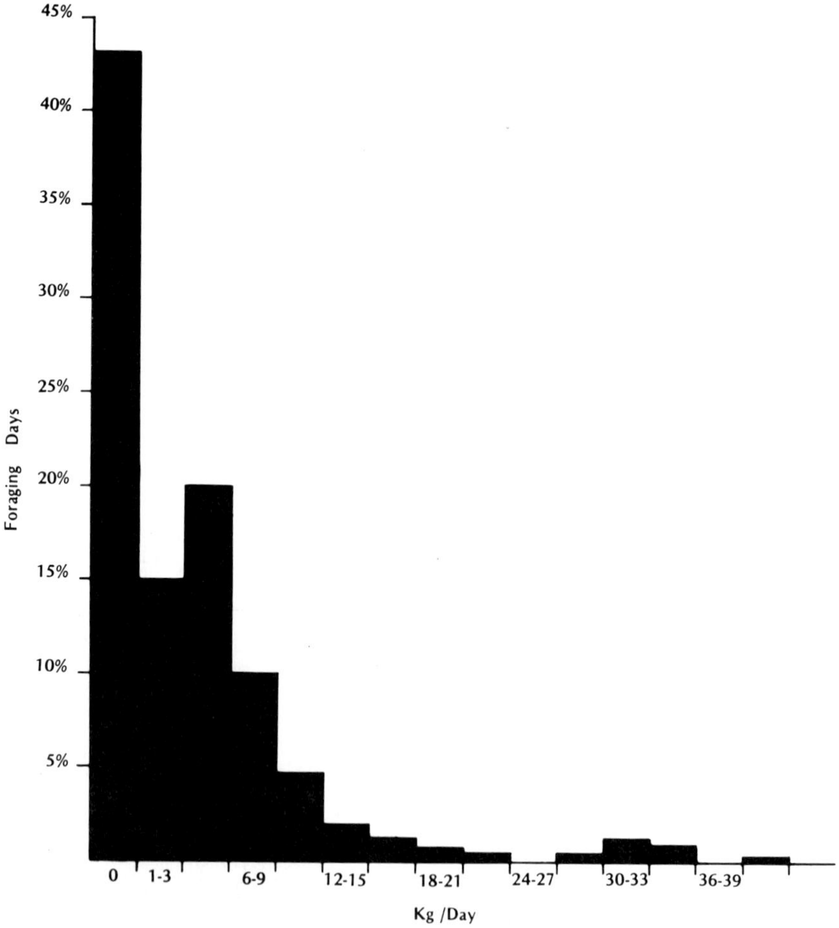

Figure 5.8 *Probability of various returns (kg/day) for an average bow hunter on any single day. All full days for individual bow hunters were sampled.*

than 100 m apart. Because, as we mentioned earlier, men frequently clump together through much of the day, we find it unlikely that a band of 40 hunters would get hunting returns much lower than those described here. Because of the higher total game take each day, they would have to move camp frequently, but it is not clear how important this difference would be to the Aché, as we observed them to move camp every day even when in small bands.

A common explanation for the minimum size of hunter–gatherer bands is risk reduction. If returns from foraging are not steady, a band of only a few families may have a high probability of days with no food. Although we have the data necessary to calculate the probability of a given sized band getting hunting returns below some limit on any given day, it is somewhat difficult to assess the important physiological limits beyond which risk should be minimized.

Figure 5.8 shows the probability of killing various amounts of game for any *individual* hunter on any one day. It represents a tabulation from the total of all man-days sampled for bow hunters, and does not take into account the fact that hunters do slightly better in smaller groups (see Figure 5.6).

Consider the extreme case: no game at all. For one hunter the likelihood of this is 43.5%. But the probability of no game falls sharply with the addition of a second hunter (18.9%, which is less than 1 day in 5 with no game) and again with the addition of a third hunter (8.2%—less than 1-day in 10). With four hunters this probability (3.6%) is just over 1 day a month. Given that gathering contributes to subsistence, we conclude that risk minimization could only be important for very small band sizes (i.e., 12–15 total members).

Pursuit Group Size

We have already mentioned that many game animals are solitary and are taken by single hunters upon encounter. As long as the spacing is such that there is little overlap in the area searched by each hunter, it is unlikely that group size will have any effect on the returns from these animals. There are, however, three game species that live in large groups and which are usually taken by multiple hunters: white-lipped peccaries, coatis, and monkeys. These animals are usually taken by groups of men who call for each other's help upon encounter. They also make up 55% of the total game taken (by weight) and, therefore, the group hunting strategy employed is quite important and has a significant effect on the overall returns from hunting.

Two alternative motivations may affect goals of individual foragers. One

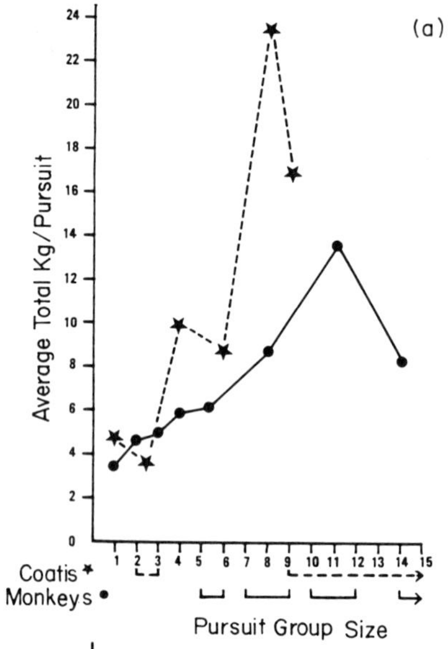

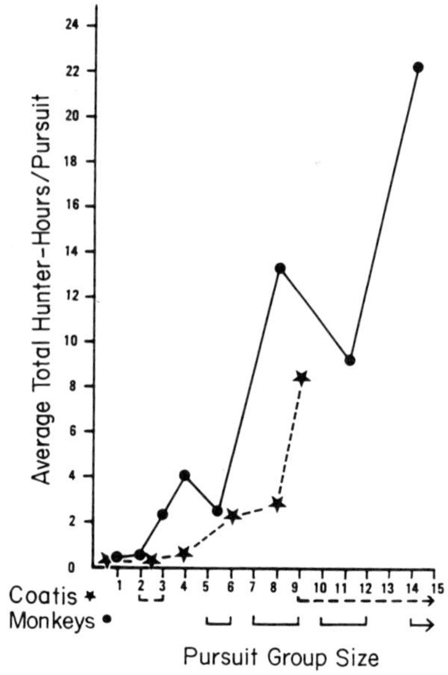

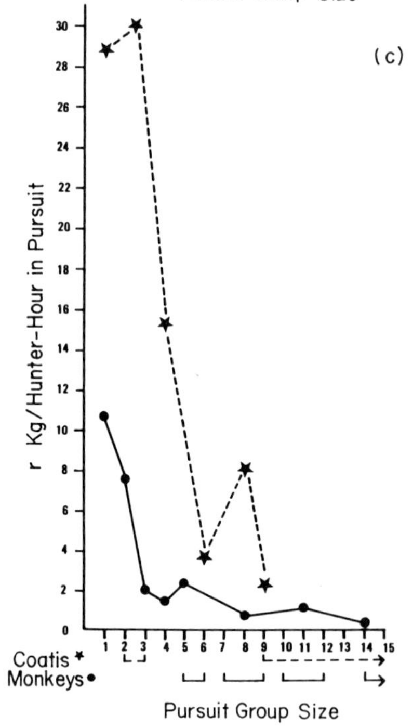

Figure 5.9 *Hunting returns according to pursuit group size (based on a sample of 21 coati and 40 monkey pursuits in which pursuit time was measured). (a) Average total returns (kg); (b) average total man-hours in pursuit upon encounter; (c) average rate of returns (kg/man-hour) upon encounter with prey.*

possibility is that each man is attempting to maximize his own personal return rate because of the possible rewards of being the best hunter. A second alternative is that since all the men eventually share all game killed, they are all trying to maximize the returns that the whole group gets each day. According to the first model, a hunter should call for help upon encounter or respond to calls only if it is likely to increase his own personal returns. If calling for another hunter or responding to a call decreases the average return *he* expects from the encounter, he will not call or respond, even if the total take is likely to be higher with more men helping (note that he may not make all the kills and thus, while more total game is killed, he may get credit for less). According to the second model, the hunter will always try to add more members to his pursuit group, or join pursuit groups, as long as overall foraging returns for the group are increased by this strategy.

Figure 5.9a,b,c shows the total kilogram returns and total man-hours spent in pursuit, for different-sized pursuit groups, upon encounter with monkeys and coatis (we do not have these data for white-lipped peccaries). It also shows the return ratio in kg/man-hour for each prey item as a function of pursuit group size.

If hunters wish to maximize the average returns of the band because they share all the game killed, the optimal strategy is to continue to add hunters to the pursuit group as long as returns remain higher than the combined returns of a smaller pursuit group plus the foraging returns of hunters not joining the pursuit. For example, if, during a 1-hour hunting period, two men can each expect 7.7 kg/hour in a joint pursuit of monkeys that lasts 0.24 hours, and a third man can expect 0.53 kg/hour (the average for hunting in general) if he continues to search for and pursue other game, their combined total will be 0.24 hours for two hunters at 7.7 kg/hunter-hour for the monkey pursuit, plus 0.24 hours for the solitary hunter at 0.53 kg/hunter-hour during this time, plus the remaining 0.76 hour for all three hunters at 0.53 kg/hunter-hour before or after the monkey pursuit. This is 4.04 kg of game. If the third man joins the others in pursuit of the monkeys, the group total will be 0.78 hours (the average pursuit time for groups of three) for three hunters at 2.2 kg/hunter-hour, plus the 0.22 hours remaining for the three hunters at 0.53 kg/hunter-hour after the pursuit. This is 5.5 kg of game. Thus, to maximize group totals, the third hunter should join the pursuit. This can be stated more generally by defining the following variables:

p = pursuit group size
r_p = the return rate/hunter at group size p
r_a = the return rate for the average hunter in general foraging (0.53 kg/hr)
h = the total number of hunters

T = the total number of hours in a hunting period
t_p = the number of hours in pursuit for group size p

Group returns during a hunting period (G) are then equal to the game taken by the hunters in the pursuit group which is $t_p \cdot p \cdot r_p$, plus the game taken by the remaining hunters during this time which is $t_p\,(h - p)r_a$, plus the game all the hunters get during whatever time this pursuit is not underway, which is $(T - t_p)h \cdot r_a$. Thus

$$G = (t_p \cdot p \cdot r_p) + [t_p\,(h - p)r_a] + [(T - t_p)h \cdot r_a]$$

Figure 5.10 shows the value of G for different-sized pursuit groups. For coati hunting, group returns are maximized in pursuit groups of 8 (Figure 5.10a). However, as Figure 5.11 shows, the distribution of actual cases does not support the hypothesis that hunters are maximizing overall returns. There is no preponderance of pursuit groups of this size among those observed. For monkeys, group returns are greatest in pursuit groups of 10–12, although the numbers are fairly constant for groups under 14 (Figure 5.10b). Again the distribution of observed pursuit group sizes (Figure 5.11) does not peak at the group maximum. The data do not support the hypothesis that Aché hunters establish pursuit groups which maximize total band returns.

An alternative hypothesis is that hunters try to maximize their personal return rates. If so, the optimal strategy depends on whether or not they are already in a pursuit group. Those in a pursuit group should continue the pursuit if $r_p > r_a$ and should exclude additional hunters if an increase in pursuit group size would decrease the returns per hunter in the pursuit group. Those outside the pursuit group should join as long as the returns/hunter in pursuit are greater than the returns for continuing to search and pursue other game. In the former case, hunters should solicit additional pursuit group members as long as $t_p \cdot r_p + (T - t_p)r_a < t_{p+1}r_{p+1} + (T - t_{p+1})r_a$. In the latter case hunters should join pursuit groups as long as $r_{p+1} > 0.53$.

Figure 5.10 also shows the individual returns during a hunting period for hunters who are members of pursuit groups of different sizes. The personal returns of hunters in the pursuit group are maximized for both coatis and monkeys when a single hunter remains alone—a pursuit group of one. As group sizes increase, this rate drops for both kinds of game. The frequency of observed cases does not match these rates (Figure 5.11). The distribution of pursuit group sizes does not support the hypothesis that the personal return rates of hunters in pursuit are maximized for either of these animals.

Hunters do not primarily form pursuit groups which would maximize either the returns for the individual who encounters the animal, or the

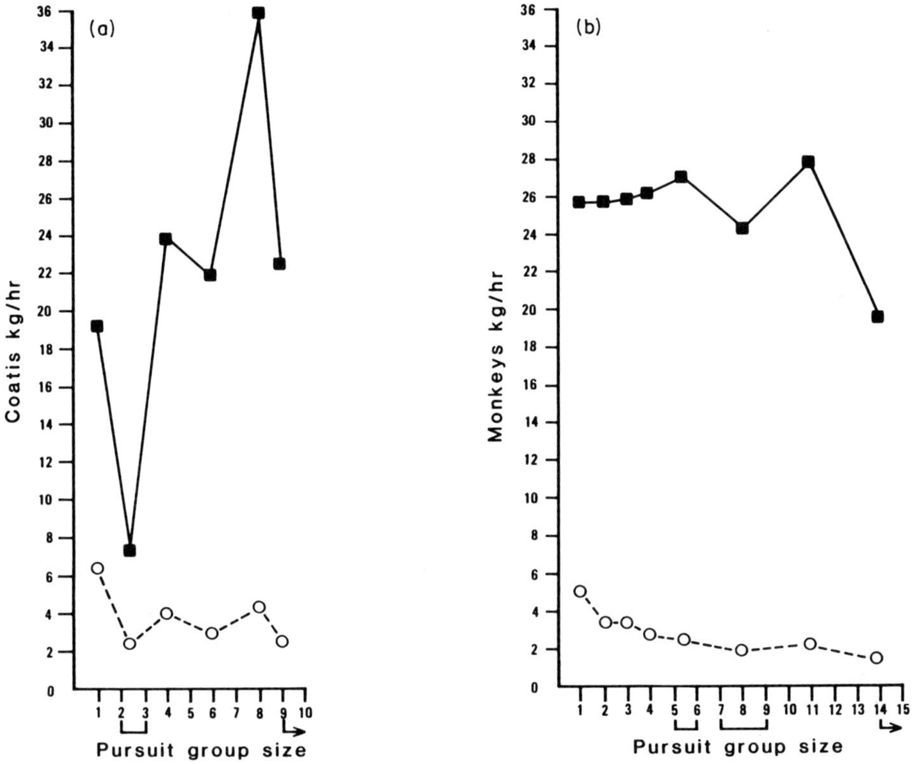

Figure 5.10 *Group return for pursuit of (a) coatis and (b) monkeys, by pursuit group size.*

Upper curve: G = total group return during a hunting period

Lower curve: I = total individual return for those in the pursuit of interest during a hunting period

Where: p = pursuit group size

r_p = return rate/hunter in the pursuit group at group size p

r_a = average return rate/hunter for general foraging (0.53 kg/hr)

h = number of hunters, here assumed to be 9 for coati hunts, 14 for monkey hunts, to simplify calculations (these are the smallest numbers which allow use of all the data)

T = total number of hours during a hunting period, here assumed to be 3 to simplify calculations (this is the smallest whole number which allows use of all the data).

t_p = number of hours in pursuit at group size p

$G = [t_p \cdot p \cdot r_p] + [t_p(h-p)r_a] + [(T-t_p)h \cdot r_a]$

$I = [t_p \cdot r_p] + [(T-t_p)r_a]$

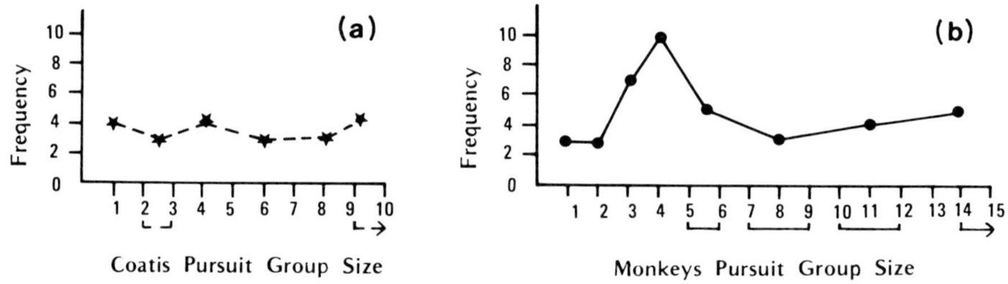

Figure 5.11 *The frequency of different pursuit group sizes over the sample of (a) 21 coati pursuits and (b) 40 monkey pursuits.*

overall foraging returns for the band. Instead, they form pursuit groups across a wide range of sizes.

There is a conflict of interest between individuals trying to maximize their personal return rates which may lie behind this wide range of band sizes. While a hunter will increase his own efficiency if he pursues the game he encounters by himself, not calling for help, a hunter will also increase his return ratio by joining other hunters in pursuit—as long as the returns for doing so are greater than his average returns for continuing to search for other game (0.53 kg/hour). Thus, while an individual maximizing his personal ratios would not want to be joined by a second hunter (or these two by a third hunter), other hunters *would* want to join him. However, very few cases were ever observed where a hunter failed to call for help upon encounter with either item.

Hunters *outside* the pursuit group increase their personal ratios by joining a pursuit as long as the rate for individuals in the group formed by their addition is higher than 0.53. This holds for coati pursuit groups of all sizes and for monkey pursuit groups of fewer than 14 hunters (Figure 5.11). In other words, individuals outside pursuit groups maximize personal returns by joining pursuits 56 out of 61 times (92%).

Spacing seems to be a major problem for hunters who wish to maximize their foraging returns. If hunters are too far apart, pursuit groups may be small because of a lack of hunters within earshot rather than because hunters fail to call for help or fail to respond to those calls. If hunters stay close together (as they often do), the area searched per man-hour decreases and thus the overlap causes a considerable increase in the search time component of foraging. Since we have shown (Table 5.4) that search time is the major component of foraging time (71.2%), increasing it by any factor would considerably reduce overall foraging returns. The hunters are therefore faced with a dilemma: spacing too far apart as to be out of hearing range decreases foraging returns for those who do not join pursuit groups, but spacing not far enough apart as to cause overlap in the area searched also decreases foraging returns. The perfect spacing needed to solve these problems is very difficult to maintain in thick jungle. This possibly explains the continual cycle of coming together and spreading out throughout the day.

Summary and Discussion

It should be emphasized that the results presented are based on observations during a single season of a single year for one hunter–gatherer group. As such, we should be somewhat cautious extrapolating the meaning of these results to the study of Aché subsistence patterns in general, more so to the study of hunter–gatherer economies, and especially to questions concerning human evolution. Nevertheless, with this cautionary note, we will explore what we consider to be some of the possible and likely implications of this study with respect to more general questions.

The primary purpose of our field work was to test hypotheses concerning the applicability of optimal foraging models to hunter–gatherer subsistence behavior. We are interested in the general patterns which may be expected to hold for all human groups and individuals faced with the same types of problems. Specifically, we are aiming to test models concerning foraging behavior of humans in order to allow us to predict what the foraging behavior must have been for groups and periods of time that are no longer observable. If the general principles of foraging behavior for humans (and other organisms) can be adduced, they will apply in all ecological circumstances and all time periods. Thus, with a knowledge about the specifics of the environment in question, we will be able to make accurate predictions about the behavior of interest, which ultimately can be tested archaeologically. This, of course, rests on a strong demonstration of the applicability of the model for currently observable behavior.

The Aché demonstrate that in the appropriate ecological setting humans are quite efficient predators and get very good returns hunting with quite simple technology. The tropical forests of this part of lowland South America seem to be well endowed with game that is easily killed by human hunters. This observation directly contradicts impressionistic statements by many earlier authors (e.g., Gross 1975; Harris 1974; Lathrap 1968; Meggers 1971), but is supported by a growing body of quantitative studies on hunting returns in lowland South America (e.g., Beckerman 1980; Gross *et al.* 1979; Hames 1979; Lizot 1978b; Vickers 1980). Hunting returns in lowland South America using traditional methods and technology range from a high of 1.35 kg/man-hour for a Yąnomamö village (Saffirio and Hames in press) to a low of 0.16 kg/man-hour for the Bari (Beckerman 1980). Aché bow hunters get returns of 0.53 kg/man-hour, and hunting with shotguns the Aché get 1.60 kg/man-hour. These numbers are strikingly similar to the returns reported by Hames (1979) of 0.56 kg/man-hour for Yąnomamö bow hunters, and 1.49 kg/man-hour for Ye'kwana men hunting with shotguns in the same ecozone. The large amount of meat that Mbuti Pygmies take daily hunting in the Ituri forest (Harako 1976, 1981; Hart 1979; Tanno 1976), the importance of meat in the Agta diet (Estokio-Griffin and Griffin 1980), and other measured hunting returns for several South American groups (e.g., Werner *et al.* 1979), would all suggest that the prevalent idea of tropical forests's being poor in game for human hunters (e.g., Hayden 1981; Lee and DeVore 1968) should be revised.

Comparison of Aché hunting to other hunter–gatherers shows a remarkably wide range of variation. There are two important differences between the Aché and the !Kung Bushmen. First, the Aché get higher returns per man-hour hunting, and receive a greater portion of their daily caloric intake from meat. Assuming that !Kung men hunt about 8 hours on an average hunting day, we calculate from Lee's data (1979:268) that !Kung hunters at Dobe got 0.33 kg meat/man-hour during July 1964. Because they hunted only 3 days a week, meat provided only 29% of the total caloric intake for the group. Wilmsen (1982) monitored hunting returns for an entire year at /xai/xai and found that only 12% of the calories in the diet for bush dependent !Kung came from meat. He estimates that hunting returns were only 0.2 kg/man-hour (personal communication).

The second important difference in hunting behavior between the Aché and the !Kung is the difference in the game species themselves. The !Kung primarily hunt large game animals, which single men stalk and shoot with poison arrows. The Aché on the other hand primarily hunt game smaller than 10 kg, and group hunting is frequent and important. These two groups therefore represent two extremes in hunting strategy due to differences in their ecological surroundings. Under the right circumstances it is likely that large game hunting would provide much higher returns than small game hunting, as the returns for the highest ranked items would be much higher. These differences point out that a hunter should not be expected to display similar behavior in all ecological circumstances. The particular aspects of the ecology of interest are of primary importance for any theory concerning probable foraging behavior, and any other behavior, including social organization which may be constrained by the subsistence pattern.

From our data we would tentatively conclude that humans, under the conditions we have described, do indeed forage optimally, for the most part. Although the adaptive importance of other conflicting behavioral possibilities may at times outweigh the importance of maximizing energy efficiency, we suggest that this parameter of fitness is likely to be important for all organisms in all environments.

The differences in time spent hunting between the Aché and other hunter gatherers is an interesting and perplexing problem. It is not clear to us why !Kung hunters for example should hunt only 3 days per week (Lee (1979) whereas the Aché hunt every day that weather permits and get better returns. This is especially difficult to understand in light of the fact that the !Kung are somewhat undernourished and show a greater need for increased calories (Truswell and Hansen 1976; Wilmsen 1979). Blurton-Jones (1978) has suggested that heat exhaustion may be a limiting factor for !Kung gathering, and this may also explain the need for !Kung men to remain in camp several days per week. There may of course be many other explanations for this difference; however, we suspect that more information is needed on what exactly !Kung men do with the time they are not hunting before a likely explanation will be found. Despite their high returns from hunting, our data for the Aché do not support the contention that hunters take the "zen road to affluence," working only a few hours a day (Sahlins 1972).

Two facts seem contradictory: that hunters hunt fewer hours the day after high returns, and that the hunters with highest average returns hunt more hours per day than those with low average returns. This is, however, primarily due to the difference in the measures used. The fact that hours hunted the day after high returns are compared to the hunter's own average is important, as a good hunter may hunt fewer hours than his own average the day after he gets high returns, and yet still hunt more hours than an average hunter. The trend for hunters with high returns to spend more hours hunting per day can be interpreted in several ways. Either the tendency for an individual hunter (for whatever reason) to hunt more hours per day has made him into a better hunter, the differences in hunting ability (for whatever reason) reinforces successful hunters to hunt more hours per day, or both are operating simultaneously. This could be tested by increasing artificially either of the two parameters. When returns were raised by allowing the use of shotguns, we observed no subsequent change in hours hunted per day over a 10-day hunting trip. This may, however, be too short a time period to produce a change in an individual's daily hunting pattern, and thus the mechanism of this correlation is still not understood.

The optimal diet model seems to predict fairly accurately when items will leave the diet (i.e., not be taken upon encounter). The analyses for capuchin monkeys can possibly account for the fact that the Aché consider these monkeys as a game item (the Aché have hunted only with bows until very recently), whereas neighboring Guaraní speaking Indians, Paraguayan and Brazilian peasants, who all hunt with shotguns, do not consider these monkeys as prey (or even edible in some cass). Although taboos based on symbolically mediated proximate mechanisms may be reported by informants as the explanation of species avoidance in hunting, it would be very useful to analyze the benefits of such beliefs carefully before concluding that they are arbitrary (Ross 1978).

There are, as we have mentioned, some apparent violations of optimality. Figures 5.3 and 5.9c show that the rank of some items can be adjusted considerably depending on the strategy used during their pursuit. Other things being equal, we expect that the most efficient strategy will be chosen. If single shotgun hunters took monkeys alone upon encounter, aiming only for large monkeys, they would still be in the optimal diet (see Table 5.7). Nevertheless, we did not observe this hunting strategy. It is possible that monkeys leave the diet of shotgun hunters for some other reason; however, we conclude tentatively that they are excluded because returns from an average encounter are lower than average foraging returns from shotgun hunting. Although much further testing is necessary, we calculate that both shotgun and bow hunters spend less than 5% of their foraging time pursuing items that should not be included in the set predicted by the optimal diet model.

Although no tapirs were killed this period, it is instructive to consider them in order to realize the effects of large game on hunting returns. Several tapir were shot during the study period, and we recorded 10 man-hours spent in pursuit of them without success. If one tapir were to have been killed during this period it would have been the number-1 ranked item in the optimal diet. The hunter who killed the tapir would then have an average hourly foraging return of more than double the next highest man during our study period, even if the hunter who killed the tapir were the man that we recorded with the lowest average returns throughout our study period. These changes that might have resulted in the hunting returns, due to one single event, point out both the impact of large game items on the picture we present (Hames [1979] observes this effect), and the weakness of data from such a short study period.

Precise measurements of Aché hunting returns should allow for more complete models concerning the economies of the pre-Columbian societies in South America. Subsistence returns for hunter–gatherers, even at the level of the Aché, who do quite well, do not compare to the caloric returns per man-hour from horticulture in South America. Although Aché bow hunters get 910 Cal/hour for their subsistence work, returns from horticulture reported in South America are much higher (Hames n.d.)

This leaves us with the difficult problem of explaining why the Aché continued as hunter–gatherers into the 1970s. We suggest that the answer to this question is likely to be found from consideration of factors other than subsistence behavior. Clastres (1968) proposed that continued warfare with neighboring Guaraní Indians drove the Aché to adopt their nomadic hunting and gathering life style.

The data presented on group size and hunting returns may explain why Aché bands rarely contain fewer than five hunters, but we cannot at this time demonstrate the relevant factors constraining maximum band size. It is also possible that precontact Aché bands moved camp less frequently than we observed during the study period, and we have some indication of the existence of large (greater than 100 people) semipermanent camps prior to outside contact. These questions will have to be further investigated before any satisfactory explanation is found.

Finally, from an evolutionary perspective, we can ask what are the implications of the individual variation in returns that we measured. It should once again be pointed out that this is a very short term study and thus the variance over a longer period more relevant to reproductive success may be quite different. Two things, however, should be considered. First, we have the impression that, over a longer time period, the position of some of the best and worst hunters may not change too much. A much longer period of observation of hunting before we began to measure returns allowed us to predict in advance who some of the best and worst hunters were going to be. This supports the contention that the results are probably representative for the extreme cases, even if there is considerable change in the hunting returns for men near the average. Second, because of the sharing pattern, which is not yet well studied, we suspect that actual differences in amount of game killed do not reach the level of individual or familial consumption. Thus, if any morphological or physiological bases of differences in hunting ability are being selected for, there must be a difference in the access to mates for hunters getting high and low returns. It seems likely that this comes from increased status due to hunting success relative to other men. Other alternatives are of course possible, but it is very difficult for us to accept the possibility that hunting ability in this society, and those which may be like it, plays no part in the evolutionary trajectory of the individuals who compose it. On the contrary, the importance of hunting success to the nutritional well-being of all the individuals in the group suggests that there should be quite heavy selective pressure to produce competent hunters, although the precise mechanism of that selection is currently obscure.

Short-term observations of variance in hunting probably consistent w/ longer term reality.

Any useful application of these data in order to address questions of larger scope requires the comparison of this case to other cases in which some of the variables have different values. We suspect that when comparable data for other hunter–gatherers, and nonhuman primates, are available, the comparative approach will make diet and general subsistence activities of our ancestors much more accessible than they can ever be using direct archaeological approaches alone. The patterns revealed *and* the questions raised by hypotheses drawn from optimal foraging theory are grounds for optimism.

Acknowledgments

We thank the Research Committee and the Biomedical Support Grant Committee of the University of Utah for supporting this research. We thank Colonel Oscar Centurion, Graciela Ocariz, and Miguelangelo Berni of INDI for their gracious assistance, and Ministro Marcial Samanego for his interest and attention. Fr. Alejandro Pytell and the staff of the Verbo Divino Mission, Chupa Pou, provided us generous hospitality. David Post gave us guidance and criticism. Eric Charnov and James O'Connell were stimulating, supportive, and full of good ideas.

Notes

[1]Before contact, in 1970, we estimate 700 ± 200 northern Aché in an area approximately 5000 km^2.

[2]During the study period, we recorded 736 man-days in the jungle, out of a possible of approximately 4800. We could not monitor all man days, as we spent considerable time on foraging trips. This means that this group of men spent a minimum of 15% of their time in the jungle foraging and we estimate the maximum for the group at about 25% of the time foraging during the study period. Several individual men spent more than 50% of the time in the jungle and the most active hunters were also most likely to be included in our sample.

[3]The seasonal *várzea* described by Hames (1980) is here divided into two ecozones in order to separate actual lagoons from a much larger area of floodland forest that is not always closely correlated with the presence of a lagoon.

[4]Not all animals are likely to be hunted, however, as some may be too small to be worth the bother. Young boys do hunt small prey (e.g., rats and birds) which are ignored by adult hunters. Various age-specific or situation-specific taboos exist but never apply to the entire population. Some food taboos elicited from informants have been observed to be broken.

[5]Two adult men did not hunt on single days because they were quite ill. Several of the men in the older age bracket also did no hunting on various occasions, due to illness.

[6]The angle chosen here was deliberately low in order to make the competition more difficult. Aché men frequently comment that their ability with the bow has gone steadily downhill as they spend more and more time at the mission. They are not able to predict on any given day who the likely winner of the competition will be, as they feel that any hunter could win if he is having an "on" day. Our impression is that some hunters do shoot consistently better, but we agree that the winner is unpredictable.

[7]Encounter rate may, however, alter the handling cost, but it is the change in handling cost, and not abundance, which will determine whether the item is included in the optimal diet.

[8]From our data, the average capuchin monkey taken weighs 2.5 kg and requires 2.4 hours pursuit to kill. Thus, (1.8 kg × 1300 cal)/(2.4 hours + 0.2 hours) = 900 cal/hour.

[9]We do recall several occasions in the camp late in the evening or early in the morning when adult hunters tried to shoot birds that were smaller than 0.4 kg; however, these attempts were accompanied by considerable laughter and joking, and we consider them to be play rather than foraging. The Aché by their own account do take many of the small birds in this area occasionally, although this seems to be primarily an activity of young boys between about 6 and 12 years old. We observed seven small birds killed by this age group.

CHAPTER 6

Shotguns, Blowguns, and Spears: The Analysis of Technological Efficiency

James A. Yost and Patricia M. Kelley

Introduction

One indication that a theory is well formed is that its concepts are defined clearly enough to permit testing. A poorly formed theoretical explanation defies quantification, often resulting in apparently unsolvable debates that eventually die out more because the adherents and the antagonists die off than because the issues debated have been solved or adequately explained. One of the attractive features of ecological models to anthropology has been that they can be stated in quantifiable terms and, hence, subjected to scientific testing.

In this vein, if Marvin Harris (1974, 1975) has been successful in nothing else, he has certainly been successful in prompting no small number of ethnologists to learn to count and calculate to test the hypothesis that a scarcity of protein was responsible for the warfare exhibited in areas like the Amazon Basin. Gross (1975:544) made the suggestion that "one of the purposes of writing and publishing this paper has been to stimulate field workers to measure such variables as dietary intake in the societies in which they conduct research. . . . The dialectic between theory and data has now brought us to the point where only more and better measurements will yield satisfactory results."

In response to such challenges, a number of quantitative studies have emerged in recent years (e.g., Chagnon and Hames 1979; Hames 1979, 1980; Hames and Vickers in press; Larrick *et al.* 1979; Lizot 1978b; Vickers

ISBN 0-12-321250-2

1979, 1980) laying a foundation for future comparative studies by supplying us with badly needed hard data and pointing out those areas where data are weak or lacking. By the mere fact of quantification, the protein debate in anthropology thus differs qualitatively from so many other debates in anthropology.

One of the issues that had to be addressed in this debate was that of the adequacy of traditional technology to provide a population with sufficient levels of protein. Chagnon and Hames (1979) demonstrated conclusively that the bow and arrow are quite adequate for the task, but to date no quantitative data have been published to demonstrate the efficiency of the blowgun and the spear in securing protein. From existing data, it cannot be determined with any degree of certainty if the blowgun and spear are equal to the task. One of the goals of this paper is to present data demonstrating the efficiency of the blowgun and spear.

At the same time it will be profitable to consider the effectiveness of the blowgun and spear as compared with modern technology—that is, the shotgun—in much the same manner that Hames (1979) has compared the efficiency of the bow and arrow to the shotgun. When Eric Ross stated "it is clear that shotguns have reduced the efficiency with which certain important animals—in particular, the terrestrial quadrupeds—can be killed" (1978:13), he sparked a reaction from several quarters. Beckerman, Good, Nietschmann and Vickers all reacted immediately in their responses to Ross's article in *Current Anthropology* (Volume 19, 1978). None of them supported his contention that traditional technology was more efficient than the shotgun. Their responses were brief and anecdotal, but in a later publication Hames responded with extensive data that "indicate that the shotgun is a far superior hunting implement to the bow" (1979:250).

The issue of the relative efficiencies of the traditional and modern technologies is not whimsical. It is important to the discussion of the relationship of the tropical forest ecosystem to various cultural features, such as warfare, infanticide, and settlement patterns in that the amount of influence environment has in the lives of men is directly related to the capacity of technology to exploit resources. If the modern technology has reduced the capability of Amazonian groups to obtain protein, which is the logical outcome of E. Ross's argument, then the relationship of the environment to cultural features is becoming more important than it was prior to the introduction of the technology.

On the other hand, if modern technology is more productive than traditional technology, then examination of environmental interactions with culture must consider what the potential for production might have been prior to the impact of modern technology. This will become increasingly difficult given the widespread tendency for Amazonian societies to adopt the newer technology.

Our fieldwork among the Waorani, a linguistic isolate in eastern Ecuador, had often raised conflicting questions in our minds as to the efficiency of their traditional technology compared to the shotguns they were beginning to accumulate in large numbers. On the one hand, the Waorani frequently make statements that support E. Ross's contention (1978:12) that a family of toucans or band of monkeys can be killed in large numbers with a blowgun before the remaining animals try to flee, but a shotgun will send the majority out of range immediately after the first bang. In addition, the Waorani also complain about the number of animals they miss because of misfires due to wet powder or malfunctioning shotguns.

On the other hand, the majority of field researchers in Amazonia have been of the opinion that shotguns are more productive (e.g., Hames 1979). Indeed, the almost ubiquitous replacement of traditional hunting implements by shotguns in Amazonia would seem to speak in strong support of the shotgun's superiority. It would be highly unusual, according to materialist explanations, for so many people in so many societies to abandon an efficient technology in favor of a less efficient one. Is there, then, any basis to the Waorani statement that blowguns are better for arboreal game, and, if so, why are the Waorani taking up the use of the shotgun when they depend so heavily upon arboreal game for protein?

The role of mentalistic factors influencing the acceptance of shotguns could not be rejected without examination. On one occasion, a Wao man complained that while hunting monkeys, many had gotten away because of the shotgun's noise. I asked him if that were so why he continued to use a shotgun for monkeys. His answer would have delighted any anthropologist seeking mentalistic explanations of behavior: "It makes such a beautiful noise!" Then he launched into a discussion of the aesthetic qualities of the shotgun—shiny, smooth metal, mechanisms that click, the polished wood stock, and that beautiful sense of power in the kick and boom—so unlike anything in the indigenous culture. Is there more to the mentalist explanation than the materialist cares to grant? Have the aesthetic qualities played a crucial role here at the expense of the need for protein?

To provide answers more substantial than generalized opinions, we began collecting data in 1977 to determine Wao protein intake and to test the hypothesis that the shotgun is a more efficient implement in protein production than the blowgun and the spear. Since technology does not operate in a vacuum independent of the user, a detailed analysis of Wao hunting behavior patterns was essential before a complete understanding of the efficiency of the various technologies could be assured. This chapter will present the results of our research. By outlining Wao hunting patterns and harvest results, we can not only demonstrate the comparative efficiencies of the traditional and modern technologies, but also suggest what it is that creates the differences in the hands of Wao users.

Relevant Environmental and Cultural Features

The homeland of the Waorani is an expanse of tropical moist and tropical wet forest (Cañadas 1977; Sanchez 1976:25) embracing the first parallel south in eastern Ecuador, a territory defined by the Napo River to the north and the Curaray and Villano rivers to the south. It falls between approximately 76 and 77°30′ west longitude. To the west, the average elevation of the floodplain is 425 m; on the eastern edge of the territory the elevation is 245 m. The terrain is highly three-dimensional and the soils are poor when compared, for example, with the territory of the Siona–Secoya to the northeast. Sourdat and Custode (1977) note that the soils covering most of the Wao area are well drained but are low in fertility. The tops of the small *mesetas* or ridges in the upper reaches of the drainage have fair soils of volcanic origin, but most of the territory is covered with ancient red clays. The ridges paralleling the streams and rivers rise an average of 185 m above the floodplain, giving the area a washboard effect when viewed from an airplane. Hunting across drainage areas is, therefore, demanding work, but paralleling the streams is easier once the top of a ridge is attained and followed.

In the upper reaches of the tributaries, the area most inhabited, the drainage basins average less than 1 km in width, but to the east the basins frequently widen to more than 3 km. The small feeder streams near which the Waorani usually locate are clear water streams, but several of the larger rivers, such as the Curaray, the Cononaco, the Tivacuno and the Tiputini are white water rivers.

The most recent census completed by Yost indicates that as of August 1980 there were 658 Waorani. At the time of European contact in 1958 there were about 500 people with access to a territory encompassing approximately 20,000 km^2 (0.025 persons/km^2), but the recent construction of roads into Wao territory has brought colonists into the area, thereby reducing the territory the Waorani had access to by approximately 15%. The resulting density (0.039 persons/km^2) is somewhat deceptive because vast areas of the territory are utilized only for infrequent expedition hunts; the majority of the population live in small villages of fewer than 100 people scattered over an area encompassing about 2200 km^2.

Culturally, the Waorani have traditionally been a people oriented to the interfluve. They lacked canoes, tabooed most of the larger fish and game from the waterways, incorporated rivers into their cosmology of things to be feared and, in general, avoided the larger rivers whenever possible. They built their houses on the tops of hills or ridges and planted their gardens on the slopes of hills away from rivers. Only in the past decade have they begun to exploit in a serious way the variety of niches available to them along the narrow floodplains.

Except for January through April when peach palm (*Bactris gasipaes*) is available in great quantities, the staples for the Waorani are manioc and plantains (*Musa* spp.) most of the year. Most of the kilocalories in the diet come from agricultural pursuits, but the quality protein in the diet comes primarily from hunting. In the past, fish figured into the diet only in a minor way, but in very recent years fish have become an increasingly important source of protein. As the hunting yields given in this paper will demonstrate, hunting in itself still produces more than adequate amounts of protein for the population; apparently, factors other than a shortage of protein are stimulating an elevation in the importance of fish in the diet.

The social and political features of Wao culture are typical of much of upper Amazonia. The Waorani are extremely egalitarian and individualistic, with concepts of rank, power, and status being foreign to them. The flexibility of local kindred organization and the continual conscious recognition of the economic importance of kin prove to be superb adaptations in the light of Wao hunting behavior. Similarly, their individualistic behavior proves to be well adapted to the style of protein procurement they have evolved. These generalizations will be explained as the discussion proceeds.

Warfare has been one of the hallmarks of Amazonian society, and the Waorani present no exception. Both internally and externally they have maintained a reputation for killing. Chagnon (1974) indicates that warfare-associated male mortality among the Yąnomamö ranges between 20 and 30%, and J. Ross (1980) reports a higher rate for the Jívaro. The data available on the Waorani are comparable to that of the Jívaro; data covering up to five generations indicate that 42% of the Waorani deaths for that period were attributable to internal spearing raids and 16% were attributable to conflicts with the outside world (Larrick *et al.* 1979:167). Even though warfare has essentially ceased in the past few years, the Waorani continue to talk of it incessantly. It is a rare day indeed if past raids, old enemies, or present threats do not dominate discussions while men jog through the forest. The way a stream takes a sudden turn, the weather on a given day, or a particular event seem to function as perpetual mnemonics of the old patterns of warfare.

Hunting Behavior and Technology

It is obvious that the interplay between the physical aspects of any technological item and the application of it in the hands of a user are what determine any tool's effectiveness. The fact that a projectile has a given velocity and penetration potential is important, but this alone does not determine whether that projectile kills animals. Numerous behavioral variables come into play in making the projectile effective or ineffective for harvesting protein. In this section we will outline both the technical capabilities of the three weapons under consideration and the behavioral patterns that surround their use.

THE BLOWGUN AND ITS USE

No small amount of energy and emphasis is given to the hunter in Wao culture, and with good reason, considering his role in survival. Adult hunters take seriously their roles in teaching small boys to develop their arms, lungs, cheeks, eyesight, hearing and endurance. Of the few rituals surrounding young boys, almost all are concerned with producing effective hunters. By the time a boy is 5 years of age he should be proficient enough with the blowgun to hit consistently targets of fruit or leaves set up by his older male kinsmen. By the time he reaches puberty, he will have an extensive knowledge of the calls, nesting habits, breeding behavior, feeding cycles, and habitats of a large number of species, and will know which animals to expect around a given species of tree, both when it is flowering and when it is fruiting.

Undoubtedly, the Wao hunter depends upon hearing more than upon any other sense to locate potential game. He learns to distinguish among the animals in the canopy by the sound they make as they move: the frequency of the movements, the loudness of the rustling leaves, the distance between movements, and the kind of tree the animal is in are all clues to the species. It is not unusual for hunters to know what kind of animal is present long before they see it or hear it call. The dense growth of the canopy often obscures animals from view, but knowledge of what species is present, combined with an understanding of that species' behavior, can make it possible for a hunter to predict an animal's next move without ever seeing the animal.

Obviously, the better the hunter understands the species' behavior, the greater his chances of success in the hunt. The Waorani are noted for the large variety of birds and mammals they keep as pets, and this is not without consequences for the Wao hunter. From the time a boy is an infant, he is around fledgling birds and infant mammals of numerous species. He hears them cry for food all day long, he watches their behavior, and he learns to distinguish their odors. All of these bombard his senses so that on a hunt he knows whether to pursue a faint squeak high in a tree or to ignore it as coming from a nongame species too small to bother with. He may even know by the sound or smell of urine dripping on leaves 5–10 m

Figure 6.1 *Gïcawæ shoulders his heavy, 3-m-long blowgun. Used to hunt the numerous arboreal species, the blowgun provides high levels of protein for the Waorani. (Photo by J. Yost.)*

away whether the animal is worth an effort. On more than one occasion, we have observed hunters making the decision to pursue or not pursue animals in the canopy on the basis of the urine falling from the canopy.

In the process of learning these things, the young hunter kills a large number of small birds and mammals. These kills have not been included in our data here, but in considering the total protein availability they should not be ignored. They are small, often including blowgunned hummingbirds or birds in the 20-gram weight-range, but their cumulative value is greater than that of insects in the diet.

Although a youth learns the use of both the blowgun and the lance at an early age, he does not become truly effective with either until his later teens. Effective blowgunning demands the development of precision through the coordination of respiratory muscle control and "intuitive" aiming. The typical blowgun has a bore 12 mm in diameter and 2.75 m long, giving a volume of 311 cc (Figure 6.1), a volume of less than 7% of the total exhalable lung capacity of an adult male, based on averages for western man (Holvey 1972:496). Obviously, the production of a constant and forceful pressure throughout the dart's travel down the bore is more a matter of controlled use of the accessory breathing muscles than of the exhalation of a large volume of air.

Aiming the blowgun is a matter of learned coordination between the strength of the impulsion behind the dart, the distance to the target, and the vertical angle of the target. Shots directly overhead require little or no correction for trajectory (Figure 6.2), but as a target comes closer to horizontal it requires increasing correction for trajectory. Generally speaking, a hunter blows each shot with all he is capable of (Figure 6.3), thereby reducing as much as is possible the number of variables that have to be taken into account in a given shot. However, on extremely long shots high in the canopy the hunter will put a little extra lung effort and body english into the shot.

The blowgun itself is made of any of several hard palms, with one known locally as *tepa* (unidentified) being preferred. The construction is composite: two matched lengths are grooved lengthwise, sealed with beeswax along the mating surface to prevent air leakage, and wrapped with the bark of a liana. Since the heavier the blowgun the less susceptible it is to movement from nervousness or heavy breathing while aiming, the Waorani tend to prefer blowguns that weigh from 2.5 to 3.2 kg. Longer blowguns are easier to aim, so the average length is 2.75 m for serious hunting. The relationship of length to power is complex: the longer the dart travels through the bore, the more velocity the hunter can put behind it, but beyond a certain point, governed mostly by his muscle control and strength, the friction between the dart and the bore diminishes the effectiveness of the hunter's effort. As long as the hunter can continue to increase the velocity of the dart as it moves through the bore it is to his

Figure 6.2 *Most game is blowgunned at a near-vertical angle. As shots approach the horizontal, range and accuracy decrease significantly. (Photo by J. Yost.)*

advantage, but at that point where he is simply maintaining the velocity or begins to lose velocity, additional length is detrimental to the weapon's efficiency.

The darts, which average 2.5 grams, are shaved from the cambium of the petiole of any of several species of palm leaves and range in length from 30

Figure 6.3 *With cheeks bulging and muscles tensed, Töbö concentrates all his energies into a shot. (Photo by J. Yost.)*

to 45 cm, with 39 cm being average. The diameter is normally about 2–3 mm. Just prior to shooting, the hunter spins enough kapok (*Ceiba* sp.) around the dart to form a tight airseal in the bore of the blowgun; he positions it on the dart at a point from 20 to 25% forward of the butt end to serve as a balance and a symmetrical airfoil while the dart is in flight (Figure 6.4). Although the surface area of the kapok is large enough to present considerable wind resistance, its tapered shape combines with the needle shape of the dart to present a streamlined projectile that is quite effective.

In defining effective range as "the greatest distance at which one will risk taking a shot," Hames (1979:227) determines the maximum effective range of the Ye'kwana blowgun to be 17 m. He points out that "all that is required of the needle-sharp dart is that it pierce the animal's flesh" for it to be effective (p. 226). Our experience with Waorani hunters suggests that this distance may not be representative of the ability of Amazonian hunters who rely heavily upon the blowgun. Instead, our data seem to be more consistent with the estimates given for other groups in Amazonia as re-

Figure 6.4 *A hunter twirls kapok onto a palmwood dart to form an airfoil and to function as a pressure seal for the dart in the blowgun bore. A hunter may carry as many as 300 darts in the quiver hanging from his neck, and sufficient kapok in the attached gourd to fletch all of those darts. (Photo by J. Yost.)*

ported in Hames' discussion (p. 227). For example, Gaba, an adult male hunter, was observed to take aim at a squirrel (*Sciurus igniventris*) from 12 m. When he shot, the poisoned dart not only penetrated the squirrel, but continued through the thorax and disappeared in the forest beyond the squirrel, leaving the kapok embedded behind in the wound. The squirrel died immediately. Furthermore, the Waorani do not hesitate to take shots at monkeys or birds in the canopy. Rarely have the authors seen woolly monkeys (*Lagothrix lagotricha*) or howler monkeys (*Alouatta seniculus*) within 17 m of hunters, yet these two species are killed in great numbers by Waorani. We have observed hunters shooting and killing monkeys from 30 m on numerous occasions.

In tests of the blowgun's range, we had men known to be proficient blowgun hunters demonstrate their capabilities, and then compared the results with those of younger teenage hunters who are not as proficient. Using an optical distance finder, we measured the distance at which the hunter could hit a monkey-sized target in the canopy two out of three shots. The acknowledged proficient hunters' average distance was 39 m; a couple of the better hunters were able to hit a target consistently at 41 m. The younger hunters had difficulty shooting accurately beyond 31 m, but were able to shoot quite accurately and with considerable force at 24 m. In all cases, part of the requirements was that the dart strike with enough force to remain in the target, which was usually a distinctive mark on a tree branch or a termite nest. It should be noted that the targets were overhead to the hunter, the normal position of game hunted with a blowgun. In tests for horizontal range, the accuracy of the experienced hunters fell off greatly when the target was moved beyond 25 m.

Although Hames used willingness to take a risk as the definition of effective range, we are unable to use this criterion in determining effective range of Wao blowguns. For shotguns or bows, this is a valid criterion in that shells are expensive and arrows in limited supply on a given hunt, so hunters tend to conserve them. However, the Waorani make no effort at all to conserve poison darts. Because when they go out on a day hunt they will typically carry over 100 darts in the quiver, and possibly as many as 300, they are willing to risk shots at almost anything, including hummingbirds 10 m away. As long as meat is carried home, it is of little significance that killing a woolly monkey may have required a dozen darts.

Of course, the critical killing power of the dart lies in the poison. For very small birds and mammals up to 80 grams or so, darts with no poison are effective, but for anything larger the poison is essential. The Waorani manufacture their own poison and have no memories of ever having traded with other groups either to obtain or to give poison. The only ingredient in the poison is the bark of the large, flat vine *Curarea tecunarum*. The watery resins are compressed from the bark, heated until a surface scum forms and then spread onto the tip of the darts as a thick, jet-black lacquer that hardens like tar. The process is not extremely time-consuming; the poison for several hundred darts can be prepared and applied in less than 2 hours, excluding the time it takes to find the vine in the forest and carry it home.[1] Once the darts are prepared, they will maintain their potency for several months if they are kept dry and reasonably cool.

Because a hunter expects to have plenty of time to shoot before game flees, he does not spin kapok around a dart until he encounters an animal. At that time he prepares the dart with the kapok and notches the tip just behind the poison so that a wounded animal will break the tip off rather than extract the dart, thereby leaving the poison in place to work (Figure 6.5). Although there are many variables affecting the time it takes for a dart to kill an animal, an adult woolly monkey will usually fall from the trees in 15 to 20 minutes.[2] Larger animals require more time and more darts; on one occasion it took over half an hour for 11 darts to bring a puma (*Felis concolor*) crashing from a tree.

Figure 6.5 *Notching the fletched dart with a pirana tooth just behind the poisoned tip ensures that, should the animal grab or swat the dart, the poisoned tip will break off, remaining in the animal to take effect. (Photo by J. Yost.)*

THE SPEAR AND ITS USE

Whereas the blowgun is used primarily for arboreal game, the spear is used almost exclusively for large terrestrial animals. Traditionally the tapir (*Tapirus terrestris*), collared peccary (*Tayassu tajacu*) and brocket deer (*Mazama americana*) were regarded as taboo, but today all the Waorani we have interviewed hunt them along with the other terrestrial species. The most common species hunted with a spear include the white-lipped peccary (*Tayassu pecari*) and the collared peccary. The size, speed, and strength of these animals make hunting them risky, particularly when the only weapon is a triangular-shaped palmwood (*Bactris gasipaes*) spear, 2.5 m long and weighing 1.5 kg.

Each man usually keeps a supply of half a dozen razor-edged, notched, double-ended hunting spears stored in the thatch of his house for quick access should someone report a herd of peccary in the area. When the alarm is sounded, he grabs as many spears as are available and heads for the area. Of all of the weapons that the Waorani use, the spear is probably the most demanding in terms of strength and endurance. To be effective, the hunter needs to get within a meter or two of the intended animal and stay there long enough to put a spear into it. That is no small challenge. The common strategy is to ram the spear into the animal, hang onto that spear and drive one or more others home before the animal can work free to escape or attack. When that animal's fate is certain, the hunter may pursue other animals, if any remain in the area, and if he has any unbroken spears remaining. Although a hunter prefers to get close enough to hang onto his spear, he will throw it if necessary, but seldom more than 2 or 3 meters.

THE SHOTGUN AND ITS USE

Hames (1979) has covered the technological aspects of the shotgun quite well, and we will not repeat that here except to note differences in the Wao use from those he describes. The Waorani obtained shotguns only recently; by January 1974 there were no more than 3 shotguns among all of the adult males. By the end of that year over 20 men had obtained shotguns, most of them by making a trip to the capital city of Quito where they requested them from government agencies. Today, there are 66 shotguns distributed among 194 hunting males. Most of the men now obtain them by walking 2 or more days to nearby towns, such as Tena, where they sell artisanry until they have enough money to buy a shotgun. Normally, 15–20 tourist-style blowguns and spears, representing about 2 months of intermittent labor, are sufficient.

Nearly 90% of the shotguns are locally manufactured 16-gauge, single-shot breechloaders. It is not at all unusual for a new one to have a firing spring so weak that it will not detonate the primer without repeated attempts. The man who buys one that lasts more than a year without repairs has made a lucky purchase; the man who gets rid of his blowgun when he buys a shotgun is a prime target for ribbing from his friends and goading from his wife.

The ammunition for shotguns is even more problematic than the weapon. It is difficult to obtain, must be guarded from moisture, and is erratic in performance. A box of 25 cartridges currently sells for the equivalent of three tourist blowguns, representing a little more than a week of intermittent labor and a minimum of 4 days of travel to make the purchase. The preferred ammunition is imported from the United States, but it is not always available. The velocity varies from a rating of 1155 fps on locally produced shells to over 1300 fps on imported shells. For game weighing less than 1 kg, hunters prefer shot 2–3 mm in diameter, but for larger species they prefer shot of 4–5 mm diameter. The Waorani also reload their own shells with primers from the United States, powder imported from England, shot produced locally, and wadding made from shredded bamboo (*Gaudua angustifolia*). These reloads produce inconsistent results: the powder and primer frequently soak up too much humidity and burn slowly and incompletely if at all; and the shot, because it is formed in molds rather than formed by dropping in a shot tower, often includes many figure-eight-shaped pellets. Consequently, patterns from a blast with these pellets are quite erratic.

In spite of the erratic performance, hunters seldom miss when the shells do fire. Hames (1979:226ff.) has pointed out most of the reasons the shotgun is so effective, including pattern size, shocking power, and range. Several comments need to be added here. First, most of the Waorani who hunt with shotguns learned to hunt with blowguns, a weapon that requires much greater skill in aiming in order to be accurate. They find the shotgun a very easy weapon to aim because of the huge pattern it makes. On the other hand, while the penetration power, that is, the high velocity, of the projectile is critical to the effectiveness of the shotgun and bow, this is much less so for a poisoned dart. It takes very little velocity to get a needle-sharp dart deep enough to kill, whereas a great deal more velocity is required to get a 5 mm diameter sphere to penetrate deep enough to kill. In addition, a shotgun pellet has a much smaller portion of the target that it must strike to result in a kill; a poison dart can be effective no matter where it strikes.

Waorani who use the shotgun to hunt terrestrial animals have not yet begun to rely heavily upon the extended range the shotgun could give them over the spear. They still try to get within 3 to 5 m of terrestrial animals before pulling the trigger, giving them sudden, dramatic kills. The shotgun does give them a higher kill rate than the spear because of its shocking power at such close range, but in the sense that the Waorani do not often try to extend the range, they are not maximizing the technology to its fullest. (They do, however, take advantage of the shotgun's greater shocking power in a herd of peccary by quickly leaving a fallen animal and immediately pursuing another.) Only in the case of deer, an animal that was taboo until very recently, do they truly rely upon the shotgun's range to accomplish a kill they would not otherwise have attempted.

In a similar vein, traditional hunting behavior patterns continue to influence their use of the shotgun in seeking arboreal prey. A group of Wao hunters was shocked to see one of the authors take a shot at a flock of geese passing low over the river early one morning. He missed, and they were confirmed in their opinions that volant game must be followed until it is sitting still. Again, they are not truly maximizing the shotgun's potential.

One other comment needs to be made regarding behavior in hunting with shotguns. One of the authors has hunted with several other Amazonian groups and observed that, when killed with a shotgun, some monkeys grip a branch with their tails and fail to fall out of a tree, and the hunters may abandon the hanging carcass if the tree is too large to chop down quickly. In circumstances like this, the blowgun would appear to be more efficient than the shotgun in that the poison is a muscle relaxant; as a monkey dies, the poison relaxes the animal's grip and it falls. The Waorani excel at climbing trees, and when a monkey gets suspended in a giant buttressed tree, they climb the lianas hanging from the tree and retrieve the game. Also, it is not uncommon for howler monkeys or spider monkeys (*Ateles belzebuth*) to hole up in a giant tree by climbing into a crotch protected from view below and to sit for hours if need be, waiting for hunters to leave the area. Frequently the Waorani will climb the tree or a nearby one with a blowgun and shoot the animals as they crouch motionless.

HUNTING PATTERNS

The Waorani distinguish between day hunts and expedition, or "sleeping," hunts. For the most part, they do not practice night hunting, primarily due to a lack of the technology to do so. On occasion a man will go into a garden at night with a flashlight to try to kill an agouti (*Dasyprocta fulignosa*) that has been ravaging his manioc, but that is a rare and very recent practice. On other occasions a man may leave an hour or so before dawn to stalk a calling nocturnal curassow (*Nothocrax urumutum*) so he can shoot it as soon as there is light enough to see it, but before it stops calling and moves into hiding for the day. Beyond these two instances, night hunting activity is almost unheard of.

Because the Waorani live very near the equator, they have approximately 13 hours of light adequate for hunting every day of the year—adjusted, of course, for cloud conditions. The normal pattern is to leave at dawn during the dry season, or after sunrise during the wet season, and to hunt until just enough time is left in the day to get home just after dark, or until the hunter judges he has killed sufficient game. "Sufficient" is usually defined in terms of the weight a hunter is willing to carry from his stopping point to the village. When a large number of peccary or a tapir is killed, the men will frequently carry what they can to the village and send the women back to retrieve the remainder the following day.

Unless a hunter is tracking a large terrestrial animal, he will almost always hunt by trail, and Wao trails almost always follow the tops of ridges paralleling the streams and rivers. Although trails crosscutting the ridges and rivers would pass through more biotopes and yield more hunting time in the transitional ecotones where, presumably, more game abounds (cf. Hames 1980:40), the extra effort necessary to cross rivers, navigate swamps, and ascend and descend hills discourages Waorani hunters from making those kinds of trails. If adequate game can be killed along the tops of the ridges, the hunters are going to be more efficient than if they have to expend the additional energy crossing 185-meter ridges all day long. Also, a hunter who must carry 20 kg of meat for several hours while returning home would expend considerable additional energy crossing ridges. Since most of the rivers in the area under study are too swift and shallow to allow good hunting by canoe, the Waorani seldom hunt from canoes. Nor do they parallel the rivers on the terraces immediately above the rivers. The terraces usually end abruptly in escarpments or excised oxbows that make them undesirable for trails.

Of the 867 hunts recorded in this survey, 547 (or 63%) were hunts by single individuals. Twenty-seven percent involved two hunters, and only 10% could be classified as group hunts with three or more hunters. Those hunts that do involve groups of hunters are usually responses to the sighting of a herd of peccary and are of a completely different nature from the normal day hunts. Instead of a single individual going off as quietly as possible in search of whatever might present itself, most of the able men and women from a village dash off in groups of three or four to the area where the herd was seen. Very frequently, the women will become the trackers for the subgroups until the herd is overtaken; then they hang back until they have carcasses to retrieve. Unless the distance to the village is in excess of 4 or 5 hours the animals are not gutted before they are carried home. Because very little of the animal is discarded (usually only 15–20 cm of the large intestine) the advantage gained in eliminating some weight by gutting is outweighed by the clumsiness of carrying a butchered carcass.

A note is in order regarding the use of dogs on hunts. Prior to peaceful contact in the late 1950s, none of the Waorani had dogs, and only within the past decade have they had them in any number. Some of the more isolated groups still do not have them. Those who use them in hunting claim that they are most effective in hunting the collared peccary, tapir, and deer. The large herds of white-lipped peccary are relatively easy to track through the forest because of the heavy disturbance they leave on the forest floor, so dogs do not give any great advantage in hunting them. The tapir, deer, and collared peccary, however, are much more difficult to track and get close enough to for a kill. Dogs give a big advantage in hunting them, particularly because a dog will pursue these animals and frequently force them to take a stance where they are vulnerable to hunters.

Many Waorani have told the authors that they never used to eat the tabooed collared peccary, deer, or tapir because they were so difficult to corner and kill rather than because of any disgust surrounding a tabooed animal. Some have confessed to eating portions of collared peccary and tapir, for example, long before the taboo on it was lifted by the majority. Tewae's response to the question "Did you eat collared peccary before you came into contact with the outside world?'' is typical: ''How in the world would I have ever eaten collared peccary! We never had any dogs!'' Dogs do increase the efficiency of the hunt; the fact that they are used to bring some of the largest protein packages in the forest into the range of the spear is highly significant. Three of the top 11 sources of protein (according to weight) are hunted by dogs and were either ignored previously or obtained only rarely. In our sample, collared peccary, tapir, and deer account for 23% of the total protein production.

Data Collection

Because most Waorani hunters continue to use both the traditional technology and the new technology, they are an ideal population in which to test the hypothesis that shotguns are more efficient than blowguns or spears. For our sample, we initially chose the villages of Tzapino, Tewaeno, Quiwado, Wamono, Dowetome, Dayono and Toñaempade. We eliminated the village of Dayono from consideration because the people there have become more acculturated, using outside food resources more than the other villages, and because we have had little contact with them and have less understanding of their hunting patterns. Eventually we eliminated Dowetome, a village of 58 people, from most of our analysis because our data on them were too thin; in more than 5 weeks of residence among them we have collected fewer than two dozen man-day-hunts, hardly an adequate representation of their hunting productivity. (When outsiders are present among them, their normal activities come almost to a standstill as they become preoccupied with observing the observers.) We also eliminated Toñaempade for reasons discussed below.

The remaining villages are all in the upper reaches of tributaries to the Curaray River. Tewaeno is the largest village, with a population of 73; Tzapino has 68; Quiwado, 65; and Wamono, 24. These numbers must be taken advisably because people are continually shifting from location to location in a cycle of rotation coordinated with manioc production (Yost 1981:102).

Each of the authors has lived among the Waorani for more than 5 years in their functions as part of the Summer Institute of Linguistics. Kelley first began her work as a literacy consultant among the Waorani in 1972, and Yost began his work as an anthropologist there in 1974. To gather the data for this paper Yost has accompanied Waorani on numerous hunts in all parts of the territory, and both authors have interviewed Waorani about hunting and have listened to their hunting stories for hours *ad infinitum*. In 1977 we began collecting data of the nature and quality needed for the comparison between the shotgun and traditional technology. Our initial approach was to interview the Waorani in the villages where we lived (Tewaeno and Tzapino) at the end of each day to determine who had gone hunting, who had accompanied them, where they went and for how long, what weapons they had used, and what their success had been. Yost made rounds daily to each of the houses within a one-half hour radius of the central village location to get this same information. In 1979 we devised a questionnaire written in the Wao language and trained at least two people from each of the sample villages (excluding Dowetome) to fill in the forms daily. We made checks on the trainees to ensure that they were reporting accurately by cross referencing their own data with ours, by cross referencing the data from two individuals from the same village, and by studying the forms for irregularities. On the basis of these checks we eliminated one village, Toñaempade, from our sample early in the study. It was obvious we could not have enough confidence in the reliability of the data collected there.

The use of the forms had a number of advantages. First, we were able to get data on days we were not physically present in a village. That meant we had the ability to cover a number of villages at once as well as the chance to get data when we had to leave for the city. In addition, once the trainees understood what we wanted, they were able to get more complete data than we could get ourselves. The trainees took a real interest in the project and listed the small species, particularly birds, about which people usually neglected to inform us.

To calculate the weight of the animals reported on the written forms, we multiplied known values of the average weight of each of the species by the number of animals reported on the forms.[3] The forms had a category indicating the size of the animal in the broad terms in which the Waorani classify them. Because our own measurements have shown that when they classify an animal as "young" it averages 60% of the weight of an adult, we multiplied any animal in the "young" category by 0.6 of the assigned adult weight.

In analyzing the 867 day hunts on which we have data, those entries that lacked information appropriate to the particular issue being examined were eliminated for that issue. For example, any entry that did not have both the time of departure and time of arrival recorded was eliminated from any analysis that involved efficiency, but for some cases where efficiency was not an issue, those data were still of value. Similarly, any hunter whose total number of recorded hunts was fewer than five was excluded from any analysis that pertained to individual hunters. The data on the hunts adequately covered every month of the year except March.

To estimate the gross annual protein production, we utilized data collected by Yost in 1974–1975 and again in 1979. These data, which sample the activities of adults on a daily basis, were collected by visiting each

Table 6.1

Rank Order of Species by Total Weight[a]

Rank	Common name	Species name	Total weight (kg)	Percentage of gross weight
1	White-lipped peccary	*Tayassu pecari*	4940.1	26
2	Wooly monkey	*Lagothrix lagotricha*	3873.5	21
3	Collared peccary	*Tayassu tajacu*	2740.6	15
4	Howler monkey	*Alouatta seniculus*	2197.4	12
5	Tapir	*Tapirus terrestris*	1314.4	7
6	Salvin's curassow	*Mitu salvini*	610.9	3
7	Cuvier's toucan	*Ramphastos cuvieri*	414.4	2
8	Capuchin monkey	*Cebus albifrons*	400.8	2
9	Spix's guan	*Penelope jacquacu*	301.7	2
10	Blue-throated piping guan	*Pipile pipile*	261.8	1
11	Brocket deer	*Mazama americana*	253.4	1
12	Capybara	*Hydrochoerus hydrochaeris*	226.5	1
13	Paca	*Agouti paca*	194.8	1
14	Coati	*Nasua nasua*	130.6	1
15	Great tinamou	*Tinamus major*	116.5	
16	Spider monkey	*Ateles belzebuth*	110.1	
17	Saki	*Pithecia monachus*	100.0	
18	Ivory-billed aracari	*Pteroglossus flavirostris*	86.1	
19	Red squirrel	*Sciurus igniventris*	82.9	
20	Gray-winged trumpeter	*Psophia crepitans*	66.5	
21	Agouti	*Dasyprocta fulignosa*	47.5	
22	Golden-collared toucanet	*Selenidera reinwardtii*	30.4	
23	Tayra	*Tayra barbara*	27.3	
24	Squirrel monkey	*Saimiri sciurea*	23.7	
25	Speckled chachalaca	*Ortalis guttata*	21.6	
26	Nocturnal curassow	*Nothocrax urumutum*	20.4	5 (ranks 15–36 combined)
27	Cayman	*Caiman sclerops*	20.0	
28	Mealy parrot	*Amazona farinosa*	19.6	
29	Red-mantled tamarin	*Saguinus fascicollis*	18.7	
30	Scarlet macaw	*Ara macao*	18.2	
31	Douracouli	*Aotus tirvirgatus*	17.6	
32	Titi	*Callicebus moloch*	17.2	
33	Squirrel	*Sciurus cocalis*	15.4	
34	Acushi	*Myoprocta pratti*	11.1	
35	Yellow-ridged toucan	*Ramphastos culminatus*	11.0	
36		Miscellaneous small birds (numerous species)	29.3	
Gross weight			18,781.1	100

[a]Includes only those species whose total weight was more than 10kg.

household near Tewaeno every other day for the months of August and December of 1974 and January and March of 1975 to interview the members regarding their activities the preceding day and the day of the interview. These data were then compared for consistency with the data collected in January, March, April, July, September, and December of 1979. The data were further validated by comparison with the number of hunting days a single adult male had in an 11-month period from May of 1980 through April of 1981. The results of all three sets of data were highly consistent with one another. Table 6.1 presents the rank ordering of the species in the kills by weight.

The Kill and Its Meaning

A brief survey of the game species the Waorani most commonly harvest reveals that they continue to rely heavily upon the blowgun in their hunting (Table 6.2). Of the 3165 animals recorded killed, 64% were killed with the blowgun. Only 3% were killed with spears, but 33% were taken with shotguns. Comparison of Table 6.2 with Table 6.3, however, demonstrates that the spear and the shotgun are more properly evaluated in terms of the total amounts of protein they produce than in terms of the number of animals killed by each. The total production of all 867 recorded hunts was 18,781 kg, of which 36% were taken with blowguns, 51% with shotguns, and 13% with spears. The reason for the difference is that blowguns are used almost exclusively for arboreal animals under 10 kg, the spear is used for the large terrestrial mammals, and the shotgun is used for both classes.

The data published by Hames (1979:234) show that bow hunters kill relatively few of the smaller animal species under 1.5 kg. Comparing the data in Table 6.2 with his, it is obvious that the Waorani use the blowgun to kill small species far more than the Yąnomamö use the bow to kill small species. Toucans, aracaris and squirrels alone account for 33% of the animals killed with the blowgun. Examination of Table 6.2 also reveals that the Waorani do not use the shotgun extensively to kill these small species. A cost–benefit analysis suggests that they do not use it for these because the yield is too low for the input required to obtain shells. A box of shells costs approximately 30 hours of labor to make trade goods, or 1.2 hours of labor per shell. In addition, the average trip to buy the shells takes 44 hours of very fast walking in rugged terrain. Usually, only one box of shells, and very occasionally two, are purchased in a single trip due to practical considerations such as the amount of trade goods that can be carried and the desire to purchase other goods with the money gained in trade. If a single box is bought, the cost of the shells soars to 2.9 hours of strenuous activity

Table 6.2

Rank Order of Species by Frequency of Kill, According to Weapon[a]

			Number killed per weapon						
			Spear		Blowgun		Shotgun		
Rank	Common name	Scientific name	*N*	%	*N*	%	*N*	%	Total killed
1	Woolly monkey	*Lagothrix lagotricha*			457	81	105	19	562
2	Cuvier's toucan	*Ramphastos cuvieri*			258	82	55	18	313
3	Howler monkey	*Alouatta seniculus*			188	76	58	24	246
4	Spix's guan	*Penelope jacquacu*			118	55	96	45	214
5	Red squirrel	*Sciurus igniventris*			178	92	15	8	193
6	Salvin's curassow	*Mitu salvini*			42	23	144	77	186
7	Blue-throated piping guan	*Pipile pipile*			61	34	120	66	181
8	White-lipped peccary	*Tayassu pecari*	32	21	4	3	116	76	152
9	Collared peccary	*Tayassu tajacu*	33	25	3	2	97	73	133
10	Ivory-billed aracari	*Pteroglossus flavirostris*			127	99	1	1	128
11	Capuchin monkey	*Cebus albifrons*			95	82	21	18	116
12	Great tinamou	*Tinamus major*			4	5	79	95	83
13	Red-mantled tamarin	*Saguinus fascicollis*			63	94	4	6	67
14	Gray-winged trumpeter	*Psophia crepitans*			36	64	20	36	56
15	Saki	*Pithecia monachus*			48	94	3	6	51
16	Golden-collared toucanet	*Selenidera reinwardtii*			42	100			42

17	Mealy parrot	*Amazona farinosa*			36	90	4	10	40
18	Squirrel monkey	*Saimiri sciurea*			28	90	3	10	31
19	Titi	*Callicebus moloch*			21	75	7	25	28
20	Paca	*Agouti paca*	15	56			12	44	27
21	Douracouli	*Aotus tirvirgatus*			19	76	6	24	25
22	Acushi	*Myoprocta pratti*	1	5	3	14	19	81	22
23	Coati	*Nasua nasua*			6	38	10	62	16
24	Scarlet macaw	*Ara macao*			16	100			16
25	Squirrel	*Sciurus cocalis*			16	100			16
26	Speckled chachalaca	*Ortalis guttata*			6	38	10	62	16
27	Blue-headed parrot	*Pionus menstruus*			14	100			14
28	Spider monkey	*Ateles belzebuth*			8	62	5	38	13
29	Agouti	*Dasyprocta fulignosa*	2	17	1	8	9	75	12
30	Nocturnal curassow	*Nothocrax urumutum*			8	67	4	33	12
31	Yellow-ridged toucan	*Ramphastos culminatus*			6	60	4	40	10
32	Tapir	*Tapirus terrestris*	3	38			5	62	8
33	Tayra	*Tayra barbara*			3	43	4	57	7
34	Capybara	*Hydrochoerus hydrochaeris*	6	100					6
35	Brocket deer	*Mazama americana*	1	17	1	17	4	66	6
36	Cayman	*Caiman sclerops*	1	100					1
37	Miscellaneous	(small birds and mammals)[b]	—	—	128	100	—	—	128
			94	3	2031	64	1040	33	3165

[a]Includes only those species whose total weight was more than 10kg.

[b]The birds and mammals of this category that are the most commonly blowgunned are: *Microsciurus flaviventer, Cebuella pygmaea, Potos flavus, Odontophorus stellatus, Heliornis fulica, Ara ararauna, A. chloroptera, A. severa, Pionites melanocephala, Amazona ochrocephala, A. amazonica, Trogon violaceus, Motmotus motmota, Capito niger, Pterglossus castanotus, P. pluricinctus, P. inscriptus, Rhamphastos culminatus, Pitangus sulphuratus, Cyanocorax violaceus, Psarocolius decumanus,* and *Lipaugus vociferans.*

Table 6.3

Output–Input Ratio (kg/hr) for the Spear, Blowgun, and Shotgun

Weapon	Total weight (kg)	Total hunts	Mean yield per hunt (kg)	Mean hours per hunt	Output–input ratio
Spear	2526.1	81	31.2	7.9	3.95
Blowgun	6772.6	433	15.6	9.6	1.62
Shotgun	9482.4	401	23.6	9.3	2.54
Total	18,781.1	915[a] (867)	21.7	8.9	2.43

[a]The total number of hunts does not equal 865 because on some hunts the hunter carried and used more than one weapon. In such instances, each weapon is credited for a "weapon hunt" in this column. But when figuring the "day hunts," these duplications have to be subtracted from the number. Thus, there were 915 weapon hunts, but only 867 day hunts. To figure the total mean yield per hunt, the 867 figure must be used rather than 915.

for a single shell. On those occasions when a man buys two boxes, he reduces the cost to 2.1 hours per shell. The authors are unaware of a man buying more than two boxes in a single trip.

On the other hand, the time invested in making poison darts is minuscule by comparison. A man can collect the materials from the forest, shape the darts, make the poison, and apply it to the darts in an average of 4.7 minutes per dart when he makes a batch of 200–300, which is normal. What is more, dart making does not demand high energy expenditure, and 68% of the time spent on dart manufacture can be "piggybacked" with other activities, such as visiting. By contrast, high levels of energy are expended in obtaining shotgun shells, and the time thus spent is not easily shared with other activities.

Ignoring these differences in strenuousness of input and assuming that the kill rate with a shotgun shell is 100% successful and that it requires four darts to kill a bird or squirrel (a high estimate), the shotgun shell is still six to nine times more costly (depending on whether the hunter bought one or two boxes of shells) than the dart to bring down the same small animal.

While the blowgun is used for maintaining a continuous supply of protein by harvesting numerous small animals, it is also quite effective in harvesting animals in the 2–10 kg range. However, as the size of the animal increases, the benefit of the shotgun also increases, placing the individual hunter in more of a need to decide which weapon he will use. A larger animal produces a higher protein return for the investment, making the absolute cost of the kill lower and, therefore, making it more justifiable to use shotgun shells. It should be noted that a larger animal killed with darts or shotgun shells is less costly in an absolute sense (i.e., in kilograms of protein returned per energy unit invested) than a smaller animal; but if four darts can kill the larger animal the relative cost between darts and shells remains as calculated above. The significant factor is that at some point the expenditure of the shell to kill a larger animal pays off well enough that the greater cost efficiency of the dart is no longer compelling.

Although larger animals may require more darts for a kill, thereby reducing the differential in cost, other factors begin to play a role. For very large animals, the shotgun is indisputably superior in killing power, and for small game the blowgun is indisputably more cost efficient. But for animals in the intermediate range (2–10 kg) the decision is not as clear cut. This is reflected in Table 6.2 which shows an increasing mixture of shotgun and blowgun use for animals in the 2–10 kg range. This trend toward increasing shotgun use is also evident for species in the 5–10 kg range but, even for animals that large, the Waorani still kill far more with the blowgun than with the shotgun. Of 1813 animals in that category 1081 or 60%, fall to the blowgun. These data are heavily influenced by the Wao preference for the blowgun over the shotgun for hunting monkeys, the most common game in the 5–10 kg range. To hunt monkeys, the silence of the blowgun is very important because it allows a hunter to kill numerous animals before a band disperses. In addition, it requires a great deal of energy to climb a tree to retrieve a shotgunned monkey, but a blowgunned animal will fall when the poison takes effect. Undoubtedly these factors combine with the higher cost of shells to encourage continued use of the blowgun.

When production with the spear is compared to production with the blowgun, the specialized nature of the spear comes into clear focus. The spear was used in only 81 of the recorded hunts, as opposed to 433 for the blowgun, but it accounted for 2526 kg, compared to 6773 kg for the blowgun (Table 6.4). In terms of numbers of animals killed, the blowgun killed 20 times more animals, but yielded only slighly over 2½ times more meat. What this means to a Wao family is a fairly constant supply of meat provided by the blowgun, punctuated by some times of feasting provided by the spear.

To understand the implications of this for a Wao family, we need to look at the social pattern of sharing and the total production for an adult hunter in a year. Of 632 sample man-days, 96 were hunting days; that is, the

Table 6.4

Mean Weight per Kill, by Weapon

Weapon	Total kills	Total weight of kills	Mean weight per kill
Spear	94	2526.1	26.9
Blowgun	2031	6772.6	3.3
Shotgun	1040	9482.4	9.1
Total	3165	18781.1	5.9

average adult male spends 1 out of every 6.6 days, or 4.5 days per month, hunting. This was corroborated by data on one individual over an 11-month period, showing him to hunt 47 days in that period, that is, 4.3 days per month. Using Tzapino as an example, we find 16 hunters among a total population of 68 persons, 27 of whom are children under the age of 10. Since a man hunts 1 out of every 6.6 days, on a given day there will be two to three men out hunting (average is 2.4 men per day). The mean weight per hunt for Tzapino is 17.28 kg, so each man is going to bring home approximately 17 kg of game. Every man, woman, and child will thus receive an average of 0.41 kg of unbutchered game/day. Because the Waorani usually eat all of the viscera, except the bladder and a small portion of the lower intestine, and all of the head, except teeth and bone, a very conservative calculation of 70% of the whole weight as edible meat means that every man, woman, and child will receive an average of 0.28 kg of edible meat per day.

An event observed by one of the authors while at Quiwado is instructive at this point. The entire village of 65 had just returned from a hugely successful peccary hunt. In less than 4 hours they had killed 12 white-lipped peccaries, carried them back to the village, and had begun cooking them. As soon as the meat was cooked, people began carting handsful of it to other houses. As one woman carried a handful of steaming boiled meat and manioc to her sister-in-law's house, 100 m away, she met her sister-in-law coming to her house with a handful of boiled meat and manioc. They did not exchange greetings or say anything, but continued on with their gifts to their destinations, returned home, and collected more to take to other households.

An aspect of this needs expansion to give a better understanding of the data presented above. In small-scale societies, particularly those in which the local kindred is the basis of social organization, sharing with a broad spectrum of people is one of the highest values (Yost 1981b:691). Because of this strict code of sharing and reciprocity, game is going to be distributed fairly evenly in the community, making it possible for a man to hunt for fewer than 5 days a month and still receive a constant supply of fresh meat. If each family had to rely solely upon its own production, it would experience extreme oscillation in its protein consumption, but the nature of the local kindred organization has a leveling effect upon consumption. In instances of hunting failure (7.4% of all hunts are unproductive) or of very low production, a family will still be provided for by the fact that other hunters in the community are likely to bring home enough meat to share. The amount of variance in hunting yields is extreme, ranging from 0 to over 200 kg for single hunts. For example, 44 day hunts were recorded for Tewaeno in one 20-day period during January and February, a time of slightly lower than average hunting yields. Three (6.8%) of those hunts were futile and 30, or two-thirds, of the hunts produced less than 10 kg of game. However, 1 hunt netted 108 kg of game, and another, 180 kg. The remaining 9 hunts in that period produced an average of 28.3 kg each, with very little variance from the mean.

What this variance in yields can mean to a family is demonstrated in the following. During this period, one hunter killed only 9.1 kg of game in five hunts, another hunter carried home 106.6 kg in three hunts, and a third netted 355.6 kg in five hunts. The meat was shared among all the families. Had the first hunter not received meat from the others, his nuclear family of 4 would have had no more than 1.42 kg of meat to share among themselves over an eight-day period. However, during his time of low production, the second and third hunters produced 155 kg of game which was distributed among 33 individuals, including the first hunter's family. Thus, through the distribution of meat along the broad lines of the kindred in a system of generalized reciprocity, lean times for one family were covered by a time of high production by two other families. This type of situation gets repeated many times over in a year with a much larger pool of families reciprocating game among themselves.

Hunts as successful as the one described above for Quiwado are quite infrequent, but they skew the appearance of the data.[4] The weight of those animals totaled almost 320 kg, and all but two were killed with spears, underlining what was said earlier about the specialized nature and sporadic production of the spear.

When we view the total data from the standpoint of successful hunts, those in which something was killed, as opposed to unsuccessful hunts, those in which no game was killed, the shotgun and the blowgun are practically equivalent, but the way the spear is used is dramatically highlighted. Of 401 hunts with the shotgun, 4.5% yielded no return; 5.7% of 433 hunts with the blowgun yielded no return, but a remarkable 32% of 81 hunts with the spear were unfruitful. When this is considered in light of the fact that the average weight per hunt with the spear, including the zero yields, is 100% larger than the average yield per hunt with the blowgun and 32% greater than the shotgun (Table 6.3), it becomes inescapable that the spear provides a punctuated bounty of meat on infrequent occasions. The spear provides an average 31.2 kg per hunt even though a third of the hunts record zero weight, but the blowgun and shotgun produce averages of 15.6 kg and 23.6 kg per hunt, respectively.

It is relevant at this point to ask what factors go into a hunter's decision to carry a given weapon on a particular day. The answer is far from simple. In some cases he may have already decided what game he wants to pursue, and chooses the appropriate weapon. For example, if a member of the community has sighted peccary or tapir tracks close to the village, the hunter may decide to go for them. For tapir he will undoubtedly select the shotgun; the same would be true for peccary, but in this instance there will usually be competition for the shotgun since large numbers of hunters often hunt peccary together and there are usually a very limited number of functioning shotguns in a community at a given time. Those unable to muster a shotgun will have to carry spears. Another instance of determination of weapon choice by a decision to pursue a particular type of game occurs during certain times of the year when hunters specialize in hunting monkeys, particularly woolly monkeys; at that time the blowgun will be the weapon of choice for reasons outlined earlier.

It is on those remaining occasions, which constitute the majority of hunting days, that hunters have to weigh the factors more carefully. A limited or exhausted supply of shotgun shells or a broken shotgun may swing the balance in favor of the blowgun. Or, if a hunter is expecting to stay close to the village and, therefore, is likely to see only small game, he will usually take the blowgun because of the cost of shotgun shells. When the hunter has no predilection for the type of game he may pursue and feels an urgency to get meat, he may optimize his options by carrying both a blowgun and a spear, or a blowgun and a shotgun. This is especially true when he plans to hunt zones distant from the village.

The data make it appear that hunters could afford to carry the spear exclusively, depending upon the mechanism of redistribution to ensure that everyone has sufficient continuous meat. What the data presented here conceal are (1) the success of the spear is too sporadic to be relied upon exclusively, and (2) the spear is selected precisely because game is almost guaranteed. Most production from the spear comes from large group hunts arranged because peccary have already been sighted in an area; such hunts are predisposed toward success. Because a number of hunters are each likely to kill two or more peccary on the same day or within 2 or 3 days, most of the spear's production is consumed in a short, intense period.[5] It would be rare indeed for a given village to experience five such hunts in a single year, scarcely sufficient to sustain people over a long period.

Thus far we have focused upon the amounts of meat the different technologies produce, but the truly significant figure is the amount of return gained from a given amount of expenditure (Hames 1980:51; Vickers 1980:16). Table 6.3 shows the average kilograms of game killed per hour (output/input ratio) with each of the three weapons. Again, what has been said about the weapons is reinforced. The spear is the most efficient weapon in terms of the gain from a given input, but the discussion above has shown that this can be misleading. All of the factors governing the weapon's use have to be taken into consideration or the wrong conclusions will be reached. If the Waorani were to try to rely solely upon the spear for protein production, their intake would drop drastically.

The blowgun is best adapted to a single ecological zone, the canopy, but it is not as restricted as the spear in that it works equally well for a large variety of species, both arboreal and volant. It continues to complement the shotgun, in spite of the shotgun's versatility, because it does some things quite well. Its value in hunting arboreal species balances it against the overall value of the shotgun.

Table 6.3 reveals that the shotgun is 57% more productive on a per-hour basis than the blowgun. Whereas it is the highly specialized nature of the spear that makes it efficient, it is the generalized nature of the shotgun that helps make it productive. The shotgun reaches into all niches, except that of animals under 1.5 kg, and thereby has the capability of supplying much of the continuous meat supply that the blowgun can supply, as well as the heavy kills the spear supplies. The way the Waorani use the shotgun, its generalized nature gives it an edge more than its range does, but its value in harvesting the large terrestrial species is significant.

One of the questions that arises is whether any of the weapons are more effective during one time of the year than another. The variation in the kill rates for nine of the more prominent species representing the arboreal mammals, the large terrestrial mammals, and the canopy and lower stratum avian species is shown in Figure 6.6. The rectangles represent the number of animals of each species that are likely to be harvested in a given hunt during that month. The values were obtained by dividing the number of kills of a species in a given month by the number of hunts for that month. Since the sample for the month of March was too small to give reliable results (five hunts for that month) the values for March have been interpolated from the values for February and April. The same has been done for Figure 6.7, which gives the overall variation in hunting efficiency on a monthly basis.

Comparison of the data in the two figures reveals several tendencies. First, hunting efficiency drops considerably during the rainy months and reaches its peak during the drier months. This is undoubtedly due to the severe difficulty of hearing animals while it is raining. Not only does the noise of the rain falling on leaves help conceal the noise the animals make in their movements through the forest, but, in addition, animals are less active during rain. Hunters who depend upon sound as the primary sense in hunting are therefore handicapped.

July appears to be contradictory to this general tendency. Every species demonstrates a low for July, making it the least productive month from any aspect, but the rainfall is not that high. It might be reasoned that this is a transitional season, accounting for the anomaly, but the more likely reason is one of a complicated interaction of various components of the ecosystem which are not apparent in Figures 6.6 and 6.7. In abbreviated form, October through January in this area are the months of highest inflorescence both in the canopy and in the lower strata. In January a number of plants are beginning to fruit, and by April the fruiting season is at its peak. During these months the animals begin to congregate, attracted by the large numbers of flowers or fruit. The game birds nest during the flowering

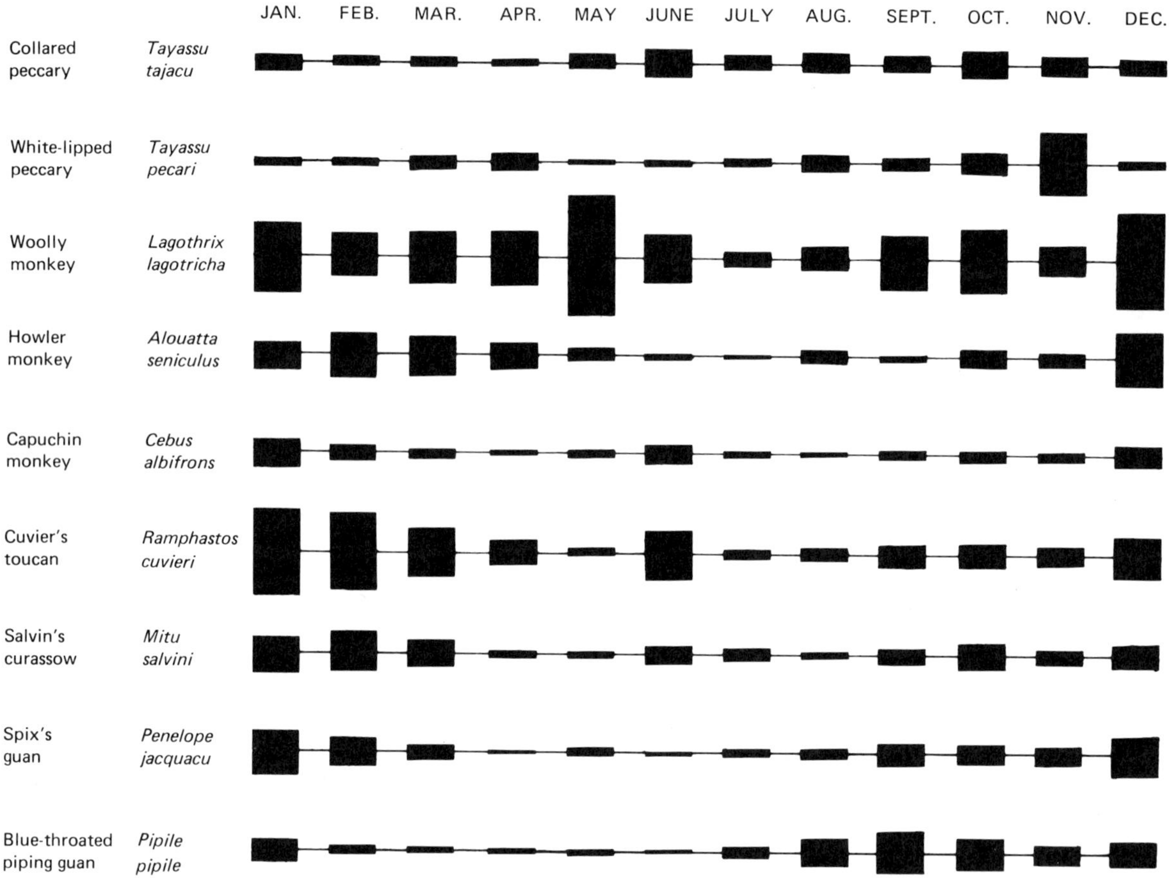

Figure 6.6 *Seasonal variation in number of animals killed, by species.*

months and, therefore, have sufficient food for their young when the fruiting season gets underway. The birthing season for the monkeys correlates with the fruiting season, coming to a peak in April when the fruiting is at a peak. The monkeys, particularly woolly monkeys, congregate in larger bands during this time, presumably providing greater defense for young and females through social mechanisms. By May the animals are fat, earning this season the designation of the "fat season" by the Waorani.

By July, however, flowering or fruiting plants are at their lowest, and animals are forced to disperse. (The Waorani note that this is one of the more difficult times of the year to try to travel great distances because there are so few foods available.) Because the animals are dispersed, the hunter has to invest more time to harvest the number of animals he would kill if he

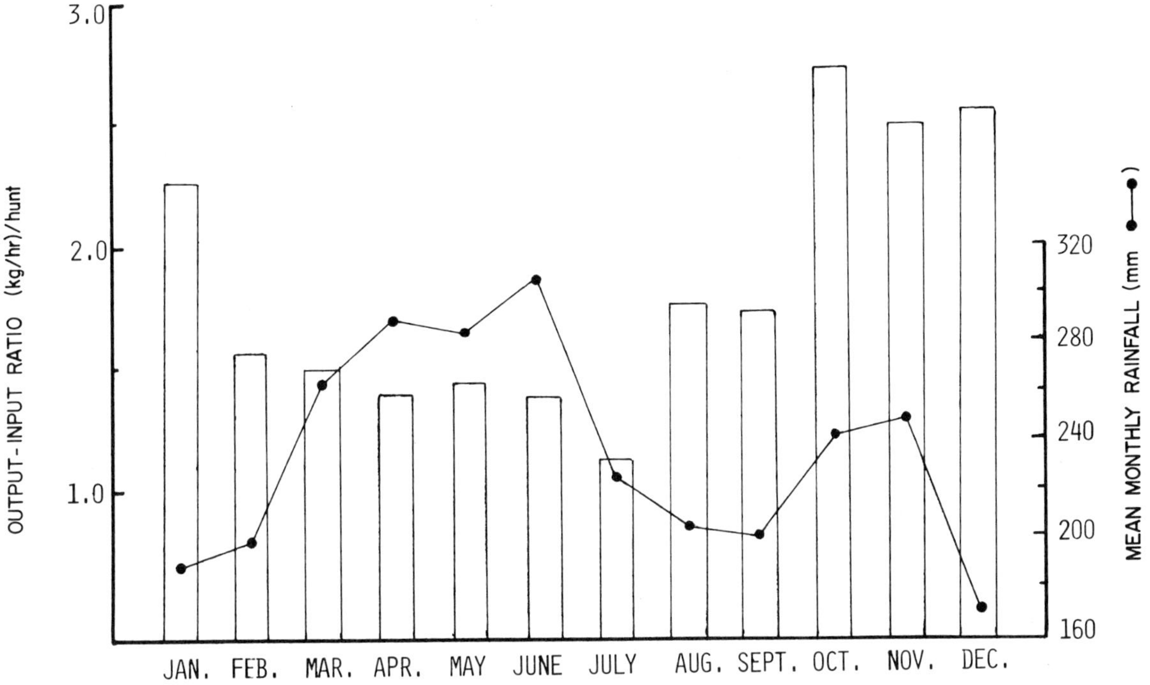

Figure 6.7 *Seasonal variation in hunting efficiency. (Based on rainfall data from Blakeslee 1980.)*

encountered a large band or flock in the fruiting season. It should be added that not only is the production per hour lower for July, the gross weight produced for this month is also lower. Fewer animals are killed, and those that are killed weigh less than at other times of the year. It is not a matter of just spending longer hours to get an equivalent amount of meat; less meat is actually harvested during this month.

Vickers (1976) has noted the relative scarcity of fats in the diet of the peoples of Amazonia and the attention the Siona–Secoya give to seeking greater quantities of fat. Figure 6.7 reveals the same phenomenon for the Waorani. Hunting efficiency is relatively low for May, meaning animals are harder to get, but during that month the Waorani kill a significantly larger number of woolly monkeys. They anticipate the fat season, particularly for the number of fat woolly monkeys they hope to harvest, and will expend considerable effort to do so. In fact, during this season they almost ignore other types of game to concentrate on the woolly monkeys.

The apparent relative evenness of monthly yields of the two species of *Tayassu* is somewhat deceptive in that for most months the yield is produced in just two or three hunts, as opposed to numerous hunts spread out over the entire month for the other species. The spikes in the production rates of the *Tayassu* species for October and November correlate with the ripening of the swamp palm (*Mauritia flexuosa*). The white-lipped peccary (*T. pecari*), in particular, returns repeatedly to forage beneath the trees as the ripened fruit falls. The Waorani, conscious of this fact, establish hunting routines that can exploit the patterned behavior of the species during that season.

In conclusion, the relationship of a weapon's efficiency to seasonality is directly related to that weapon's effectiveness in hunting any given species. That is, if monkeys are the species being hunted at a given time of year, the blowgun is going to be the most effective weapon for that season; thus, the seasonal patterns of a species become the determinant of the seasonal efficiency of the weapon. Therefore, it is best to view the above data from the perspective of the weapon most commonly used to kill each of the species listed in Figure 6.7.Because the blowgun is most efficient at taking arboreal and volant species, it appears that the most productive months for the blowgun are December through February, and in May. The shotgun, however, tends to be effective over a wide range of months, again because of its capacity to reach into a broader spectrum of ecological niches. It can be employed regardless of the animals most likely to be harvested in a given season. The spear is most effective during October and November because of the higher predictability of the behavior of the two species of *Tayassu*..

One further explanation is in order here. Earlier we alluded to the fact that we have recorded very few hunts for March. The lack is not due to sampling error, but due to the fact that March is the height of the *chonta* season. Cogi, an astute Wao man, stated it succinctly: "During the *chonta* season, Waorani do not hunt or move around much. We stay home, drink *chonta*, and give *chonta* feasts. So the game comes right up to our door. It's not frightened off by our hunting or moving around. Then when the fat season comes [April–June], the game is right here. In big bands! and fat!" The Waorani do not consume much meat during March; instead they rely almost exclusively upon the huge quantities of *chonta* they cultivate near their homesites.

Impact of the Shotgun

Given that the Waorani are using shotguns more frequently than before and that the shotgun is more efficient than the traditional technology, it is reasonable to expect that the Waorani are either hunting fewer hours now to meet their protein needs, or they are consuming greater amounts of meat than in the past. Examination of the data from a diachronic perspective reveals that the latter is the case. The Waorani have not significantly changed the amount of time they spend hunting, but they have increased the amount of meat they bring home in an average hunt by a third. In Tzapino, the mean yield per hunt for 1977–1978 was 14.2 kg. At that time there was only one shotgun in the village, and shells were at a premium. By 1980, the mean yield had risen to 21.2 kg per hunt, there were several shotguns in the village, and shells were more readily available. In spite of the 33% increase in the per-hunt yield, the hunters were hunting an average of only 3% less time per hunt and were hunting as many days as they had in 1977. The question, of course, is how long the area can sustain the increased pressure.

By comparing the Ye'kwana with the Yąnomamö, Hames (1979:247) shows that among the Ye'kwana the shotgun has had its greater impact upon the arboreal and volant species. Our data do not demonstrate these trends. Because of the Wao preference for the blowgun over the shotgun when hunting arboreal species, over four times as many kills of arboreal species are made with the blowgun as with the shotgun. Only selected volant species, curassows (*Mitu salvini*) and guans (*Pipile pipile*), show any marked impact from the shotgun (see Table 6.1). The greatest impact is upon the terrestrial quadrupeds, *Tayassu tajacu* and *T. pecari*, those species which both E. Ross (1978:13) and Hames (1979:248) cite as less affected by the shotgun. As used by the Waorani, the shotgun kills three to four times as many of these species. It is likely that the differences in the traditional technologies explain the differences between Hames's data and ours. Bow hunters find it more difficult to kill arboreal species than blowgun hunters do, and blowgun and lance hunters find it more difficult to kill the large terrestrial species than bow hunters do. Because bow hunters are killing few arboreal species to begin with, when the shotgun is introduced it enables them to expand their kills among arboreal species rather dramatically. The same holds true for blowgun and lance hunters expanding into terrestrial niches more effectively with shotguns.

It is worthwhile to extrapolate an estimate of how much the shotgun has affected the total meat intake of the Waorani. Assuming that the 33% increase in yield in Tzapino is due only to the shotgun, we can reduce the current daily average intake by one-third and arrive at a conservative estimate of the meat intake prior to the shotgun. The resulting figure, 0.19 kg, is the average amount every man, woman, and child would have had prior to the use of the shotgun, an amount very consistent with the 0.173 kg/day reported by Chagnon and Hames (1979:911) for the bow-hunting Yąnomamö.

Comparison of the Blowgun to the Bow

Using Hames's data, it is possible to give a rough estimate of the bow's efficiency as compared with that of the blowgun. This can be done first by comparing his relative efficiency of the bow with the shotgun and then by comparing our relative efficiency of the blowgun with the shotgun. In doing so, we are using the shotgun as a rough constant to which the other weapons may be compared, recognizing that there may be differences in the shotguns available to the Waorani and those available to the Ye'kwana. This is not intended to be a precise statement of the relationship between the blowgun and the bow, but can give us a rough estimate of general tendencies.

After controlling for confounding variables in location and hunting technique, Hames (1979:245) demonstrates that the Ye'kwana obtain an output/input ratio of 1.49 kg of game per hour of hunting with shotguns and compares this to Yąnomamö yields of 0.45 kg per hour of hunting with the bow and arrow. This means that the shotgun is 3.3 times more efficient than the bow and arrow. Our data demonstrate that the shotgun is 1.57 times more efficient than the blowgun, which would make the blowgun about twice as effective as the bow. Perusal of the species hunted with the two weapons indicates that the bow is most productive for terrestrial animals, but the blowgun's adaptation to arboreal species gives it the edge. However, the fact that the Waorani do not maximize the shotgun's potential must not be forgotten. When they become more specialized in their use of the shotgun, its efficiency will undoubtedly increase, thereby broadening the differential between it and the blowgun.

Conclusion

The intense level of Waorani warfare and the exiștence of infanticide (5% of the deaths cited in Larrick, *et al.* 1979) make the Waorani an ideal testing site for the hypothesis, propounded by Harris (1974, 1975) and Gross (1975), among others, that warfare and infanticide are a competitive response to scarce resources, particularly protein. This hypothesis has come under serious question in research by Chagnon and Hames (1979), Vickers (1979) and Lizot (1978b), and is further discredited by the data presented in this paper. Waorani use shotguns along with traditional technology to supply an average as high as 250 g of edible meat per person per day; using traditional technology alone they could provide 190 g per person per day. These yields fall within the range of levels published for eight other groups in Amazonia, as cited in Chagnon and Hames (1979:212). The Waorani exhibit very high levels of health and nutritional status (Larrick *et al.* 1979) partly because they are consuming protein at levels in generous excess of those established by nutritionists as adequate (National Research Council 1974, cited in Chagnon and Hames 1979:912). Since the Waorani in this study are at least 10 times more densely populated than ever before in their history, it appears highly unlikely that competition for scarce protein resources was the unique sustaining factor in their intense levels of warfare. This research not only establishes that there was no scarcity of protein available to the Waorani, it also demonstrates that the traditional weaponry, the blowgun and the spear, was adequate to exploit that protein.

E. Ross's contention (1978) that the shotgun is less efficient than traditional hunting technology is shown not to be accurate as it pertains to the blowgun and spear. The generalized capacity of the shotgun to reach into a broad variety of niches makes it more efficient than the more specialized blowgun or spear. Although this efficiency has given the Waorani sufficient reason to adopt the shotgun for what it is best suited, they are not adopting it wholesale for application in all circumstances. They are utilizing it in a balanced relation to the traditional weapons, particularly the blowgun. A number of individuals say that at one time they had opted to hunt with a shotgun exclusively, but after a couple of years decided it wiser to use the shotgun selectively, particularly while hunting monkeys.

If the blowgun alone is compared to the shotgun, the shotgun is over 1½ times more efficient, but if the blowgun and the spear are viewed as a complementary package, the shotgun's efficiency is not nearly so much greater. The combined blowgun–spear output–input ratio is 2.07 kg/hour, which makes the shotgun 1.22 times more effective. Such differences seem to be adequate reason for Amazonian hunters to take up the shotgun, particularly when it is adopted to complement the traditional technology, not supplant it.

Acknowledgments

We are deeply indebted to Margaret Scharr of the University of Texas, Arlington, Computing Center and to Max Cobbey of the Summer Institute of Linguistics for their expert consultation and time in getting the data computerized. We also want to thank our colleagues Glen Turner, David Turner, Janet Morris, Carolyn Dijk, Lori Belden, Barbara Walsh and Kathie Yost of the Instituto Lingüístico de Verano, Ecuador, for their long hours helping us with data tabulation, typing, and chart preparation. James W. Larrick, Raymond Hames, William Vickers, Catherine Peeke, Carolyn Orr, and Omar Johnson all gave helpful suggestions to strengthen this paper. We thank Wade Davis of the Botanical Museum of Harvard University for many of the plant identifications. His identification of *Curarea tecunarum* follows the reclassification given by Barn and Kruk. We are indebted to Taëbaëta, Dägaë, Wïdï, Wëtae, Yïiwa, and Pegöca for their enthusiastic participation in collecting much of the data for this paper. Their work not only increased the amount of data, but added to the quality of the study.

Notes

[1]When a man finds the vine he will frequently collect sections to transplant in areas near his home. Thus, the plants have been dispersed over the entire Wao territory, and there is never a problem of shortage.

[2]Experimental and chemical analysis of the poison is in progress, but Jim Larrick (personal communication) indicates that preliminary tests suggest that Wao poison may act through a novel mechanism different from other types of "curare."

[3]The values used for this paper are those published in Blake (1977), Chiarelli (1972), Hames (1979), Walker (1968), and from Vickers (personal communication) and Tallman (personal communication) as well as our own. For those values that were not our own we used a priority system: since Vickers's and Tallman's work has been done very close to the Wao territory, their values were used first, then those by Hames, followed by the other published values.

[4]This hunt is not included in our data because Quiwado was not being sampled at the time.

[5]Following a successful peccary hunt, even though the meat is distributed throughout a village, most people will end up with such large quantities of meat that for the next day or two there will be a great deal of joking about the number of people with diarrhea, nausea, and even vomiting resulting from overeating fresh meat.

CHAPTER 7

Why Do the Mekranoti Trek?

Dennis Werner

Introduction

Trekking is a common practice among many South American Indian groups. The Sirionó (Holmberg 1950:42–43), the Bororo (Cook 1909:390), the Nambikwara (Levi-Strauss 1948:38–39), the Kaingang (Métraux 1946:452; E. Hoerhan, personal communication), and the Kayapó (Banner 1961) are all well known for the lengthy trips villagers undertake to gather products from the forests and savannas surrounding their villages. Various researchers have given treks an important role in explaining native social structures or adaptations to their environments. Gross (1979) sees the seasonal breakups of village communities to go out on trek as responsible for the many nonkin sodalities (age-grades, ceremonial groups, men's societies, etc.) of Gê groups. Turner (1979) views trekking as important for the exercise of power by male heads-of-households among the Kayapó. And Meggers (1971) regards trekking as an adaptation to a tropical forest that limits population densities and political development. But despite this interest in trekking, we have few data on actual expeditions in specific groups, and we know little about why groups trek in the first place.

Using information gathered among the Mekranoti–Kayapó Indians of Central Brazil (Posto Indígena Mekranoti), I attempt in this study to analyze some of the possible reasons for trekking in lowland South America. The Mekranoti are a group of 285 Kayapó-speaking Indians living in a single village in the southern part of the Brazilian state of Pará. Originally part of a larger village, the Mekranoti split into two groups in 1953 shortly

ADAPTIVE RESPONSES
OF NATIVE AMAZONIANS

ISBN 0-12-321250-2

after their first contact with Claudio Villas Boas near the Xingu River. The group remaining at the Xingu River is now known as "Txukahamãe." The other group is now called "Posto Indígena Mekranoti" or simply "Mekranoti." This latter group remained isolated in the tropical forest until 1966 when a missionary moved into their village.

At the time of my fieldwork in 1976–1977, the Mekranoti depended entirely on shifting agriculture, hunting, fishing, and gathering for their subsistence. Although their lands had not been officially demarcated, they lived in an otherwise seemingly uninhabited territory about the size of Belgium. Compared to other Indians in Brazil, they exploited a relatively fertile and productive environment (Gross *et al.* 1979; Werner *et al.* 1979). Working through FUNAI (the Brazilian Indian Foundation), the Mekranoti traded Brazil nuts and handicrafts in exchange for medicines, shotguns, metal pans, and other articles of Western manufacture. But only one Indian spoke any Portuguese, and none could deal well with money.

The Mekranoti spend a great deal of time on trips away from their village, including lengthy journeys into the forest by the entire village travelling as a unit. Many of these trips would be called "treks" by anthropologists. But some of them would not be included under the "trekking" rubric. Presently, the Mekranoti, with the help of FUNAI or missionaries, sometimes travel to Belém or Brasília to receive medical treatment, buy goods for the villagers, or participate in meetings of the Summer Institute of Linguistics. Young men also go off for part of the year to collect Brazil nuts to sell to FUNAI, and individual families sometimes go off for a few days to live near their gardens or to hunt in the forest. In the past, the Mekranoti sometimes deserted their homes entirely to escape enemies after raids or to relocate near new garden sites. While all of these different trips may be important to Mekranoti life, they do not constitute treks in the sense intended by most anthropologists. In this study I restrict trekking to overland trips into the forest or savanna, lasting several weeks, and involving large numbers of people with the intention of returning home to a base village. During my stay among the Mekranoti, the Indians spent about 22% of their time on treks. (They spent another 1% of their time on short family trips into the forest.) I accompanied the Indians for portions of three different treks for a total of about 8 weeks, during which time I carried out studies of time allocation, dietary intake, and hunting and fishing productivities.

All Mekranoti treks are similar in certain respects. And the Mekranoti refer to them all with the same expression, *mẽ djy kôt tẽ*. Usually treks involve daily or almost daily moves of campsites from one area of the forest to another. They also require that people carry around large quantities of garden produce to feed themselves. On trek, the young men generally spend most of their time hunting, while the women (and some elderly men) transport produce and occasionally do some gathering of wild foods.

But there is a great deal of variation in the seasonality, duration, composition, and organization of treks. I have described some of these variations elsewhere (Werner 1978).

Some anthropologists see Kayapó trekking as exclusively a dry-season activity (cf. Meggers 1971:67). But the Mekranoti also trek during the rainy months, albeit not as often. Table 7.1 shows that for the year of my fieldwork people spent a greater proportion of their time on trek during the dry season. Although my historical notes are unsystematic and far from complete, they also suggest a possible dry-season bias. Of the treks mentioned in my notes, 37% took place in the dry season that constitutes only 25% of the year.

Trekking is undoubtedly more difficult during the rainy season, when the Mekranoti must cut paths through large stretches of flooded land and must construct shelters against the rain. Consequently, new campsites are generally set up only 30–45 minutes away from previous encampments. During the dry season, the Mekranoti can travel longer distances over solid land, and can sleep in the open air on leaves spread on the ground. But the distance between campsites is never more than a few hours walk. Part of the reason is that the Mekranoti must transport garden produce from one site to the next. This requires frequent return trips and could not be accomplished if distances were too great. As it is, Mekranoti adults on trek average about 2 hours a day simply transporting garden produce from one campsite to the next or from home and forest gardens to the trekking village.

The duration and composition of treks also vary. Some treks last several months; others are completed in a few weeks. Sometimes the entire village travels together as a single unit; at other times only the men go out, or the village splits into two separate treks. During my stay among the Mekranoti, half of the village left for a long informal trek from June to September, and the other half made a shorter trek in June and July. In December, and again in May, the entire village participated in formal ceremonial treks. In October, the men trekked without women and children.

The organization of treks depends to some extent upon the people who participate. If only the men go out, they are likely to divide themselves into their two men's societies and go off in different directions. If both men and women trek, organization depends on whether the trek is ceremonial or nonceremonial. Nonceremonial treks are informal and may involve only part of the village. Nuclear families locate themselves anywhere they wish within a campsite, often varying their relative placement from one day to the next. Ceremonial treks are much more rigidly organized.

Most Kayapó ceremonies center on the bestowal of special names to children. These ceremonies (such as *Bẽp, Ṫakàk, Nhàk,* or *Kôkô*) last several months. In the early stages of the festivities, men (or women, depending on the ceremony) must sing songs every morning and evening in the central plaza. They are rewarded for their efforts by gifts of food provided by the parents of the children who will receive names. For the final days of a ceremony the Mekranoti need tortoises to feed the assembly, and they must trek for several weeks to obtain a large enough supply.

If only the men do the trekking, then they divide themselves into their men's societies and go off in different directions. But when both men and women go on ceremonial treks people generally travel together as a single unit. Housing arrangements are decided on the days before departure. The fathers of the children receiving names (*mẽ kra rêr mêx*), and their "helpers" (*mẽ kuka tyk*) go to the men's house and "choose up teams" consisting of their real and fictive brothers with their respective wives and children. Virtually everyone is chosen at this point. During wet-season treks each team builds a long lean-to with sections set aside for its nuclear family units. Each team also constructs its own *atykbê,* a dormitory for boys and young men who do not sleep with their families. And in front of each team dwelling a clearing is prepared and covered with leaves to serve as a men's house (*ngà*) for nightly men's meetings. Most ceremonies require wet-season treks, but there are also ceremonial treks during the dry season. The organization is much the same, although shelters need not be prepared.

Of the 1449 observations of people on trek that appeared in my time allocation studies, 816 (or 56%) were of nonceremonial treks, 608 (42%) of ceremonial treks, and 25 (2%) of male-only treks.

Mekranoti treks vary widely, but they all require a great deal of work transporting garden produce. Why do the Mekranoti leave the comforts of their village to engage in such pursuits?

Explanations for Trekking

NATIVE EXPLANATIONS

When I asked (informally and unsystematically) why the Mekranoti went on trek, they gave various answers. They may want to collect forest products such as game, fish, tortoises, palm oil for hair, tree resin for body painting, or wood for bow staves. One informant mentioned an incident in the past when the Mekranoti trekked to avoid enemy attacks. Others mentioned reasons such as escaping illness in an area, or avoiding intravillage conflicts. Indians elsewhere in Amazonia offer a similar variety of interpretations for trekking. Maybury-Lewis (1974:44) even records a case when the Shavante went on a trek while their beans were growing in order to avoid the temptation to eat unripe green beans. All of these native explanations may be valid for particular treks. But they do not explain why trekking exists in the cultural repertoire of some Indian groups but not others. For example, many groups probably suffer from the temptation to eat unripe beans, but only some groups trek. To explain why trekking exists in some cultures but not others, we need to turn to more comprehensive theories.

PROTEIN SCARCITY

One explanation for trekking regards the practice as an adaptation to scarce protein resources in areas distant from large rivers (Gross 1975:535; Meggers 1971:101). I accepted this explanation in a previous article (Werner 1978). The basic argument is that some South American Indians may be forced to go out on trek in order to meet minimum protein requirements because the areas in which they live are short in fish and game.

Recently analyzed data do not support the protein scarcity view. First, a comparison of protein capture in four Central Brazilian societies—the Mekranoti, the Shavante, the Bororo, and the Kanela—shows that the groups that trek most are those with the highest, not the lowest protein availability (Werner *et al.* 1979). The Shavante and the Mekranoti trek a great deal and also have relatively high amounts of protein in their diets. The Kanela trek very little even though their protein supplies are very low.

An analysis of the seasonality of Mekranoti trekking also casts doubt on the protein scarcity argument. Table 7.1 shows that the Mekranoti are most likely to trek during the dry and intermediate seasons. But Table 7.2 shows that these are the times when hunting and fishing returns are greatest. According to the Mekranoti, hunting is easier during the dry season because animals must come together at this time around smaller streams and waterholes. Also, game is easier to chase over dry land than through flooded swamps. Thus, trekking seems to be most popular when game and fish are most plentiful, not when they are scarce.[1]

SOCIAL PROBLEMS

In arguing that resource scarcity cannot account for trekking among South American Indians, some anthropologists (e.g., Bamburger 1979; Turner 1979:177) suggest social, rather than economic, origins for the practice. Speaking about the Kayapó (including the Mekranoti), Turner (1979:177–178) argues that trekking exists primarily to allow male heads-of-households to express their dominance in matrilocal societies. According to this argument, men cannot "test" their control over their daughters in normal matrilocal situations because the daughters always remain in their father's household, making a test of control impossible. But the organiza-

Table 7.1

Seasonality of Trekking

Season	Month	Rainfall[a] (mm)	Percent of time spent on trek[b]
Dry	June	98.9	62
	July	0.0	54
	August	32.5	30
Intermediate	May	125.5	33
	September	106.5	24
Rainy	October	310.5	3
	November	356.5	7
	December	459.3	50
	January	231.0	0
	February	159.5	—[c]
	March	351.6	4
	April	230.5	9

[a]Data come from FUNAI (Fundação Nacional do Indio—National Indian Foundation) records taken daily with a rainfall gauge installed in the village by a hydroelectric company.
[b]Data come from time allocation studies based on randomly timed visits to random households (see Gross *et al.* [1979] or Werner *et al.* [1979] for a description of the technique).
[c]No information.

tion of treks around males allows male heads-of-households to demonstrate their control by requiring that their daughters and sons-in-law remain with the household head regardless of the collective group affiliation of the sons-in-law. Turner argues that this demonstration of control is important, not only to the individual household head, but also to the "consistency and integration of society at large" (Turner 1979:178). Turner is not very clear about why this elaborate test of control is so important to social integration, but the Mekranoti data do not support the argument in any case. First, not all treks are organized around males. Second, even in those that are, heads-of-households do not necessarily attempt to remain with their sons-in-law on trek. When the men divide themselves into their respective men's societies to go on all-male treks, young men follow their men's society regardless of their household affiliation. When the entire village goes on a ceremonial trek, teams are chosen based on the choices of the ceremonial sponsors. My daily records of living arrangements on these treks showed that sons-in-law are often separated from their fathers-in-law.

Another social reason for trekking may be the avoidance of conflict. In her discussion of treks and village splits, Bamburger (1979) argues that conflict can be handled either by "voice" or by "exit." That is, individuals can either discuss their problems and search for a solution, or they can simply leave the troublesome situation altogether by moving away. The Kayapó, according to Bamburger, have chosen the "exit" option. Bam-

Table 7.2

Hunting Success by Season[a]

Season	Successful hunting–fishing trips (%)	Mean yield of undressed game or fish (kg/hour)
Dry	72	0.49
Intermediate	56	1.62
Rainy	60	0.45

[a]Data come from the random sample of hunting and fishing trips that appeared in the time allocation studies. When someone was reported out hunting or fishing, I checked back later in the day to see what was obtained. If possible I weighed the returns. Otherwise I used standard figures to determine the weights of reported kills.

burger is not entirely clear about which kinds of society would choose which option. But one possibility is that the choice is related to the amount of territory people have to move into. Unfortunately, this argument does not account very well for the South American data on trekking. Some groups (such as the Kuikuru or the Kamayura in the Xingú) have large territories, but they do not trek.

Another possibility is that trekking might be related to the amount of conflict in a society. Presumably, groups with more internal fighting might have a greater need for the "exit" option. But this argument also fails to account for the South American data on trekking. Some groups with a great deal of warfare (e.g., the Jívaro [Harner 1972], or the Y̨anomamö [Schuster 1956:10]) do not trek as much as more peaceful groups (e.g., the Nambikwara [Levi-Strauss 1948:38–39], the Sirionó [Holmberg 1950:42–43], or the Bororo [Cook 1909:390]) which are known for their lengthy journeys. Finally, many Mekranoti treks do not result in village split-ups at all, and so could not be attributed to the need to avoid intravillage conflict. Of course, these critiques do not mean that individuals cannot take advantage of some treks to deal with conflict. Indeed, the Mekranoti mentioned instances in their past when they went on trek specifically to avoid enemies or to solve internal disputes. But the reason trekking exists in the Mekranoti's cultural repertoire in the first place must be something other than internal conflict. After all, not all Mekranoti treks resolve conflict, and the evidence from other South American groups suggests that conflict alone does not necessarily lead to trekking practices.

EASE OF GATHERING

Although protein scarcity may not explain Mekranoti trekking, it is still possible that the practice may result from other dietary considerations.

Although he did not actually accompany the Indians on forest trips, Posey (1982) emphasizes the importance of wild foods to trekking Gorotire–Kayapó in years past. In the days before steel axes, gardening may have been more difficult, and gathering a more attractive way of obtaining food. For certain groups, the ease of gathering wild foods (such as pine nuts among the Kaingang) may, indeed, explain why these Indians trekked. But the explanation is unlikely among the Mekranoti. Table 7.3 shows that the amount of time spent gathering on trek is minimal—less than an hour a week. And in the food consumption studies, gathered foods like Brazil nuts, honey, and *açai,* formed a minuscule part of the diet, usually amounting to no more than a handful of nuts or fruits a day. Garden produce provided the overwhelming bulk of calories.

Parenthetically, I also think a word of caution is in order about reconstructions of past Kayapó treks. In discussing past expeditions, the Mekranoti insisted that they, too, lived entirely off gathered foods. But on eliciting further details, I discovered that the Indians, in fact, consumed garden produce from outlying forest gardens of Kayapó and non-Kayapó cultivators. Thus, statements about past treks may not be entirely reliable among the Kayapó.

MAXIMIZING PROTEIN

If the Mekranoti trek for food, it is not wild fruits they are after. More likely, they want extra meat, even if they are not short of protein. Individual Mekranoti sometimes told me that the *village* was going on trek for

Table 7.3

Adult Mekranoti Activities on Trek and at Home

Activity (from 6:00 A.M. to 8:00 P.M.)	Males (hr/week)		Females (hr/week)	
	On trek	At home	On trek	At home
Hunting	29.6	8.3	0.0	0.2
Fishing	1.4	3.7	0.0	0.4
Gardening	0.0	4.7	0.7	16.7
Gathering	0.0	0.9	0.7	1.8
Domestic animal care	0.0	0.5	0.0	0.2
Market labor[a]	0.0	7.1	0.0	0.8
Other work[b]	26.1	22.8	55.5	29.7
Nonwork[c]	40.9	50.0	41.1	48.2

[a]Includes manufacturing handicrafts for sale and the collecting of Brazil nuts.
[b]Includes housekeeping, child care, food preparation, manufacture, gathering firewood.
[c]Includes eating, sleeping, ceremonial activities personal hygiene, conversation, recreation, sitting idly.

some other reason—such as collecting tortoises—but they would add that they, *personally*, were going along because they wanted to eat lots of meat. Dietary studies of a few families (Table 7.4) showed a clear difference in the amount of meat consumed while on trek and while in the village. On trek, people averaged almost three times more meat or fish a day than at home. The extra meat is due primarily to an increase in the amount of hunting men do while on trek. Table 7.3 shows how much time Mekranoti adults spend on different activities while on trek and in the village. On trek men hunt on the average 29 hours per week, compared with only 8 hours per week at home.

If the Mekranoti are not short of protein, then why do they bother with all of this hunting? As argued elsewhere (Werner *et al.* 1979), the Mekranoti could live with less meat than they do, but they would probably not live as well. Extra protein helps in conferring greater resistance to disease, in allowing optimal use of other nutrients, and in protecting against manioc poisoning (Axelrod 1964; Lowenstein 1973; Osuntukun *et al.* 1969; and Spath 1971). I think it is the relative ease of gardening and hunting in the Mekranoti environment that makes the pursuit of a higher-quality diet worthwhile. Elsewhere in Amazonia people may be forced to accept lower dietary standards. They may be too busy with gardening to allow time for hunting, or they may not have enough land for trekking. On the other hand, in some places (such as the Xingú Park) fishing may be so productive that hunting is not worth the trouble. In summary, I suggest that people will be willing to go on trek whenever trekking is an easy way to improve one's diet. That is, trekking is primarily a way to optimize diet. The Mekranoti obtain sufficient calories with relative ease from their gardens (Werner *et al.* 1979), leaving them plenty of time to improve their diet through hunting. I shall return to this argument later in order to suggest possible cross-cultural tests. But first, I need to answer a possible objection to the "trekking for protein" argument.

Some anthropologists might argue that extra protein can be obtained more easily through vegetable than through animal products. This would

Table 7.4

Meat Consumption on Trek and at Home[a]

	On trek	At home
Person-days with meat or fish	38	25
Person-days without meat or fish	3	15
Total	41	40
Meat or fish per person per day (kg)	0.92	0.32

[a]Data come from dietary studies of a few families during different seasons of the year. (See Werner *et al.* [1979] for a description of techniques.)

make extra hunting for protein a wasteful activity. Beckerman (1979) argues that in some areas of Amazonia, Brazil nuts (*Bertholetia excelsa*) might be an especially rich source of protein. Since the Mekranoti environment contains large numbers of Brazil nuts, which the Indians export for sale, I decided to check whether Brazil-nut gathering would be a better way to obtain extra proteins than would hunting. Table 7.5 shows that the Mekranoti must spend more time gathering Brazil nuts than hunting to obtain an equivalent amount of protein. This suggests that the Mekranoti are wise to devote more time to obtaining meat than to Brazil-nut gathering. But fishing is less productive than either of these two other pursuits. Yet the Mekranoti spend more time fishing than gathering Brazil nuts. This latter finding does not seem to support the protein maximization argument. But on closer inspection the apparent problem disappears.

Table 7.5 shows that Brazil nuts are, indeed, high in proteins. But they are also high in calories. Wu Leung (1961) reports that Brazil nuts contain 640 kcal for every 100 g edible portion. For meat, the figures would be closer to 150 kcal, and for fish, closer to 100 kcal per 100 g edible portion. Mekranoti individuals currently consume an average of 63 g of animal protein a day. For them to acquire an equivalent amount of protein from Brazil nuts, they would also need to consume 3055 kcal in Brazil nuts alone, compared to only 455 kcal from the animal proteins they now consume. This would not leave much room for any other kind of food, and is hardly the best way to balance a diet. During my stay in the field, one Indian contracted scurvy and needed to be taken to Belém for treatment. Thus, I am particularly

Table 7.5

Return Rates for Game, Fish, and Brazil Nuts[a]

	Game	Fish	Brazil nuts
Mean yield/trip (kg, dressed or shelled)	3.78	1.10	2.18
Mean number of hours/trip	5.5	5.50	5.0
Mean yield/hour (kg, dressed or shelled)	0.69	0.20	0.44
Number of hours spent hunting, fishing, or gathering/adult–day	0.87	0.21	0.05
Consumption/person–day (kg)	0.29	0.02	0.01
Protein/person–day (g)[b]	58.6	4.1	1.40

[a]Data come from studies of hunting, fishing and gathering productivities. (See Werner *et al.* [1979] for details.)

[b]Estimates for protein content are based on Wu Leung and Flores (1961):
Protein/100 g edible portion game is 20 g.
Protein/100 g edible portion fish is 20 g.
Protein/100 g edible portion Brazil Nuts is 13.2.

interested in the loss of vitamin C that might come about if the Indians were to depend more on Brazil nuts for their protein. As it turns out, Brazil nuts are particularly low in ascorbic acid; even bitter manioc has more (Wu Leung 1961). It seems the Mekranoti are better off if they depend on meat or fish for their protein supplies and consume more vegetables for other nutrients. In short, they must not simply maximize protein but also limit the number of calories in their diet in order to meet minimum requirements for other nutrients. In short, they must optimize their diet.

Cross-Cultural Tests

As I have suggested, the explanation for trekking in lowland South America may lie in the concept of dietary optimization. I expect Indians to go on trek whenever trekking is the best way for them to improve the quality of the food they eat. This hypothesis sounds rather vague, but I think it can be tested. First, we must show that native South Americans do, in fact, attempt to optimize their diets. And second, we must evaluate whether trekking would be the best way to do this. Let us look at each of these propositions in turn.

Are people diet optimizers? Some anthropologists have protrayed humans in technologically simple societies as interested primarily in minimizing the work they do rather than in improving the quality of their diets or in acquiring additional goods. That is, they see humans as energy minimizers rather than as diet optimizers. A classic example is Sahlins's (1972) attribution of "Zen" affluence to primitive peoples. According to Sahlins, by simply desiring less, people can feel satisfied with what they have and enjoy abundant leisure. Allen Johnson (1980) formalized this argument in his linear programming analysis of Machiguenga data. In this study, Johnson attempted to account for how the Machiguenga allocate their work activities by assuming that these Indians are trying to minimize the amount of work they do while still acquiring minimum basic nutrients and avoiding too many calories. Johnson's linear programming model suggested that the Machiguenga ought to devote 100% of their subsistence labor to gardening and none to hunting and gathering. In reality, however, the Machiguenga devoted about one-fourth of their working hours to hunting and gathering. I suggest that the discrepancy between the model and reality may be due to the incorrect assumption that the Machiguenga are energy minimizers rather than diet optimizers. The linear programming model ought to be conceived differently. Instead of attempting to find a solution that minimizes the amount of work people do, the linear program ought to find a solution that optimizes diet (e.g., maximizes protein) while holding constraints on the number of hours people are willing to work. That is, hours of work may be the constant, while quality of diet varies.

Table 7.6 helps to make this point clearer. It shows the amount of time the Mekranoti work at different village tasks in the different seasons. The last column shows the total amount of work people do between the hours of 6:00 A.M. and 8:00 P.M. Monthly figures vary somewhat, but the averages for the different seasons are remarkably similar. It seems people have decided to limit their work to about 50 hours/week regardless of the season. There may be brief spurts of work or leisure (chi-square tests of work versus nonwork observations showed significant variations from the 50-hour mean for the months of June, July, August, and January). But the variations seem to even out in a relatively short space of time (chi-square tests show that work inputs do not differ significantly by season). Except for brief periods, it seems people have allowed themselves only a limited amount of work, or stated differently, they have allowed work to expand to fill the space available.

Table 7.6

Adult Mekranoti Village Activities, by Season[a]

Month	Nonwork[b]	Gardening	Other subsistence[c]	Nonsubsistence work[d]	Total work
June	37	11	11	41	63
July	63	13	7	18	38
August	43	29	3	26	58
Mean of dry season	48	18	7	28	53
May	50	17	6	27	50
September	45	19	11	24	54
Mean of intermediate	48	18	9	26	52
October	46	10	9	36	55
November	56	7	7	30	44
December	49	12	10	28	50
January	65	3	4	28	35
February	—	—	—	—	—
March	41	4	7	47	58
April	46	18	12	24	54
Mean of rainy season	51	9	8	32	49

[a]In percentage of time between 6:00 A.M. and 8:00 P.M.

[b]Includes eating, ceremonial activities, personal hygiene, conversation, recreation, sitting idly, and sleeping.

[c]Includes hunting, fishing, gathering, and domestic animal care.

[d]Includes market activities, child care, food preparation, housekeeping, manufacture, and gathering firewood.

This figure of approximately 50 hours/week of work also appears in other studies of time allocation in lowland South America. Using the same techniques as for the Mekranoti, Flowers and Ritter calculated work at 50.9 hours/week for the Xavante, and 50.4 hours/week for the Kanela, respectively (Werner *et al.* 1979; also cf. Beckerman, this volume). Using different protocols, but the same basic random sampling technique, A. Johnson (1978) arrived at a figure of 54.6 hours/week of work for the Machiguenga. The Bororo, who suffer from a high rate of invalidism, work somewhat less—44 hours/week (Werner *et al.* 1979). Although these constitute only a small sample of South American societies, the figures are impressive. Possibly, people at vastly different technological levels may work very different numbers of hours (indeed the whole question of what constitutes work becomes difficult with such broad comparisons). But within a given technological level (e.g., the combination of gardening, hunting, and gathering practiced by most South American Indians), there may well exist a set amount of time people are comfortable working. The major variations between these groups may not be in the total number of hours they work, but rather in the allocation of work to different tasks. The economic question facing the Indians is not how to minimize the number of hours they work but, rather, how to allocate their labor to get the best results. (In a hierarchy of needs, basic nutrients would have to be met first, of course, regardless of labor costs. But if these needs can easily be met, as seems to be the case among lowland South American Indians, then people may begin to work to improve their diets.)

The question of diet optimization in Native South American might be examined further by comparing time allocation studies for the total amount of work done in different societies, and by carrying out linear programming analyses. If the hypothesis is correct, we should expect most groups to work roughly the same number of hours, although the amount of time they devote to different tasks should vary. (The variance across groups in overall work should be lower than the variance for specific tasks.) Also, linear programs that assume people are optimizing diet should give results closer to reality than linear programs that assume people are minimizing labor. Testing the hypothesis is beyond the scope of this study, but is entirely possible given the data now available on these societies.

If the "diet optimization" hypothesis is supported, we can begin to check whether trekking is a means of improving diet. For this, we need data from different societies on the activities people pursue while on trek and the productivities of subsistence tasks on trek and in the village. Such data may be difficult to find given the diminishing importance of trekking in South American societies. But the few groups that still trek might be examined and contrasted with groups that do not trek. Cross-cultural tests may be difficult, but not impossible.

Conclusion

In this review of trekking among the Mekranoti–Kayapo, I have argued that protein maximization accounts better for the Mekranoti data than do alternative explanations that stress absolute shortages of protein, conflict resolution, social control by elder males, or the ease of gathering foods. I have also suggested some possible cross-cultural tests of a broader, "dietary optimization" hypothesis for trekking. Probably I have raised more questions than answers. The problem of trekking is a complex one, involving basic assumptions about the economic motivations behind human actions. If the questions raised here dealt only with trekking, they would probably not be worth the trouble answering, but I think more important issues are involved. We need to arrive at a better understanding of why people work, not only to understand cross-cultural differences, but also to understand our own society. The problem of trekking is closely related to these basic questions.

Note

[1]One question that might arise concerning the seasonality data is whether the higher hunting yields in the dry season are due simply to the fact that the Mekranoti trek more at this time and bag more game while on trek. The data do not support this view. Hunting yields are the same on trek as in the village. The failure of trekkers to bag more game may be due to two reasons. First, because many more men go out hunting while out on trek, there is more competition over the day's individual hunting area than in the village. Second, younger and inexperienced men do more of the hunting while on trek, and their hunting productivity may be lower. Were it not for these drawbacks, hunting productivity would presumably be higher while on trek. Unfortunately, my data are too limited to control for all of these factors. In any case, the recording of higher hunting yields for the dry season is not due to data gathered while on trek.

CHAPTER 8

Cocamilla Fishing: Patch Modification and Environmental Buffering in the Amazon Várzea

Anthony Stocks

Introduction

The purpose of this paper is twofold. First, I will draw attention to a hitherto unreported relationship between humans and fish in the Cocamilla Indian community of Achual Tipishca in the Huallaga River Basin in northeastern Peru. The fishermen of Achual Tipishca exploit a *várzea* (Brazilian term for the Amazon floodplain) lake and one aspect of their adaptation is to enrich its waters with garbage and human feces. In so doing, they are practicing a form of "patch modification" in which their lake resource is enhanced. Second, I will attempt to demonstrate through an analysis of fishing data that the fertilized zone of the lake, that part nearest the community, becomes an important resource during the times when the lake is flooded, thus buffering the effects of an annual fluctuation in fish availability. During the course of the analysis and discussion of the data on Cocamilla fishing, I shall also touch on four other subjects: (1) whether the Cocamilla fishing patterns can be explained by optimal foraging theory, (2) whether the Cocamilla are "time-minimizers" or "energy-maximizers" in Schoener's (1971) sense, (3) whether the view of Cocamilla fishermen as optimal foragers precludes a view of them as "managers" of their lake resources, and finally (4) whether the data on *várzea* lake fish production and Cocamilla fishing patterns can shed light on prehistoric Amazonian populations.

ADAPTIVE RESPONSES
OF NATIVE AMAZONIANS

ISBN 0-12-321250-2

Theoretical Background

Optimal foraging theory is a label that has come to be applied to a variety of concerns centering around the choices predators make about where to forage and what food items to eat. Emlen (1966) and subsequent theorists (Charnov 1973, 1976; Charnov *et al.* 1976; MacArthur 1972; MacArthur and Pianka 1966; Orians and Pearson 1979; Pearson 1974; Pulliam 1974; Schoener 1971) have all treated, in essentially similar ways, a mathematical consideration of cost–benefit ratios in determining how such decisions are made for nonhuman predators. MacArthur and Pianka (1966) initiated studies of patchy, or "rough-grained," environments. They used assumptions of economic behavior to determine how predators that spend large amounts of time searching in an environment in which prey were not evenly distributed should select their prey. Optimal foraging and related theories have been extended to human foragers by a number of anthropologists very recently (Behrens 1981; Feit 1973; Hames and Vickers 1982; Hawkes and O'Connell 1981; Heffley 1981; McCay 1981; O'Connell and Hawkes 1981; E. Smith 1981; Winterhalder 1981a, 1981b. See also Beckerman, this volume; Hill and Hawkes, this volume).

Pyke *et al.* (1977:140) classify most optimal foraging models into categories reflecting four decisions predators make in order to forage efficiently: (1) decisions about which patch types to visit, (2) decisions about how long to stay in each patch, (3) decisions about which food types to eat, and (4) decisions about which foraging path to employ in each patch. At the same time, they recognize another area within optimal foraging concerns that has remained largely speculative and qualitative, that is, patch modification. Patch modification refers to actions taken by foragers that directly affect the environment in such a way that optimal action at a later time is modified. There are two possibilities for such an effect; the predator may either depress or enhance the resource. Resource depression was treated by Charnov *et al.* (1976) in a paper that recognized the importance of "behavioral depression," a phenomenon in which behavioral changes in the prey induced by the predator—improved hiding, changed foraging schedule, etc.—may outweigh resource depression due to direct exploitation and may also enhance the resource for competing predators. Alternatively, the predator may enhance the resource. This possibility is considered in Pyke *et al.* (1977) who predicts that such a phenomenon is most likely to occur with a predator that has more or less exclusive use of a patch. Whether exclusive use referred only to conspecifics is not made clear; "If an animal has such exclusive use of an area, then it could 'manage' its resources for 'sustained yield' rather than maximize the initial yield at the cost of poorer yields later" (p. 140).

This chapter argues that the fertilization of the lake by the Cocamilla at

Achual Tipishca constitutes patch modification as described above. That is, the fertilization of the lake produces a situation in which the fertilized zone becomes the most efficient zone of the lake to fish at times when the lake is flooded. Thus, the optimal action "at a later time" is changed by virtue of a direct effect that the Cocamilla have upon the environment. Whether the Cocamilla practices constitute "management" is another issue. Certainly the yield of the lake is sustained for the Cocamilla by their cultural practices of fertilization, but the question whether the Cocamilla are "maximizing initial yields" is open. I will argue in the concluding section of the chapter that management, like strategy, may be defined either cognitively or behaviorally (see Behrens 1981), and that behavioral management is quite compatible with a conscious strategy of maximizing fishing efficiencies.

The following parts of the chapter treat separately the method of data collection, a brief presentation of the cultural and historical context, a discussion of fishing resource areas and the tools and techniques used to exploit them, and a discussion of the data as they apply to optimal foraging theory and our notions of Amazon riverine populations now and in the past.

Data Collection

The data on which this chapter is based were gathered during the course of 22 months of fieldwork among the Cocamilla Indians of the Huallaga River *várzea* in eastern Peru. Three main methods of data collection were employed. The first method consisted of random visits. Following a method developed by A. Johnson (1975), random visits were made to village households over a 12-month period after an initial 6-month period to refine the procedures and the coding. The visits were made five times each day at times selected at random in 15 minute units between the hours of 5:30 A.M. and 8:00 P.M. The household to be visited was also randomized. During each visit, the activity of each member of the household was ascertained and coded. This method resulted in approximately 15,000 coded observations of the daily activities of persons of both sexes and all ages, which resulted in a computer printout showing time allocation by sex and age for several hundred activities.

The second method utilized records kept by informants. Three families were selected to fill out reports on their daily activities. Reporting included the location of daily work activities (field, lake, stream, forest, etc.), the time spent getting there, the personnel involved in the activity, the time spent working, the techniques of work (net fishing, spear fishing, hook-and-line fishing, etc.), and the type and weight in kilograms of the product brought back to the village. This method was used by Bergman (1974) in work among the Shipibo Indians. Informants were given watches, weighing containers, and scales. Reports were collected, coded, and checked daily for a 12-month period. This system resulted in approximately 5000 separate observations of work activities and provided the basis for calculation of work efficiency in terms of time costs/product benefits. The system also served as a check on the reliability of the random visit method.

The third method employed was that of participant observation. Daily interaction with community residents over a period of 22 months provided a check on the accuracy of quantitative methods employed and a thorough understanding of the qualitative meaning of the quantitative data. This method was employed in the dietary studies reported in this paper. In order to collect the data, my wife and I spent a week at each of three different times of the year actually living in the household of the subject of each study, weighing and measuring everything the family consumed. During the course of research, a good deal of time also went toward examining documentary materials on the ethnohistory of the Cocamilla and the general historical background of the region. This information has been published elsewhere (Stocks 1976, 1978, 1981).

Historical Background of the Cocamilla

The Cocamilla Indians are a subgroup of the larger Tupian-speaking Cocama. They are the westernmost Tupian Indians in South America. Although the Cocama are known to have inhabited a large portion of the lower Ucayali River aboriginally, the Cocamilla migrated to the neighboring lower Huallaga River sometime before initial European contact around 1540. Figure 8.1 shows their approximate location at contact. Intensive contact with Europeans in the form of Jesuit missionaries and soldiers did not take place until about 1640 when the Cocamilla were placed in missions. The Cocamilla have continued to occupy an ecological niche as *várzea* fishermen and slash-and-burn horticulturalists under European and Peruvian domination ever since. Until recently, their main economic function for the developing upper Amazon economy was as canoemen for exploration and commercial trade, peon labor in fields, and suppliers of riverine mammals and fish for local *haciendas* and urban areas. As recently as 1935, however, few of them spoke Spanish. Today they are in the economic position of peasants, involved in the cashcropping of plantains, rice, and jute (*Malva urena*). At present, fishing in the study community is mainly for subsistence; although the Cocamilla theoretically could sell more fish to the markets than they do, they are concerned with maintaining the production of the lake for subsistence. More highly capitalized

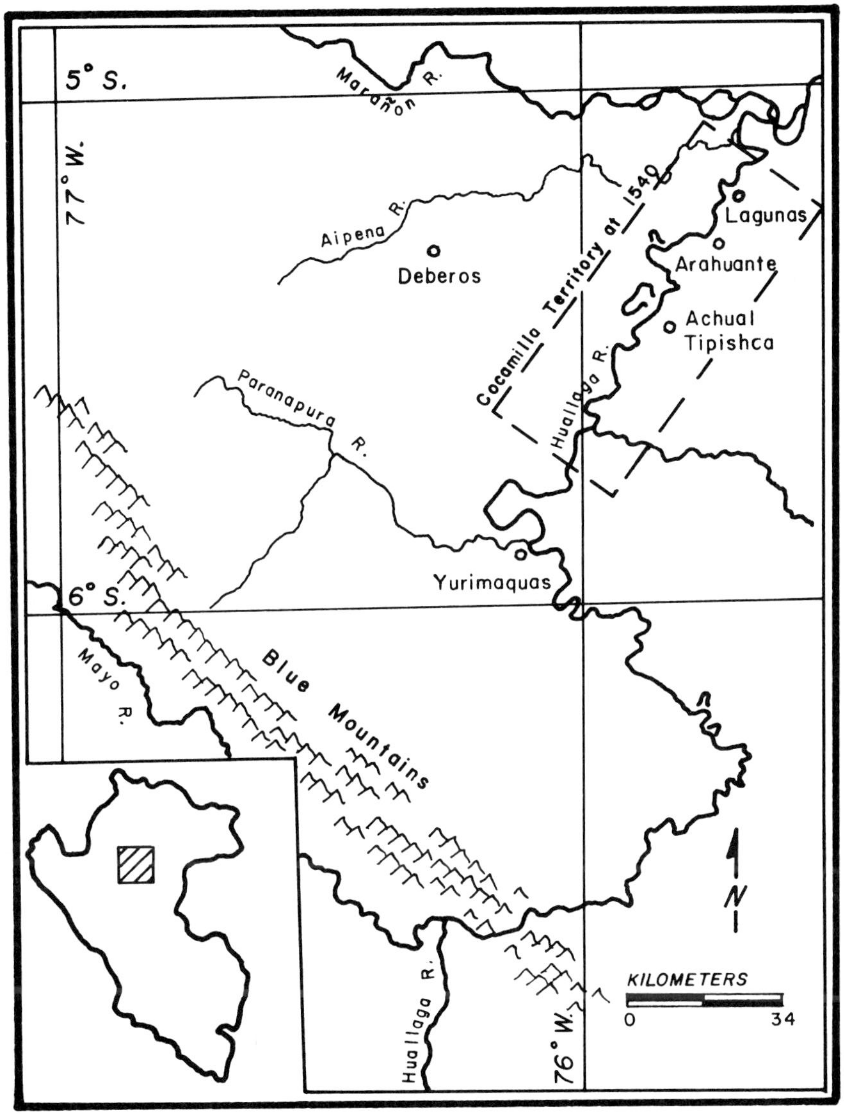

Figure 8.1 *Lower Huallaga River, Peru.*

fishermen have begun to supply the urban markets and the Cocamilla often work for such entrepreneurs as crewmen or, more often, in a form of debt peonage in which the entrepreneur advances salt and other fishing supplies (the *adelante*) and the individual Cocamilla sells salt fish back to the *patrón* (*habilitador*) at rates set by the latter. However, during the course of my fieldwork with the Cocamilla in the village of Achual Tipishca there was little commercial fishing on their *várzea* oxbow lake because the Cocamilla resisted such incursions into their lake domain, and the commercial fishermen were unsure what power I might have to keep them out. Therefore, the data on fish production should be a reasonably good, if conservative, estimate of typical annual production for subsistence purposes. It does not include fish caught clandestinely by commercial fishermen or by the neighboring villages.

Aboriginally, Cocamilla riverine subsistence was probably based much more on the exploitation of river turtles (*Podocnemis unifilis* and *P. expansa*) and aquatic mammals such as the manatee (*Trichechus inunguis*) than it is at present. Such is the case for all Tupian groups on the main Amazon that are known historically. These animals were largely destroyed over the main course of the Amazon River during the nineteenth century when their meat and oil were commercialized (N. Smith 1979, 1981c). An additional resource, present during precontact times but largely absent today, was the black caiman (*Melanosochus niger*). Like the freshwater dolphins today, the caiman was a direct competitor with humans for fish, and there is some evidence that caiman feces fertilized the lake in the past (Fittkau 1970, 1973). These animals were destroyed along large portions of the Amazon River system during the present century when their hides became valuable. It is likely, therefore, that the carrying capacity of the *várzea* ecosystem was greater in precontact times than it is today.

Although the Cocamilla are known to have used nets and weirs aboriginally, the nets were probably not large, and there were no casting nets as are currently in use (see section on Fishing Sites and Techniques in this chapter). I have not observed the Cocamilla to use weirs in fishing although they report trying to dam up the outlet to the lake on two different years, both times unsuccessfully.

Since 1979, with the formation of a tribal political entity, the Cocamilla Federation, the Cocamilla have made a concerted attempt to obtain titles for their lands, still communally held, and to secure guarantees for their lake as a subsistence resource. They have become militant about protecting their fishing grounds. The Peruvian government has dragged its heels in the matter of protecting Cocamilla rights, but the terms of the Cocamilla struggle are set for some time to come.

There is no completely accurate census of the Cocamilla, but it is estimated that they number about 4000 to 7000 individuals, about 3000 of whom live on the lower Huallaga River *várzea* (Stocks 1976). The remainder either carry on a rural peasant lifestyle on the Marañon River and its tributaries to the north, or have become wage laborers or fishermen in cities such as Iquitos and Yurimaguas. In either case, their self-identification as Cocamilla is often relatively weak. The study community of Achual Tipishca is composed of 304 residents, of whom 50 are healthy adult males, between 20 and 50 years of age, who do the bulk of the fishing.

The Huallaga River population of Cocamilla is fairly diverse socially. The description of fishing techniques and behaviors in this paper is perhaps applicable to an understanding of the Cocamilla as representatives of Tupian *várzea* lake inhabitants in the past, with the exception of changes in subsistence resources noted above, but applies only to the seven communities of the Federation that exist on *várzea* lakes. The large Cocamilla *barrio* in the district capital of Lagunas remains the largest concentration of ethnic Cocamilla, but the *barrio* population as a whole is not representative of the fishing way of life that the upriver Cocamilla carry on.

The Ecological Context

Várzea lakes are unique. The *várzea* lakes of the upper Amazon River system constitute a biome that is in a permanent state of violent flux. The rise and fall of the Huallaga River is continuous, with a peak flood season in May, an extreme low-water period in August and September, and a lesser flood peak in November. Because "living" *várzea* lakes (i.e., those not cut off from the river during low water) are connected by canals to the river, the rise and fall of the river is reflected in the lakes.

Figure 8.2 shows weekly fluctuations in the level of the lake at Achual Tipishca over a 12-month period. The total elevation difference between high and low water is more than $7\frac{1}{2}$ m. A comparison of Figure 8.2 with Figure 8.3 will indicate that the peak rainfall and the peak water levels are closely related.

The weekly water level averages do not give a totally accurate impression. A 24-hour period may see the river and its associated lakes rise or fall as much as 2 m. As a constantly disrupted environment, the *várzea* lake and river system has selected for numerous species of fish that multiply and grow rapidly. This feature is predictable in heterotrophic systems (Odum 1971:252). Many of the species of fish seem to be what Pianka (1970) has termed *r*-selected species, species whose survival strategy revolves around maximizing reproductive capability (r) by rapid growth to maturity and weedlike egg dissemination. Nights on a *várzea* lake are often full of the sounds of fish croaking under water or pattering on the surface. Lucky catches with nets, even during the flood season, may yield hundreds of kilograms of fish. *Arapaima gigas*, one of the larger and better-known denizens of the lake systems, grows to a full 2 m in 5 years or less. The heterotrophic aspects of the *várzea* can also be seen in the plant communities along the river, communities that have large monospecific stands of grasses and cane and are in sharp distinction to the high diversity of the tropical forest away from the rivers.

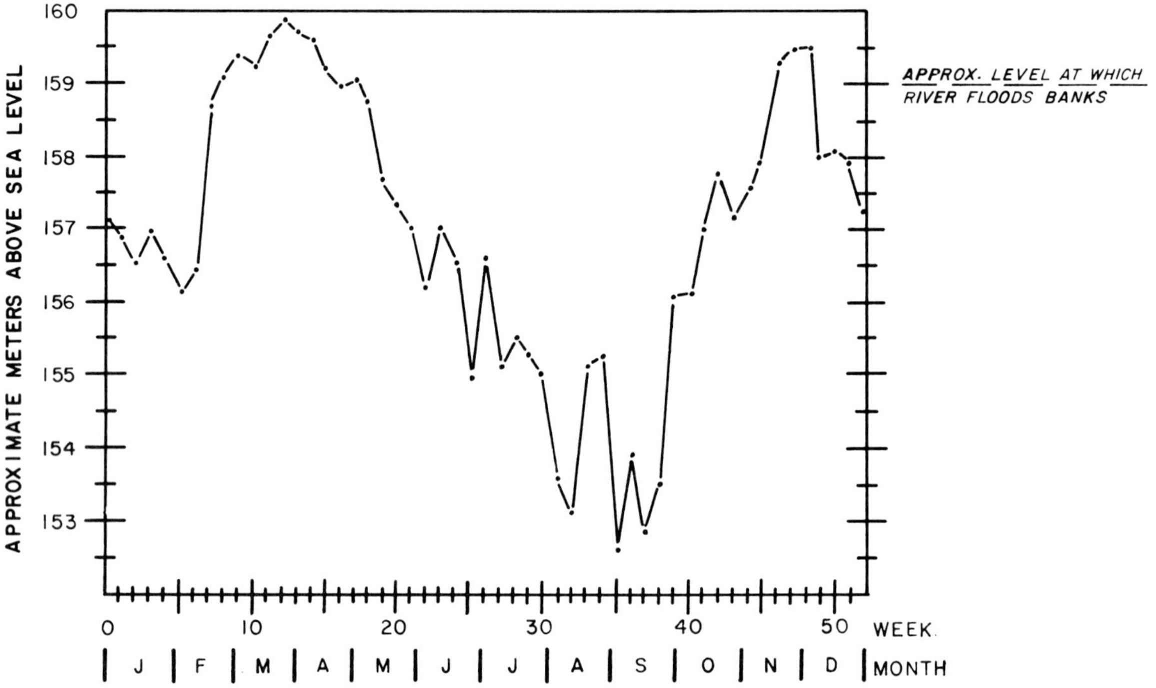

Figure 8.2 *Weekly average of lake-water level, January 1977 to January 1978.*

For all the high primary and secondary productivity of the lake and river system, the lake-dwelling Cocamilla annually are presented with a fish shortage. This is not to say that the protein problem reaches crisis proportions in the sense that other protein sources simply are not available.[1] But when the lake floods, fish "go into the trees," as the people say. Dietary studies made during the fieldwork show that protein intake drops from 75 to 50 g/day/person during the flood season. Fish intake drops from 17.2% to 4.3% of the total weight of food consumed. Thus, there is a relative and consciously perceived strain on subsistence when the peak flood season comes from March to May and again in November. I believe this strain is sufficient to account for the cultural practices related to garbage disposal and personal hygiene described in this paper, but because the Cocamilla themselves do not make the connection in the way that I do, it is difficult to substantiate this point. Figures 8.4 and 8.5 show that the *efficiency* of fishing peaks during low water, whereas the *intensity* of fishing peaks during the floods. Together, the two figures support the assertion that the problem in catching fish during high water is sufficient to be perceived by the Cocamilla.

During high water, the minnows of fish that spawn in the rivers (locally

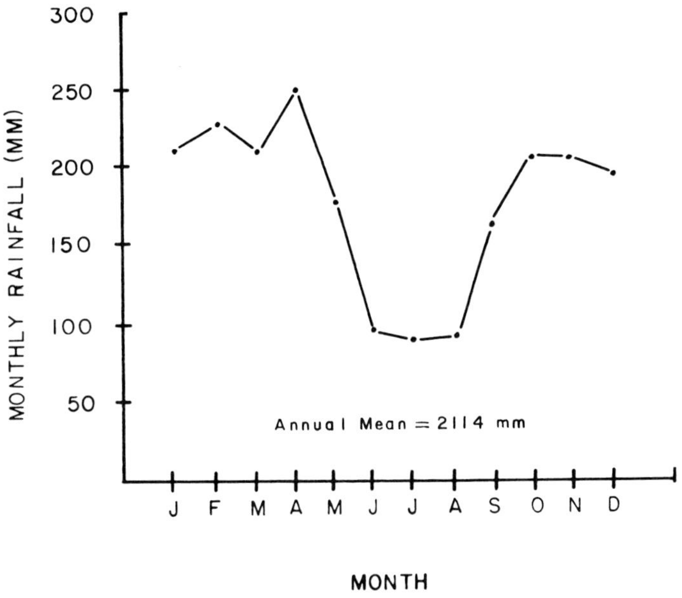

Figure 8.3 *Twenty-one-year average rainfall in Yurimaguas. (Data from Soil Science Department, North Carolina State University, 1974.)*

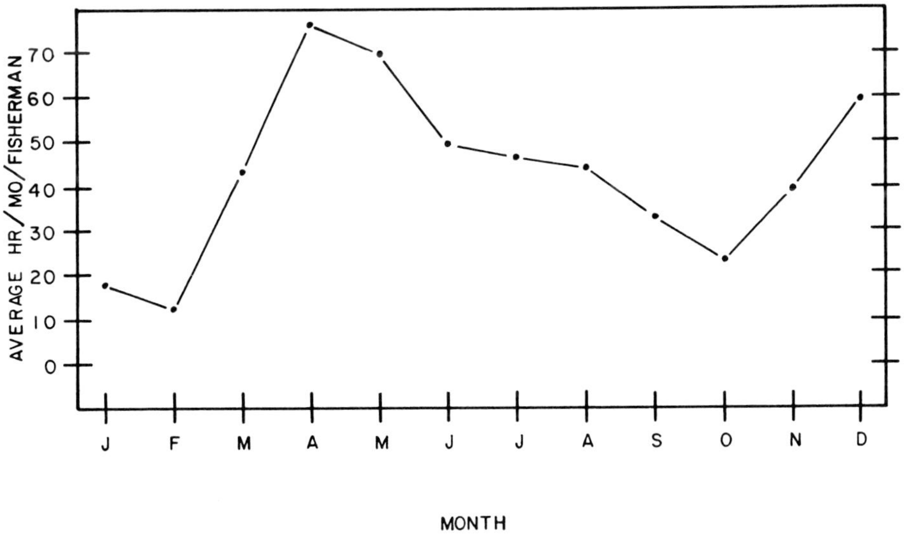

Figure 8.4 *Average intensity of fishing effort (hrs/mo/fisherman) in all fishing areas combined. (Data from records on three families.)*

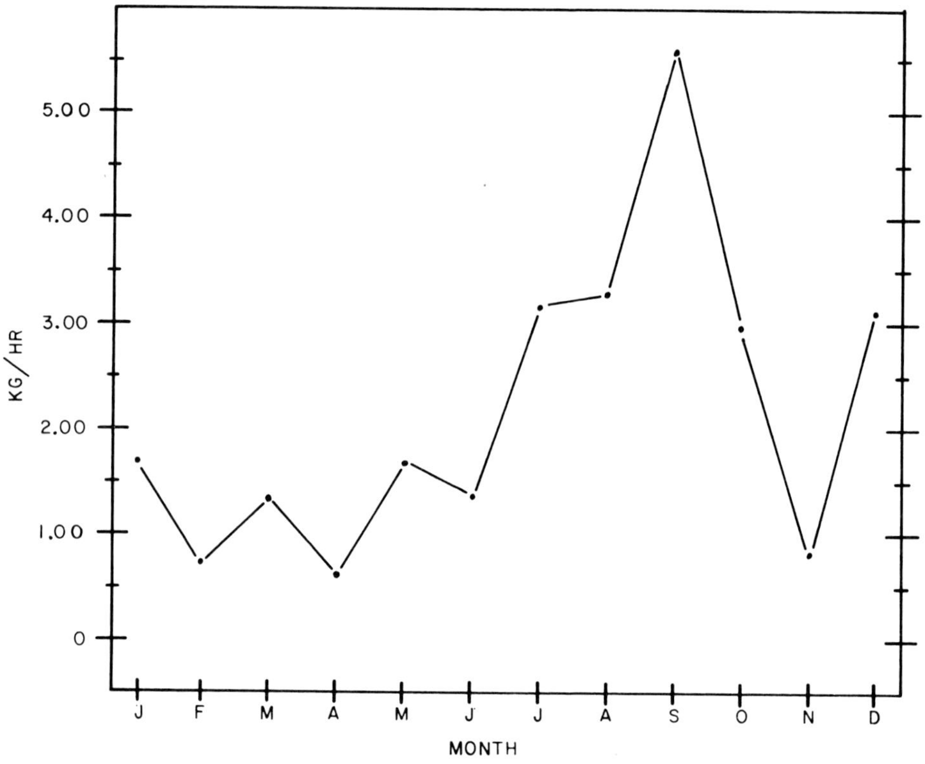

Figure 8.5 *Average efficiency of fishing in (kg/hr) in all fishing areas combined. (Data from records on three families.)*

referred to in the aggregate as "white fish," prominently including members of Prochilodontidae, Characidae, and Pimelodidoe) are found in the lake and in the flooded forest where they fatten and grow. As the water level drops in June, a phenomenon called the *mijano* (a Quechua (?) word that glosses as "fishrun") occurs when these fish leave the lakes and migrate to the rivers to spawn. The fish remaining in the lake (nonmigrating "dark" fish including many members of the Loricaridae, Doradae, and Cichlidae families, as well as some Characidae and Pimelodidae) are often caught in abundance during the low-water period. Lake fishing is relatively easy and productive during this time.

Fishing Sites and Techniques

The major Cocamilla fishing locations at Achual Tipishca may be divided into five general types; four of them are relatively unimportant to overall subsistence in quantitative terms. The types are the following: (1) local streams that originate inland and empty into the river or the lake, (2) drying lakes left in shallow depressions in the forest when water levels fall during low water, (3) dying oxbow lakes, often choked with vegetation, (4) the main river channel, (5) the lakes of the *várzea* that are still connected to the river by canals (*caños*) throughout the year. Figure 8.6 shows the area of Achual Tipishca with the locations of typical water features.

The streams are normally fished only during the dry season from June to October, and are not heavily exploited during the rest of the year. During the dry season, the streams are sluggish and fish can easily be caught in the nearly stagnant pools. The normal fishing technique involves the use of *barbasco* fish poison (*Lonchocarpus nicou* or *Jaquine* sp.), the roots of which are crushed at the fishing site and mixed with the water when the stream is low. The fish, finding their oxygen supply cut off by the drug, rise to the

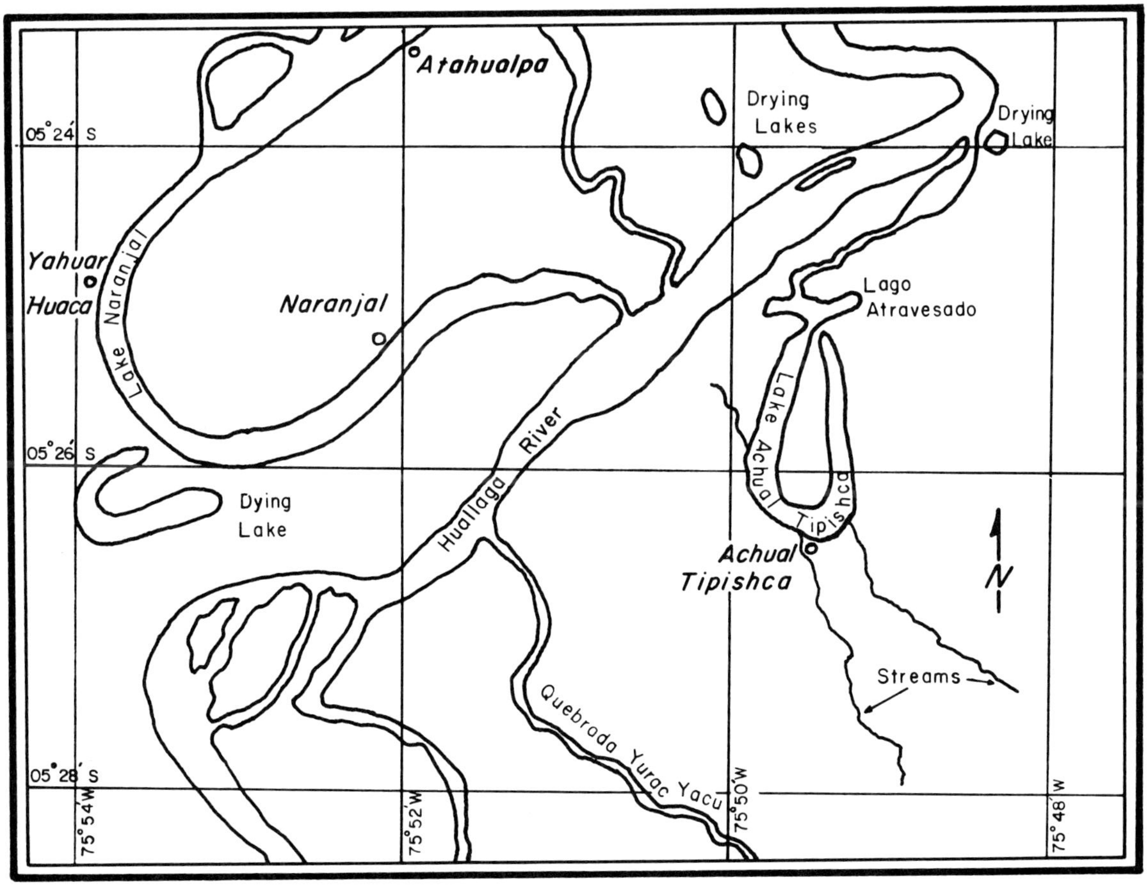

Figure 8.6 *Achual Tipishca area.*

surface. Smaller fish, stunned by the drug, are usually ignored. The larger fish thrash about in the water and are speared, hacked with machetes, or netted in baskets. In areas of low population density, a poisoning party composed of men, women, and children (about six people seems to be the mode) can be surprisingly successful at harvesting a 200–500 m section of stream, netting as much as 0.1 kg/m^2 of stream in an hour or two. The streams running into the lake of Achual Tipishca are generally less productive due to use of the streams by the Cocamilla and exploitation by mestizo settlers who have recently appeared in the inland areas. Less than 1% of all fishing involves the use of fish poison (see Table 8.1 for a breakdown of the frequency of techniques).

Other fishing resources of relatively minor importance over the annual cycle are the lakes not directly connected to the river during the low-water periods. These include two types that I have called "dying" lakes and "drying" lakes. The drying lake is a shallow depression filled with water left over from the flood. Because these lakes tend to be seasonal, they are not filled with aquatic vegetation. When they are discovered, if they are accessible, they are intensively fished with gill nets in comprehensive deployment patterns that lead to a rather complete harvest. Often they are sources of river turtles for the brief period of their exploitation. Occasionally, large *paiche* (*Arapaima gigas*) can be harpooned in such lakes with metal-tipped, detachable-head harpoons. Dying lakes are old oxbows that have become cut off from the river, choked with vegetation, and deoxygenated. Their exploitation at present is almost exclusively for *paiche*, an air-breathing species. These lakes seem to be the main breeding ground for this fish, perhaps because they are not heavily exploited due to the difficulties of maneuvering canoes in the choked water. In the past, special ritual precautions were necessary before fishing in such lakes. They were thought to harbor protective spirits, often the spirits of departed shamans, in the form of giant anacondas.

Table 8.1

Percentage of Time Spent in Various Fishing Techniques, for Males 20–50 Years Old[a]

Fishing technique	Percentage
Hook and line	29.9
Gill nets	29.3
Circular casting net	23.8
Spear or harpoon	16.3
Fish poison	0.7
	100.0

[a]Data from time allocation study (random visits) over a 12-month period.

The Huallaga River's main channel is exploited by the Cocamilla in areas along sandy or muddy beaches, or in areas where there are secondary shallow channels. Fish are usually speared with metal tridents or captured with circular casting nets in those areas. The Cocamilla lack, on the whole, large deep-nets (seines called *honderos*) capable of harvesting fish from the main channel in deep water. Their *honderos* are much smaller than those of most commercial fishermen, and are useful in parts of the lake but not in the river. Needless to say, river fishing is most successful during low-water stages, especially during the late June *mijano*. For Cocamilla who fish mainly for subsistence and who have access to lakes, therefore, the river is of relatively little importance in terms of their total fishing effort. It is significant that the main concentrations of people along this part of the Huallaga River are on lakeshores, not on the riverbank, despite the fact that the river is the main avenue of transportation and communication. Because the lakes are somewhat inland, the riverbank may often appear relatively unpopulated, a fact that might cause a reconsideration of the reports in Spanish chronicles of large stretches of "unpopulated" riverbank.

The exploitation of the lake is complex and important. About 94% of the total fishing done by the Achual Tipishca fishermen takes place either on the lake or in the canal that connects the lake to the river. Hence, the remainder of this chapter will concentrate on lake fishing. I would like first to describe the techniques and the personnel involved in lake fishing and fish storage, then move on to a description and analysis of the cultural methods used to buffer the annual fluctuations of the water level.

Techniques and Personnel Involved in Lake Fishing

As Table 8.1 indicated, the bulk of the fishing done by the Cocamilla today is with hook and line (*anzuelo*) relatively small gill nets (*tramperos*), and circular casting nets (*tarafas*). Most of the fishing with hook and line is done by men and boys from canoes along the shoreline of the lake during the daytime hours. The shoreline is fairly regular except for small bays where streams empty into it, and is deeply shaded from the trees and mangroves that crowd the banks. The bays are preferred fishing spots. The fisherman sits in the prow of the canoe, often guiding the canoe with a paddle held in one hand while he fishes with the other. The paddle never leaves the water. Fishpoles, made from a hard willowlike bush (*tortuga*, species unknown), are normally used. These do not exceed 2 m in length. The lines are rarely more than 2 m long, and the preferred landing technique is to jerk the fish directly into the canoe when a bite is felt. Fishermen know exactly which variety of fish to expect from the context of the fishing hole and often from the wave pattern that a fish will make when it is startled. The shoreline of the lake seems to be zoned horizontally and vertically. Different fish species are expected at different depths and at different distances from the shoreline. On a good day, production from hook-and-line fishing can be quite high, with the fisherman harvesting as much as 2 or more kilograms of fish per hour of fishing time.

Gill nets are of three types, the *hondero* or deep net, the *trampero* (trap), a shallower net, and the *arrastrador,* a purse seine that is deployed from the shore with long handlines on each end. The *trampero* is the most common of the nets used by the Cocamilla and seems to have been part of the aboriginal subsistence kit. The net may be as long as 30 m, but is usually smaller. It is rarely more than 3 m wide. *Tramperos* are sometimes deployed by pairs of canoes working together and sometimes in stationary positions in the forest during the flood season to catch the large *gamitana* (*Colossoma macropomus*). When pairs of canoes work together, the net is anchored at one canoe while the other canoe lays out the other end of the net in a semicircular pattern. The men in both canoes then rapidly haul in the handlines on the end of the net, standing up in the canoes to do so and skidding the canoe sideways across the water. The *hondero* is sometimes employed in this manner, but is just as often left overnight to entangle passing fish. The *arrastrador* is most often used, by those few Cocamilla that have them, to fish drying lake bodies or river beaches. They are often anchored to the shore while men wade out into the water to deploy the other end. The danger of fishing in this fashion is that of stepping on a stingray, a hazard that is extremely common in most of the Amazon Basin rivers. To avoid this possibility, the fishermen probe carefully in front of themselves with a spear as they proceed in order to drive the stingrays away.

The use of the circular casting net often involves both men and women. The role of the woman is to paddle the canoe, and she sits in the stern to do so. The casting net weighs as much as 25 kg and it takes strong arms and shoulders to fling it from the bow of the canoe far enough out so that it can open up. The thrower positions himself in a standing position at the extreme front of the canoe and balances himself precariously with one foot in front of the other. When fish are seen, the net is thrown and hauled in by a centrally attached line which causes the net to close. The bottom (outer edges) of the net is heavily weighted with lead to drag the bottom while the net closes. The use of the circular casting net is relatively new to the Cocamilla, and there is some evidence that the disturbance of the bottom by the drags is ecologically destructive. Figure 8.7 shows the lake of Achual Tipishca. Zone 4, Punta, was formerly the site of much activity with circular casting nets because it is the shallowest portion of the lake. It is seldom used today and both the fish and the fishermen seem to shun it.[2]

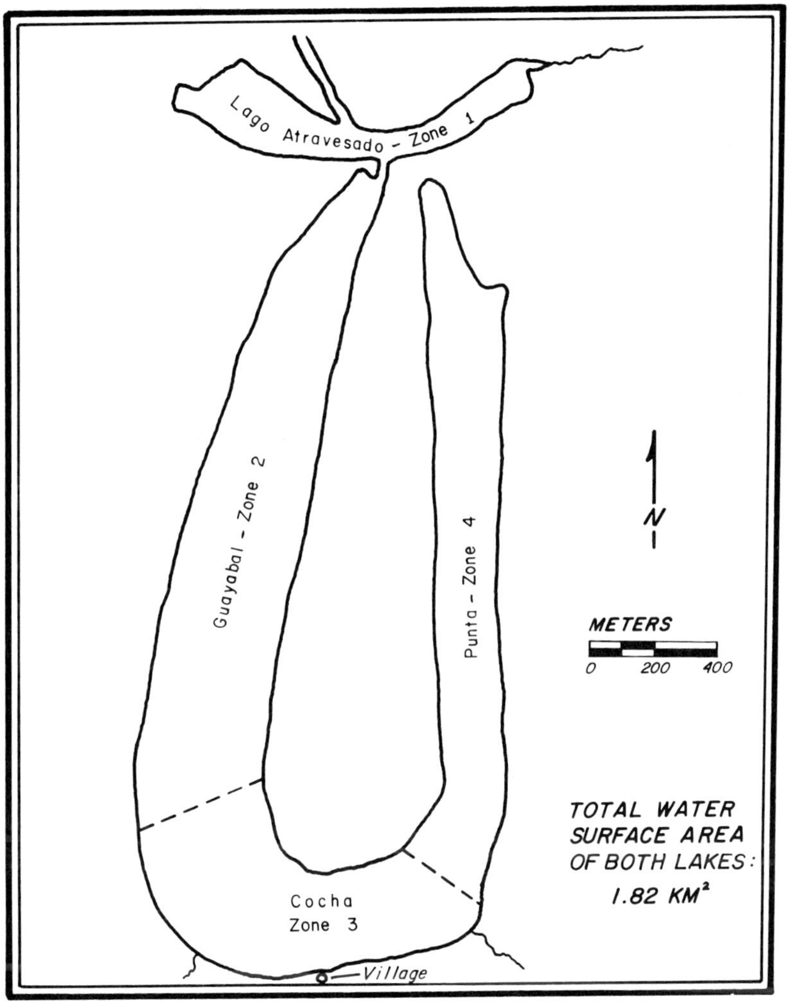

Figure 8.7 *Fishing zones of Lake Achual Tipishca.*

The use of the spear in the context of lake fishing was probably much more common aboriginally than it is today when the Cocamilla have access to nylon for large and durable fishnets. Fishing with spears is usually carried on when the lake level is falling rather than rising. Fish are sometimes speared from canoes during the daytime during low-water periods. The use of the spear requires much dexterity in order to compensate for light refraction in the water and in order to hit sometimes distant targets. It is a skill which young male Cocamilla begin to acquire at an early age. The spear fisherman usually fishes alone and maneuvers the canoe along the lakeshore using one hand for the paddle. Larger fish such as the *fazaco* (*Hoplias malabaricus*), the *tucunaré* (*Cichla ocellaris*), and the *turushuque* (*Oxydoras niger*) are preferred, although the latter is often large enough to require a harpoon. During the extreme low water period, Zone 1 in Figure 8.7 (Lago Atravesado) is sometimes fished with spear when the fisherman cannot see the bottom. The spear is simply pushed into the mud at the bottom of the lake on the chance that a fish is there. Population densities of the *carachama* (*Pterygoplichthys multiradiatus*), a bottom-feeding loricarid that can live for long periods out of the water and seems to be adapted to surviving in drying lakes by "walking" to another water source, are quite high in the shallower muddy areas and the success rate is considerable. Spears are also used at night. Fishermen always prefer moonless hours of the night to fish with any technique. Spear fishing at night is done today with flashlights. The technique involves probing the shoreline with the canoe about 1 m offshore. Many fish inhabit the extreme edge of the lake at night, perhaps for the cooler water to be found there, and they are explosed by the flashlight and speared. Most spearfishing takes place at night, but only during periods when the water is low and falling.

The remaining technique used in lake fishing involves the use of fish poison. Traditionally, the entire community paddles canoes to a part of Lago Atravesado during December for a communal fish hunt with poison. A small bay of the lake is closed off with nets and the poison is put into the water. I witnessed one such hunt in which most people speared fish from the canoes, but smaller boys and girls perched on logs to collect fish that were hard to get from the watercraft. Some fish were later collected from the nets used to seal off the bay.

Storage of Fish

The aboriginal technique was to sun-dry the fish on platforms constructed for the purpose. According to testimony, only the larger fish such as the *paiche* (*Arapaima gigas*) and the *gamitana* (*Colossoma macropomus*) were sun-dried. Filets not more than 4 cm in thickness were carved out of the fish for drying. Today, the Cocamilla use salt purchased from rafts that ply the Huallaga River from the Andean foothills. They now dry and salt all sizes of fish, especially for the season in which commercial boats come into the lake to purchase fish. The salt resource may have been available aboriginally, at least on a sporadic basis, but this is uncertain (see Varese 1973 for a description of the historically known salt trade). Turtles, however, were certainly stored in precontact times in pens along the riverbank and lakeshores until they were needed. As noted earlier in this chapter, the river turtles were largely eliminated from the upper Amazon region during the nineteenth century by capitalist exploitation.

Buffering Environmental Fluctuations

The Cocamilla have adapted to the annual fluctuations in lake-water level in a highly interesting way. The method used is essentially that of fertilizing that part of the lake closest to the community (see Figure 8.7, Zone 3, Cocha) with garbage, human feces, and fish and animal entrails, a practice that seems to maintain a dependent population of fish in that particular area. This population is harvested most intensively during the two flood peaks in April–May and in November.

The explanation offered in the preceding paragraph amounts to a hypothesis and obviously needs support, particularly because the Cocamilla do not consciously recognize the systematic nature of their complex relations with the lake beyond an admission that they do indeed tend to fish close to the community during high-water stages because fish are hard to catch elsewhere. The lakeshore that fronts the community is a repository for an enormous quantity of organic material. Each of the 41 residences in the community daily throw plantain, banana, and manioc peels onto the shore. The traditional location for defecation in the past was about 25 m up a small stream that empties into the lake just east of the community. The stream is called, with admirable simplicity, *Capichiru* (''shit creek''). While this custom has diminished for those not living next to this stream, a surprising number of adults and children defecate in the area of their port, close to the water. Older people often defecate from canoes offshore or go to the shore opposite the community. The preliminary reaction of the foreign observer tends to be negative and visceral inasmuch as many community activities—bathing, washing clothes, toothbrushing, visiting, and, often, water collecting—take place at the lakeshore. Furthermore, the unused portions of the entrails of animals such as wild peccary, deer, and domestic pigs are thrown into the water at the port. This practice is supported by a belief that if village dogs eat such material it will ruin them for hunting. Fish are usually cleaned at the lakeshore and their entrails thrown to the piranha. Even fish cleaned in private residences often have their entrails carried to the lake for disposal.

Fish can often be observed nibbling at garbage and human stools; the *piranha* are particularly noticeable. Because water levels fluctuate daily and even feces deposited on land are close to the water, most organic material is incorporated into the lake within a few days of deposit.[3] If we, as a ballpark estimate, assume that 50% of the feces end up in the water and that the 203 adult male equivalents produce 0.25 kg/day of solid waste/day (exclusive of water content) we can calculate that approximately 9 metric tons per year of feces are added to the lake each year. Add to this the approximately 5.5 metric tons of animal and fish entrails and the 30 metric tons of garbage deposited annually at lakeshore; the total organic material annually fertilizing the port area through active cultural practices is a not inconsequential 44.5 metric tons per year, including at least 100 kg of

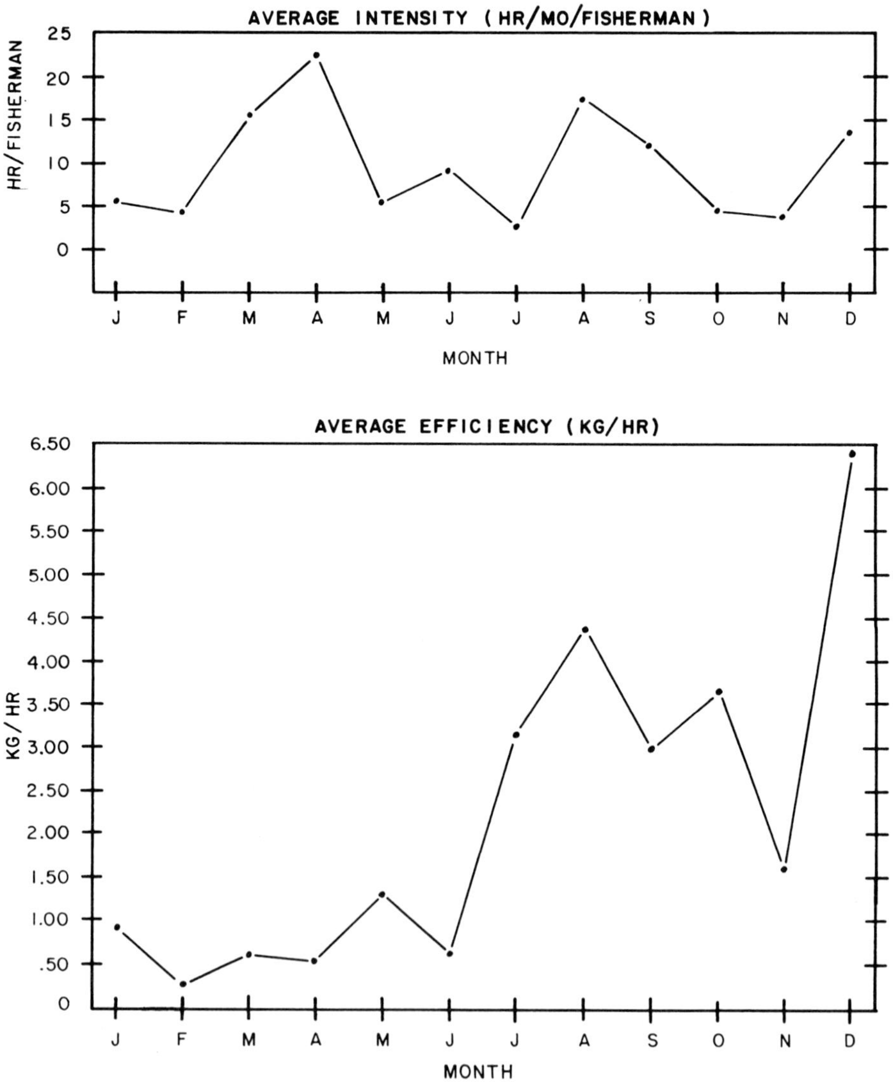

Figure 8.8 *Intensity and efficiency of use in Zone 1, Lago Atravesado, over a one-year period. (Data from records on three families.)*

nitrogen. This figure does not include the urine deposited, which may add an additional 900 kg of nitrogen, nor does it include the effects of having the entire community flushed annually by the floodwaters.

Figures 8.8–8.10 show (a) the average time/month/fisherman plotted against (b) the average efficiency in kilograms of undressed fish/hour for

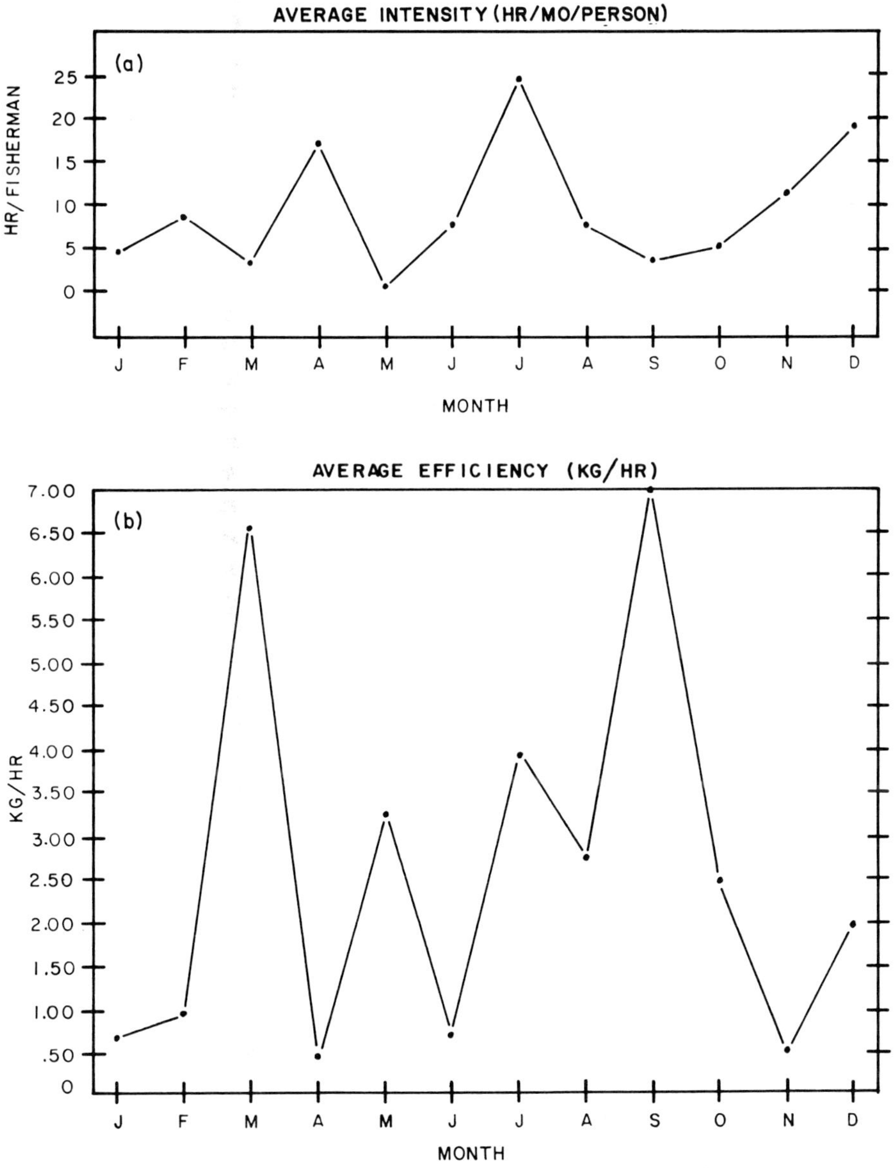

Figure 8.9 *Intensity and efficiency of use in Zone 2, Guayabal, over a one-year period. (Data from records on three families.)*

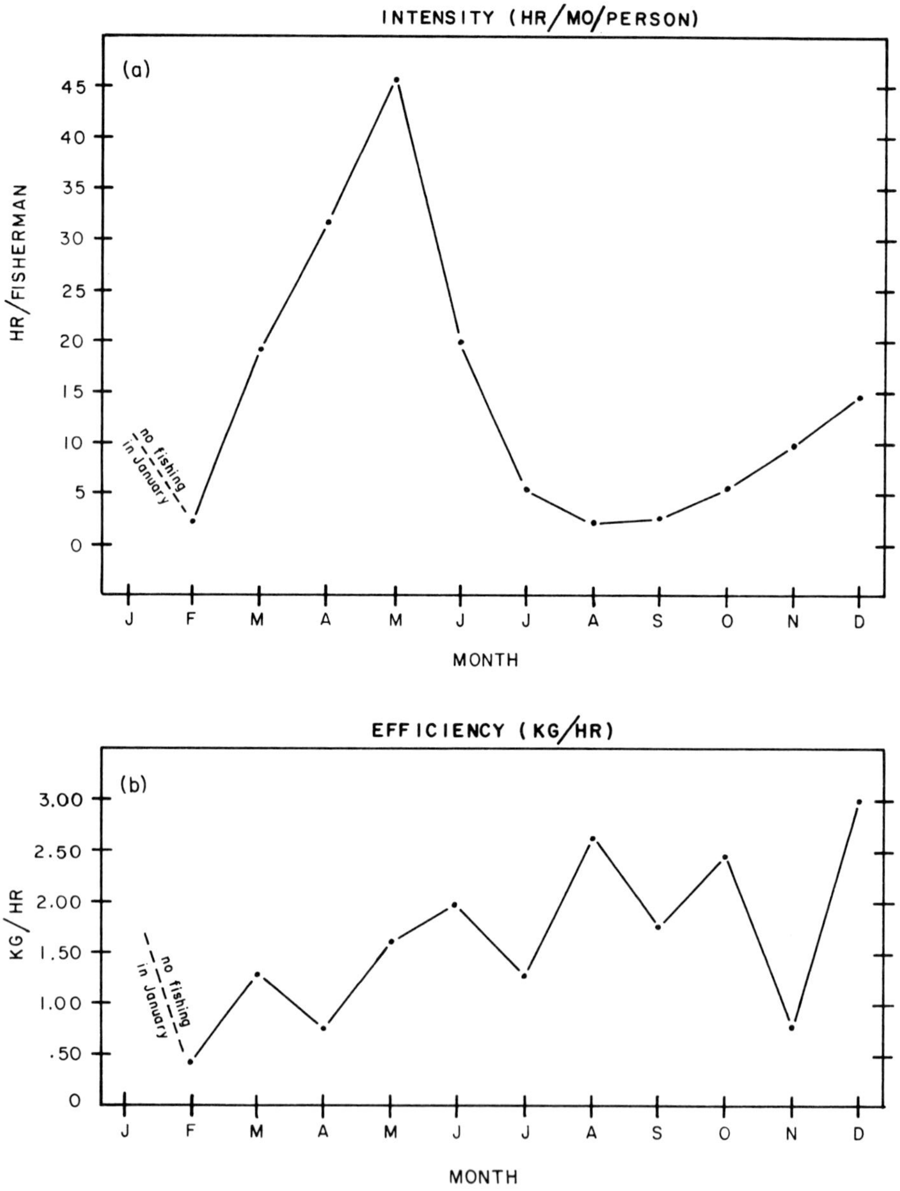

Figure 8.10 *Intensity and efficiency of use in Zone 3, Cocha, over a one-year period. (Data from records on three families.)*

each of three fishing zones shown in Figure 8.7. Data for two of the three families list only the hours for the male head of household because they fished alone or with other males. The catches show only the share of the fisherman. The hours for the third household include both the work of the head of hosuehold and his son, who often fished with hook and line while his father worked in the fields. Approximately 25% of the household fishing was done by the boy.

In Figure 8.8, which shows the intensity (hrs/mo/fisherman) and efficiency (kg/hr) of exploiting Lago Atravesado, a shallow, abandoned river channel, we see considerable use through the year with minor peaks in April, August, and December. The efficiency of use of this area rises sharply during low-water periods in July, August, and late December. The extremely sharp December efficiency peak is nearly entirely due to one enormous lucky net catch when one informant went out with several other men and a large borrowed net.

Guayabal (Zone 2 in Figure 8.9), which is located at the outlet of the oxbow lake, shows extreme variation in both intensity and efficiency. Again, the sharp peaks of efficiency in March and in September are due to single extraordinary catches which were surprising even to the Cocamilla fishermen. Little consistency is noted in the pattern for this area when considered alone.

Cocha (Zone 3 in Figure 8.10), which is the fertilized zone nearest the community, shows a highly interesting pattern. The intensity of use for this area is a duplicate of the pattern of the rise and fall of the lake. Use increases dramatically as the lake rises and drops just as dramatically as the

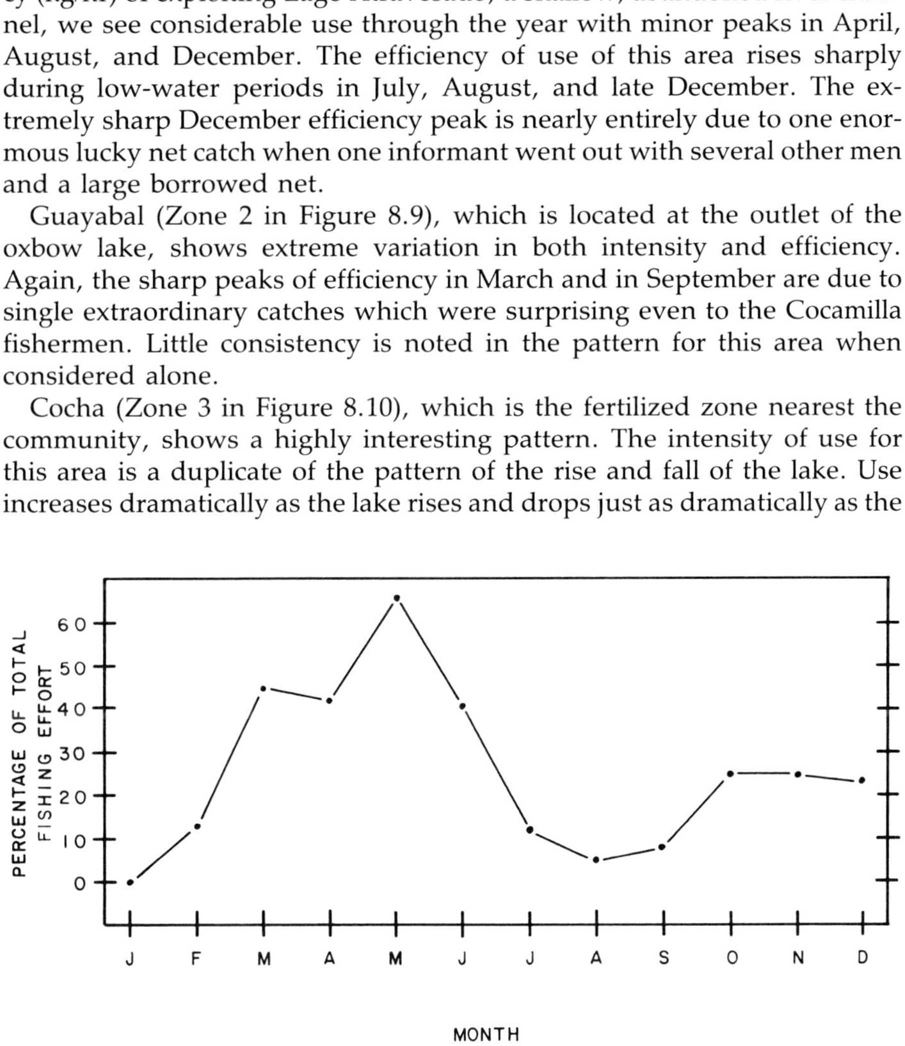

Figure 8.11 *Time spent in Zone 3, Cocha, as percentage of total time spent fishing in all areas (streams, rivers, and lakes). (Data from records on three families.)*

lake falls. Efficiency varies less than in other areas during the year. This area occupys a considerable amount of time in the high-water period even when its use is calculated as a percentage of total fishing time in all areas, as Figure 8.11 shows.

To explain these patterns, we must look to a slightly different organization of the data, a comparison of the efficiencies and intesities among the three areas on a month-by-month basis. Table 8.2 shows this comparison; it reveals that the Cocamilla are, for the most part, behaving in a manner predictable by optimal foraging theory; that is, during 9 months of the year they are fishing that part of the lake where efficiencies tend to be the highest. I have indicated the values in which maximum efficiency and maximum intensity coincide in any given month and have eliminated from consideration those efficiencies that are unduly influenced by one catch. One can clearly see that the fertilized zone of the lake becomes the most efficient part of the lake during the entire high-water period of March through June. As the water in the lake falls, there is a shift in fishing effort toward the lake outlet as the river-spawning fish leave the lake. During low water, Zone 1, the shallowest zone, is intensively fished.

Table 8.2

Comparative Efficiencies and Intensities of Lake Zones 1, 2, and 3

	Zone 1 (Lago Atravesado)		Zone 2 (Guayabal)		Zone 3 (Cocha)	
Month	Efficiency[a]	Intensity[b]	Efficiency[a]	Intensity[b]	Efficiency[a]	Intensity[b]
January	0.93	6.3[c]	.52	4.9	—	—
February	.29	4.5	.91	8.8[c]	.44	2.2
March	.62	15.8	6.55[d]	3.4	1.29	19.2[c]
April	.54	22.9	.49	17.0	.76	31.7[c]
May	1.31	5.8	3.29[d]	0.6	1.61	45.7[c]
June	.54	9.3	.72	7.8	1.98	20.0[c]
July	3.19	3.0	3.95	24.6[c]	1.29	5.4
August	4.39	17.9[c]	2.76	7.7	2.63	2.3
September	3.00	12.5[c]	7.00[d]	3.7	1.77	2.6
October	3.64	4.6	2.49	5.1	2.45	5.7
November	1.60	3.9	.51	11.5	.78	9.9
December	6.42[d]	13.6	1.96	19.2	3.00	14.6

[a]kg/hr.

[b]Hours/month/fisherman.

[c]Values in which the maximum efficiency and the maximum intensity coincide.

[d]Values are the results of one particularly good catch in an otherwise undistinguished month. They are not considered, even by the Cocamilla, to be representative of the overall efficiency for that area in that month and are eliminated from consideration on that basis.

Table 8.3

Comparison of Fishing and Hunting in Terms of Intensity and Efficiency[a]

Month	Intensity (hours/person)		Efficiency (kg/hour)	
	Fishing	Hunting	Fishing	Hunting
January	18.1	12.1	1.69	1.39
February	17.4	6.2	0.72	1.07
March	43.5	23.4	1.34	0.37
April	76.1	11.7	0.62	0.09
May	69.3	1.7	1.68	0.00
June	49.3	0.0	1.35	—
July	47.1	0.0	3.18	—
August	44.8	9.8	3.30	0.55
September	32.9	2.1	5.59	0.00
October	22.9	14.2	2.98	0.01
November	39.8	11.7	0.84	0.41
December	59.7	5.4	3.11	0.48

[a]The fishing shown here represents the effort in all areas, not only the lake. The data are based on the three families' records. The hunting data are based on the same three families. One family did not hunt at all during the year, while another hunted regularly. The third hunted sporadically.

The puzzling feature of Table 8.2 is that during October, November, and December, the Cocamilla are not concentrating on that part of the lake most efficient to exploit. One might suppose from data on other eastern Perurian fisherfolk, the Shipibo Indians (Bergman 1974), that as the water rises, hunting becomes more efficient than fishing and less pressure is therefore placed on the fishermen to follow efficiency in their fishing. This is not the case, however, as Table 8.3 reveals. Hunting efficiencies in October are close to zero, and are only 0.41 kg/hr in November and 0.48 kg/hr in December, all below the expected efficiencies of fishing in any part of the lake at those times.

In summary, the data on fishing intensity and efficiency in various named parts of the lake show that the effect of fertilizing the lake near the community is to create a favorable habitat for fish even during the high-water period. This fish population is reflected in the shift in fishing intensity and efficiency to the fertilized zone during high-water times. Everything happens "as if" the Cocamilla were "fish managers" using a complex pattern of nutrient subsidies and exploitative patterns in order to buffer environmental fluctuation in water levels and fish availability. This systematic aspect of their behavior is not, however, conscious.

Gross Fishing Production

Table 8.4 shows the results of the annual production for three Cocamilla fishermen who reported their own activities for 12 months. When their production is applied to the village as a whole—taking into account the distribution of nets in the village—a conservative figure of 33.8 metric tons can be extrapolated for the village as a whole. Approximately 32.0 metric tons were produced by the lake. The lake is 1.82 km^2 in size. The productivity of the lake over 12 months was therefore .032 kg/m^2, or about 17.6 metric tons/km^2. From ethnohistorical information that I have published elsewhere (Stocks 1978, 1981), it is known that the community has remained at its present location since the lake was formed nearly 100 years ago. The size of the community was precisely the same in 1864 as it is today and, while it has been considerably larger at times, it has never fallen below its present size. Oral accounts of the population increase during the *barbasco* "boom" (when native fish poison was exported to use as a central ingredient in commercial insecticides) of the 1940s through the 1950s state that the fish population was severely depleted when the population of Achual Tipishca reached 500. Over the past 10–15 years, however, the population has stabilized at its 1864 figure and fish production seems to be stabilized as well.[4]

Calculating a 15% wastage of fish, the net weight of dressed fish available for consumption is 28.73 metric tons/year. Fifteen percent of the fish are sold to outsiders in a normal year, leaving 24.42 metric tons for the community. Using Gross's (1975:540) formula for estimating protein per capita yield, I conclude that the fishing production yields a conservative

Table 8.4

Summary of Annual Fishing Production for Three Fishermen

	Number of trips	Total time (hours)	Mean time per trip (hours)	Total catch (kg)	Mean catch (kg)	Mean efficiency (kg/hour)
Segundo[a]	138	609.71	4.42	838.2	6.07	1.37
Javier[a]	112	508.59	4.54	650.0	5.80	1.28
Eleuterio[b]	91	444.93	4.89	1827.0	20.07	4.11

[a]Segundo and Javier fish mainly by traditional methods—hooks, spears and small casting nets—and only rarely with small gill nets. They produce for families of six and five, respectively. The time shown is their own and does not include the time of other family members who may be aiding in steering the canoe.

[b]Eleuterio fishes often with a deep gill net (*hondero*) and fishes in conjunction with his brothers. His production is his share of the catch and feeds eight people. He sells the surplus both in the village and to outsiders.

estimate of 44 g of protein per person per day.[5] This figure is supplemented by minor additions from hunting activities (about 3% of the overall weight of the diet) and from vegetable sources.

Discussion

The data presented in this paper have implications for optimal foraging theory, for theories of carrying capacity in the Amazon *várzea* lakes, and for more general discussions of Amazon ethnohistory and populations. Optimal foraging theory predicts that predators will be either of the time-minimizing or energy-maximizing types (Schoener 1971). Time minimizers are predators with alternative uses for time and fixed energy needs. If an animal is of this type, it should, theoretically forage only until its specific energy requirement is met. Although the Cocamilla data are not conclusive, the overall month-by-month correlation between fishing intensity and efficiency is negative ($r = -0.15$). This correlation is not significant with only 12 data points (calculated from the data shown in Figures 8.4 and 8.5), but it remains clear that intensity and efficiency are not positively correlated. The tentative conclusion is that the better the fishing, the less the Cocamilla tend to fish. They would be closest to Shoener's time-minimizers because, on the whole, they do not fish for longer than it takes them to secure dietary requirements.

The data also show that short-term efficiency considerations cannot always predict where the Cocamilla will fish. The months of October, November, and December find the Cocamilla fishing where it seems *least* efficient to do so. The Cocamilla, if they are proper optimal foragers, should concentrate on hunting in February because only in February is hunting efficiency higher than that of fishing. Yet, from Table 8.3 we see that little hunting was done by the informants in February. Actually, much of the year's somewhat sporadic hunting is accounted for by two factors, both in the consciousness of the Cocamilla. The first of these is that they desire variety in their diet. The second conscious motive for hunting is that wild game is the only acceptable offering for community festivals, a tradition that has been passed down to them from their Catholic past. It was also the custom of Spanish priests and governors, until fairly recently, to request wild game on a daily basis from Indians designated as *mitayeros* (from the Quechua *mita*, "labor assignment"). Wild game is still called *mitayo* in most of the old mission areas of the Peruvian Amazon region. What we can see in the comparative hunting and fishing patterns is an effort to obtain at least some meat throughout the year for reasons both of a preference for variety and of the demands of the festival cycle, and apparently without too much regard for the relative efficiencies involved. The Cocamilla persist in old patterns of hunting even though their area of the Huallaga River seems to be severely depleted in terms of game. In other areas of the lowlands, hunting seems to be a good deal more successful (Behrens 1981; Bergman 1974).

Whether the Cocamilla can be said to "manage" their lake or whether we wish to view the nutrient cycle outlined here as merely a latent function of cultural practices evolved for other reasons, the effect is that of patch enhancement and sustained annual production which buffers environmental fluctuation. I do not believe that these cultural practices are merely coincidental because the Cocamilla support these disposal customs with an array of folklore that suggests that the customs might have some practical value. For that reason, I wish to make an argument that the Cocamilla manage their lake resources "in effect," or behaviorally. It is a most provocative fact that while their conscious behavior seems to be governed by a logic of efficiency and least effort with little thought to long-term production, they are at the same time carrying out behaviors that ensure such long-term sustained production. Moreover, it is this deeper level of patch modification management that determines in the end where it will be efficient to fish during the stressful part of the year. Their adaptation, in short, has both "surface" and "deep" structural aspects, one conscious and supported by a largely economic rationale, the other unconscious and supported by folklore and magical beliefs. What I am suggesting briefly, since I do not have the space to elaborate the position here, is that a view of humans simply as economic persons may miss a great deal of the point. If native South Americans were not good managers at some level, it is difficult to conceive how the *várzea* could have been teeming with aquatic life at contact as reported historically, given the dense human populations reported in the middle Amazon region.

The human carrying capacity of the Amazon River ecosystem has been debated for some time (Denevan 1976; Dobyns 1966; Moran 1973; Steward and Faron 1959). The various estimates are critically examined by Denevan (1976) who produces his own estimate of 8 persons/km^2 in the floodplain habitat. None of the estimates, however, are based on any real data on animal or fish production in that biome. Estimates of populations based on the productivity of riverine biomes have not used reliable or applicable data. Although Gross (1975:528; see Beckerman 1979 for a critique) estimated that the upper Xingú River might reliably yield 5 metric tons of fish/km^2/year, the data in this paper show that a *várzea* lake on the Huallaga River can reliably produce over three times (17.6 metric tons/km^2) that amount. The reliability of this figure is attested by the long-term stability of the Achual Tipishca population, a population dependant on the resources of one *várzea* lake.[6] Because a close examination of Cocamilla subsistence patterns shows that their adaptation is to the *várzea* lakes rather than to the river, one way to estimate aboriginal population densities for natives known to exploit lakes as their main protein source might be to calculate the total lake surface of the *várzea* and multiply that figure by a constant which would represent known lake carrying capacity. From the Cocamilla data, it appears that such capacity cannot be less than 170 persons/km^2 of lake surface, and is probably higher, both in the past and at present. It is apparent from the Cocamilla data that agricultural land, the highly reputed rich alluvial soils of the floodplain, is not, in fact, the determining factor for populations. The Cocamilla leave many areas, close to the river but distant from lakes, relatively unexploited. If this feature of *várzea* adaptation is understood, then the large gaps in populations reported by the early Spanish explorers become comprehensible, not necessarily in terms of warfare and "buffer zones" (DeBoer 1981), a factor that is possible but difficult to quantify precisely, but in terms of measurable resources and their uneven distribution. It is easily possible that there may have been thousands of natives, living on *várzea* lakes that were not visible from the rivers, who were never even guessed at by the early explorers, a point also recently made by Beckerman (1979). The Omagua were undoubtedly highly visible precisely because their section of the Amazon River lacks a wide floodplain and forced their adaptation to the point-bar formations of the islands. One might argue that population densities might well have been similar in the upper Amazon and the narrower middle Amazon valley.

A more general point might also be made here. On the basis of an analysis of another upper Amazon riverine group, the Shipibo of the central Ucayali River, Bergman (1974) has argued that upper Amazon populations were probably never dense enough to have evolved sophistication in recycling or resource management. I would argue to the contrary, based on the data from this paper on fish production and the facts of Cocamilla demographics. The man–fish relationship described here indicates populations in close balance with lake resources and a demographic pattern of strong out-migration and flux to balance the natural rate of population increase. While I would endorse Bergman's view that a subsistence system is as wasteful as it can afford to be, the Cocamilla evidence indicates that the Tupians probably could not afford to be wasteful of their lake resources. In addition, is should be pointed out that there is reason to believe that the Panoans of eastern Peru, as represented by the Shipibo of Bergman's study, have origins in the *tierra firme* Gran Chaco well within the last millenium (Lathrap 1974:136–142), a fact which, if true, should lead to caution in using them to exemplify *várzea* adaptation for the main Amazon. The general ethnohistorical picture might be, therefore, not one of an upper Amazon Basin gradually and sparsely filling until the Conquest, as Bergman suggests, but rather one of the ebb and flow of perhaps quite dense riverine populations over millenia, populations which may have evolved a number of cultural practices that acted to manipulate energy flows in an environment-enhancing way.

Summary and Conclusions

This chapter has attempted to fill in some gaps in our understanding of Amazon fishing ecology. The central point regards the evolution of cultural behaviors among upper Amazon Tupians that operate, quite beyond their consciousness, to buffer the annual fluctuation in fish availability occasioned by the Amazon flood. The analysis of Cocamilla fishing patterns presents a case in which a human predator handles a resource in such a way that optimum production is sustained throughout the year. Furthermore, this paper has argued that optimal foraging theory can predict many but not all Cocamilla fishing and hunting patterns. Historical facts and preference for variety in diets must also be utilized in a complete explanation of these foraging patterns. It was also argued that the carrying capacity of *várzea* lakes is much higher than has been previously supposed. Finally, it was argued that our picture of upper Amazon ethnohistory must be adjusted to account for population densities impelling the evolution of cultural behaviors that operate to recycle large amounts of organic material produced by humans in *várzea* lakes.

Acknowledgments

The fieldwork that forms the basis for this paper was conducted under grants from the following agencies and institutions: The Social Science Research Council supported research with a doctoral dissertation grant and an extension for additional fieldwork and writeup. The National Science Foundation made a grant for improvement of doctoral dissertation research (BNA76-09554). The Department of Health, Education, and Welfare supported me under a Fulbright–Hays grant for doctoral dissertation research for 12 months. The Tropical South American Research Program at the University of Florida under Dr. Charles Wagley supported a preliminary ethnographic census among the Cocama–Cocamilla in 1975 and supported postdoctoral research among the Cocamilla in 1978–1979. Idaho State University supported further research during the Summer of 1981 with a Faculty Research Grant that allowed me to check the accuracy of some specific observations in this paper. Special thanks go to Kathleen Stocks who organized, coded, and supervised the collection of much of the random visit data. The Centro Amazónico de Antropologia y Aplicación Práctica in Peru provided both institutional support and some funding during the course of research. The Universidad Pontificia Católica in Peru was helpful in providing institutional affiliation. Special thanks goes to Alejandro Camino in helping to arrange the details. This paper has benefitted from critical reading by Allen Turner, Raymond Hames, and William Vickers.

Notes

[1]Nutritional studies were made at three points during a 12-month period covering low, mid-, and high water. Each study was a week long and involved the physical presence of me or my wife in the study household all day long during waking hours. All food consumed by each family member was weighed and garbage ratios were determined. The daily per-person protein intake during high-water states drops from 75 to around 50 g/person/day. The caloric consumption remains relatively stable at around 2000 kcal/person/day. During the summer low-water period, fish provide 17.2% and game 0.6% of the total weight of food consumed. The peak flood season changes the ratio to 4.3 and 8.3% for fish and game respectively, while gathered products (snails, palm fruits, etc.) increase in importance from 1.4% of the low-water diet to 12% of the high-water diet. Thus, it is evident that the Cocamlla, within the constraints on hunting and fishing time placed by agricultural activities, have a sharp annual drop in fish protein intake for which they try to compensate by increasing hunting and gathering activities. Whether it is more efficient for them to do these other activities is discussed elsewhere in this paper.

[2]The possibility exists, of course, that the Cocamilla are practicing some sort of "zone rotation" (Hames 1980) with the various named zones of the lake. If this is indeed the case, there seems to be nothing in their consciousness about such rotation.

[3]One factor which is substantially changing the feces–lake–fish–human nutrient cycle outlined here is the recent addition in Achual Tipishca of large numbers of domestic pigs which, to the disgust of some Cocamilla at least, are allowed to forage throughout the community. These pigs compete directly with dogs for the fecal and garbage resources. It is doubtful in 1981 that very much, if any, fecal material that is not deposited directly into the water is incorporated into the lake except during high water stages when the community is flooded and the ground on which it sits is flushed by the floodwaters. However, the customs surrounding garbage, fecal material, and fish and animal entrails are still in force.

[4]Fieldwork during the summer of 1981 reveals, in fact, that the lake production is probably higher than in the period of the major fieldwork. All fishermen interviewed agreed that the lake was undergoing a renewal since the residents closed off fishing to outsiders definitively in 1980.

[5]An earlier estimate published in my dissertation (Stocks 1978) of 56 g/person/day should be revised to this new figure. The method of calculation here is more refined and reflects both a more accurate judgment about which households should be assigned the sampled production values and my increasing inclination, as I grow older, to err, if error is inevitable, on the conservative side of estimates.

[6]A central problem, when proceeding on an assumption that population stability has some direct relation to the carrying capacity of the lake and river system, is that of specifying how environmental information is communicated to the population and what they do about it. In recent years of familiarity with the community of Achual Tipishca, I have found only very slight growth in the population (304 in 1976, 309 in 1979, 312 in 1981). At the same time, the turnover (flux) in population has approached 15%/year with emigration and immigration considered with births and deaths. The causes of such migration relate, as often as not, to subsistence, particularly fishing. Information is constantly exchanged on the quality of fishing and hunting in the various Cocamilla settlements. When, as in recent times, urban migration is an option, the quality of hunting and fishing is only one of several motives for moving. Considering births and deaths alone, the natural rate of increase of the Huallaga Tupians is 4.5%/year, a figure which is from two to three times as high as I have measured in one inland group. The picture, then, is one of a population with a high natural rate of increase which handles it by population flux and emigration, a picture in tune with what is historically known of Tupians from archaeology (Lathrap 1974).

CHAPTER 9

Carpe Diem: *An Optimal Foraging Approach to Bari Fishing and Hunting*

Stephen Beckerman

Introduction

The concern of this chapter is the examination of a feature of Bari protein foraging (i.e., fishing and hunting) strategy that apparently contradicts one of the central tenets of optimal foraging theory. The Bari are a group of tropical forest horticulturalists who live in the southwesternmost lobe of the Marcaibo basin; pertinent subsistence behavior is sketched in this chapter. Basic ethnography is covered in Pinton (1965), while a good many fishing and hunting data relevant here are presented in Beckerman (1980). Good bibliographies on optimal foraging theory appear in Charnov and Orians (1973), Pyke, Pulliam, and Charnov (1977), and Orians and Pearson (1979).

The tenet of optimal foraging theory of interest here posits that a forager should exploit a particular resource only when he cannot, on the average, get a better return for his time by exploiting some other resource(s) (MacArthur 1972). That is, it is hardly an efficient use of time and labor to hunt at an average return of 100 g of meat per man-hour when one can fish at an average return of 400 g of meat per man-hour. The problem is that the Bari do just that—they continue hunting even in those seasons when fishing returns are several-fold higher than hunting returns.

My effort to explain this unexpected behavior is organized in four parts: first, this introduction, gives a brief overview of the Bari, particularly focusing on subsistence activities, and specifies the questions and problems that will be explored. Second, a development of the particular aspects of opti-

ADAPTIVE RESPONSES
OF NATIVE AMAZONIANS

ISBN 0-12-321250-2

mal foraging theory to be tested here, along with some suggestions as to later modifications is given. The third part is a presentation of the data, which include day-by-day fishing and hunting records and day-by-day rainfall records: in the port, a tentative resolution of the apparent contradiction of theory and data is proposed, based on the proposal of a set of hypotheses about what the Bari are doing. To test these hypotheses will require the gathering of additional data.

SUBSISTENCE ACTIVITIES OF THE BARI

The Bari live in longhouses that ordinarily hold a local group of about 50 people. Usually a local group has two to five longhouses, often spaced along an altitudinal gradient, and occupied in turn throughout the year. Subsistence activities conform well to what has come to be seen as the "typical" tropical forest pattern. The vast majority of kilocalories in the diet come from "sweet" manioc (*Manihot esculenta*) cultivated by using the slash-and-burn technique. Most protein is acquired by fishing and hunting, the former supplying about 75% of the meat in the diet.

By far the most common kind of Bari fishing expedition involves the building of two stone weirs from the upstream and downstream ends of an island to adjacent spots on the shore, and the spearing of the fish trapped between. This kind of fishing trip usually involves all the adult members of a local group. Less common is the use of spears and sometimes arrows in an undammed stretch of water. Often, although not always, men participate in this kind of fishing. There are a number of rare techniques, one of which will be discussed in following sections. For the first technique just mentioned, the number of possible fishing spots is limited to the number of islands; and for the second, to the number of stretches of water shallow and tame enough for a spearman to walk easily through them, yet deep enough to be acceptable habitats for fish. Recently, hook-and-line fishing has been added to the traditional repertoire of fishing techniques. Known fishing spots do not seem to be as critical to this style of fishing as to the older methods.

Although I have previously calculated returns per man-hour of fishing and hunting expeditions by taking into account the time expenditure of all the adult participants in the expedition in question (Beckerman 1980), for present purposes my "per man-hour" figures will refer to adult male labor only. The justification for this approach, which eliminates women's contributions to the double dam fishing trips that supply most Bari protein, is just this: optimal foraging is a matter of making decisions. To the best of my knowledge, fishing and hunting foraging decisions among the Bari are made by men only. From their point of view, the women's labor is an unearned resource. That is to say, as far as I know, when men make decisions about where, when, and how long to fish, they do so on the basis of considerations of the returns they will get to their own labor only.The complementary situation seems to apply in regard to women's decisions about their gathering activities, which are usually accompanied by at least one man. The attitude seems to be that the "specialists" know best. Adoption of this "men only" rule for foraging also makes comparison between hunting, which is a strictly male activity, and fishing, which is usually done by both sexes, straightforward on a per man-hour basis. A brief discussion of women's role in fishing appears in Beckerman (1980).

For traditional Bari fishing methods, the season is an important variable. Returns per man-hour correspond closely to the depth and opacity of the rivers, which are in turn controlled by rainfall. A formula relating average monthly returns per man-hour to average monthly rainfall is given in the section entitled "Methods and Findings."

Although traditional Bari hunting used only the bow and arrow, some Bari have recently acquired shotguns (with results to be discussed subsequently). Although hunting returns per man-hour also show some tendency to vary seasonally, the correlation with rainfall is not reliable in this case. Hunting parties are also less consistent in composition than fishing parties. They vary from one man to all adult males in the local group. However, large game is shared throughout the longhouse if taken by a subset of the local group.

In addition to seasonal variation, both hunting and fishing returns tend to vary with proximity to major rivers. Fishing returns go down, of course, because travel time to the larger rivers goes up, and the smaller rivers have a poorer ichthyofauna than the main streams. Hunting returns often (although not always) go up—possibly because the areas farthest from the major rivers often are occupied for the least amount of time in the annual migration cycle and are therefore subject to less hunting pressure. (It will not do to insist too much on this explanation, however, because this "cycle" is quite irregular.)

FOCUS OF THIS CHAPTER

I hope to show in the following that an explanation of *why* the Bari hunt is to be sought in an understanding of *how* they fish, and to this end most of the apparatus of optimal foraging theory developed in the next section is informed by the goal of producing tools to elucidate how the Bari decide when to fish, how they decide what spots (what island or shallow water locations) to fish, and how they decide how long to stay at it once they are fishing. Most of the formulation of the optimal foraging questions of "where," "when," and "how long," then, is guided by the idea of people going off to fishing spots. I hope I have made it plain when other ideas of foraging are also in play.

I must also caution that the data used here were gathered in 1970–1972, without any thought of using them to test optimal foraging theory, which was then in its infancy. Collected as they were with another purpose in mind (the study of energy flow) these data do not permit the building of an unassailable case for or against the theory. What they do permit is the building of a plausible case, and the specification of a set of hypotheses to test that case, along with the specification of the additional data needed.

Theoretical Background

The fundamental idea of all optimal foraging theory is that the forager should maximize the ratio of resources procured to time spent in procuration. The assumption is that thus an individual (or, in the case of food-sharing creatures such as people, a group) will have the greatest amount of time and energy to devote to alternative activities such as constructing and maintaining shelter, defending against enemies, seeking mates, caring for offspring, and so forth, which also contribute to fitness. In most optimal foraging work, resources are measured in units of food energy, and the problem is modeled as that of maximizing the number of kilocalories obtained per hour. In the Bari case, the question is the amount of meat (fish or game) rather than the amount of energy obtained in foraging, but this substitution does not affect the mathematics of the argument.

The branch of theory that is relevant to Bari fishing and hunting is called "central place optimal foraging theory" (Charnov and Orians 1973; Kramer and Nowell 1980; Orians and Pearson 1979). It deals with foragers who, like the Bari, return with their prey to a fixed location. It attempts to predict what type of resources the forager should exploit, how far he should go to get them, and how long be should stay at his foraging—all with a view to maximizing his rate of return. Resources are usually construed as particular prey species in the optimal foraging literature. However, there is another branch of the theory that is concerned with the selection of "patches," particular areas that are more or less uniform in the abundance of a given array of prey, from the point of view of the predators, and which are distinguishable from other areas in that their prey abundance is either greater or lesser. It is with issues around the selection and exploitation of patches (see Charnov [1976] and McNair [in press] for bibliographies) that this section is concerned.

The theory makes a fundamental distinction between time spent in travel *to* a patch (T_t) and time spent foraging *in* a patch (T_f). This distinction was anticipated by Schoener (1971) who early recognized the difference between Type I, or "sit and wait," predators who do their hunting while engaged in some other activity, be it sunning themselves, searching for mates, or whatever, and thus lose to the hunt only that time in which they are actually pursuing, subduing, and consuming prey; and Type II, or "Nimrod," predators who, when they hunt, do so to the exclusion of all other activities, and thus lose to the hunt the time they spend searching for and traveling to prey, as well as the pursuit, subduing, and consuming time. I shall return to this distinction in the "Discussion" section.

For central place foragers such as the Bari, the overall ratio of meat procured to time spent in procuration for a fishing trip is equal to the amount of meat procured, divided by the round trip time to the patch (fishing spot), plus the time spent in the patch: $M/T = M/(T_t + T_f)$. If we visualize this ratio (Figure 9.1) with time as the abscissa and meat as the ordinate, then we can think of various ratios of meat procured to time spent in procuration as a set of straight lines of varying slopes. The optimal forager will try to achieve the line with the highest possible slope, the greatest *rate* of meat acquisition.

Such a family of slopes describes the overall M/T (rate of return) ratio for a whole trip or the sum of several whole trips. It does not represent the actual progress of any single trip, which is more complex. Figure 9.2 shows the general form of the M/T_f function, illustrating how the acquisition of meat proceeds once the forager is in the patch. Unless the patch is very large, the amount of meat obtained will rise at a decreasing rate to a limit

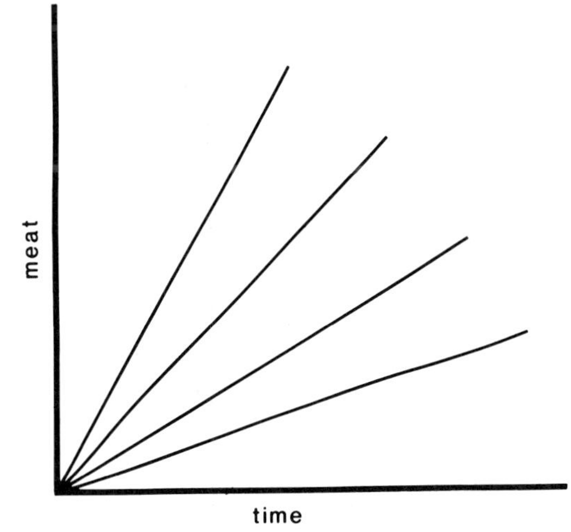

Figure 9.1 *Various overall rates of return in amount of meat per unit time are shown by the rays from the origin.*

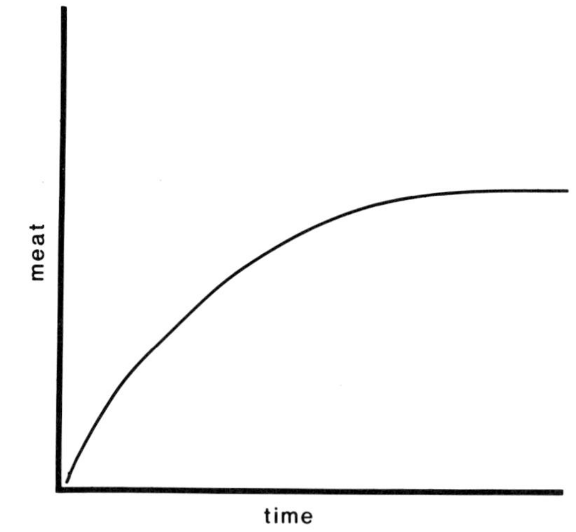

Figure 9.2 *The rate of acquisition of meat once a forager is in a patch declines with time as the patch is exhausted.*

given by the amount of meat initially present, because the forager will deplete the patch by his foraging.

How should a forager behave once he is in a patch? How long should he stay there? The question can be visualized in Figure 9.3. Adhering to the assumption that T_t, travel time to and from the patch, does not, by definition, produce any meat, we place travel time to the left of the origin. The forager's round trip travel time (to and from the patch, but not in it) is the

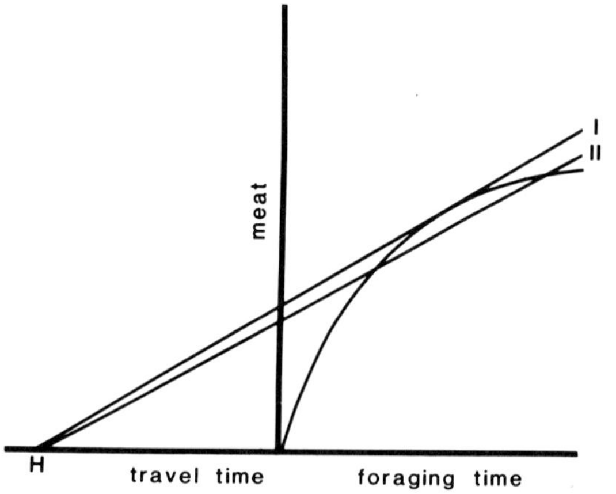

Figure 9.3 *The optimal time to spend in a patch is that time which will produce the highest overall rate of return, given that the yield of the patch declines with time. The optimal foraging time is marked by the point at which line I is tangent to the yield curve of the patch.*

distance from point H to the origin. His time in the patch is measured along the abscissa to the right of the origin. His objective is to achieve the highest possible *M/T* ratio, the fastest rate of return, for the entire trip—a problem which is equivalent to choosing the *M/T* line with the greatest attainable slope. Clearly there is only one such line, I, and it is the line from H that is tangent to the curve M/T_f. Any other attainable line (any other line which cuts the curve of actual meat procuration), such as II, will have a lesser slope. As shown by line II, staying in the patch too long is equivalent to leaving too soon. Both mistakes produce a lower overall *M/T*.

We have just seen that the amount of time spent in a patch is related to the amount of time needed to get there and back. Therefore the total amount of meat taken home from a patch is related to both T_t and T_f. The rate at which meat is extracted from the patch is a function of the richness of the patch and the time spent foraging there, whereas the decision to leave the patch, and bring T_f to a close, is a function also of T_t.

Clearly, then, there is an interaction between distance from home to patch and correct length of stay in patch. This relationship is shown in Figure 9.4, in which the optimal, tangent lines are drawn to the M/T_f curve for a hypothetical patch and two central places, one of which has a round-trip travel time to patch, which is much less than that of the other. As is intuitively satisfying, the longer the travel time, the longer the optimum time to spend in the patch.

Up to this point I have adhered to the usual assumption in the optimal foraging literature that no food is encountered outside patches. In the case of the Bari, however, this assumption does not hold. In order to get to a fishing spot, whether to fish it at once or just to evaluate it for future

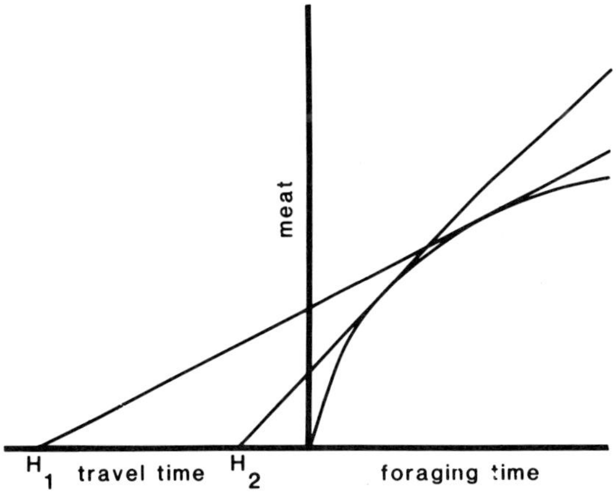

Figure 9.4 *The optimal time to spend in a patch is related to the distance of the patch from the central place. The longer the travel time, the longer the optimal time to spend in the patch.*

fishability, a Bari fisherman must walk through the forest. There is game in the rain forest.

Although some investigators (Hames 1980; Hames and Vickers 1982) have treated the various hunting grounds of tropical forest Indians essentially as a set of individual patches, the position I take here is that, from the point of view of the Bari and for present purposes, the jungle within the hunting radius of a given house is one big patch. Obviously there must be some variation in faunal richness throughout the jungle, but in the absence of any good data on this issue, it seems a reasonable first approximation to assume that the centrifugal force of hunting in driving game away from the vicinity of the house is roughly balanced by the centripetal force of the gardens (especially the abandoned gardens) in drawing game toward the vicinity of the house. (This rough balance becomes quite plausible in view of the importance of notorious garden pests such as agoutis and peccaries in the Bari game inventory [Beckerman 1980]). There is thus a situation of embedding: small, potentially rich patches are geometrically contained in a large, less productive patch. If the jungle is one big patch, then there is no real prospect of depleting it in a single hunting expedition. It is so large that a day's foraging by the Bari is a barely noticeable component in the other birth and death processes that regulate its animal populations. This situation contrasts sharply with the depletion by fishing, which regularly, although only temporarily, exhausts a fishing spot after a single expedition. It means that the rate of return of a hunting trip, unlike that of a fishing trip, does not vary predictably with time. More precisely, it means that the *average* return from a hunting trip is simply some constant times the amount of time spent foraging: $M = kT_f$, with k depending on the richness of the area around a given house. Furthermore, there is no travel time associated with hunting, because the central place is located in the rain forest. All time is foraging time.

What does this reevaluation of T_t for fishing as T_f for hunting do to the predictions of optimal time to spend in a patch? Figure 9.5 shows the resolution of this problem. For a trip in which no game can be taken in the round trip from H to the fishing spot, the actual progress of meat acquisition over time is marked by the line $H0_1$ and the curve 0_1L_1, to give an overall m/T ratio equal to the slope of the line HL_1. However, if we allow game to be taken on the trip to and from the fishing patch, then the progress of meat acquisition over time is shown by the line $H0_2$ and the curve 0_2L_2. The effect is to lift the curve of fishing returns, because one now starts not with no meat when one begins to fish, but with a certain amount of game, both that taken on the way to fish and that anticipated on the way home. The overall M/T ratio for the trip is now given by the line HL_2. This line is obviously steeper than the line HL_1. Further, the optimum time to leave the fishing patch is now sooner than would be the case if no game could be taken along the way. In a central-place optimal

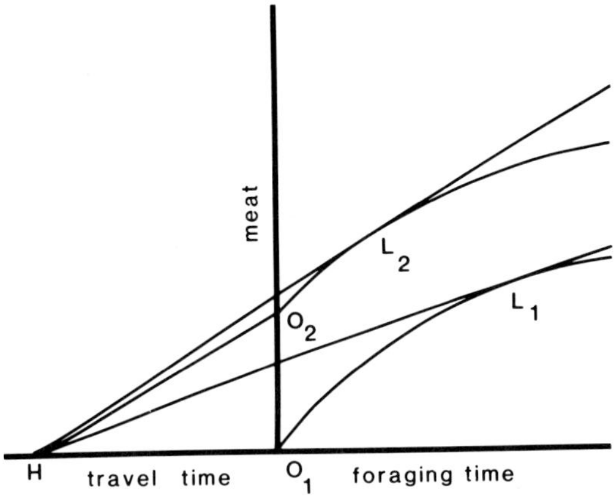

Figure 9.5 *The tangent line HL_1 shows the optimal meat to time ratio for the patch characterized by the curve 0_1 L_1, if no game can be taken on the way to the patch. The tangent line HL_2 shows the optimal meat to time ratio for the same patch, if game is taken on the way to the patch, at a rate given by the slope of the line HO_2. The effect of allowing hunting on the way to fish is to dispace the origin of the fishing curve upward, and to move the point of tangency of a ray from H to the left.*

foraging sense, allowing returns on the way to the patch has brought the patch closer.

Now, this style of theorizing was developed, in the sources already indicated and those cited in their bibliographies, to deal with cases in which there exists only one patch, or one patch type. What about cases in which there are many different patches, of varying richnesses and distances from the central place? How should the predator choose among them? Charnov (1976), in dealing with this problem, developed what he called the "marginal value theorem." Briefly, this theorem posits that choice of patch is not really important, as long as the predator avoids the really impoverished ones. What is important is that the predator should leave whatever patch he is working at the point at which the instantaneous capture rate (the slope of our M/T_f curve) drops to the overall capture rate, including travel time, for the whole habitat (our aggregate M/T ratio for the summation of a large number of trips). At this point, the patch the forager is currently working is so depleted, that he can do better by moving to another patch. Thus, if there are many different kinds of patches, the forager might find himself leaving one of them, a relatively poor one, before the optimal moment, considering that patch alone, in order to improve his overall habitat standing by returning to the central place and then setting out again for yet another patch. He might also stay longer than the optimal individual time in a rich patch, since his immediate return would still be above the habitat average. Figure 9.6 depicts this situation.

But is this the appropriate model for the Bari? Implicit in the marginal

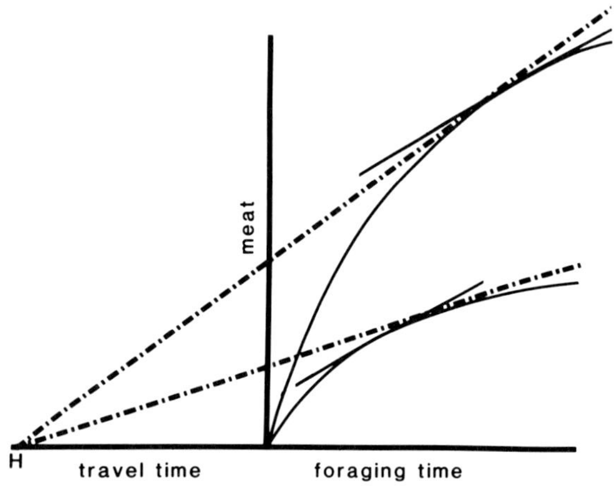

Figure 9.6 *The points of tangency of the dot–dash lines with the solid curves indicate the optimal times for foraging in the patches represented by the solid curves,* considering those patches in isolation. *The parallel solid tangent lines show what the foraging times in those patches might be under the rules of the marginal value theorem, considering all the patches in the habitat.*

value theorem is the idea that the forager abandons one patch in order to exploit another as soon as he can get there. In terms of Bari fishing, this strategy would imply leaving a fishing spot, depositing the catch back at the longhouse, and setting off immediately for another fishing spot. While the Bari do sometimes make more than one hunting expedition in a single day, or follow a hunting expedition with a fishing trip, they almost never fish more than once a day. Indeed, a successful fishing trip is usually followed with a day of rest or hunting. And, as stressed in the preceding, fish is the most productive kind of foraging the Bari do, and it provides most of their meat. A theory of Bari foraging must address itself first to their fishing behavior.

I believe that at least a plausible case can be made that if the Bari are foraging optimally, then as well as trying to optimize per month (or whatever period is taken as the standard for the overall habitat rate) they are also trying to optimize per day. Optimization per day implies, for fishing, optimization per patch. If such is the case, then we can associate with each patch its particular optimal value, given by its richness at a specific moment and its distance from the central place, as shown in Figures 9.3, 9.4, and 9.5, and inquire again into the problem of patch choice.

Let a "patch type" be a collection of patches that are similar in that their optimal M/T ratios are the same, even though the individual values of T_t and the M/T_f curves may be different, as shown in Figure 9.7. In the concrete situation of Bari fishing spots, a particular location will belong to many different patch types over the course of time, as the water rises and

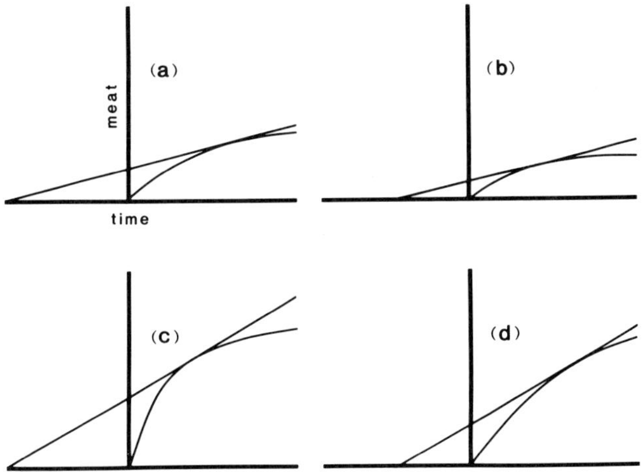

Figure 9.7 *A "patch type" is a collection of patches which have the same* M/T *ratio, regardless of their distances from the central place or fish abundance on a single day. In this example, (a) and (b) belong to the same patch type, even though they have different* M/T_f *curves and different distances from the central place; (c) and (d) also belong together, but in another patch type. A single location, for instance that represented by (a) and (c) or (b) and (d), belongs to many different patch types over the course of time.*

falls, clears and muddies, and the fish enter and leave; this last change in particular is quite unpredictable. All these changes will, of course, affect the M/T_f curve, and thus also the optimal M/T associated with the location.

The overall M/T ratio for a month's time, for example, will be the summation of the individual M/T ratios for each of the patch types exploited, weighted by the proportion of the time each type is exploited. For notational convenience, let R_i stand for the optimal M/T ratio for patch type i, and let the proportion of the time that i is exploited be P_i. Then the overall habitat M/T ratio will be $M/T = \Sigma P_i \cdot R_i$, assuming that the total time T is large enough that we do not have to worry about fractional trips, which must be disallowed under this formulation.

An interesting conclusion as to the changing value of information follows immediately from this trivial formulation. It is clear that in order to achieve the highest overall M/T ratio one should manipulate P_i in order to concentrate one's activity on the patch types with the highest R_i, since one cannot manipulate the R_i itself, except to one's disadvantage (one can always leave a patch at a suboptimal time).

Suppose in a month's time there are N hours in which fishing is possible (essentially, these would be the daylight hours). Suppose also that fishing takes place for only Q hours ($Q < N$) during that month. And suppose finally that during the N possible fishing hours the three patch types, 1, 2, and 3, listed in descending order of their R values, are available for fractions of N equal to P_1, P_2, and P_3.

Now, if the hours of Q are randomly distributed with respect to those of N, the ratio of the proportions of occurrence of patch types 1, 2, and 3 will also be the ratio in which they are exploited. However, since Q is smaller than N, there is enough patch type 1 time available to fill a proportion of Q which is equal to $(N \cdot P_1)/Q$.

(For instance, if $N = 400$ and $Q = 100$, and the proportions of occurrence of patch types 1, 2, and 3 are $P_1 = 0.2$, $P_2 = 0.5$, and $P_3 = 0.8$, then if the hours of Q are random with respect to N, the ratios of exploitation of the three patch types will be $\frac{2}{15}$ of the time for patch type 1, $\frac{5}{15}$ of the time for patch type 2, and $\frac{8}{15}$ of the time for patch type 3. However, patch type 1, the richest type (highest R) is actually available for (0.2) (400) = 80 hours. It could fill a proportion of Q equal to (0.2) (400)/100 = 0.8.)

Given the information as to where and when the best patch types occur, one can do considerably better than chance, but one must acquire the information. That will cost time, but at what price? Must the time be expended just in scouting with no return except information? It is clear that the answer is no if we recall the embedding situation (discussed previously), which increases the M/T ratio for each patch by allowing returns from the surrounding area when travelling to and from the patch. In fact, the search for information as to where the good fishing spots are can be seen as an activity almost incidental to hunting. Thus, the result of being able to forage in a large, diffuse patch while on the way to small, concentrated patches, whether to forage in them immediately or simply to evaluate them for future foraging, is at least threefold. It raises the R value of the small fishing patches; in a sense it brings them closer; and it decreases the cost of discovering where the good small patches are because this discovery not only costs time but also pays (on the average) meat. We have already looked briefly at the first two effects (Figure 9.5). Let us now examine the third a bit more closely.

The expense of discovering where the really good patch types are at any particular moment most be evaluated in terms of the value of that information. The expense is the amount by which the extra T from an average hunting–scouting trip will lower the overall M/T ratio for both hunting and fishing for the whole habitat over a period of time (and lower it, it will, on the average, because hunting–scouting on the average is less rewarding than fishing). The value of the information is the amount by which the extra M of a well-timed and -placed fishing trip will raise the overall M/T slope for the habitat. The ability of information to raise the habitat M/T ratio will depend on three things.

The first is the disparity of R_i for each of the patch types. If these values are quite close, then there is not much to be gained by choosing the very best—second or even third best will do nicely. The cost of information gain will be greater than its reward. The second is the value of the ratio N/Q. If foraging hours are a very large fraction of all hours, then there is not much to be gained from precise information, because patch types will have to be exploited in something close to their natural proportions in any case. However, if N is considerably larger than Q, then stacking foraging efforts at the high R end of the patch type range can produce substantial improvement. The third is the proportions of all time in which the various patch types occur (P_i). If the very best patch types occur so infrequently that, even with perfect information, they cannot make up more than a tiny fraction of total foraging, then (unless these patch types are spectacularly high yielding) the expense of finding them will be greater than the reward of fishing them. Contrariwise, if the very best patch types occur so frequently that even by chance they will dominate the foraging itinerary, then there is little to be gained and something to be lost in spending time at a lower rate of return just to locate a few more representatives of the best types. Information will be more useful in the middle case (which I venture must be the common one) in which improvement in overall habitat M/T can be substantial with information, but will not come automatically without it.

The major hypotheses generated in this section, then, have to do with the value of information. All other things being equal, this model predicts the following: (1) the value of information increases as the proportion of foraging time in all available time decreases; (2) the value of information increases as the disparity of meat return to time invested ratios (R_i) for different patch types increases; and (3) the value of information will be greatest when the proportion of all time in which the highest return patches are available (P_i of the most valuable patches) is neither very large nor very small.

Methods and Findings

The data base used in this chapter consists of the foraging records for a single small but otherwise very traditional Bari longhouse at four different time periods. I believe the records for this single longhouse are the appropriate data set for testing the arguments made in the preceding section, and, even though my field notes contain foraging records from three additional houses at four other time periods, I have chosen to refer to these additional records only when they amplify or contradict the testimony of the records of the single house. Aside from considerations of economy, this decision is based on the following criteria: it is important to keep location and, as much as possible, personnel constant in testing predictions such as those made by optimal foraging theory because of the possibility that differences between places and people might not reflect optimality or suboptimality in foraging behavior, but simply variation in local or personal style. It is, of course, wise to use the largest possible sample, and it is paradoxical that for this small house my sample of foragers is largest. (Because this house had such a small population—usually about 10 people—I was able to keep a reasonable check on what everyone was doing most of the time. The size of my forager sample—all adult males—varies from 4 to 6. For the larger houses I could keep track only of my immediate hearth group, and the size of my forager sample is mostly only 2 in those cases.) Finally, the records from this house are the most complete and reliable of my foraging records. They represent, in sum, the changing activities of a largely unchanging group of traditional Bari men, as they respond to seasonal changes in foraging returns in the same locale.

Because it is unfortunately the case that day-by-day fishing and hunting return data are still rather rare for the tropical forest, it will be helpful to specify the composition of the group to which the records presented in Tables 9.1–9.4 pertain, in order to maximize their utility for other workers. As is true for most tropical forest longhouses, this one, initially derived from the fragmentation of a larger local group, had its comings and goings. For the time periods covered in Tables 9.1–9.4, the population varied as follows. From November 1 through 22, 1970, there were nine people resident—an elderly man, his teenaged son, his daughter, his daughter's husband, and their toddler-aged daughter; as well as another young man, his wife, and their infant; and another teenaged boy not related to anyone else. The two young men and the two teenaged boys were the usual foragers; the elderly man had a bad case of leishmaniasis and very seldom left the house—only once to hunt and three times to fish. On November 22, there arrived another adult man and his wife; they stayed through the following February.

From January 24 through February 21, 1971, the population of the house had returned to nine, but a somewhat different nine. The teenaged boy who had no kin in the house had gone, along with the father of the infant, to visit another fraction of the group from which this longhouse had split: the infant and her mother remained. By this time the elderly man was feeling a bit better and regularly joined in fishing trips, which were not terribly strenuous at this time of low water.

By August, 1971, the population of the house had stabilized. The father of the infant girl had returned, though without his teenaged companion, and the couple who had arrived in November had left. The composition of the house remained thus through September, 1971. The elderly man fished but did not hunt much in August, but in September, his leishmaniasis well on the way to eradication, he engaged actively in hunting as well.

In addition to changes in population, there were also changes in tool kits during the time covered. On my arrival at this house in the last week of October, 1979, I distributed a large number of fishhooks, which had previously been unavailable here, and which were immediately put into use.

Table 9.1

Foraging Returns, November[a]

Date	Traditional Hunting			Traditional Fishing			Hook and Line Fishing			Shotgun Hunting			Total Foraging		
	kg	mh	kg/mh	kg	mh	kg/mh	kg	mh	kg/mh	kg	mh	kg/mh	kg	mh	kg/mh
Nov. 1							3.75	18.25	.205				3.75	18.25	.205
2							1.50	4.25	.353				1.50	4.25	.353
3	0.20	9.00	.022				0.50	6.00	.083				0.70	15.00	.047
4	2.60	18.50	.141				1.00	3.25	.308				3.60	21.75	.166
5	2.00	6.00	.333				2.50	2.75	.909				4.50	8.75	.514
6	0.00	2.25	0.00				2.75	12.00	.229				2.75	14.25	.193
7															
8	0.25	11.00	.023				0.50	5.25	.190				0.75	16.25	.046
9															
10	0.00	15.75	0.00				0.50	2.50	.200				0.50	18.25	.027
11	0.00	20.00	0.00	2.75	14.50	.190							2.75	34.50	.080
12	3.00	14.50	.207										3.00	14.50	.207
13	0.25	14.00	.018										0.25	14.00	.018
14	0.10	2.00	.050	8.50	25.00	.340							8.60	27.00	.319
15	0.00	4.00	0.00				0.37	2.75	.135				0.37	6.75	.055
16	0.10	5.75	.017				1.25	6.25	.200				1.35	12.00	.113
17	0.00	4.00	0.00				1.00	1.00	1.000				1.00	5.00	.200
18	4.00	12.25	.327	1.50	15.50	.097							5.50	27.75	.198
19	0.00	0.50	0.00										0.00	0.00	0.00
20	0.50	13.00	.039				1.50	3.00	.500				2.00	16.00	.125
21	3.00	13.50	.222	7.50	33.50	.224							10.50	47.00	.223
22	0.00	24.25	0.00										0.00	24.25	0.00
23										12.25	30.87	.397	12.25	30.87	.397
24	0.75	1.00	.750	6.00	11.25	.533							6.75	12.25	.551
25	0.25	0.75	.333							3.00	5.25	.571	3.25	6.00	.542
26							1.00	1.00	1.000				1.00	1.00	1.000
27				2.00	22.50	.089				1.25	2.00	.625	3.25	24.50	.133
28							6.50	1.00	6.500	17.00	23.63	.734	23.50	24.63	.954
29	0.00	34.25	0.00				0.00	1.00	0.00				0.00	35.25	0.00
30	0.25	6.75	.037										0.25	6.75	.037
Totals:	233.00	17.25	.074	28.25	122.25	.231	24.62	70.75	.350	33.50	61.75	.543	103.62	487.75	.215

[a]In kg of fish and game, butchered weight. Rounding is to nearest 100g and nearest ¼ hour.

Table 9.2

Foraging Returns, January–February[a]

Date	Traditional Hunting			Traditional Fishing			Hook and Line Fishing			Shotgun Hunting			Total Foraging		
	kg	mh	kg/mh	kg	mh	kg/mh	kg	mh	kg/mh	kg	mh	kg/mh	kg	mh	kg/mh
Jan. 24				16.50	29.00	.569							16.50	29.00	.569
25							1.00	1.00	1.000				1.00	1.00	1.000
26	0.50	13.25	.038	3.25	10.75	.302							3.75	24.00	.157
				21.00	23.25	.903							21.00	23.25	.903
28										1.00	2.63	.381	1.00	2.63	.381
29	0.10	6.75	.148										0.10	6.75	.148
30				5.50	12.00	.458							5.50	12.00	.458
31				14.00	23.00	.607							14.00	23.00	.609
Feb. 1				0.00	1.87	0.00							0.00	1.87	0.00
2	0.00	3.25	0.00	0.50	.75	.666							0.50	4.00	.125
3				0.25	5.00	.050	2.00	2.50	.800				2.25	7.50	.300
*4	0.10	0.25	.400	4.40	19.75	.223	4.00	12.50	.320				8.50	32.50	.262
5				2.75	10.25	.268							2.75	10.25	.268
6															
7				1.00	4.00	.250							1.00	4.00	.250
8				8.75	14.38	.609							8.75	14.38	.609
9				7.25	14.50	.500							7.25	14.50	.500
10				31.00	23.00	1.348							31.00	23.00	1.348
*11	0.75	6.75	.111	1.50	28.87	.052							2.25	35.75	.063
12	0.00	4.25	0.00	3.25	13.25	.245							3.25	17.50	.186
13				5.50	10.25	.537							5.50	10.25	.537
*14	1.00	0.25	4.000	4.75	12.75	.373							5.75	13.00	.442
*15	0.10	0.25	.400	17.50	25.50	.686							17.60	25.75	.683
16															
17										2.75	2.00	1.375	2.75	2.00	1.375
18				0.50	13.00	.039	0.50	1.00	.500	1.00	7.12	.140	2.00	20.13	.099
19							3.50	6.50	.538	2.75	3.00	.917	6.25	9.50	.658
20	3.00	18.63	.161										3.00	18.63	.161
21				24.50	32.00	.766							24.50	32.00	.766
Totals:	5.55	53.63	.103	173.65	327.13	.531	11.00	23.50	.468	7.50	14.75	.508	197.70	419.01	.472

[a]In kg of fish and game, butchered weight. Rounding is to nearest 100g and nearest ¼ hour. Dates on which a fishing trip also procured game on the way to or from the fishing spot are marked with an asterisk (*).

Table 9.3

Foraging Returns, August[a]

Date	Traditional Hunting			Traditional Fishing			Hook and Line Fishing			Shotgun Hunting			Total Foraging		
	kg	mh	kg/mh	kg	mh	kg/mh	kg	mh	kg/mh	kg	mh	kg/mh	kg	mh	kg/mh
Aug. 14	0.00	11.50	0.00										0.00	11.50	0.00
15	1.00	3.25	.308										1.00	3.25	.308
16	0.00	7.50	0.00										0.00	7.50	0.00
17	0.10	9.50	.011										0.10	9.50	.011
18	3.15	20.00	.158										3.15	20.00	.158
19	2.00	34.88	.057										2.00	34.88	.057
20	0.00	7.12	0.00	1.50	13.00	.115							1.50	20.12	.075
21	0.00	6.12	0.00	7.50	20.00	.375	0.00	4.00	0.00				7.50	30.12	.249
22															
23	2.50	12.00	.208	0.00	15.00	0.00							2.50	27.00	.093
24				104.00	64.00	1.625							104.00	64.00	1.625
25	0.35	1.75	.200										0.35	1.75	.020
26															
27															
28	0.50	19.50	.003										0.50	19.50	.003
29															
Totals:	9.60	133.12	.072	113.00	112.00	1.009	0.00	4.00	0.00				122.60	249.25	.492

[a]In kg of fish and game, butchered weight. Rounding is to nearest 100g and nearest ¼ hour.

Table 9.4

Foraging Returns, September[a]

Date	Traditional Hunting			Traditional Fishing			Hook and Line Fishing			Shotgun Hunting			Total Foraging		
	kg	mh	kg/mh	kg	mh	kg/mh	kg	mh	kg/mh	kg	mh	kg/mh	kg	mh	kg/mh
Sep. 4	0.10	12.00	.008										0.10	12.00	.008
5	0.30	10.25	.029				2.00	6.00	.333				2.30	16.25	.142
6	0.20	9.00	.022										0.20	9.00	.022
7	2.45	23.75	.103	1.00	4.00	.250							3.45	27.75	.124
8	3.60	2.62	1.371				1.00	2.50	.400				4.60	5.12	.898
9	0.15	10.50	.014										0.15	10.50	.014
10	0.40	41.00	.010										0.40	41.00	.010
11	0.00	44.00	0.00										0.00	44.00	0.00
12				6.75	47.00	.145	1.00	1.88	.533				7.75	48.88	.159
13							0.25	1.00	.250				1.00	0.25	.250
14															
15							0.50	2.50	.200				0.50	2.50	.200
*16	0.10	6.75	.015	0.10	28.75	.003	0.00	1.00	0.00				0.20	36.50	.005
17	0.92	14.25	.065				0.38	1.00	.380				1.30	15.25	.085
Totals:	8.22	174.12	.047	7.85	79.75	.098	5.13	15.88	.323				21.20	269.75	.079

[a]In kg of fish and game, butchered weight. Rounding is to nearest 100g and nearest ¼ hour. Dates on which a fishing trip also procured game on the way to or from the fishing spot are marked with an asterisk (*).

The remains of this gift were still being used the following September, but much less often, possibly as much in response to their shortage (one loses fishhooks at a great rate in these rivers) as to foraging optimization returns. More interestingly, the man who arrived in November brought with him a shotgun, which was again put into operation immediately. He was using it only occasionally by the following February, again, perhaps, as much in response to running low on shells as to concerns of optimality.

A few words must be said as to the coding and reliability of these data, and as to what information is and is not expressed in the tables. For each category of foraging activity—traditional fishing, traditional hunting, hook and line fishing, and shotgun hunting—the total number of man-hours and the total amount of meat procured (butchered weight) are recorded for each day. No indication is given as to whether the man-hours represent the work of one man or several, whether, if they do represent the work of more than one man, the men were hunting together or separately, or whether the hours transpired in a continuous track or were separated by intervals of time in which other activities were pursued. Similarly, no indication is given as to whether the meat obtained represents one large prey item or several small ones, or whether, in the case of multiple expeditions, it came from only one or is the total catch of two or more expeditions. I intend eventually to publish the entire data base for all houses and time periods in its disaggregated form, but for the present concerns these aggregated data will suffice.

Weights of fish and game are almost all quite reliable, representing actual weighings on a pocket fish scale, or known figures for the weight of an individual of a particular species. Rounding conventions will, I hope, be obvious. Man-hours are less reliable. In several cases I did not notice someone leave the house, but did see him return, or notice later that he was missing, and had to guess when he went out. More rarely I did not notice when someone came back in. Additional caveats and caveats and conventions as to man-hour figures are specified below under each category of foraging activity.

TRADITIONAL FISHING

The recorded man-hours represent the total time spent by all adult male members of the fishing expedition, including the trip to and from the fishing spot. In those cases, indicated by asterisks, in which game was also taken on the trip, a token 15 man-minutes are allocated to the traditional hunting category, along with the weight of the game, and the time is deducted from the fishing man-hours. These 15 man-minutes are a rough average of the amount of time taken from a trek to a fishing spot by the pursuit, killing, and stashing of an encountered game animal.

TRADITIONAL HUNTING

All time spent out in the woods with hunting gear, except that used in travel to and from fishing spots (but see previous discussion) and in travel to collect specifically nongame products such as matting materials, is included in hunting time, whether an expedition brought back any game or not. (This formulation is in distinction to that used in Beckerman 1980.) A possible disturbance of the accuracy of hunting times is noted below.

HOOK-AND-LINE FISHING

The figures here include travel time to and from fishing spots, but this time was often minimal, as a great deal of hook and line fishing was done just in front of the house, which overlooked a major river. However, it is quite possible that a number of hunting expeditions that I did not accompany, particularly in November, included some hook and line fishing. If that fishing happened to be unsuccessful, I would have had no way of knowing without specifically inquiring—which I did not usually do—whether it had taken place. It is thus possible, and perhaps likely, that my man-hour figures for traditional hunting are somewhat high and those for hook and line fishing are somewhat low.

SHOTGUN HUNTING

All time spent by all the men present on any expedition on which a shotgun was taken are coded as shotgun hunting, even though there was only one shotgun in this house. The reason is that for game unused to firearms, as was the fauna around this house in November when most shotgun hunting took place, hunting success was more a result of discovering game, a process to which all members of the expedition contributed, rather than shooting, an event in which only the owner of the shotgun participated. The unwary animals, mostly birds, tended, when discovered to fly only out of bow range, not out of shotgun range.

If the data in Tables 9.1–9.4 and Figures 9.8–9.11 are to be used to test the predictions made in the preceding section, it is necessary to establish that the data are consonant with the stated assumptions. The first matter that must be dealt with is the seasonal variability of fishing returns, which has so far only been discussed in general terms. I have previously shown that the average monthly returns per man-hour for fishing for the *bocachico*, the most common fish the Bari catch, vary quite closely with average monthly rainfall (Beckerman 1980). Figure 9.12 gives the average monthly rainfall data. The relationship of average returns per man-hour for *all* species of fish to the average rainfall is $Y = -0.911X + 545$, where Y is the

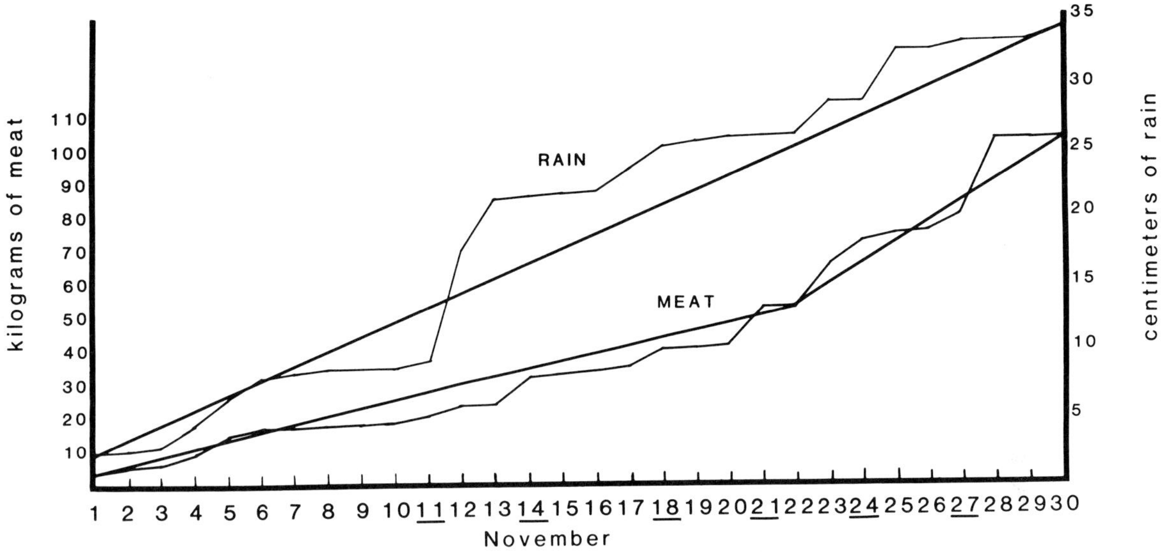

Figure 9.8 *Cumulative returns of meat from all kinds of foraging are plotted alongside cumulative rainfall figures for the November observation period. Mean rates of accumulation of both meat and rainfall are shown by the straight lines connecting the origin and terminus of each cululative plot. For this observation period, the slope of the average meat return line changes on the date on which a shotgun was introduced to the house. Dates on which fishing took place are underlined.*

number of grams of fish taken per man-hour and X is the average number of millimeters of rainfall for the month in question. The correlation coefficient is -0.853 ($r^2 = 0.727$). This formula is derived from the batch records of all houses, not just the one examined here; it does omit one aberrant, indeed unique, fishing expedition, which is described subsequently. Traditional fishing, then, does manifest strong variation in returns along the seasons.

Hunting returns per man-hour show no such regular relationship to rainfall, as a comparison of Tables 9.1–9.4 with Figures 9.8–9.11 will demonstrate. In fact, the most likely influence on hunting returns for this house is simply a slow hunting-out of the immediate environs (see, e.g., Vickers 1980), as these people remained in one longhouse longer than Bari usually care to.

In view of the regular variability of traditional fishing returns, and the irregular change in traditional hunting returns, it is striking that the average number of man-hours devoted to all types of foraging should be as inelastic as it is. In November, considering that the elderly man hardly participated (see previous section) the average amount of time spent in all foraging was 3.86 hours per man per day (118 hours per 30.5-day month). In January–February, with this man now participating (almost all foraging

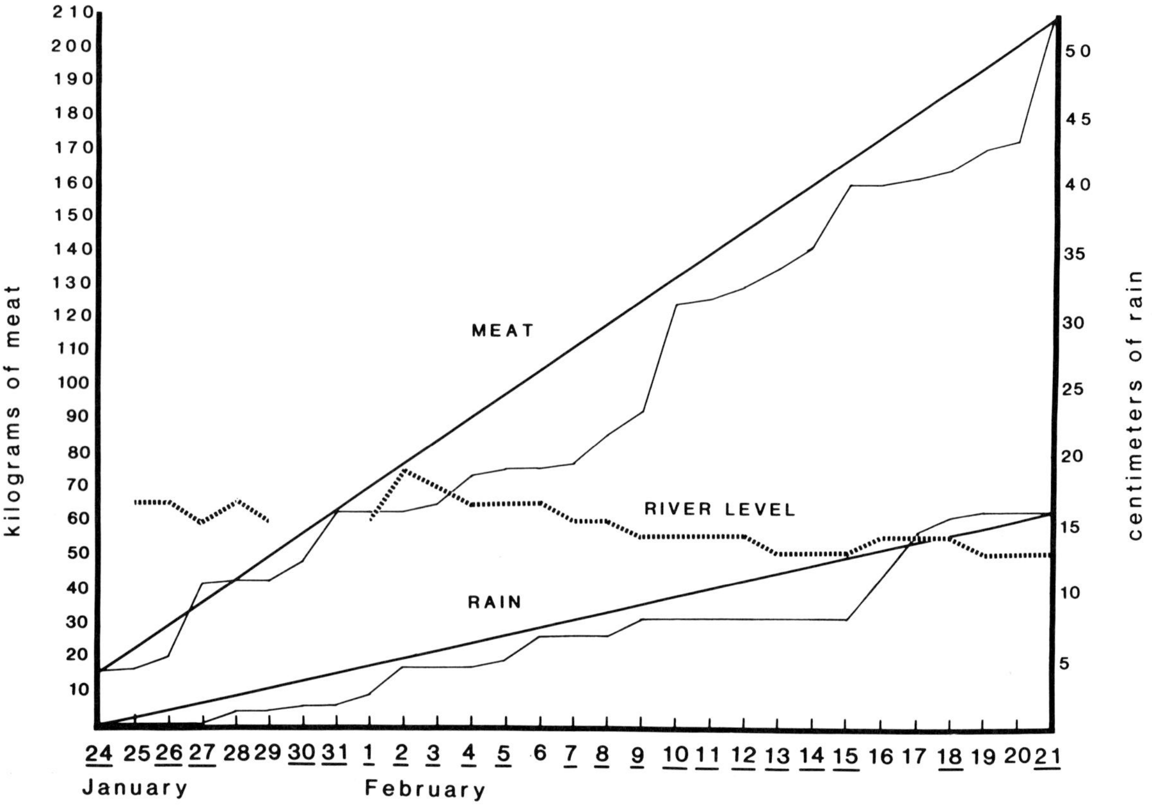

Figure 9.9 *Cumulative returns of meat from all kinds of foraging are plotted alongside cumulative rainfall figures for the January–February observation period. Mean rates of accumulation of both meat and rainfall are shown by the straight lines connecting the origin and terminus of each cumulative plot. The depth of the river, as measured at an arbitrary point, is given for those dates on which the data were recorded, by the dashed line. The scale on the left of the figure can be read in centimeters to find the depth at this point. Dates on which fishing took place are underlined.*

in January–February was fishing, which is less strenuous than hunting) the equivalent figure was 111 hours per man per month. In August, with the old man now working his way up to full participation in hunting as well as fishing, the figure (counting him as ½ forager) was 136 hours per man per 30.5-day month. And in September, this man now participating fully in hunting as well as fishing, the monthly figure was 142 hours. (These figures do not change much no matter how the elderly man is counted, as the reader is invited to verify.)

This surprising stability of the average number of hours per man per month devoted to all types of foraging is a result I neither looked for nor expected to find in these data. Checking it against the records for the other

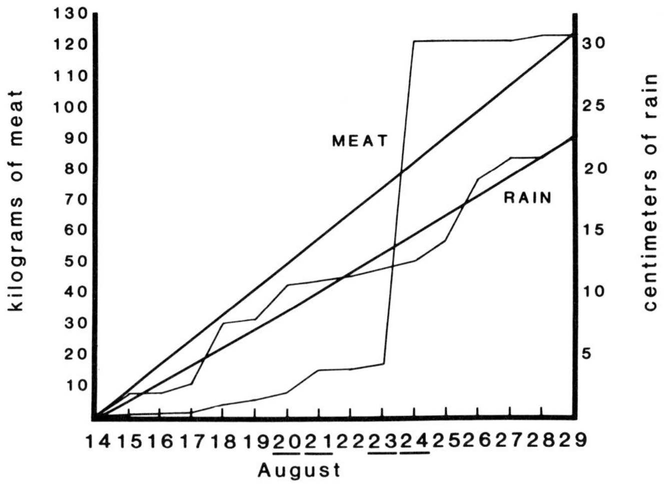

Figure 9.10 *Cumulative returns of meat from all kinds of foraging are plotted alongside cumulative rainfall figures for the August observation period. Mean rates of accumulation of both meat and rainfall are shown by the straight lines connecting the origin and terminus of each cumulative plot. Dates on which fishing took place are underlined.*

longhouses and time periods in the total data set confirmed it quite remarkably. For a house of about 20 inhabitants (I was only able to keep track of my own hearth groups here), the figures were 107 hours in October and 129 hours in March. For two full-sized longhouses of about 50 inhabitants (again observing only my own hearth group) the figures for the one house were 143 for March and for the other house, 131 for May–June. It thus seems well established that the Bari have a remarkably standard behavioral pattern which sets 125 ± 20 hours as the average amount of time a man should devote to all types of foraging during a month. It would be very

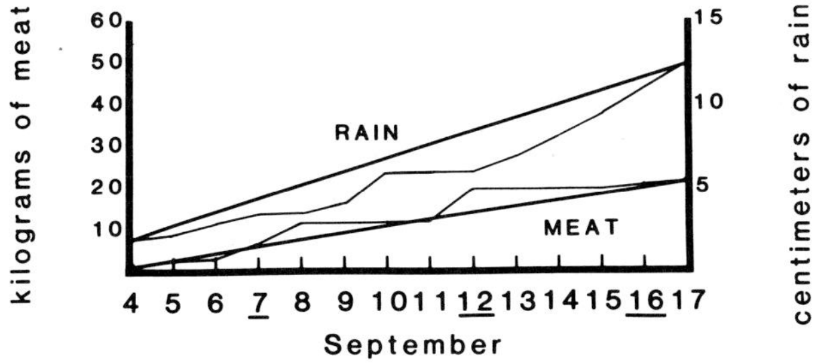

Figure 9.11 *Cumulative returns of meat from all kinds of foraging are plotted alongside cumulative rainfall figures for the September observation period. Mean rates of accumulation of both meat and rainfall are shown by the straight lines connecting the origin and terminus of each cumulative plot. Dates on which fishing took place are underlined.*

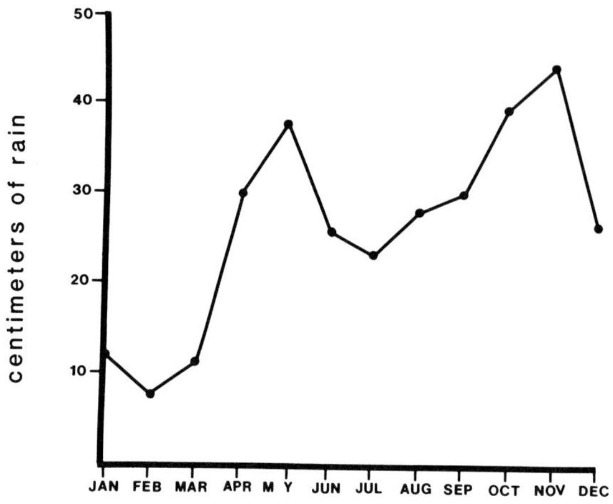

Figure 9.12 *Average rainfall figures for each month of the year in the Bari area.*

interesting to compare this figure with similar figures for other tropical forest horticulturalists.

This inelasticity of the total number of foraging hours gives some minimal plausibility to the hypothesis that the Bari are foraging optimally. The number is a fairly small one, barely more than a quarter of the approximately 427 daylight hours (assuming 14 hours of daylight and a 30.5-day month) available to them. Thus, it seems unlikely that they are so hard pressed for meat that they must scour the environment for every bit of fish and game that comes their way. General ethnographic observation leaves no doubt that the Bari do have "free" time which could be dedicated to foraging if the need were felt.

Nevertheless the case is far from conclusive. Ordinary considerations of optimal foraging theory would hold that given two alternative resources with different *M/T* (meat to time) ratios, say fishing and hunting, the forager should concentrate exclusively on the one with the higher ratio. Not only does this situation not occur, but, as Figure 9.13 shows, in most months the Bari spend most of their time in traditional hunting, which is the least productive of their foraging options.

Are the Bari then deliberately, almost perversely, foraging in a resolutely suboptimal way, or is there an explanation for this behavior? A close examination of Figures 9.8–9.11 suggests a possible resolution. In these figures, the comulative daily acquisition of meat is shown by the irregular lines. The *average* rate of meat acquisition for the period in question is shown by the straight lines. Rainfall is plotted in the same way. Additionally, in the one case in which I have the information, river depth at an arbitrary point is also shown (in Figure 9.9).

Whenever the slope of the irregular line is less than that of the straight

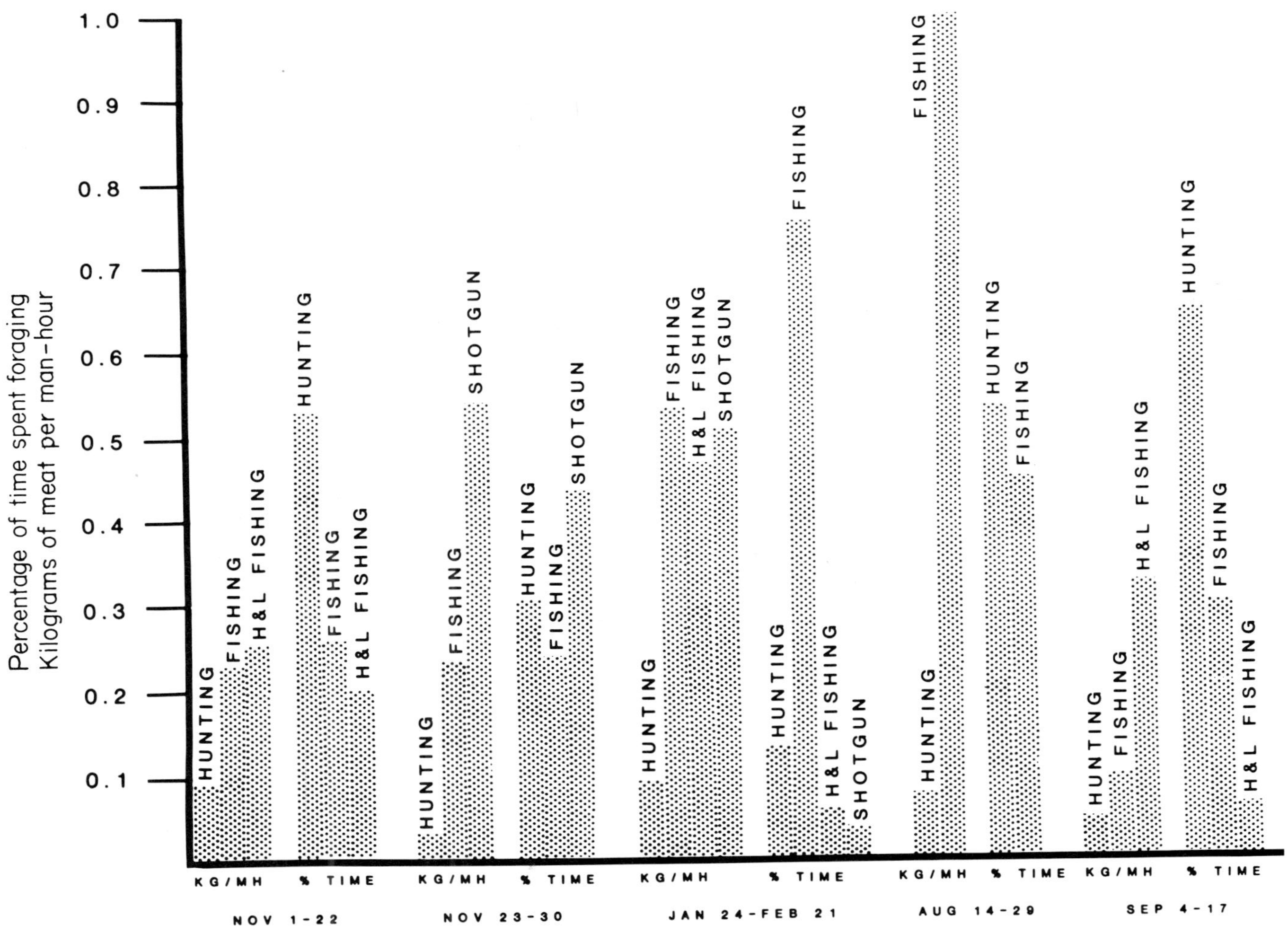

Figure 9.13 *Distribution of foraging time among the major foraging activities is plotted beside the return per man-hour for each activity, for each recording period. November is divided into two periods because a new foraging option opened, and overall rate of return rose, when a shotgun was introduced to the house in the last week of the month. Hook-and-line fishing is omitted for those periods in which the absolute amount of time dedicated to it was so small that average return figures would be meaningless. The scale along the left edge of the graph can be read as kilograms of meat acquired per manhour, for the foraging return bars, and as the proportion of time dedicated to each foraging activity (1.0 = 100%) for the percentage time bars.*

line, the house is entering a period of meat acquisition deficit, relative to the average rate for the month, and the same is the case for rainfall. Since rain almost always falls between midnight and dawn, the meat acquisition activity, the foraging, for a particular day takes place after the rain has fallen.

Now it can be seen that in most cases the days on which the meat acquisition exceeds the monthly average are those days on which traditional fishing is practiced. These days are underlined on Figures 9.8–9.11. (The major exception to this rule is the last week in November when the arrival of the shotgun raised the overall rate of meat acquisition dramatically—hence the two average slope lines for that month.) Clearly it is reasonable that after a day of successful fishing the Bari could tolerate a deficit day or two, during which time they can eat the fish acquired on the surplus day. But why not fish more on the days when they have clearly used up any surplus they may have accumulated, on days such as November 13 or September 10?

Comparing the cumulative meat lines with the cumulative rainfall lines, it is clear that most fishing takes place on days of little or no rainfall. Given the known negative correlation between average monthly rainfall and average monthly fishing success, it is reasonable to suppose that this relationship might hold on a daily basis as well. Because so little fishing takes place on rainy days, the data to test this assumption are not as complete as one might like; nevertheless, to the degree to which they exist, they support this view.

In November, there is only one day, the 18th, on which fishing is recorded following substantial rain. On this date the return per man-hour, the *M/T* ratio, is 0.097 kg, as compared with a monthly average of 0.231 kg (Table 9.1). In January–February, there are four days, February 1, 2, 9, and 18, on which fishing follows significant rain. For these days the average return is 0.274 kg per man-hour, as compared with a monthly average of 0.531 (Table 9.2). Even this figure is spuriously high. As shown on Figure 9.9, the rain of February 9 has no effect on the depth of the river. (Apparently, in this region of highly localized rainfall, we were on the very upstream edge of the storm. All the water fell downstream of us and the river level at our location actually dropped on that date.) Subtracting the expedition of February 9, then, on the grounds that we know that this storm essentially missed the longhouse location, lowers the average yield of the remaining rain days to 0.064 kg per man-hour. August figures are distorted by the enormous haul of one aberrant fishing expedition (as discussed subsequently). However, the one fishing trip in this month that follows a heavy rain, on August 20, achieves a return of only 0.115 kg per man-hour; it is unrealistic to compare this figure with the monthly average, distended as that latter figure is by the one freakish expedition. Of the three fishing trips in September, the one which follows a night of heavy rain, on the

16th, has a yield of only 0.003 kg per man-hour, as contrasted with a monthly return of 0.098 (Table 9.4).

Perhaps, then, the Bari are behaving as completely reasonable optimal foragers—hunting on those days when hunting is likely to give a greater return per man-hour than fishing, fishing (except for a few "mistakes") on those days when fishing is likely to give a greater return than hunting. But is that all there is to it, and can good fishing returns be assured simply by the absence of rainfall in the preceding day or two?

The answer to the second question is clearly no. The fishing expeditions of November 11 and February 11, to take two examples, are clearly wash-outs, despite major investments of man-hours, on days with little or no rain on the date in question or the days preceding it. In addition to having a rainless river, one must know where the fish are.

The *bocachico* (Characidae, *Prochilodus reticulatus*) the fish that alone supplies about half the meat in the Bari diet, is an algal browser. It moves in schools up and down the rivers, nibbling the algae from the cobbles of the sunlit stretches of stream. There is a yearly migration cycle, down to the swamps in the lower reaches of the river system to breed, up to the head-waters to grow fat, but apparently not all fish migrate (Beckerman 1975), and even within that cycle there is scant day-to-day predictability of what stretch of river will have a school of *bocachico,* unless one has actually seen a school moving in a particular direction within the previous few hours.

We are back to the value of information. I argued previously that the value of information, the improvement it could make possible in the overall *M/T* ratio for a whole habitat by raising the *M* in that ratio, would depend on three things: (1) the ratio of foraging time to all time, (2) the spread of patch type *M/T* ratios, and (3) the proportion of the time that particular patch types are available. Two of these things can be specified now.

Foraging time is 125 ± 20 hours per month per man. The amount of time in which traditional fishing could conceivably be done is more or less the number of daylight hours in a month—about 427. Thus, the ratio of foraging time to all time is at least three. In fact, it is considerably greater than that in all the records presented here. It varies from 4.6 in January– February when 75% of all foraging time (taking the average 125 hours as the standard) is spent in fishing, to 17.1 for September, when only 20% of all foraging time is spent in fishing. Clearly there is room for substantial improvement in the matter of the ratio of times.

Nonzero patch *M/T* ratios range from 0.089 to 0.533 in November (a change by a factor of 6); from 0.038 to 1.476 in January–February (factor of 38); from 0.115 to 1.625 for August (factor of 14); and from 0.003 to 0.250 for September (factor of 83). Again the spread of R_i for each patch type (the individual *M/T* ratios of the patches) indicates that there is considerable room for improvement. (It is probably significant that the lower values do not change as much as the upper ones. What happens in the good months is that more valuable patch types become available, not that less valuable ones disappear.)

The last thing that affects the value of information, the proportion of all time in which particular patch types are available, cannot be extracted from these data. Indeed, it would be very difficult to obtain this information from anything less than a complete day-by-day ichthyological and limnological survey of all the fishing spots within walking distance of a given longhouse. Nevertheless, given the monthly and daily variability of *M/T* returns, it seems quite unlikely that either of the cases that would minimize the value of information obtains year-round. That is, it is hard to believe that in all seasons the availability of very high *R*-value patches is so constant that there is nothing to be gained in looking for them. Contrariwise, it is equally hard to accept that high *R* patch types are so rare year-round that it is more trouble than it is worth to find them.

Provisionally accepting this line of argument, let us go through the data month by month and suggest what is happening.

Discussion

In the first place, it seems likely that in all months the return of fishing trips is augmented by casual hunting done on the way to and from the fishing spots. Examples of this phenomenon are represented by the asterisked days in January–February and September (Tables 9.2 and 9.4), and probably do not occur in the other time periods simply because the sample of fishing trips is too small for them to have turned up. The hypothesis here is that, while on their way to fish as Type II foragers (foragers whose travel and search time is dedicated exclusively to foraging), the Bari are also Type I (opportunistic) foragers with respect to hunting. This hypothesis predicts that (as discussed in the section entitled "Theoretical Background") the Bari should leave a fishing spot somewhat sooner than what the optimal time for that spot would be if no hunting took place on the way to and from it, as shown in Figure 9.6. This case should obtain in all seasons. The other hypotheses proposed here are seasonally variable.

In the very rainy month of November (Figure 9.12) the hypothesis is that as much fishing as possible is going on. Traditional hunting is pursued in order to bring home the bacon on days when fishing is not going to be profitable (high-water and muddy-water days) and to keep track of the location of the fish for those days when fishing is possible. Because there are so few good fishing days, and because hunting returns are so low, information is at a premium. When the shotgun arrives, shotgun hunting to some extent replaces traditional hunting. The returns are much higher and the information gain is similar. However, and this is a crucial point, traditional hunting is not abandoned entirely. Shotgun hunting is practiced on only a few days. Apparently it is considered important to monitor the environment constantly, so that traditional hunting takes place on those days when shotgun hunting does not, even if only for less than an hour. The percentage of time devoted to both types of hunting together actually increases above the percentage devoted to traditional hunting alone before the arrival of the shotgun. This view seems to me the most plausible explanation of why people continue to hunt traditionally when the shotgun is exceeding their wildest dreams of hunting success. A good day's fishing can still be in the M/T range of the fabulously successful hunting that the shotgun brings, as shown by the figures in Table 9.1. The value of information is high because the proportion of the time in which good fishing patch types are available, the P_i for patch types in the R_i range of 0.5 kg per man-hour and above, is low.

In the dry season months of January–February, the hypothesis is that the value of information is low, because the P_i for patch types in the range at and above $R_i = 0.5$ kg per man-hour is quite high. During this season the best patch types are so common that it is a waste of time to look for them. Even after the bad days of February 5 and 7 nobody bothered to go scouting. A good day will come along unsought—as indeed it did on February 8.

In August, somewhat more than half the time is spent hunting, even though the returns from fishing are spectacularly higher. However, these high returns pertained exclusively to one fishing trip—a trip that is unique in my field experience in timing, technique, and returns (and is thus excluded from the calculation above of monthly fishing returns as a function of rainfall, for these reasons and for others which will become clearer in the following discussion). This single fishing trip stretched over 2 days. On the afternoon of the first day a dam of poles and palm fronds, with a single gap left in the center, was constructed completely across the width of a small tributary to a major river. Just after midnight the following morning, we all returned to the dam from the longhouse, where we had been napping, with torches. We closed the gap in the dam and waited. In no more than $\frac{1}{2}$ hour, large catfish, up to 5 kg each, began literally throwing themselves at the dam, trying to get back to the main river. After these monsters had been speared, the expedition continued upstream in the tributary, spearing *bocachico,* which were also abundant in this creek. Clearly, such an expedition needs, in addition to a detailed knowledge of natural history, the specific information as to what creek the catfish are ascending, where in the creek they are, and when they are likely to try to get back to the main channel. It is clearly no accident that most of the hunting hours in this month are in the days preceding this expedition. The hypothesis here is that the P_i, the proportion of time in which it occurs, for this particular patch type is very small indeed, but it is compensated by the spectacular R_i, which attaches to it.

In rainy September the situation is quite similar to November and the hypotheses are the same: that information as to where and when fishing will be profitable is at a premium, and thus hunting goes on at a great rate. Even though, in this month, neither hunting nor fishing is giving returns of appreciable magnitude, fishing is still the more profitable of the two, and the one to which least time is dedicated. Hunting fills in with some meat on those days on which fishing is prima facie hopeless (days when all the rivers are high and muddy), and provides the information as to where fishing is going to be best on the days when the rivers are clear enough to be worth a try.

Before considering the testing of these hypotheses, one must consider the puzzle of hook-and-line fishing. It regularly yields substantial returns (Tables 9.1–9.4, Figure 9.13); why is it never more than a minor component in the foraging time budget? I can offer only guesses here. The first is that, although the M/T ratio for this type of fishing is high, both the M and the T of that ratio are small in absolute terms. Certainly it was very seldom that anyone brought home enough hooked fish to keep the house stocked for several days—yet this should be the optimal ideal of a fishing trip, enough meat to permit the luxury of choosing the next fishing trip for the highest possible R_i. Perhaps the absolute value of the M for hook-and-line fishing cannot be easily increased simply by adding more T, because the relatively few fish in these rivers that will take a hook (the *bocachico* will not) are sparse enough that once one has taken the few in a given location, one has fished out that location as far as these species are concerned. Another possibility, alluded to in the preceding is that more hook-and-line fishing than I was aware of actually did go on. If, say, a man often in the course of a hunting–scouting trip threw his line into the water, and gave up and tucked the line back in his waistband if he did not get a strike within the first few minutes, then hook-and-line fishing could have a considerably lower M/T ratio than my figures indicate. Again, it is possible that the husbanding of fishhooks, rather than concerns of optimization, caused the low proportion of time dedicated to this type of foraging. I simply do not know.

However, the way to solve this problem is the same as the way to test the hypotheses suggested above: ask. When the fieldwork reported here was conducted, my concern was with energy flow parameters. The goal was to record all intakes and expenditures of energy as exhaustively as possible, and with as little observer effect as possible. I was very concerned—in retrospect, overconcerned—with influencing subsistence behavior with anything I said or did. Consequently, I tended to avoid asking even questions such as "What are we going to do tomorrow?" for fear that someone might ask me what *I* thought we should do (the Bari are extremely hospitable) and thus put me in the position of influencing subsistence behavior.

The questions that need to be asked are simple and straightforward. The hypotheses claim that the information gathered in hunting informs decisions as to where and when fishing should take place, in seasons when good fishing spots are rare. It is a reasonably simple matter to ask hunters what they have learned after a hunting trip, and fishermen why they are fishing a particular spot at a particular time. Also, the hypotheses claim that hunting is not needed during the dry season good fishing months, because rewarding fishing spots are so abundant there is no need to spend time seeking them out. It is only a little more difficult to ask fishermen, after a failed fishing trip, whether they are going to hunt the next day, and if not, why not. And it is quite simple to ask people who are not eating meat at all on a particular day, what they are going to do about that unfortunate situation on the morrow—whether they are going to fish (and if not, why not) or hunt (and if not, why not), and if so, why and where, and if not, what else they are going to do and why.

If the hypotheses are correct, the answers to these sorts of questions will show a close concern for collecting fishing information in the bad fishing months, and a relative lack of interest in the same information in the good months. They will also show an approach to foraging dominated by the strategy that one fishes as much as one can, and hunts as much as one must, within the 125-hour limit, of course, and modified by the exception that the spotting of an item of large game near the house raises the likely M/T ratio of an immediate hunting trip to something higher than can ever be obtained in a fishing expedition. Specifically, responses as to why fishing is not done on a particular day should stress the lack of attainable fish, or lack of knowledge as to where they might be found, in the bad months, and a surplus of fish, making a trip unnecessary, in the good months. Answers which do not conform closely to this pattern will disconfirm the hypotheses.

It is in this ability to ask our subjects what they are doing, if nowhere else, that human ecologists have the advantage of our biological colleagues.

Acknowledgments

The fieldwork forming the data of this chapter was financed by The Ford Foundation, The Society of Sigma Xi, The Smithsonian Institution Urgent Anthropology Fund, the Explorers' Club, and the University of New Mexico. Kristen Hawkes got me interested in optimal foraging theory, Donna Howell kept me interested, and James McNair kept me from making a number of stupid mistakes. Renata Wulff provided all sorts of support during the writing of this chapter.

CHAPTER 10

Adaptation and Ethnozoological Classification: Theoretical Implications of Animal Resources and Diet of the Aguaruna and Huambisa

Brent Berlin and Elois Ann Berlin

With the Collaboration of James L. Patton, John P. O'Neill, Roy W. McDiarmid, and Camm C. Swift

Introduction

The research of workers in the field of ethnobiology over the past 20 years has demonstrated unequivocally that nonliterate peoples possess an impressively comprehensive, scientifically accurate knowledge of their environments. Concurrently, the past decade has seen a growing interest in ecologically oriented research that includes a focus on native systems of adaptation. There exists a widespread underlying conviction in both of these areas of research that the comprehensive ethnobiological knowledge of native peoples must reflect an intense ecological awareness fundamental to their adaptive responses.

In this chapter, we address both of these issues. Our ethnozoological studies show that the Aguaruna and Huambisa, two closely related Jivaroan groups residing in the montaña region of northcentral Peru, have extensive, detailed knowledge of the animal resources in their environment, as exhibited by the number of named species and the system of classification of these named species. Our dietary records indicate that

ADAPTIVE RESPONSES
OF NATIVE AMAZONIANS

ISBN 0-12-321250-2

these native populations sample widely from the diverse fauna characteristic of the region. An unanticipated finding of our research, however, has been that, although both of these groups possess a complex knowledge of hundreds of different species of animals and although many of these species are recognized as food sources, only a small fraction of them contribute significantly to diet. The question thus confronts us: Why do these people have such a detailed classification of animals and yet systematically utilize so few of them?

We outline an argument that claims that the extreme reliance on but a few species is explicable in terms of the relative availability and ease of capture of these species, as well as a reflection of adaptation to resource availability in specific environmental settings. The explanatory value of this functionalist approach to native knowledge is unquestionably powerful. It is limited, however, in that it leaves unanswered the reasons for naming and classifying perhaps 90% of the remaining fauna, and we examine alternatives to utilitarian explanations for the recognition of this major portion of the animal kingdom.

The Setting

The Aguaruna and Huambisa, two of the five recognized ethnolinguistic groups of the Jivaroan language family, occupy the greater upper Marañón River drainage, an area of Peru's northcentral montaña (see Figure 10.1). It is an area of sharply rising ridges cut by relatively fast moving streams, especially in the western edge of the territory. Elevations range from 200 to 2000 m. Mean annual temperatures are around 25°C and annual precipitation is nearly 3000 mm (Guallart 1964).

Approximately 20,000 Aguaruna (Uriarte 1976) are found in the area, with communities beginning around the Pongo de Rentema and extending below the Pongo de Manseriche (on the Marañón River proper), as well as along the river's major tributaries—the Cenepa, Comainas, Numpatkaim, lower Santiago, Apaga, and Potro. Some Aguaruna are to be found on the upper Mayo River, a tributary of the Huallaga (M. Brown 1980).

Huambisa is spoken by approximately 5000 people (Chirif and Mora 1977). Huambisa villages begin at the community of Galilea on the lower Santiago River and extend northward along the upper reaches of the Santiago and its tributaries to the Ecuadorian border (Figure 10.1).

Both the Aguaruna and Huambisa practice swidden horticulture, and sweet manioc and plantains and bananas are the staple crops. Major animal food resources come from fishing and hunting, although both groups

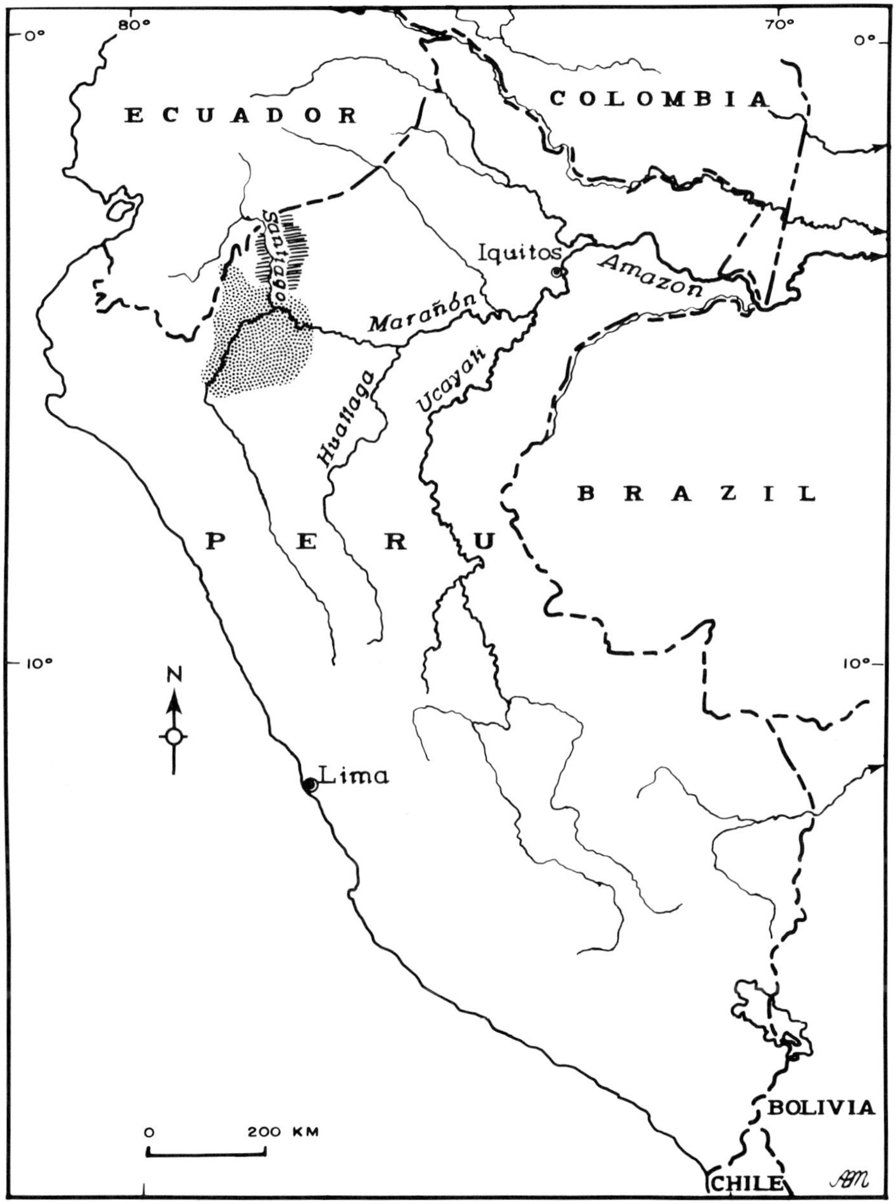

Figure 10.1 *Approximate distribution of the Aguaruna (stippled area) and Huambisa (lined area) of Amazonas, Peru.*

are turning increasingly to animal husbandry (E. Berlin, in press, Berlin and Berlin 1978, Berlin and Markell 1977).

Research Sites

The data reported here are drawn from three research sites (see Figure 10.2). One is a Huambisa community, the remaining two are Aguaruna settlements. The Huambisa village occupies the banks of a small, meandering tributary of the Santiago, about 4 hours by canoe from the confluence of the stream with the major river. Huambisa populations along the broad Santiago valley are widely dispersed, although many are now grouped into villages. Most of the Huambisa at our study site reside in the immediate vicinity of the main settlement, with a population of about 50. The creek as a whole is quite sparsely settled with long stretches uninhabited.

The two Aguaruna communities lie on the middle reaches of the Cenepa River, about 100 km southwest of the Santiago. The Cenepa flows much more rapidly than the Santiago and is broken in places by short rapids. The first Aguaruna site, which we will refer to as the "riverine Aguaruna," lies at the confluence of a small creek with the Cenepa itself. The village was established 15 years ago, when a school was founded and part of the Aguaruna occupying the upper portions of the creek were induced to settle downstream at the creek's mouth. The population was 180 in 1973 and had expanded to 265 by 1975.

The second Aguaruna site is located approximately 20 km upstream from the riverine Aguaruna. We will refer to this group as the "interior Aguaruna." People here live more traditionally than the riverine village. The estimated population for this site is about 75 people.

Data Base

The data base on which the analysis rests consists of weighed dietary intake information, zoological collections, and ethnobiological elicitation for each of the three groups, as well as less systematic participant observation. The first set of dietary data was produced by five riverine Aguaruna families over a 10-month period in 1975. A second set of food intake material comes from nutrition records made by three interior Aguaruna families for a period of 5 months in 1977. The last set of records is from six Huam-

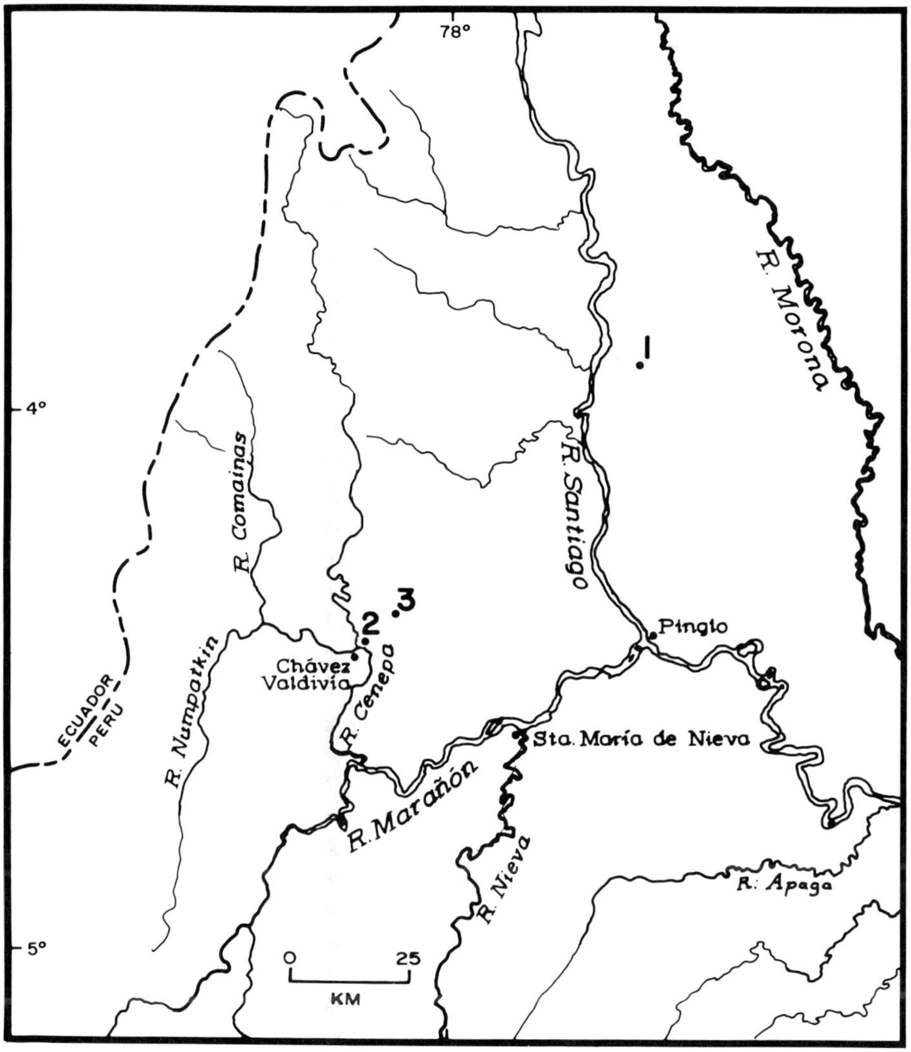

Figure 10.2 *Location of research sites: (1) Huambisa; (2) riverine Aguaruna; (3) interior Aguaruna.*

bisa families in our third research site during a 5-month period in 1979 and 1980.

The zoological information relating to Aguaruna and Huambisa biological resources and ethnobiology are drawn from results of zoological and ethno-zoological data collection over three field seasons in 1977, 1978, and 1979–1980.

Sketch of Aguaruna and Huambisa Ethnozoology

The data we have collected thus far indicate that both of these peoples possess a highly systematic, empirically based, conceptual organization of the animal world that is substantively similar to that of western science. (This is not the place to provide a detailed treatment of Aguaruna and Huambisa ethnozoological systematics, and only the barest outlines will be presented here.) Approximately 80% of the 110 mammalian species of the region are recognized and named (Berlin and Patton 1979). Of these species, all are classified as edible, except the carnivores, sloths, bats, anteaters, nutrias, most opossums, one of the armadillos, and rats and mice (Patton *et al.* in press).

All of the readily observable bird species in the area (nearly 300) are recognized with folk generic names that almost invariably correspond in a one-to-one fashion with scientific species (Berlin *et al.* 1981), a finding quite in line with the results of other detailed ethnoornithological studies of non-Western societies (Diamond 1966, Feld 1980, Hunn 1977, Saenap and Bulmer 1979). Of these 300 or more species, all are deemed edible with the exception of hawks, vultures, nightjars, potoos, and owls.

The analysis of our ethnoherpetological data is as yet incomplete, although our initial findings indicate an astute recognition of biologically valid species, genera, and families (Berlin and McDiarmid in preparation). Thirty of the 85 recognized reptiles and amphibian species are eaten. Snakes, some lizards, poisonous toads, and for some individuals, the caiman, are not consumed.

Some 150 distinct species of fish are recognized linguistically by the two Indian groups. All fish are considered edible except the small parasitic canero (*Henonomus* sp., *Plectrochilus* sp., *Pseudostegophilus* sp.). The fresh water dolphin (*Inia* sp.) is not considered to be edible, nor is it classified as a fish (Berlin and Swift in preparation).

Invertebrate classification is, as might be expected, less fully developed, although more than 250 perceptually distinctive groupings are recognized. The major portion of invertebrates are considered inedible, the notable exceptions being the prized palm grubs (probably of the Pasallidae), some beetles, snails, crabs, clams, and shrimp.

Animal Consumption Patterns

Although nearly 500 species of animals are conceptually recognized and classified as food, the Aguaruna and Huambisa depend upon an extremely limited number of species as primary contributors to diet. This is readily seen by an examination of the relative dietary contributions of these resources, species by species. (Invertebrates, reptiles, and amphibians contribute so little in terms of absolute bulk in our dietary records that they will be ignored in the following description). It will be useful to discuss animal consumption patterns in terms of the three major groups of vertebrates—mammals, birds, and fish—according to their bulk contribution to diet.

MAMMALS

The proportional contribution of mammal species to total animal consumption in each of the three research sites can be seen in Table 10.1. The weight figures represent the animal at time of preparation or edible portion plus bone.

Overall contribution (by weight) of mammals to diet is 30% for the Huambisa, 25% for the riverine Aguaruna, and 31% for the interior

Table 10.1

Contribution of Mammal Species to Total Animal Consumption

Scientific name	Common name	Native name	Percentage contribution (by wt.)
Huambisa			
Agouti paca	Paca	káshai	9
Dasyprocta fuliginosa	Agouti	ká̱yuk	8
Nasua nasua	Coati	kúshi	3
Tayassu tajacu	Collared peccary	jigkipík	2
Alouatta seniculus	Howler monkey	yakúm	2
Dasypus novemcinctus[a]	Nine-banded armadillo	shushuí	2
Tayassu pecari	White-lipped peccary	páki	1
Lagothrix sp.	Wooly monkey	chúu	1
Sciurus spp.	Tree squirrels	kunám	
Potos flavus	Kinkajou	kúji	
Sus scrofa	Domestic pig	kúchi	
Mazama americana	Brocket deer	jápa	2
Ateles sp.	*Maquisapa*	wáshi	
Pithecia hirsuta	*Asaki*	pentsepéts	
Saimiri sciureus	Squirrel monkey	tséem	
Callicebus moloch	Titi monkey	sugkamát	
			30

(*continued*)

Table 10.1 *Continued*

Scientific name	Common name	Native name	Percentage contribution (by wt.)
Riverine Aguaruna			
Tayassu pecari	White-lipped peccary	páki	5
Agouti paca	Paca	káshai	3
Mazama americana	Brocket deer	jápa	3
Dasyprocta fuliginosa	Agouti	káyuk	2
Sus scrofa	Pig	kúchi	2
Potos flavus	Kinkajou	kúji	2
Dasypus novemcinctus	Nine-banded armadillo	shushuí	
Tayassu tajacu	Collared peccary	yugkipák	
Nasua nasua	Coati	kúshi	
Alouatta seniculus	Howler monkey	yakúm	
Sciurus spp.	Tree squirrels	waiwásh	8
Aotus trivergatus	Night monkey	butúch	
Cebus albifrons	Capuchin monkey	machíg	
Sassaricyon alleni	Olingo	nuásak	
Tapiris terrestris	Tapir	pamáu	
Callicebus moloch	Titi monkey	sugkamát	
Hydrocaeris hydeochaeris	Capybara	ugkubiú	
Sylvilagus brasiliensis	Rabbit	wapujúsh	
Myoprocta pratti	*Achuchi*	yugkíts	
Bos taurus	Cow	báka	
			25
Interior Aguaruna			
Agouti paca	Paca	káshai	6
Sus scrofa	Pig	kúchi	6
Dasypus novemcinctus	Nine-banded armadillo	shushuí	5
Dasyprocta fuliginosa	Agouti	káyuk	3
Alouatta seniculus	Howler monkey	yakúm	2
Tayassu tajacu	Collared peccary	yugkipák	
Sciurus spp.	Tree squirrel	waiwásh	
Mazama americana	Brocket deer	jápa	
Nasua nasua	Coati	kúshi	
Tayassu pecari	White-lipped peccary	páki	9
Potos flavus	Kinkajou	kúji	
Cebus albifrons	Capuchin monkey	machíg	
Tapiris terrestris	Tapir	pamáu	
			31

[a]There are possibly two species of *Dasypus* (*D. novemcinctus* and *D. cappleri*) in the region. We are as yet unable to differentiate them in the dietary records for this group.

Table 10.2

Contribution of Bird Species to Total Animal Consumption

Scientific name	Common name	Native name	Percentage contribution (by wt.)
Huambisa			
Gallus gallus	Chicken	atásh	4
Nothocrax urumutum	Rufous curassow	iwáchi	2
Ortalis guttata	Speckled guan	wakáts	1
Tinamus major	Great tinamou	wáa	1
T. guttatus	White-throated tinamou	waják	1
Mitu salvini	Salvin's curassow	máshu	1
Psophia crepitans	Trumpeter	chíwa	1
Ramphastos cuvieri	Cuvier's toucan	tsukagká	1
Aburria pipile	Blue-throated piping-guan	kúyu	1
(24 remaining species, each less than 1%)			3
			16
Riverine Aguaruna			
Gallus gallus	Chicken	atásh	15
Steatornis caripensis	Oil bird	táyu	
Geotrogon montana	Ruddy quail-dove	tsamáu yámpits	
Ramphastos culminatus	Yellow-ridged toucan	kéjua	3
Leptotila rufaxila	Grey-fronted dove	shugkáem	
Cyanocorax violaceus	Violaceus jay	kíjuancham	
Phloeoceastes melanoleucos	Crimson woodpecker	tátasham	
Ramphastos cuvieri	Civier's toucan	tsukagká	
Ortalis guttata	Speckled guan	wakáts	
Trogon viridis	White-tailed trogon	táwai	
(49 remaining species, each less than 1%)			2
			20
Interior Aguaruna			
Gallus gallus	Chicken	atásh	8
Steatornis caripensis	Oil bird	táyu	3
Tinamus major	Great tinamou	wága	3
Psarocolius sp.	Oropendola	chúwi	1
Ramphastos cuvieri	Cuvier's toucan	tsukagká	1
Phoenicircus sp.	Continga	pétsa	
Ramphasos culminatus	Yellow-ridged oucan	kéjua	2
Ortalis guaa	Speckled guan	wakáts	
(24 remaining species, each less than 1%)			18

Aguaruna. Each group is similar as regards the most frequently taken mammals. Paca (*Agouti paca*) ranks first as the major mammal species for both the Huambisa and the interior Aguaruna, placing second among the riverine population. Not far behind, the agouti (*Dasyprocta fuliginosa*) is ranked second in the Huambisa village, placing fourth for both Aguaruna groups. The nine-banded armadillo (*Dasypus novemcinctus*) is a major contributor to diet among the interior Aguaruna (ranked third) and is in or near fourth place with the other two groups.

The major differences between the three study sites in mammal consumption are seen in the relative importance of deer and domesticated pig. Deer contributes less than 1% for both the interior Aguaruna and the Huambisa, but *M. americana* is the third most important contributor for the riverine Aguaruna. Pigs are the second most important mammal for the interior Aguaruna and tie for fourth for the Riverine group. This animal does not appear in the dietary records of the Huambisa for the time period sampled. The relative importance of white-lipped peccary (*Tayassu pecari*), which ranks first for the riverine Aguaruna, is probably skewed by sampling period. Reasons for the difference between its high occurrence for the riverine group and its low contribution for the other two study sties will be discussed in detail below.

BIRDS

Birds contribute 16% of total weight of animals consumed for the Huambisa, 20% for the riverine Aguaruna, and 18% for the interior group. It can be noted that chickens are ranked in first place by all three groups. This is most marked for the riverine Aguaruna for whom 15% of animal consumption (a full 75% of all birds consumed) is derived from this single source. While 55 species of game birds were taken during the time period under study, no single species accounts for more than 1% of the total for this site. Chickens play a large role for the interior Aguaruna as well, and are comparatively less significant for the Huambisa. The 3% contribution of oil birds (*Steatornis caripensis*) for the interior Aguaruna is due to the presence of a cave in the area that is regularly raided for the fat young birds of this species. Their abundance in diet is seasonal, however. Of the other wild birds, only curassows (for the Huambisa), guans (all groups), and tinamous (for the Huambisa and interior Aguaruna) are noteworthy. Overall, wild game birds provide little to total animal consumption for the Huambisa, riverine Aguaruna, and interior Aguaruna (9, 5 and 9%, respectively).

FISH

Fish contribute more than half of all animal bulk in the diet of each of the three study groups, with 54%, 55%, and 51% for the Huambisa, riverine

Table 10.3

Contribution of Fish Species to Total Animal Consumption

Scientific name	Common name	Native name	Percentage contribution by wt.
Huambisa	—		
Characids			
(*Hydrolychus* sp. *Rhaphiodon* sp. *Serrasalmus* spp. *Triportheus* spp. *Astyamax* spp. *Creagratus* spp. *Phenacogaster* spp. *Serrasalmus* spp.		mamayák, champerám, pani, tsapaúm, tsarúr	20
Loricariids			
(*Ancistrus* sp., *Panague*, sp., *Chaetostoma*, *Hypostomus* spp. *Reneloricaria* spp	*Carachama* Armored catfish	pútu shári, shácham, karachám, shúwi	10
Prochilodids	*boquichico*[a]	kágka	7
Pimelodids			
(*Pimelodela* spp., *Rhamdia* spp. *Pimelotus* spp.	catfish	ámiag, kushám, kumpáu	5
Erythrinids			
(*Erythrimus* sp, *Hoplias* sp.)	—	kugkuí, kanímu	3
Anostomids			
(*Abramites* sp. *Leporinus* spp.	—	katísh, kúwi	2
Curimatids			
(*Curimata* spp.)	—	kúum	2
Cichlids	—	kántash, chúwi, wapurús	1
Gymnotoids, Ageniosidids, Callichthyids	—	wáncha	1
(remaining 14 folk genera each less than 1%)			3
			54
Riverine Aguaruna			
Pimelodids and doradids			
(*Pseudodoras* spp., *Pterodoras* sp., *Pimelodina* sp., *Pimelotus*, spp., *Pimelodela* spp.)	Catfish	tugkáe, agákaem, buta, tujushíku, kumpáu, kúnchi, kushám, manitúja, putúsh, sáwi, wachicháyu, yutuí	22

(*continued*)

Table 10.3 *Continued*

Scientific name	Common name	Native name	Percentage contribution by wt.
Prochilodids			
(*Prochilodus* spp.)	*Boquichico*	kágka	14
Characids			
(*Serrasalmus* spp.,	—	tsajún, pani, cham-	9
Creagrutus sp.		pegkám, mamayák,	
Astyanax spp.		kamít, misáchu,	
Salminus sp.		wámpi	
Loricariids			
(*Ancistris* spp.	Armored catfish	pútu, kaejám, nayúm,	6
Chaetostoma spp.		shúwi, shácham,	
Hypostomus spp.		tsutsúm	
Reneloricaria spp.)			
Loricaria spp.			
Plecostomus sp.			
(remaining species in eight families, each less than 1%)			4
			55
Interior Aguaruna			
Loricariids, (especially *Hypostomus* sp.)	*Carachama*, Armored catfish	nayúm, shúwi, tsútsum, shácham	41
Characids		mamayák, misáchu	4
(*Astyanax* spp.,	—	samík, chumakái,	
Creagutus spp.)		tsapakúsh, tsajúm	
Pimelodids			
(*Pimelotus* spp.	Catfish	búta, kumpáu, putúsh	1
Pimelodela spp.)			
Prochilodids			
(*Prodilodus* spp.)	*Boquichico*	kága	1
(remaining species in four families, each less than 1%)			4
			51

[a]The term *boquichico* is found throughout Peru. *Boca chico* is an alternative spelling in some other Spanish-speaking countries of South America.

Aguaruna and interior Aguaruna respectively. Table 10.3 indicates that the three groups depend primarily on distinct fish faunas. The Huambisa draw heavily on small characids, loricariids, prochilodids, and pimalodids, which among them account for more than 40% of total animal consumption and 84% of all fish eaten.

The riverine Aguaruna are dependent on the large, free-swimming catfish (doradids and pimelodids) as well as the ubiquitous *boquichico* (probably *Prochilodus amazonensis* in the main). Fish species of these three families account for one-third of the total animal consumption for this group and represent more than two-thirds of all fish eaten.

Characids and loricariids follow in relative importance for the riverine Aguaruna. The interior Aguaruna are strongly focused on loricariid species, especially the *carachama* (*Hypostomus* sp.). The Loricariidae account for more than 40% of all animals eaten and 80% of all fish taken for the interior Aguaruna.

Comparative Patterns of Animal Utilization

A fuller picture of Aguaruna and Huambisa utilization of animal resources can be seen by combining the inventories of the primary animal species consumed by each of the three groups. Beginning with the Huambisa, we can see in Table 10.4a that two fish families, the Characidae and the Loricariidae, rank first and second, followed directly by the two game rodents, paca and agouti. These groups account for almost half of all Huambisa animal food resources. Fifty-four pecent of all animal resources consumed is reached by adding the pimelodid catfish. Chickens bring the total to 58%. The coati contributes 3% and each of the remaining major species adds 2% to diet, to total 73% of all animal foods. Only a single game bird appears in this fraction of the total. The remainder is accounted for by a large number of faunal species, none of which contributes more than 1% to diet.

The riverine Aguaruna present a different picture. The most important group is represented by the pimelodid and doradid catfish, followed by chicken and *Prochilodus* spp. These three groups represent more than half of all animal species in the diet (22, 15, and 14%, respectively). Two fish families, the characids and the loricariids, bring the total to 66%. The white-lipped peccary follows but, as stated earlier and discussed below, is unlikely to be a consistent contributor to diet. Five species of mammals divide the following 12% for a total of 78%. Thus, we find the populous riverine Aguaruna depending heavily on domesticated chicken and river fish, with mammals playing a more minor role in overall consumption; wild birds do not approach 1% in any instance.

Finally, the interior Aguaruna appear intermediate between the relatively conservative Huambisa and their more acculturated riverine cousins. For this group, the 40% mark is reached in one fell swoop by the loricariid armored catfish, followed by the second-ranked chicken (49% cumulative). Paca, domesticated pigs, and armadillo reach the 66% (cumulative) mark. Agouti, oil bird, and great tinamou contribute 3% each (75% cumulative) and howler monkey, characids, (2% *each*) bring the total to 81% of animal

Table 10.4

Rank Order and Percentage Contribution of All Vertebrate Groups to Total Animal Consumption

Scientific name	Common name	Native name	Percentage contribution (by wt.)
Huambisa			
Characids (Various spp., see Table 10.3)	—	mamayák, champerám, páni, tsapaúm, tsárur	20
Loricariids (Various spp., see Table 10.3)	*Carachama*, Armored catfish	pútu, shari, shacham, karachám, shúwi	10
Agouti paca	Paca	káshai	9
Dasyprocta fuliginosa	Agouti	ka̲yuk	8
Prochilodids	*Boquichico*	kágka	7
Pimelodids (Various spp., see Table 10.3)	Catfish	ámiag, kushám, kumpáu	5
Gallus gallus	Chicken	atásh	4
Nasua nasua	Coati	kúshi	3
Erythrinids (*Erythrinus* sp., *Hoplias* sp.)		kugkuí kanímu	3
Tayassu tajacu	Collared peccary	jigkipík	2
Alouatta seniculus	Howler monkey	yakúm	2
Dasypus novemcinctus	Nine-banded armadillo	shushuí	2
Nothocrax urumtum	Rufous curassow	iwáchi	2
Remainder:			
mammals (9 spp.)			4
birds (31 spp.)			10
fish (14 spp.)			9
			100
Riverine Aguaruna			
Pimelodids and doradids (Various spp., see Table 10.3)	Catfish	tujushiku and several folk genera, see 10.3	22
Gallus gallus	Chicken	atásh	15
Prochilodids (*Prochilodus* spp.)	*Boquichico*	kágka	14
Characids (Various spp. see Table 10.3)		mamayák and several folk genera (see Table 10.3)	9
Loricariids (Various spp., see Table 10.3)	*Carachama*, Armored catfish	nayúm and several folk genera, see Table 10.3)	6
Tayassu pecari	White-lipped peccary	páki	5
			4
Agouti paca	Paca	káshai	3
Mazama americana	Brocket deer	jápa	3

Table 10.4 *Continued*

Scientific name	Common name	Native name	Percentage contribution (by wt.)
Dasyprocta fuliginosa	Agouti	ká̲yuk	2
Sus scrofa	Pig	kúchi	2
Potos flava	Kinkajou	kúji	2
Remainder:			
mammals (14 spp.)			8
birds (9 spp.)			5
fish (12 spp.)			4
			100
Interior Aguaruna			
Loricariids, especially *Hypostomus* sp.	*Carachama*, Armored catfish	nayum and several other folk genera, (see Table 10.3)	41
Gallus gallus	Chicken	atásh	8
Agouti paca	Paca	káshai	6
Sus scrofa	Pig	kúchi	6
Dasypus novemcinctus	Nine-banded armadillo	shushuí	5
Dasyprocta fuliginosa	Agouti	ká̲yuk	3
Steatornis caripensis	Oil bird	táyu	3
Tinamus major	Great tinamou	wága	3
Alouatta seniculus	Howler monkey	yakúm	2
Characids	—	mamayák and various folk genera (see Table 10.3)	2
	—		2
Remainder:			
mammals (4 spp.)			9
birds (24 spp.)			4
fish (4 families)			6
			100

foods. For this study group, traditional patterns of fish harvest and the relatively commonly cropped paca and armadillo continue to represent major animal resources, but chickens and pigs have begun to assume a major role as well. As with the other two groups, wild game birds—with perhaps the exception of oil birds and tinamou—contribute but an insignificant amount to the total.

Discussion

Plausible explanations for the extreme reliance of the Aguaruna and Huambisa on a very few animal species are suggested by considerations of ecological constraints in combination with factors of relative availability and conditions of social change, although these latter are supported primarily by inferential reasoning. We will first examine some factors that we believe to be important in the heavy utilization of individual species of animals and then move to a discussion of factors relevant to the recognition of those numerous remaining species of animals whose dietary importance is insignificant.

Focusing first on mammals, three species occur among the first six wild mammals in each of the three study sites—the paca, the agouti, and the nine-banded armadillo. The behavioral habits of these creatures must be considered as relevant to their high contribution to diet. Pacas are frequent visitors to manioc gardens, where they dig up manioc and other roots and tubers. They are hunted at night with shotgun and flashlight and the feeding animals are easy targets. Our native assistants volunteered comments that the meat of pacas is highly prized.

The second terrestrial rodent, the agouti, is hunted during the day and can be drawn to the hunter, who produces high squealing sounds (which informants say is typical of a young agouti in distress) by blowing over the edge of a tightly held leaf.[1] Our informants state that the flesh of the agouti is as highly valued as that of the paca.

Nine-banded armadillos, while not attracted to gardens as are the two caviomorph rodents, nonetheless are found relatively near to human habitations. The species is secretive, but indigenous ethnoecological knowledge of the habits of these animals allows for the rapid identification of burrows from which it may be easily captured. The smoked meat of armadillo is excellent and is especially savored for its fat content. Our informants' activity diaries indicate that in no case did the hunter travel far from home to hunt these three animals. This pattern is typical of some other regions of Amazonia (but cf. Vickers 1980; and Hill and Hawkes, this volume). Comparative data taken from several other semisedentary lowland South American aboriginal groups that corroborate the relative importance of paca, agouti, and armadillo as major game species are as follows: in an extensive literature review, E. Ross has observed that "smaller vertebrates [such as the paca, agouti, and armadillo] obviously provide the most secure basis for obtaining game within a certain range of settlement size" (E. Ross 1976:270). Ross continues by quoting Beeke's experience with agoutis: "the most abundant mammal in the research area is the agouti . . . my Indian hunter and I have killed one hundred and thirty-five for eating, and yet we hardly ever take a walk through the jungle without seeing or hearing one of these red-rumped rodents" (Beeke 1925: 180, quoted in Ross 1976:270). Beckerman, writing about the Barí of Columbia, states that, "Agoutis are probably the most reliable kind of game, certainly, the most reliable mammal. Their population is comparatively dense and they seem to be well dispersed throughout the forest" (Beckerman 1980:95). Finally, it is worth noting that

Linares, in her Cerro Brujo excavations in Panama, reports "that the Cerro Brujo inhabitants had strong hunting preferences for certain species. Agouti (*Dasyprocta punctata*), paca (*Cuniculus paca*) and nine-banded armadillo (*Dasypus novemcinctus*) together constitute 80.9% of all animals taken" (Linares 1976:338).

Two larger terrestrial animals, the white-lipped peccary (*Tayassu pecari*) and the brocket deer (*Mazama americana*) figure somewhat prominently in the diet of the riverine Aguaruna, in contrast to the other two research sites. On the basis of our current data, we suggest that the relative importance of *T. pecari* among the riverine Aguaruna is probably serendipitous. These animals, traveling in large herds, may, from time to time, wander near enough to a village to be discovered. Here, all shotguns of the community are put to work at once, and as many kills are made as possible. One such event occured during the 1975 period of record keeping for this group. However, in our experience, white-lippeds do not represent a reliable and regular source of animal meat.

Comparable observations have been made about the serendipitous nature of *T. pecari* kills throughout lowland South America. During Ross's stay among the Achuara, only two white-lippeds were taken in any of the three sites where he collected data and both of these animals were "killed in a single morning . . . brought in by the same hunter" (E. Ross 1976:263). Kiltie, who has carried out one of the few field studies on the behavior of *T. pecari*, supports our suggestion of random harvest. Kiltie states "probably the white lippeds' best defense against people is simply their extreme mobility, which when coupled with the rarity of herds overall, makes their appearances infrequent and short-lived at any particular place . . . white-lipped peccaries [are] extremely unreliable in their appearances" (Kiltie 1980b:542). Of the 30 white-lipped peccaries taken during Silverwood-Cope's 9-month stint among the Colombia Maku, 27 were taken in just three hunts—4 on one, 16 on another, and 7 on a third (see Silverwood-Cope 1972:77). Bergman, reporting on fieldwork among the Shipibo, describes a single hunt in which 22 animals were killed in $1\frac{3}{4}$ hours as a herd passed by the community in which he was working. Significantly, Bergman qualifies this extensive kill as follows: "this was a unique episode in the history of the village. No one had heard of it happening before." (Bergman 1974:215).

Deer (both *Mazama americana* and *M. gouazoubira*) have traditionally been tabooed as edible game among all Jivaroan groups. The taboo is beginning to break down among the riverine Aguaruna, as seen in its relative contribution to total mammal consumption. Ross has suggested the "commercial contacts . . . provide a powerful incentive for at least killing deer that were chanced upon" (1976:297). It is true that deer is occasionally sold to military personnel and missionaries by the Aguaruna residing on the Cenepa. Alternatively it would be easy to suggest that "overall acculturation" might account for the growing acceptance of deer as food, although the precise mechanisms involved are essentially impossible to specify. It is also possible that deer consumption among the riverine Aguaruna might be related in some way to population density and concomitant game scarcity. None of the above explanations is entirely satisfactory, however.

Domesticated animals, especially pigs and chickens, are seen to be major contributors to diet among the two Aguaruna groups. This development is also difficult to account for with a simple, unambiguous analysis. Possible factors to consider are, again, increased population density, leading to lowered abundance of game in the immediate region either because of migration away from populated areas or because of overhunting. An additional, equally plausible hypothesis is a recognition by both groups of the constant and increased productivity of animal husbandry over hunting (cf. A. Johnson 1977 for something comparable among the Machiguenga.) Efforts by outsiders, especially missionaries (both Protestant and Catholic) and Spanish development volunteers, to encourage animal husbandry is clearly an additional incentive for the local populations on this river and must also be considered as a potential factor in accounting for the reliance on domesticated animals. As with the decreased significance of the taboo on eating deer, however, no undebatable explanation suggests itself to us given our current information.

Although the overall importance of fish for the three study groups is about the same, the individual species of fish taken appears to be governed in large part by the ecological constraints present at each of the sites.[2] The significance of these constraints is highlighted by reviewing the geographical locations of the three groups: the riverine Aguaruna reside directly on the Cenepa River itself; the interior Aguaruna inhabit the headwaters of a creek leading to the Cenepa; and the Huambisa lie about midway up a meandering tributary of the Santiago. These slight ecological differences apparently have given rise to relatively distinct fish faunas from which the representative groups can readily sample. The Riverine Aguaruna's heavy dependence on the widespread, free-swimming catfish, the river doradids and pimelodids, as well as the *boquichico,* is partially explained by the abundance of these species in this specific area.

Although the interior Aguaruna and the Huambisa are both creek dwellers, the creeks they occupy may represent distinct microhabitats. The bottom-scavenging, small, armored loricariid catfish species are the major fish resource for the interior Aguaruna. These species frequent the rocky bottoms and shores of the creek on which the interior Aguaruna live and are easily captured.

Finally, the sandy-bottomed, slow-moving tributary that flows eastward across the broad, flat valley of the Santiago may offer yet a third microhabitat for fish. Here, one finds the Huambisa harvesting small characids, the most abundant family of fish, prochilodids, pimelodids, and loricariids, of various species.

To summarize to this point, we have shown that a small number of species of mammals, birds, and fish contribute significantly to total animal consumption for the Aguaruna and Huambisa. We have suggested that the selection of these particular species might be explained as some combination of factors relating to availability, ease of capture, and, particularly in the case of fish, to specific ecological features of individual habitats. The question remains, if the Aguaruna and Huambisa utilize so few species of animals, why do they continue to maintain such a detailed zoological classification of the remaining faunal diversity, whose absolute numbers constitute more than 90% of all known species?

Two competing explanations have been suggested in the literature. One, which we will call intellectualist, is prominently associated with Lévi-Strauss (1966). Hays (1982) recently examined a second position that he has called "utilitarian–adaptationist." This view has most recently been espoused by Hunn (n.d.), Bulmer (1974) and is implicitly held by Diamond (1966). For clarity of presentation, we will first briefly discuss each of these positions, and then consider a third alternative.

Lévi-Strauss, in a detailed review of the literature on several systems of ethnobiological classification (1966, especially Chapter 1), aruges that the complexities of native systems of biological knowledge are not to be explained on the basis of need, in contrast to the functionalist position of Malinowski, who claimed that primitive peoples' concern with "plants and animals was inspired by nothing but the rumbling of their stomachs" (Lévi-Strauss 1966:3). Instead, for Lévi-Strauss, "the universe is an object of *thought* at least as much as it is a means of satisfying *needs*" (1966:3, emphasis added). Further, primitive peoples' "extreme familiarity with their biological environment . . . cannot relate to just practical needs" (1966:6,8), because the detail found in such systems of classification" can scarcely be of much practical effect [because] its main purpose is not a practical one. It meets intellectual requirements rather than . . . satisfying needs [because] *classifying . . . has a value of its own*" (1966:9, emphasis added).

Recently, the intellectualist position has been challenged by the development of an explicit utilitarian–adaptationist position. This view, as summarized by Hays in his critique is as follows: The utilitarian–adaptationists see the environment "as a setting in which people must satisfy their physical needs, that is, a setting to which they must *adapt*. Folk biological classification would then be regarded as a way by which people systematically organize, store, and retrieve environmental information which will enable them to accomplish this adaptation. Thus, a utilitarian/adaptationist perspective would seek to identify the practical consequences of folk conceptual and terminological distinctions, and ultimately explain folk taxonomies as structures which are motivated by a concern with these consequences" (Hays 1982).

Hunn has espoused this position most forcefully, arguing that all ethnobiological distinctions made within any particular folk system must have "practical significance." He claims that terminological distinctions between organisms are maintained solely because they have different functional utility. For Hunn, "no two folk taxa, if their practical significance were adequately described, would be found to be precise functional equivalents . . . [one must] describe the practical content of folk biological knowledge in detail sufficient to discriminate each taxon from every other" (Hunn n.d.: 20).

Both the extreme intellectualist and utilitarian–adaptationist positions are inadequate in our view and cannot account for the patterns of utilization and classification of animals that we have described for the Aguaruna and Huambisa. We will explore some shortcomings of the utilitarianist model first.

It will be recalled that although only a handful of animal species make significant contributions to Aguaruna and Huambisa diet, large numbers of animal species are nonetheless considered as potentially edible. It might be argued that if an animal species is considered to be edible, this will be grounds for its continued linguistic recognition regardless of its minuscule contribution as food. In light of the data we have presented, it seems implausible to imagine that edibility alone is sufficient reason for recognition when this utilitarian value is so low. If potential edibility were the sole consideration for recognition, such species would be highly likely to be lost from the classificatory system over time simply on the basis of their infrequent occurrence in diet.

Obscure nutrient needs are sometimes cited as the reasons for inclusion of a wide variety of foods in diet. The evolutionarily adaptive response of humans to dietary resources seems to have been the ability to manufacture those nutrients not consistently available and the inability to synthesize those nutrients that have been relatively constantly available throughout our evolutionary history (Stini 1980). Humans must obtain, for example, the B vitamins and some amino acids from dietary sources. Current nutritional knowledge would suggest that the nutrient content of animal flesh is relatively uniform. (Plant species are much more likely to vary widely in nutrient content). Amino acids or proteins are the primary contribution of animals to diet. Infrequent or rare consumption of a given animal species would not be expected to alter significantly the nutritional status of the consumer.

A utilitarianist might then counter that animal species will be recognized for many other functional reasons than their edibility, such as possession of esthetic properties useful in ornamentation, music, dance, and other creative activities of human life. The upper and lower tail coverts of certain species of toucans, for example, are prized by the Aguaruna and Huambisa for their use in the production of native crowns and these features no doubt play some part in their linguistic recognition. Nonetheless, the feathers of most of the more than 250 species of recognized birds are decidedly unimportant as decorative plumage and yet their linguistic recognition is as solid as their more colorful companions.

If animal species cannot be said to have importance as a dietary item or for the production of products that are important in some area of human material culture, it might then be argued that some species, such as venomous snakes or dangerous cats, will be recognized because of their negative or noxious characteristics. Negative evaluation no doubt plays a role in the identification of some species of animals but, as a trait, has relatively minor importance for understanding the diversity of the system as a whole.

Myth and ritual are yet another instance of posited functions contributing to the importance of species in an ethnobiological classificatory system. In two volumes that deal comprehensively with Aguaruna mythology, Chumap and Garcia-Renueles (1979) mention 162 folk genera (approximately 20% of the named fauna), although broad generalizations such as "all of the animals we don't eat" also occur. Most stories, in fact, are instructional in the behavioral habits of animals and would seem to reinforce recognition of common species rather than increase recognition of uncommon ones. Few, if any, animals are used in ritual.

Simply stated, diet should comprise the largest category of usefulness in a utilitarian model (63% of named species are considered to be edible), with adornment, myth, aversion, and so on adding smaller numbers of species. The meat, skins, and bones of animals have a limited number of uses. If consistent use in diet explains roughly only 10% of the classificatory system, it is unlikely that a list of uses can be generated that is great enough to account for the remaining 90% of named animals. The ad hoc, checklist of attested, pragmatic man–animal relationships for all recognized animal species rapidly fades as the full faunal inventory is considered.

A more subtle utilitarian–adaptationist argument is broadly ecological (hence, again functional) and has been elaborated by Bulmer (1974) and Reichel-Dolmatoff (1976). This view defines the notion of utility more generally than simple, direct relationships of animal (and plant) species to human beings. Bulmer's words outline this view concisely:

> If one sees individual plant and animal categories solely in their direct relationships to man, there are many which appear irrelevant, neither utilized nor noxious. However, if the relationships between different kinds of plants and animals are recognized as relevant, then a great range of additional forms will very usefully be identified and classified. If hunting or collecting of wildlife are economically or socially important, or even if these activities are unimportant but accurate observation of wildlife is still culturally relevant for other reasons (e.g., signs of seasonal change, omens) then because animals exist in significant relationships with other animals and plants, there is a considerable impetus to classify these forms also (Bulmer 1974:12–13).

The difficulty with such an open-ended conception of utility is its scope, which allows each and every possible conceivable relationship to be treated as functional no matter how indirect the relationship or how many organisms are involved. Practicing field ecologists are well aware of the difficulties of specifying quantitatively even the most simple relationships between *two species* in relatively uncontrolled (i.e., natural) settings. At best, total ecological dependency arguments take on the character of elaborate Rube Goldberg cartoons. They are, in the final analysis, untestable (cf. Hays 1982).

If strict functionalist explanations appear not to take us very far in exploring the totality of Aguaruna and Huambisa animal classification, the intellectualist views of Lévi-Strauss are also difficult to support in their entirety. Like the functionalist arguments that he rejects, Lévi-Strauss's notions suffer as well from their ad hoc application. The propensity of primitive peoples to classify is proclaimed to be due to an inherent "intellectual need" to do so, to some kind of internal, cognitive motivation for a "demand for order" (1966:10). We are asked to accept this alternative to the utilitarianist position on the basis of the beauty of the argument alone.

We propose a third alternative, one that is close to that of Lévi-Strauss, but one that differs in several important respects concerning the nature of biological reality. The view we outline is not new, and has its basis in the theory of natural classes.

We begin by reconsidering again a portion of the question, why have primitive peoples developed such a detailed classification of animals? (cf. Diamond 1966). There is something peculiar about this question, one motivated perhaps by a bit of unintentional western ethnocentrism. Rather than ask, Why does the primitive classify his environment in such detail? we might inquire, Why not?

This turn of the query results from the work of comparative studies in ethnosystematics over the past several years that have shown individual systems of plant and animal classification to be formed from the same mold. The overriding similarities that have been observed in small-scale, nonliterate societies' view of the world of plants and animals are now so apparent that one is allowed to conclude that human beings living in close, direct contact with their natural environment are constrained by essentially the same kinds of biological facts. The so-called "detailed" classification that westerners have marveled about refers primarily to the smallest groupings of animals (and plants) that can be easily recognized with minimal cognitive effort. These minimal, perceptually salient classes have been referred to as folk genera (Berlin 1972, 1973, 1976; Berlin *et al.* 1973, 1974). What is of importance here is the growing body of comparative evidence that shows these fundamental folk genera to be in close correspondence with recognized biological species and groups of closely related species of the same biological genus. This is not the place to review the extensive literature that has developed on the subject of the correspondence of folk and scientific systems of biological classification, (but see Berlin 1973; Berlin *et al.* 1974; Bulmer 1974; Diamond 1966; Dwyer 1976; Hunn 1976,

1977). For the purposes of this paper, a cross-cultural comparison of the correspondence of folk generic taxa and scientific species in two vertebrate groups, the birds and mammals, will suffice to demonstrate our main point. In Table 10.5, we compare our recent data from Aguaruna mammal and bird classification with that of Tzeltal, a Mayan society of highland Chiapas, Mexico, recently studied by Hunn (1977), and Kalam, a society from highland New Guinea that has been the subject of long-term investigation by Bulmer (Bulmer 1970, 1974; Bulman and Menzies 1972–73; Bulmer and Tyler 1968; Majnep and Bulmer 1977). These groups are selected because each has been given intensive ethnobiological study, for which we now have comprehensive descriptions, based on thorough zoological collecting methods rather than on restricted faunal sampling.

Table 10.5 indicates the proportion of folk generic categories in each of the three societies that correspond in a one-to-one fashion to zoological species of birds and mammals. Considering the numbers of species involved and the differing ecological settings of each group, the similarities of the proportional figures are striking. For the three groups combined, more than two-thirds of all mammal folk genera and more than 70% of all bird folk genera correspond exactly with recognized zoological species.[3]

This essential matching of folk and scientific lower order categories suggests that the biological systematists—be they self-trained ethnobiologists or western trained biologists—do not *invent* a conceptual order for the world of animals (or plants). Rather, the similar classifications of locally restricted, nondimensional species (Atran 1981; Diamond 1966; Mayr 1969) results from simply recognizing and codifying the natural order inherent in

Table 10.5

Zoological Correspondences of Folk Generic Taxa with Scientific Species in Three Societies for Comprehensive Faunal Inventories of Mammals and Birds

	One-to-one correspondence (%)	
	Mammals	Birds
Aguaruna[a]	63	80
Kalam[b]	63	70
Tzeltal[c]	76	64
	$\bar{X} = 67\%$	$\bar{X} = 71\%$

[a]Number of mammal species is 110; number of folk mammal genera is 51, of which 32 correspond to zoological species; number of bird species, 300; number of folk bird genera, 259, of which 206 correspond to zoological species.

[b]Number of mammal species is 43; number of folk mammal genera is 35, of which 22 correspond to zoological species; number of bird species, 140; number of folk genera, 105, of which 74 correspond to zoological species.

[c]Number of mammal species is 63; number of folk mammal genera is 45, of which 34 correspond to zoological species; number of bird species, 149; number of folk bird genera, 106, of which 68 correspond to zoological species.

the biological world. In this sense, both systems are strongly perceptually based, although the westerner has developed an evolutionary theory to explain explicitly the similarities and differences so readily obvious to the careful observer. One need not, then, invoke utilitarian motivations for either system, any more than one must enlist intellectualists' drives for conceptual order. Rather, it would seem that the physical and behavioral characteristics of localized species of a given geographic region are sufficiently transparent that not to recognize them is rendered difficult.

Conclusions

The Aguaruna and Huambisa, in their daily interactions with their tropical forest environment, are confronted with natural discontinuities of the biological world that express themselves in a multitude of ways to the human observer. It is true that the desire to fulfill the basic needs of sustenance (with least effort?) will intensify human interaction with some of these species and many species may be killed and evaluated as potentially edible. But the satisfaction of these needs and uses is surely not the primary mover in the recognition and classification of these species in the first place. Prior to these functional assessments, all of these species will be seen or heard.

We assert that the structure of the biological environment that the astute Aguaruna and Huambisa recognize is fundamentally given, and that cognitively, as a result of constant interaction with this environment, they come to act as codifiers of its inherent structure.

Acknowledgments

This research has been supported by a generous grant from the National Science Foundation (BSN 76-17485, "Field Research in Ethnobiological Anthropology, Brent Berlin and James L. Patton, Co-principal Investigators). Additional financial support has come from the National Institute of Mental Health (NIMH Grants 22012 and 25703); the Quantitative Anthropology Laboratory, University of California, Berkeley; the Museum of Vertebrate Zoology, University of Calfornia, Berkeley; the Museum of Natural Sciences, Louisiana State University; the US Fish and Wildlife Laboratory; the US Museum of Natural History; and the Departments of Ichthyology and Herpetology, Natural History Museum of Los Angeles County. This support is gratefully acknowledged.

Our investigations have been carried out with the direct collaboration of the Museo de Historía Natural "Javier Prado," Universidad de San Marcos, Lima, and the Ministerio de Agricultura, Dirección de Forestal y de Fauna, Lima, Peru.

We have benefitted especially from the comments of Terrence Hays, whose cautionary notes on the limitations of adaptationist models (Hays 1982) served as a crucial impetus for the present paper. We also appreciate the helpful criticisms of the volume editors, Ray Hames and William Vickers. The sustained support of José María Guallart, S. J., natural historian to the Aguaruna, has been important to us. Several of the ichthyological determinations used in the paper derive from Guallart (1976).

The orthography used for Aguaruna and Huambisa names is one developed by Mildred Larson of the Sumer Institute of Linguistics, (Larson 1966). Native terms are pronounced as in Spanish (roughly), where j is [h], g is [ŋ], and e is "uh" as in "uhuh." Underlined vowels, (e.g., u) are nasalized.

The work reported here could not have been accomplished without a thorough survey of the diversity of the vertebrate fauna in the upper Marañón area. We are especially grateful for the support of the four zoologists who have coordinated the collections and determinations of vertebrates: James L. Patton (mammalogy), John P. O'Neill (ornithology), Roy W. McDiarmid (herpetology), and Camm C. Swift (ichthyology). These team leaders have been ably assisted by John Cadle (herpetology and ichthyology), Mercedes Foster (ornithology), Thomas Justice (ichthyology and herpetology), Ted Parker (ornithology), Carol Patton (mammalogy) Marty Ruggles (ichthyology and herpetology), Douglas Stamm (ichthyology), and Donald Stewart (ichthyology).

We are highly indebted to the following field assistants: Ernesto Anguash, José Asunción, Nicolas Chigkún, Juan Magkásh, Feliz Domínguez, Hernán de la Vega, Hernán Jempe, Manuel Jempe, Alejandro Nestor, Alfredo Nunigo, Maximo Pati, and Juan Yampits.

Notes

[1]Silverwood-Cope reports a similar hunting technique among the Colombian Maku. He notes that "the hunter goes to the manioc garden at dawn or at dusk when the agouti comes to steal manioc, or he goes to a swamp which the agouti frequents. He rolls a cone from a leaf. He puts the tip of the cone in his mouth and draws in air making a squeaking sound like a young agouti. At this the agouti approaches the hunter and is killed" (1972:73).

[2]Despite the fact that fish species contribute half of the animal flesh consumed by all three groups, fish is not nearly as valued as the meat of mammals. *Boquichico* (*Prochilodus amazonensis*) and armored catfish are favorite fish, but only mammals elicit the response of a gourmet delight.

[3]We are the first to admit that the *specific* perceptual and biological constraints at work in the folk linguistic recognition of any *particular* species are essentially unknown at the present time, although initial efforts to describe some of these constraints, following leads first suggested by Hunn (1977), have been made (Berlin, *et al.* 1981). Our initial observations lead us to believe that vertebrates, in general, are more likely to be given explicit recognition than are invertebrates, and that among the vertebrates, individual species of birds and mammals are more likely to be named than are the less readily observable species of reptiles and amphibians. Our data on the precise correspondence of folk generic categories and invididual species of fish are not yet analyzed. The full characterization of the precise factors involved, leading ultimately to the prediction of which species will be named in any particular folk system of biological classification, represents one of the most challenging problems facing modern ethnobiology.

III
Nutrition

CHAPTER 11

Nutrition in the Northwest Amazon: Household Dietary Intake and Time–Energy Expenditure

Darna L. Dufour

Introduction

Most native Amazonians have traditionally relied on low-protein high-bulk cultigens such as manioc (*Manihot esculenta*) and plantains (*Musa* spp.) as caloric staples. High-quality protein was obtained from fish, game, and smaller fauna, such as insects and other invertebrates. A variety of wild vegetable products were collected, including palm fruits and hearts, and the fruits, nuts, and seeds of dicotyledonous trees. In contrast to cultigens, these wild vegetable foods are characterized by high caloric density and relatively small edible portions.

Although the adequacy of the diet of native Amazonians, particularly in relation to dietary protein, has been debated (Beckerman 1979; Chagnon and Hames 1979, 1980; Gross 1975; Lizot 1977), there are relatively few quantitative evaluations of either dietary intake or nutritional status. Berlin and Markell's (1977) study of the Aguaruna is the only detailed information available. They report that the average Aguaruna diet of manioc and fish is more than adequate in all nutrients, and that the community as a whole is well nourished. Their findings are in general agreement with the reports of good nutritional status for other groups pursuing traditional modes of subsistence (Ayres and Salzano 1972; Black *et al.* 1977; Fagundes-Neto *et al.* 1981; Glanville and Geerdink 1970; Layrisse and Layrisse 1980:160; Neel 1970; Niswander *et al.* 1967). Indeed, as Ayres and Salzano (1972:175) point out, patent malnutrition is rare in relatively unacculturated groups who maintain their traditional way of life. On the other hand, acculturation can

ADAPTIVE RESPONSES
OF NATIVE AMAZONIANS

ISBN 0-12-321250-2

be accompanied by a decrease in nutritional status. Dricot-D'ans and Dricot (1977, 1978) found that the nutritional status of men in the more traditional Machiguenga villages was better than in the more acculturated ones. For the Xavante, Flowers (this volume) found that fluctuations in food supply have probably increased with the change from a hunting–gathering to a horticultural mode of subsistence.

This essay presents data on the food intake of Tatuyo Indians and the time and effort expended to maintain their traditional diet. The subsistence pattern of the Tatuyo is similar to that of other Amazonian groups in that it is based on rootcrop horticulture, hunting, fishing, and collecting. The dietary staple, manioc, is under continuous cultivation, but the availability of wild foods varies seasonally.

Here I will attempt to show that (1) food intake at the household level appears to be adequate in calories and protein, although protein intake can fluctuate widely; (2) because there is little tendency to store foods, the adequacy of the diet depends on daily procurement activities; and (3) the time and energy devoted to food acquisition is relatively high, and is a direct reflection of food resource availability and processing.

Background

Research was conducted in the Colombian Vaupés between November 1976 and April 1978, in the Tatuyo village of Yapú, in the headwaters of the Papurí River (see Figure 11.1).[1]

The Tatuyo are one of the various exogamous patrilineal groups of Tukanoans in the Colombian Vaupés (see Bidou 1976; C. Hugh-Jones 1979). Like other Tukanoans, they have been in contact with Colombian society since about the turn of the century. They traditionally lived in small, dispersed longhouse settlements in the headwaters of the rivers Papurí, Tiquíe, and Pirá–Paraná. The present-day settlement pattern in the Papurí includes both traditional longhouses and multi-house villages. The latter, of which Yapú is one, have formed in response to outside influence. To the east, downstream of Yapú are more acculturated villages, and to the northeast is the Catholic mission post of Acaricuara. To the west is the Pirá-Paraná River, still an area of longhouse settlements. The entire region is sparsely settled, with a population density of about 0.2 persons/km^2.

The present village of Yapú is located on a traditional Tatuyo longhouse site which had been inhabited earlier in the century. The village was formed during the early 1960s by Tatuyo who were then living in small local descent groups along the stream Yapú. By 1977 the village included

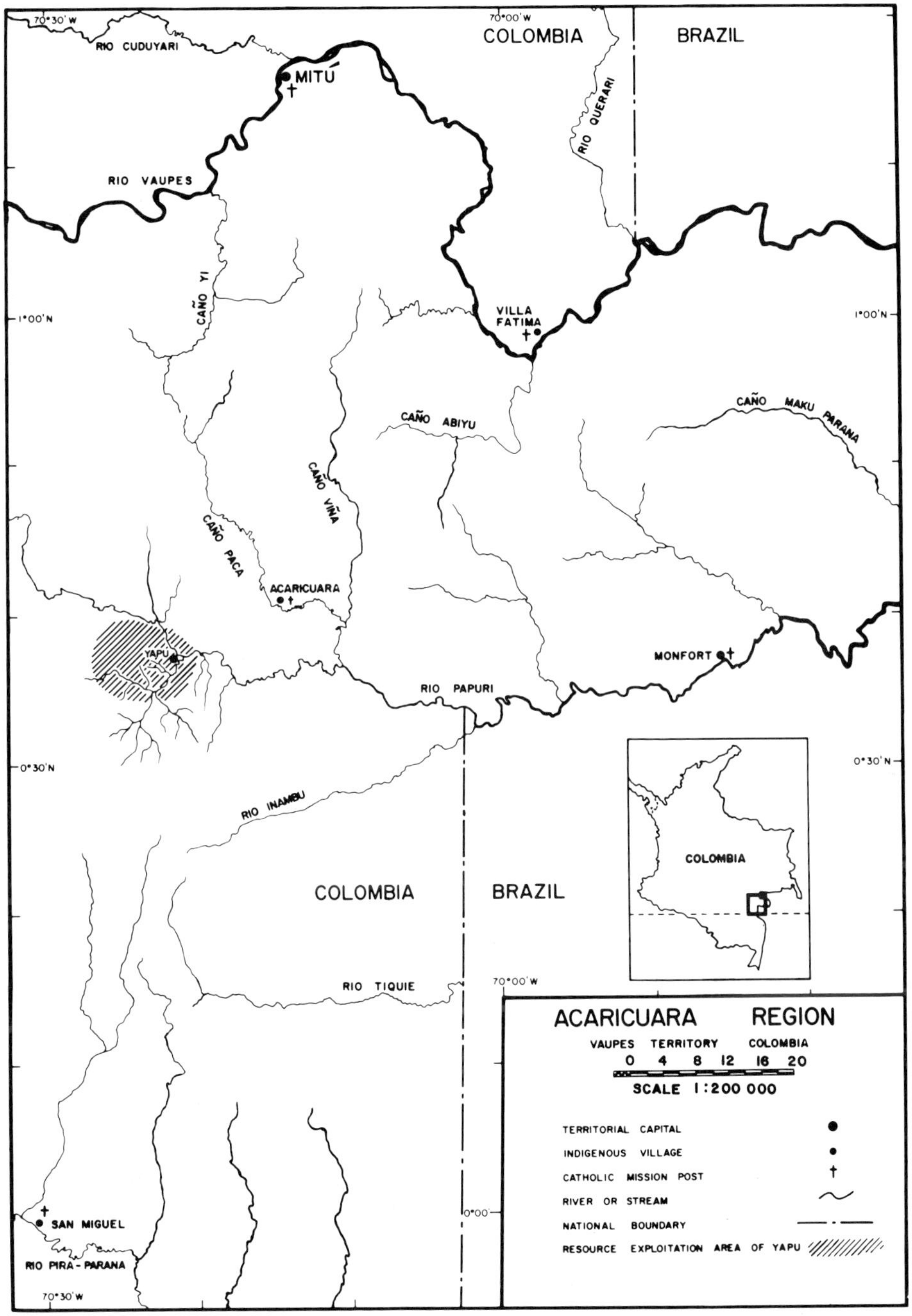

Figure 11.1 *Yapú fieldsite, Colombian Vaupés.*

members of some other exogamous groups and had a population of 140 persons.

Residents of Yapú were living in some 25 household groups, occupying 18 or so individual houses. Households are formed by both nuclear family and extended family groups. The minimum membership of a household is one adult man and one adult woman, who are a conjugal pair, or a parent and an adult child. The figure of 25 includes all households that claimed the village as their primary residence. However, since both individuals and households were quite mobile, the actual number of people present in the village at any one point in time fluctuated considerably. In a random sample of 60 days the mean population of the village was 108 people.

PHYSICAL ENVIRONMENT

The upper Papurí is an area of low elevation, about 200 m above sea level. The terrain is gently undulating and intersected by numerable small black-water rivers and streams. Mean annual temperature is relatively high at about 26°C, and rainfall abundant at about 3500 mm/year. Seasonal differences in temperature and rainfall are not well marked. There is a dry season of slightly less rainfall from November to February, but during this period rainfall does not usually drop below 150 mm/month. During the rainy season, from March to July, it can reach 500 mm/month.

Black-water rivers are characteristic of the region. These rivers are relatively poor in fish and invertebrates (Marlier 1967; Roberts 1973). The abundance and distribution of fish in the Yapú area are further limited by the area's location upstream of the large rapids on the Papurí River. These rapids appear to restrict the movements of fish. For example, the rapid at Yapara, just a couple of hours downstream of the village, traps at least one large species (of up to 10 kg) below it. Fish above Yapara, in the streams near the village, are small, with no species weighing more than 1 kg.

In headwater areas, such as the upper Papurí, the rivers respond rapidly to local fluctuations in precipitation and hence show a great deal of variation in water level. In general, however, they are lowest in the dry season and highest in the rainy season. The productivity of fishing correlates roughly with water level, being best at times of low water.

The forest cover in the upper Papurí area has been classified as dense forest with savanna and hills (PRORADAM 1979:232). It is a very heterogenous forest in which the relative abundance of any one species is no greater than 7% (PRORADAM 1979:276). Within the forest are naturally occuring zones of low savanna vegetation, or *caatinga* (compare Figures 11.2 and 11.3).[2] These are areas of infertile sandy soil (Klinge and Medina, 1979) which are not suitable for gardening. Furthermore, in comparison to the forest, they are considered poor in both game animals and wild food

Figure 11.2 *Dense tropical Forest along the main trail between Acaricuara and the Papurí River, downstream of Yapú. This is an area of relatively small trees and open canopy. The logs placed across the trail in the foreground indicate that the area is often wet.*

Figure 11.3 *Yapú woman and child standing in a zone of* caatinga *vegetation on their way to the garden.*

plants. The total surface area covered by *caatinga* is difficult to estimate, but there are several large patches of it near the village. In recent years, gardens have been cut beyond these areas of *caatinga* at a distance of about 1 hour's walking time from the village.

FOOD RESOURCES

The Tatuyo cultivate a number of crops, but rely on bitter manioc (*Manihot esculenta*) as the dietary staple.[3] It is not eaten simply boiled or roasted as are other rootcrops, but rather is elaborated into a number of products. In the ordinary diet the most important of these are *manicuera* (boiled manioc juice), *mingao* (starch and water drink) and *casabe* (manioc bread). *Fariña*, a toasted manioc meal, is also prepared.[4] It is the staple food on trips away from the village, but is not very important in the diet of village residents.

Cultigens of secondary importance include rootcrops such as sweet manioc (*Manihot esculenta*), taro (*Colocasia* spp.), sweet potato (*Ipomoea* sp.), arrowroot (*Maranta ruiziana*) and malanga (*Xanthosoma* sp.), as well as breadfruit (*Artocarpus communis*), bananas and plantains (*Musa* spp.), sugarcane (*Saccharum officinarum*), coconut (*Cocos nucifera*), pineapple (*Ananas sativa*), *naranjilla* (*Solanum* spp.), *guama* (*Inga* sp.), papaya (*Carica papaya*), maize (*Zea mays*), peach palm (*Bactris gasipaes*), *uvilla* (*Pourouma cecropiaefolia*), and hot peppers (*Capsicum annuum*). Of these crops, peach palm fruits, *guama* and *uvilla* are only available during a limited harvest season. The other crops are harvested throughout the year as they are needed. Women do most of the horticultural work, as well as the harvesting and processing of vegetable foods, but the gardens are cut and burnt by the men.

Coca (*Erythroxylum coca*) is also cultivated and is here considered both a food and a drug. As in the rest of the Vaupés, coca is prepared as a dry powder and is mixed with leaf ash (usually *Cecropia* sp.). It is used principally by adult males, and it is they who cultivate and prepare it.

A number of seasonally available wild vegetable products are collected in sufficient quantity to form the basis of one or more family meals. The most important of these are certain nuts and oil seeds (such as *Erisma Japura*, *Caryocar amydaliferum* and *Caryodendron orinocense*), palm fruits (particularly *Jessenia* sp. and *Euterpe* sp.), a pulse (*Monopteryx angustifolia*), and a leaf (*Phytolacca rivinoides*). Other wild foods such as mushrooms, tree fruits, and berries are collected in smaller quantities and are used primarily as snack foods. Most wild vegetable foods are collected by both men and women. However, men are more likely to harvest the palm fruits and on occasion collect large quantities of nuts and oil seeds.

Fish, game, and insects are the sources of dietary animal protein. Almost all fishing is done by adult men and boys. It tends to be an individual pursuit using hook and line, traps, and poison. The most frequently caught species include *Leporinus alternus*, *Hoplias malabaricus* and *Aequidens latifrons*, all of which weigh less than 1 kg. Very small fish, which weigh less than 200 g, make up a significant portion of the catch. Frogs and shrimp are seasonally important riverine resources that are collected by men, women, and children.

Game animals are a valued source of meat, but hunting is less frequent an activity than fishing. Most hunting is done by individual men with shotguns and within a half day's travel of the village. The animals most frequently killed are the two large rodents, agouti (*Dasyprocta* sp.) and paca (*Agouti paca*). Paca are hunted at night along the rivers, and most agouti are killed during daylight in and around gardens. The latter are often cornered in hollow logs by the dogs that women take to the garden.

Insects can also provide a significant amount of dietary animal protein. Some 20 species of edible insects are collected by men, women, and children (Dufour 1979). The most important of these are palm grubs (larvae of *Rhynchophorus* spp.), ants (*Atta* spp.), termites (*Syntermes* spp.), and caterpillars (several species of Lepidoptera). They are generally available all year-round, except for the caterpillars whose availability is strictly seasonal.

Some domestic animals, such as chickens, and occasionally pigs, are kept by individual households, but their meat is sold rather than consumed. At the time of this fieldwork the village also had three head of cattle, newly acquired through a promotional program of the Catholic mission. The intent was to raise them for resale, and at this time they were not in any way a significant part of the subsistence system.

Data Collection

As part of a larger study, detailed records of household food production, consumption, and time and energy expenditure were kept on four households during November 1977. These households are reasonably representative of village households. Their age and sex compositions are shown in Table 11.1. Household I was a young conjugal pair with a 2-year old child. Household II was a middle size nuclear family of four children under 10 years of age. Households III and IV were both large, mature domestic groups of six and seven children, respectively. In both of these households the older children participated in subsistence activities.

Household food production and intake were measured for a 4-day period. Within this 4-day period, records of the time and energy expenditure of the adult male and female heads of household were kept for 3 days. General observations on the activities of children were also recorded. The studies were done in sequence within a 21-day period, thus eliminating any ques-

Table 11.1

Age, Sex Composition, and Estimated Daily Energy Requirements of Four Households[a]

Households and members	Age (yrs)	Weight (kg)	Daily energy requirement (kcal)
I Father	20–25	55.0	2,960
Mother	20–25	52.5	2,607
Son	2.25	9.0	909
Household total			6,476
Per capita			2,159
II Father	30–35	56.5	3,014
Mother	30–35	47.5	2,773
Son	9.0	20.0	1,560
Daughter	7.0	17.0	1,204
Son	4.5	14.0	1,326
Daughter	2.0	8.0	808
Household total			10,755
Per capita			1,792
III Father	40–43	50.0	2,691
Mother	35–40	46.5	2,176
Daughter	17.0	43.0	2,113
Son	13.0	34.0	2,824
Son	11.0	29.5	2,301
Son	7.0	18.0	1,404
Daughter	5.0	18.5	1,684
Daughter	3.0	14.0	1,414
Household total			16,617
Per capita			2,077
IV Father	35–40	57.0	3,086
Mother	35–40	44.0	2,609
Daughter	14.0	43.0	2,168
Son	12.0	29.5	2,094
Daughter	9.0	21.0	1,638
Son	7.0	19.0	1,142
Son	4.0	12.5	1,138
Son	1.5	8.5	858
Household total			14,733
Per capita			1,842

[a]Estimated energy requirements are based on FAO/WHO standards (FAO/WHO 1973) for "very active" individuals. The estimated requirement for pregnant and lactating women was increased by 150 and 550 kcal/day, respectively. These increments were applied for pregnancy (first trimester) in Household I and lactation in Households II and IV.

tion of seasonal differences between the households. The methods used are discussed in detail in Dufour (1981), and are here summarized briefly.

Food production was recorded per individual as the weight of all food harvested, gathered, fished, hunted, or otherwise obtained, that was brought back to the village. The edible portion of foods was determined by weight, and energy production was calculated for that portion only.

Total household food intake was estimated for a 4-day period, using a modified larder method in which a continuous record was kept of all food inputs into the household and all food outputs. Household intake was taken as the difference between food inputs and outputs after controlling for food stored in household larders. Food composition values were obtained from standard tables (principally Wu Leung and Flores 1961 and Instituto Nacional de Nutrición 1967), and the biochemical analysis of some Yapú foods (Dufour n.d.).

Time expenditure was assessed using 3-day time–motion studies. These are a continuous record of the nature and duration of each activity a subject engages in during a specified interval. Records were kept in activity diaries similar to those presented in Hampton (1960).

Energy expenditure in important subsistence activities was measured through direct calorimetry on a sample of 10 village women. In this method, the calories used in performing a task are calculated from the volume of oxygen consumed. These values, in conjunction with the time–motion studies, were then used to calculate total daily energy expenditure and energy expenditure in specific activities. Rates of energy expenditure for activities not measured were taken from the literature and adjusted proportionally to the mean body weight of Yapú men and women.

Food energy and protein requirements were estimated from age, sex, and body weight according to the recommendations of the 1973 Food and Agriculture Organization/World Health Organization (FAO/WHO) committee. These recommendations are used here as one indication of the adequacy of household energy and protein intakes. However, it must be kept in mind that the FAO/WHO energy requirement is an average population value, and as such is of limited usefulness when applied to individuals as has been done here.

The FAO/WHO model allows for the adjustment of adult requirements to four activity levels. By these standards, Yapú adults tend to be "very active" (Dufour 1981:216), and the correction factor for that level of activity has been applied to individuals 13 years and older. An energy supplement for pregnancy and lactation has been included where applicable (Table 11.1). However, this supplement may overestimate the energy needs of Yapú women because several studies have failed to find increased energy intakes during pregnancy and lactation in non-Western women (Hipsley and Kirk 1965; Norgan *et al.* 1974:322; Prentice *et al.* 1981:2790). The energy requirements for children up to 12 years of age were adjusted for body weight because Yapú children are lighter and smaller for age than are reference children. No allowance was made for catch-up growth because this does not seem to occur. The sum of the requirements of household members was taken as the household requirement (see Table 11.1).

Results

The household case studies were done during November, the early part of the dry season. It was the time when new gardens were being cleared by men and those gardens planted in the earlier part of the year were ready, or almost ready to harvest. Households with an adequate supply of manioc were weeding, but not harvesting, their new primary-forest gardens because the manioc tubers in these gardens were still quite small. Tuber growth was adequate in secondary-forest gardens and these were being harvested.

In November, river levels fluctuated considerably but were generally low and falling. Fishing conditions were more or less average, neither especially good nor as difficult as they can be. Men were taking advantage of the fluctuations in water level by alternately opening and closing the mouths of small inlets with fish weirs. An unusually sharp rise in water level triggered spawning in *boteka* (*Leporinus alternus*), behavior which is characteristic of the early rainy season. Men were also fishing with hook and line in daylight and with set hooks at night. Toward the end of November, some shrimp were being collected from the detritus at the river's edge, but water levels were not low enough to make it very productive. In the area of the village, some *barbasco* (fish poison) was being used in small forest streams. There was considerable movement in and out of the village as individual men, as well as entire households, moved to the headwaters, where river levels were lower, to fish for a few days or longer.

The only wild vegetable product of significance in the diet was *batí* (*Erisma Japura*), an edible oil seed collected in May and June. It is prepared as a butter and allowed to ferment in underground storage pits (see Dufour and Zarucchi 1979). In November, a number of households still had stores of *batí* butter. The only wild food plants collected in November were small quantities of *poá* (*Parkia* sp.) and *au* (*Phytolacca rivinoides*). The edible portion of *poá*, a long seed pod, is extremely small, probably less than 5%, and it is eaten principally by children. *Au*, a leaf, is one of the first weeds to colonize newly burned gardens.

HOUSEHOLD FOOD SUPPLIES

Household food production data provide one measure of the dietary energy and protein potentially available to the household. It cannot, however, be equated with food consumption because production fulfills social as well as biological needs. Some daily production is given to other households, some is used to exchange for trade goods, and some is given to domestic animals. Households also receive gifts of food from one another.

They do not, however, purchase food (except salt) or use domestic animals as a source of food.

In this study, food inputs into the household were defined as coming from three sources: household larder stores, household food production, and foods from other households. Four kinds of outputs were distinguished: food remaining in the household larders, food gifts given to other households, food given to domestic animals or wasted, and food sold or exchanged for trade goods. Net energy flow, the difference between total food energy input into the household and total food energy output over the 4-day period, is taken as household energy intake. Similarly, net protein flow, the difference between total dietary protein input and output, is taken as household protein intake.

The pattern of food energy and protein flow within each of the four households during four days is summarized in Table 11.2. The larder stores in all households contained the equivalent of a 3–5-day supply of calories. These stores were almost exclusively manioc products and thus were relatively low in protein. In Household III, the food stores were unusually high because *fariña* was being prepared for sale and the raw ingredients, as well as some of the toasted *fariña,* were in the larder.

Food energy production was greater than intake (net flow) in Households I, II, and III, and less than intake in Household IV. Food energy production was particularly high in Household III because the women were accumulating manioc to toast as *fariña.* Protein production was greater than intake in Households II and III, and less than intake in I and IV. The production was highest in Household II, partly in response to an opportunity to sell fish.

The gifts of food received were less than those given away in all four households, although the differences were not very great in Household I. The large gift output in Household IV, an amount equivalent to over one-third of their total production, represents the hospitality extended to a visiting family. Relatively little food energy or protein was wasted, and little given to domestic animals, except in Household IV. This household was keeping a full-grown pig, which consumed prepared manioc products equivalent to about one-sixth of the production of the household.

The most important component of household net energy and protein flow is daily food production. This is clearly seen when the net flows are partitioned into the inputs from larder stores, food production and gifts (Figure 11.4).[5] Gifts were a minor source of food energy, and the contribution of larder stores varies from minimal to about 23% in Household III and 28% in Household IV. The reliance on larder stores in Household III resulted from the allocation of manioc production to *fariña* rather than to daily use. On the other hand, the use of larder stores in Household IV was the result of relatively low production in relation to the food energy needs

Table 11.2

Food Energy and Protein Flow in Four Yapú Households During 4 Days

Category	Households I	II	III	IV
	ENERGY			
Input (kcal)				
Larder	30,249	44,625	139,245[a]	92,028
Production	28,508	52,899	81,523	60,092
Gifts	2,240	883	3,784	6,076
Subtotal	60,997	98,407	224,552	158,196
Output (kcal)				
Larder	30,017	37,528	144,381	62,840
Waste/Animal	619	0	2,220	9,142
Gifts	2,697	13,417	11,557	22,188
Cash Crop	0	12,386[b]	0	0
Subtotal	33,333	63,331	158,158	94,170
Net energy flow (kcal)	27,664	35,076	66,394	64,026
	PROTEIN[c]			
Input (g)				
Larder	144(05)	205(30)	230(20)	423(25)
Production	243(40)	1,795(78)	981(33)	1,042(60)
Gifts	211(88)	24(67)	146(88)	233(92)
Subtotal	598(49)	2,024(73)	1,357(36)	1,698(56)
Output (g)				
Larder	137(00)	169(06)	241(00)	187(00)
Waste/Animal	5(00)	0(00)	17(00)	98(00)
Gifts	11(11)	228(45)	313(10)	238(26)
Cash Crop	0(00)	244(88)	0(00)	0(00)
Subtotal	153(01)	641(51)	571(05)	523(12)
Net protein flow (g)	445(65)	1,383(83)	786(59)	1,175(75)

[a]Included in the larder were about 91,770 kcal of manioc prepared as *fariña* that was being set aside for sale to the mission.
[b]This included 6210 kcal of *fariña*, 4964 kcal of cultivated crops, and 1212 kcal of fish that were used to pay a debt.
[c]Values in parentheses are the amount of animal protein as a percentage of the total protein for each category.

of the household, their obligations of hospitality, and the needs of their pig.

With the exception of Household I, most of the net protein flow was from production. The contribution of larder stores was relatively small, but that of gifts was significant in three of the households. Of the protein received as gifts, over 80% was from animal sources. These gifts had the greatest impact on the smallest household, Household I. The effect is striking if Households I and IV are compared. Each received similar amounts of protein as gifts (211 g and 233 g respectively), but whereas gifts repre-

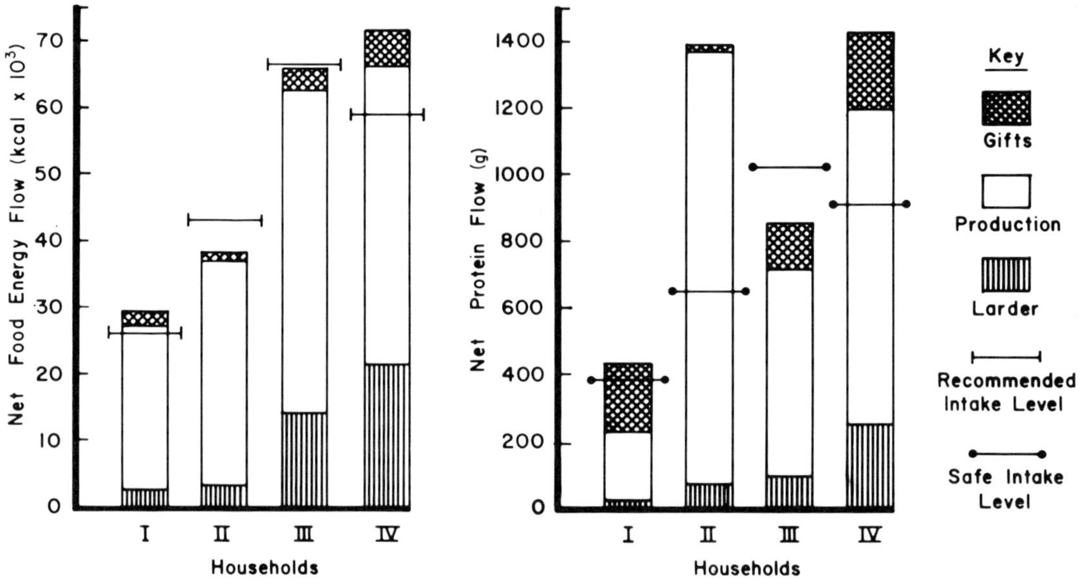

Figure 11.4 *Net energy and protein flow during 4 days for four households. Each flow is partitioned into inputs from gifts, productive activities, and larder stores. The recommended level of energy intake and the safe level of protein intake for each household for 4 days are indicated by the horizontal line segments.*

sented 47% of the net protein flow for Household I, they represented only 20% for Household IV.

SOURCES OF FOOD ENERGY AND PROTEIN

The types of food consumed in each of the households are shown in Table 11.3. In all of the households manioc supplied the greatest percentage of dietary energy, between 72 and 92%. Fish were second in importance as a source of energy, supplying between 3 and 18% of the total. Cultivated vegetable products accounted for an average of 3.8% of the household energy intake. They were, in order of relative importance, plantains, corn, *naranjilla,* yams, sugarcane, manioc leaf and lemons. Wild plant foods, the oil seed *batí,* and the leaf *au,* contributed 1%. Coca accounted for another 1% of the mean intake.

Fish supplied the greatest percentage of dietary protein in each of the four households, between 49 and 81%. Manioc was the second most important source of protein and supplied more than either meat or insects. Of the total household protein intake, between 59 and 83% was from animal sources.

Table 11.3

Sources of Food Energy and Protein Consumed in Four Yapú Households over Four Days

Source	Households I	II	III	IV	Mean
			ENERGY (%)		
Garden Crops					
Manioc	85.7	72.2	92.3	83.2	83.3
Others[a]	2.3	4.1	1.5	7.4	3.8
Coca	0.1	3.7	1.0	0.0	1.2
Wild Plants[b]	0.9	0.8	1.1	1.2	1.0
Fish	4.4	18.1	3.4	5.2	7.8
Game	0.0	0.1	0.5	3.0	0.9
Insects	6.6	1.0	0.2	0.0	2.0
Total	100.0	100.0	100.0	100.0	100.0
			PROTEIN (%)		
Garden Crops					
Manioc	29.3	8.3	29.2	15.7	20.6
Others[a]	4.9	2.0	3.9	8.5	4.8
Coca	0.2	6.4	4.4	0.0	2.8
Wild Plants[b]	0.6	0.3	3.5	0.8	1.3
Fish	48.9	81.3	50.5	50.0	57.7
Game	0.0	0.3	6.7	25.0	8.0
Insects	16.1	1.4	1.8	0.0	4.8
Total	100.0	100.0	100.0	100.0	100.0

[a]Includes all cultivated vegetable foods except bitter manioc, that is, other tubers, sugarcane, corn, plantains, bananas, etc.
[b]Food in this category was derived from: (1) *batí* (*Erisma Japura*), an oil seed which had been harvested the previous May and June; and (2) *au* (*Phytolacca rivinoides*), the leaf of a garden weed. Household I also consumed a small amount of *poá* (*Parkia* sp.) as a snack. It was excluded from the calculation of household energy consumption for lack of food composition data. It is assumed that this involves an error of less than 1%.

ADEQUACY OF ENERGY AND PROTEIN INTAKES

The food energy intakes of the household over 4 days compares favorably to their estimated needs, based on FAO/WHO recommendations (Table 11.4). The mean daily intake of Households I, III, and IV were 107, 100, and 109% of the recommended levels, respectively, but intake in Household II was only 82% of the recommended level.

The adequacy of protein intake is more difficult to assess because the value of dietary protein varies with the food source. Standards for safe level of intake, as set by FAO/WHO (1973:40–73), are expressed in terms of egg and milk, proteins of high biological value. For the dietary data pre-

Table 11.4

Comparison of Estimated Energy Requirements and Consumption During Four Days

Household	Energy requirement[a] (kcal/capita/day)	Observed energy consumption (kcal/capita/day)	Percentage of FAO/WHO requirement
I	2159	2305	107
II	1792	1462	82
III	2077	2075	100
IV	1842	2000	109

[a]FAO/WHO (1973:34) recommendation. See Table 11.1 and text for explanation.

sented here, there is no information as to protein quality, and data are insufficient on the amino acid composition of component foods to estimate it. In such cases, FAO/WHO (1973:73) recommend an arbitrary correction for quality, and they present correction factors for diets having protein qualities of 60, 70 and 80% relative to milk or eggs. The low relative quality of 60% is for diets in which 70–80% of the protein is derived from foods such as manioc and maize, and virtually none from animal products (FAO/WHO 1973:73). Because the data indicate that over 50% of the total dietary protein is obtained from animal sources (Table 11.3), it is assumed that the diet has a protein quality of greater than 60%, and probably between 70 and 80% relative to milk or eggs.

With the exception of Household III, the total protein intakes compare favorably with the safe level of intake for diets with a protein quality of 70% relative to milk or eggs (Table 11.5). The mean daily consumption of the

Table 11.5

Comparison of Household Protein Consumption and Safe Levels of Intake Adjusted for a Protein Quality of 70% Relative to Milk or Eggs

Household	Safe level of intake score 70 (g/capita/day)	Observed consumption (g/capita/day)	Percentage of score 70[a]
I	32	37	116
II	27	58	215
III	32	24	75
IV	28	37	132

[a]This diet is assumed to have a protein quality of at least 70% relative to milk or eggs, as defined by FAO/WHO (1973:73). The safe level of intake per household is calculated as the sum of the safe level of intake of household members divided by the number of members.

Households I and IV was 116 and 132% of the safe level, respectively, and of Household II was 215% of the safe level. Household III had the lowest intake relative to estimated needs with only 75% of the safe level.

TIME AND ENERGY EXPENDITURE OF HOUSEHOLD MEMBERS

The mean time and energy budgets of the adult men and women in the four households are shown in Table 11.6. A time–energy budget is the way in which time and energy are allocated to various activities; that is, the way in which time is spent, and energy expended. Records of these budgets were kept for 3 consecutive days for the adult males and females in the four households, and thus represent 12 man-days and 12 woman-days.

Men. The mean daily energy expenditure of the adult males was 2802 kcal. All of the men were in the process of clearing new gardens and worked at it at least 1 of the 3 days for which time–motion data were kept. The general pattern was to work at garden clearing for most of the morning every other day, and then do some fishing in the afternoon (Figure 11.5). About 25% of the total time budget and 50% of the total energy budget were spent in

Table 11.6

Mean Time and Energy Budgets of Four Adult Males and Four Adult Females During 3 Days

	Adult males				Adult females			
	Time (min)		Energy (kcal)		Time (min)		Energy (kcal)	
Activity	$\bar{X}$	%	$\bar{X}$	%	$\bar{X}$	%	$\bar{X}$	%
Food acquisition								
gardening	217	15.2	821	29.3	224	15.6	600	28.6
collecting	4	0.3	7	0.2	5	0.3	13	0.6
hunting	21	1.4	58	2.1				
fishing	158	10.9	599	21.4				
Food preparation[a]	38	2.6	125	4.5	125	8.7	272	13.0
Materials acquisition: fuel, water, and forest products	19	1.3	81	2.9	23	1.6	58	2.8
Miscellaneous activities	481	33.4	704	25.2	499	34.6	752	35.8
Sleep	502	34.9	407	14.5	564	39.2	403	19.2
Total	1440	100.0	2802	100.0	1440	100.0	2098	100.0

[a]For males, the category "food preparation" refers to coca preparation.

these two activities. Little time or energy was used in either hunting or gathering.[6]

The mean daily time and energy devoted to gardening was slightly greater than fishing, although both were done on 8 of the 12 days observed. Most of what is labeled gardening was the work of clearing new gardens, although a small amount of harvesting is also included. The energy cost of garden clearing is high because of the high rate of energy expenditure in the actual felling. This cost is reflected in the mean daily energy expenditures. For the 6 days on which garden clearing was done, the mean daily energy expenditure was 3119 kcal, compared to 2481 kcal for the other 6 days. However, although the energy expenditure of individual men tended to be higher on days that included garden clearing, the highest daily values calculated for 2 of the 4 men were for nongardening days.

Among the men in the four households, there was considerable variation in the amount of time devoted to the acquisition of animal protein foods, and in the degree of success achieved. The male head of Household II fished intensively in order to have fish to sell. During 3 days he devoted 18% of his total time budget (13.5 hours), and 31% of total energy budget to fishing, and produced over 2.5 kg of fish. The male head of Household I

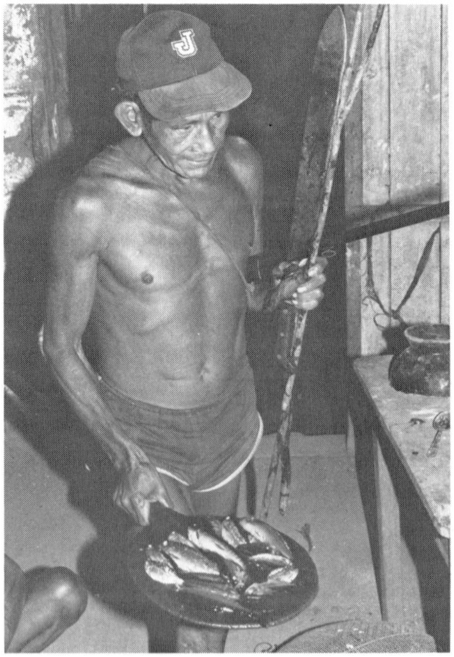

Figure 11.5 *Yapú man returns home carrying a paddle of fish and a fishing pole and net.*

spent a roughly similar proportion of his time (18%, or 12.7 hours) and energy (36%) budget in hunting, fishing, and collecting. However, in hunting he was unsuccessful, and in fishing he produced only 0.14 kg of small fish. This low production was offset somewhat by the collection of palm grubs (1.3 kg of *Rhynchophorus* sp. larvae). These grubs are a kind of food cache in that particular palms are cut in anticipation of a later harvest. In this instance, the male head of household harvested the grubs after returning empty-handed from a short hunting trip.

The male head of Household IV devoted a more moderate amount of his time (13%, or 9.5 hours) and energy (22%) to fishing, and produced some 1.7 kg of fish.[7] In contrast to the other two men who spent a minimum of 3 hours/day in hunting, fishing, and/or collecting, his time input into fishing was zero on the first day, just over 1 hour on the second, and over 8 hours on the third. Considering only the days on which he fished, his mean time input was 4.8 hours, a daily average similar to the other two men.

The male head of Household III did virtually no fishing at all (only about 2% of his time budget). He was bothered with an arthritic-like pain in his legs, and except for his garden work at clearing, remained quite sedentary. His sons, aged 13 and 11 years, supplied the only fish (about 1.75 kg) produced by the household during the 4 days.

Relatively little time or energy was devoted to coca preparation (listed as food preparation), or to materials acquisition. The latter materials included some thatching palms to repair a roof, and the fuel needed to toast coca. About one-third of the mean daily time budget was classified as miscellaneous activities around the village. During daylight hours, this time was spent primarily in basketry, fish net manufacture, and visiting. In comparison to periods of resource acquisition outside the village, the time spent in the village is sedentary and the rate of energy expenditure low.

Women. The mean daily energy expenditure of the four adult women was 2098 kcal. The women gardened on 10 out of the 12 days for which the time–energy budgets were kept, a proportion of gardening to nongardening days which is representative of the time of year. Of the 10 gardening days, 8 were routine manioc-harvesting and -processing days and exemplify the ordinary work schedule of women. On these days women leave for the garden early in the morning. They return to the village with the morning's harvest around midday to rest and eat. The afternoon is largely taken up with manioc processing (Figure 11.6). In the late afternoon, firewood is collected, the fresh manioc juice is boiled, and a manioc bread often is baked. The evening is spent relaxing and visiting.

For the 12 woman-days considered here, gardening was the most important food acquisition activity. Very little time was devoted to the collection of wild foods. Gardening accounted for about 16% of the mean time budget and about 29% of the mean energy budget. Almost half of this time and

Figure 11.6 *Yapú woman and adolescent girl working grated manioc mash in a basketry sieve to express the starch and liquids from the more fibrous portion.*

energy was spent traveling to the garden and back. Travel time averaged almost 2 hours/day for these women.

Food preparation accounted for 9% of the total time and 13% of the energy budget. It included both the extensive processing of manioc and the cooking of manioc and other foods. The manioc processing itself accounted for almost half of the total time and energy expended. The days of observation only included routine manioc processing and cooking. The making of beer and the preparation of *fariña* involve somewhat more lengthy processes. Materials acquisition, which in the case of these women was limited to fuel and water for cooking, used a relatively small proportion of the time–energy budgets: 1.6 and 2.8% respectively. Both of these resources are very accessible.

The acquisition and preparation of food, along with the fuel and water for cooking, accounted for 26% of the mean daily time budget and 45% of the mean daily energy budget. If daylight in the tropics is taken as 13 hours, then these activities accounted for 48% of the total daylight time.

There was very little difference in the time budgets of the women. The adult female of Household I was an exception. Because her young son was sick, she spent 2 full days at home rather than going to the garden as she would normally have done. The only other noteworthy distinction in the time budgets of the women is that the adult female head of Household III did very little food preparation because her oldest daughter did most of it.

Children. Although many of the time inputs of children in the four households could not be quantified, the subsistence activities in which they participated are clear. They are defined, for the most part, by the sexual division of labor exemplified by adults. Girls worked at gardening and food preparation, and boys at garden clearing and fishing.

There are important differences between the households in the age of the children and hence in their participation in subsistence activities. In Household III, the oldest daughter was in late adolescence and was able to assume a share of the gardening and food preparation work almost equal to that of an adult woman. In the day-to-day operation of the household, she assumed somewhat more than half of the woman's work. The 13-year-old daughter in Household IV was capable of carrying out much of the routine work of an adult woman, but her strength, endurance, and motivation seemed to be important limitations. Younger daughters, aged 10 years or less, in Households II, III, and IV helped to varying degrees with the lighter chores.

The male members of the household do not work together as consistently as do the females. This is in part a reflection of the fact that hunting and fishing tend to be individual endeavors. This does not detract, however, from the important contributions made by boys. In both Households III and IV, the sons over 10 years of age fished every day. Their daily returns, which varied from 20 g to 1 kg, were not particularly great, but their contribution did ensure a daily input of animal protein into the household. This was particularly important in Household III where the adult male was ill.

Discussion

In the data presented here, manioc was the most important source of food energy in the Tatuyo diet and fish the most important source of dietary protein. Vegetable foods other than manioc were relatively minor sources of dietary energy and protein. For cultivated crops, this is essentially true during the rest of the year as well. Both coca and wild plant foods were minor sources of dietary energy and protein, but probably provided significant amounts of fat and other nutrients. Coca powder, although relatively unimportant in the intake of the household, could be significant in the diets of individual men considering its relatively high nutritional value: 316 kcal and 15.9 g protein/100 g edible portion (Dufour n.d.). The small contribution of wild plant foods to the diet is partly a reflection of the time of year. Many wild plants fruit in the early to mid-rainy season, and it is at this time that they are most important in the diet. Virtually no wild plant foods were collected in November, but stored supplies of *batí* were still being included in the diet some 5 months after the harvest.

Animal foods, other than fish, were of limited importance. Meat made a significant contribution to the dietary intake only in Household IV, and insects were significant only in Household I. Meat is generally less available than fish, and insects tend to be collected irregularly. As was illustrated by Household I, the latter are often gathered when neither fish nor meat are available.

Although there are seasonal variations in the availability of wild foods, and hence their inclusion in the diet, overall energy intake appears to remain constant. This observation is based on the fact that the body weights of adults remained relatively stable during the period of field work. Most of the individual changes in body weight were due to illness.

In diets in which a single food provides an overwhelmingly large proportion of the total energy, that food's characteristics are particularly important. As a staple food, manioc is a valuable source of carbohydrate, but contains very little protein and virtually no fat. In manioc, as little as 1.3% of the total energy is provided by protein as compared to over 7% for cereal crops (United Nations Educational Scientific and Cultural Organization/United Nations Environment Programme/Food and Agriculture Organization 1978:355). It is a significant source of protein in the Tatuyo diet because it is consumed in such quantities. Young manioc leaves are a better source of protein (Terra 1964:99) and are frequently included in the diet. Not only is the quantity of protein in manioc low but its quality, as determined by amino acid composition, is relatively poor (Albuquerque 1969:119–128). Thus, supplementary foods, particularly animal foods, are important sources of dietary protein.

A second characteristic of manioc is its toxicity. It has been known for some time that raw manioc tubers contain potentially toxic concentrations of hydrocyanic acid (Coursey 1973:28). This is well recognized in the Papurí area, and the elaborate processing applied to the tubers and leaves is intended to make them palatable and to reduce their toxicity. The work of processing is done by women and occupies almost half of the total time and energy they devote to food preparation. Their processing technique includes: (1) fine grinding of the raw tubers; (2) mixing and washing of the ground mash to separate the starch, fiber, and liquids; (3) boiling of the liquids into a drink; and (4) the overnight storage of the starch and fiber. Seigler and Pereira (1981:362) have shown that fine grinding, mixing, and the precooking storage period are the steps most effective in reducing the cyanogen levels in cooked manioc foods.

Although the effectiveness of the processing techniques used in the Papurí area is not known with any degree of certainty, there are no reports or evidence of manioc toxicity in the human population. There are, however, occasional reports of the accidental poisoning of domestic animals. Even though the processing techniques appear adequate, the possibility that small amounts of cyanogens are present in the processed foods can not be overlooked. When small amounts of cyanogens are included in the diet, they are detoxified metabolically (Oke 1973:97–103). It has been suggested that the efficient functioning of this metabolic detoxification is dependent on a generally adequate diet, and in particular, an adequate supply of animal protein (Osuntokun 1973:135).

In comparison to their estimated needs, household energy intakes appear generally adequate, although Household II was low in comparison to the standards. Dietary protein intakes were adequate in Households I, II, and IV, but were only 75% of the safe level in Household III.

The relatively low energy and high protein intakes in Household II are difficult to interpret. There is nothing to suggest that the supply of food energy was any more limiting in this household than it was in any of the others, and nothing to suggest it was inadequate. As was true of all four households, they had a considerable supply of manioc stored in the larder throughout the period. It is entirely possible that the energy needs of the household were less than their estimated requirement. As noted previously, the FAO/WHO recommendations for energy are average values. The total protein intake was high in Household II but, in comparison to the others, daily intake fluctuated more. There was very little animal protein available on the first 2 days, a moderate amount on the third, and a surplus on the fourth. There was a tendency in all of the households for the daily intake of protein to fluctuate much more than did energy. This is an important characteristic of the diet not shown by the mean values presented here.

The close interrelationship between the level of energy intake and the utilization of dietary protein must be kept in mind. When energy intakes are deficient, dietary protein is used to provide energy (FAO/WHO 1973:19). Therefore, in diets in which protein intakes are variable, and sometimes low, an adequate energy intake ensures that the protein supplied will be used as protein.

The only other group of native Amazonians for which household dietary intakes have been reported are the Aguaruna Jívaro. The Jívaro are manioc cultivators inhabiting an area of upland tropical forest in northcentral Peru. They rely on manioc, bananas, and plantains for about 75% of their total caloric intake, and fish for most of their dietary protein (Berlin and Markell 1977:70). They thus share with the Tatuyo a broadly similar diet, although they live in a different ecological zone. Berlin and Markell (1977:78) found household dietary intakes 136–214% above FAO/WHO recommendations for energy intake, and 213–448% above the safe level of protein intake. They note that the intakes recorded may represent a seasonal high because their data were collected when fishing was particularly good (Berlin and Markell 1977:77).

In this study the adequacy of household protein intake was associated with the time–energy input of men in wild-food acquisition. Dietary protein intakes were adequate in Households I, II, and IV, households in which the adult male spent an average of over 4 hours a day in hunting, fishing, and/or collecting. In contrast, the adult male in Household III spent no time in wild food acquisition. He was not well, and did only a minimal amount of fishing. Although his sons provided some fish every day, the dietary intake of the household was only 75% of the safe level.

The relationship between dietary protein adequacy and the time and energy invested in wild-food acquisition is not a simple one. Time spent, especially in hunting and fishing, does not always yield a return. Even though the male heads of Households I and II invested similar amounts of time and energy in wild-food acquisition, the male head of Household I produced much less. He returned empty-handed after several hunting and/or fishing trips, a frequent occurrence among village men.

The productive activities of adult men are usually the major, but not the only, source of animal protein for the household. In large, mature family groups the older male children also provide animal foods. The two oldest sons of Household III contributed virtually all of the animal protein produced by the household. In Household IV, the oldest son fished every day and provided the household with as much fish as his father. The fish caught by the sons was extremely important, but not sufficient to meet the needs of their respective households. However, it would have been sufficient to meet the needs of a smaller household, such as Household I.

Gifts of food are another source of protein for the household. Although in multihouse villages, food is not shared as frequently as it is in longhouses, the gifts that pass from one household to another are still important. These gifts contain a high percentage of animal protein and are more important in ensuring the adequate flow of protein than of energy. Further, they seem to have a greater impact on the smaller households which have relatively lower dietary protein needs.

Characteristic of the November period was the clearing of new gardens. This is the period of peak agricultural work for men. Garden clearing is relatively hard work, and therefore their daily energy expenditure was probably higher than it would have been at other times of the year. It was, however, about 400 kcal lower than the mean value for Machiguenga men over the course of the entire year (Montgomery and Johnson 1977:104). In this study, it is difficult to say how much of an effect the time and energy devoted to garden clearing had on wild-food acquisition. The men were not working so intensively at clearing as to exclude other activities. Three of them devoted time to fishing almost every day. They were, however, restricting their fishing to the immediate vicinity of the village and choosing not to travel to the very headwaters where early dry season fishing is often better.

The productivity of fishing reported here appears low, and it is considerably lower than that reported by Århem (1976:39) for Makuna longhouses on the Pirá–Paraná. There are at least two reasons why the productivity of fishing may be lower near the village than in other areas of the upper Papurí. First, because of village size, riverine resources are more heavily exploited than they are near smaller settlements. Second, Yapú is in a headwater area where no large fish are found. Århem's (1976:34) data support the observation that both the diversity of fish and the mean weight of fish caught are greater in downstream regions.

In comparison to animal foods, manioc appears to be a consistent and abundant source of food energy in the household. Because of its relative abundance, the time and effort needed to maintain adequate levels of production are easily overlooked. However, the data presented here indicate that the production of manioc foods accounts for a large proportion of the women's time and energy budgets, a proportion of the time budget equivalent to almost half of the available daylight time.

The time and energy inputs most closely linked to the production of manioc foods are food preparation, gardening, and fuel and water acquisition. Of these, the first two are the most significant. The inputs into food preparation reflect the elaborate processing techniques characteristic of manioc-based systems and are probably common in Amazonia. Montgomery and Johnson (1977:104) found that Machiguenga women spend about 10% of their time and 12% of their energy budgets on food preparation, inputs very similar to those reported here. On the other hand, Yapú women expend considerably more time and energy on gardening than do Machiguenga women. The higher values for the Yapú women could be a reflection of the actual work done in the garden, that is, the tasks of harvesting, weeding, replanting, and so forth, or they could reflect differences in the time and energy used in traveling to and from the garden. For Yapú women, garden travel time is high, almost 2 hours/day, and it alone accounts for almost half the time and energy expended in gardening. This time commitment appears to be near the upper limit if manioc is to be harvested and processed the same day.

The time devoted to garden travel is greater in Yapú than it would be in traditional longhouse settlements because of the greater size and permanence of the village. This time is further increased by the presence of a band of *caatinga* vegetation which limits the amount of cultivatable land in the immediate vicinity of the village.

The time–energy budgets of the women did not differ appreciably (if we allow for the unusual circumstances in Household I), although their total energy production was quite different. This supports the observation that women allocate a fairly standard portion of their daily time and energy budgets to routine food-acquisition and -preparation activities. Their pattern of work results in a rate of energy flow from vegetable foods that is high and constant.

In contrast, the flow of protein of animal origin is considerably lower and more irregular. Although the time–energy budgets of men show a high degree of commitment to the acquisition of animal foods, the return on this input is variable, and there are rarely larder stores to cover fluctuations in production. Further, when the necessary commitment is not made, or when there are complicating circumstances such as illness, household protein flow is likely to be inadequate.

Acknowledgments

Financial support for the fieldwork in the Vaupés was provided by a dissertation fellowship from the Social Science Research Council, a grant from the National Science Foundation to Dr. M. A. Little (No. BSN 75-20169), and a predoctoral fellowship from the Organization of American States. I am grateful for the facilities and technical assistance provided by the *Instituto de Ciencias Naturales* in Bogotá under the direction of Dr. Polidoro Pinto E. Grateful acknowledgment is also made to Paul N. Patmore for his assistance with the fieldwork in the Vaupés and the preparation of the figures which accompany this text.

Notes

[1]In this paper I refer to the inhabitants of Yapú as Tatuyo, that is, speakers of the language known by that name. This is somewhat of an oversimplification because several other groups are also represented. There are almost as many Bara in the village as there are Tatuyo. However, Yapú is considered a Tatuyo village, and the core members of the village, politically and socially, belong to the sib of Tatuyo speakers referred to as Owa Huna. Bidou (1976) provides a valuable account of the Tatuyo speakers in the Pirá–Paraná region.

[2]The possibility that human populations could have played a role in the formation of *caatinga* vegetation has been suggested (Klinge *et al.* 1977:274).

[3]In this chapter, the distinction made between "bitter" and "sweet" manioc follows the

Tatuyo distinction between *kii* and *makasera*. I use the word "manioc" to refer only to bitter manioc (*kii*), varieties of the crop that are considered toxic and are the dietary staple. Sweet manioc (*makasera*) is not considered toxic. It is a cultigen of minor importance and is prepared by boiling or roasting like other rootcrops such as yams.

[4]*Manicuera, mingao, casabe* and *fariña* are the names used in Vaupés Spanish. In Tatuyo, boiled manioc juice is *ñuka*, a starch and water drink is *okobarique*, manioc bread is *naulo*, and toasted manioc meal is *poka*.

[5]For all foods except manioc the three inputs into net flow, production, larder stores, and gifts are easily distinguished. *Fariña* as well as the yellow manioc tubers harvested to prepare it are also easily distinguished. However, for white manioc the relative contributions of production and larder stores to everyday use are difficult to distinguish because some products of the daily processing are used immediately and others stored for variable lengths of time. The storage is part of the processing because a slightly fermented product is desired. In partitioning the net energy and protein flow derived from white manioc, I have assumed that: (1) the white manioc produced during the 4-day period was used (i.e., consumed, given away, etc.); (2) the white manioc stored in the larder was used only when production was insufficient to meet household needs.

[6]The food acquisition categories of fishing, hunting, and gathering are somewhat artificial as a single trip away from the village can easily include all three types of activities. In this study, trips away from the village were coded for the explicitly stated purpose of the trip, and the time spent traveling was assigned to that category. Time spent in other activities was then coded as appropriate. For example, on a fishing trip upstream, the collection of forest fruits along the way would be coded as "gathering," and the remainder of the time as "fishing."

[7]The 1.7 kg of fish was caught on a father–son fishing trip and so some of it was actually caught by the son.

CHAPTER 12

Seasonal Factors in Subsistence, Nutrition, and Child Growth in a Central Brazilian Indian Community

Nancy M. Flowers

Introduction

In this chapter I look at the subsistence practices, diet, and the growth of young children in a community of recently sedentarized Xavante Indians of Central Brazil to ascertain some of the effects of a transition from foraging to food production. Until about 15 years ago, this group was largely dependent on wild foods, as they were nomadic during much of the year. They spent less than a month on cultivating the maize, beans, and pumpkins, which were the only crops they grew (Maybury-Lewis 1967: 48). Physical examinations made in the community in 1962 by a scientific team that included physical anthropologists and physicians (Neel *et al.* 1964) showed that the nutritional and general physical status of the group at that time was good. Maybury-Lewis (1967), who studied the same community in 1958 and accompanied the Indians on their seasonal treks, found no evidence that they experienced a season of hunger. This would indicate that they effectively countered seasonal variation in availability of food by moving about to exploit different resources.

For hunters and gatherers, all subsistence labor is essentially harvest labor. There is little or no delay between expending effort to procure a food resource and consuming it except in northern latitudes where food storage is common. Some hunting and gathering peoples, however, have a "hungry season" because fewer resources are available at certain seasons, and stored supplies may give out before a new season brings abundance again. This is especially true in the Arctic, where a fluctuating environment and

ADAPTIVE RESPONSES
OF NATIVE AMAZONIANS

ISBN 0-12-321250-2

lack of vegetable foods make the failure of any game resource a potential disaster (Balikci (1968:62). Even in latitudes where plant resources are abundant, thus providing greater security through diversity, there is seasonal variation, with many species ripening at one time of the year, while at another there is a relative dearth of food. For Great Basin groups, the season of hunger was winter (Steward 1938), whereas for the G/wi Bushmen of the Kalihari Desert, Silberbauer (1972) found it to be the dry season.

Since hunters and gatherers have to contend with fluctuation of resources, a number of authors have suggested that with the advent of agriculture food supplies became more abundant and reliable. Butzer (1971:314) makes a typical statement of this viewpoint:

> The invention and adoption of new tools and a new economic subsistence would inevitably promote a great increase in population made possible by the increased and more reliable food supply. Food production per unit area was much greater, and even during a bad crop year a certain amount of food would be available. There would not be complete dependence on the seemingly erratic movements and biological cycles of wild game. Life and death were no longer so precariously balanced; birth rates increased and human mortality declined.

However, recent investigation, including fieldwork among contemporary hunters and gatherers, suggests that populations dependent on cultivated crops may be more vulnerable to environmental fluctuations than are foragers living in a similar environment. Lee (1968:39–40) found that the flexibility of !Kung Bushmen and their wide knowledge of alternative vegetable foods allowed them to cope better with a drought than did their neighbors who depended on herding and agriculture. In a study of skeletal remains from two North American populations living in similar habitats, Cassidy (1980) found that the distribution of Harris lines (growth arrest) in long bones showed that in the hunting and gathering population growth arrest episodes occurred at regular intervals, indicating periodic mild food shortages, perhaps in late winter; in the agricultural population such episodes were of longer duration and occurred randomly, indicating that they were caused by disease or crop failure.

With dependence on agriculture, a new factor enters into the scheduling of time and labor. Agriculture requires that people spend time planting and caring for crops that they will not eat for several months. In regions with pronounced seasonality, planting must be done at a specific time, either in the spring or at the beginning of the wet season. Even in tropical forest areas where the rainfall distribution permits planting to be done at any season, agricultural tasks may conflict with wild food collecting. Hames (1980:53), for example, notes that the Ye'kwana and the Y̨anomamö, at seasons when garden-related tasks require their attention, hunt more in the depleted game areas near their villages even though this is less productive than long-distance hunting. In native lowland South America, where animals are not domesticated for food, protein in the diet may fall to a low level if people do not have time to hunt and fish.

Does dependence on agriculture give the Xavante an increased and more reliable food supply compared to their former mode of subsistence? I attempt to answer that question by observing whether, with the use of agriculture, their present food supply appears to be seasonally stable or fluctuating.

The health and growth of young children is a sensitive indicator of the nutritional status of a community (Jelliffe 1966). Growth in stature reflects the effect of long-term conditions, while growth in weight is more variable and fluctuates in response to short-term nutritional changes. (Gross and Underwood 1971). Therefore, seasonal changes in the rate of weight growth of young children should reflect seasonal fluctuations in the food supply.[1]

If the rate of children's weight growth is even year-round, this would indicate little variance in daily food intake from season to season and would therefore suggest that increased reliance on agriculture has augmented, or at least not diminished, Xavante subsistence security. If their rate of weight growth varies considerably by season, this may suggest that the Xavante are experiencing stress as they attempt to reconcile conflicting demands on labor for maximizing crop production while still exploiting the wild food resources on which they continue to be partially dependent. Such stresses may well be similar to those experienced by prehistoric populations during the transition from wild food procurement to food production.

The Study Community

This study was carried out between May, 1976, and July, 1977 in a Xavante Indian village with a population of about 250. It is the only village on a reservation (Pimentel Barbosa) of 205,000 ha, bordering on the Rio das Mortes.

Some 4000 Xavante Indians now live in the eastern part of North Mato Grosso State in Central Brazil, between the Rio das Mortes and the upper affluents of the Xingu River. They occupy nine villages, each of which either has a Brazillian government Indian post, or is attached to a Salesian mission. The Xavante language belongs to the Central Gê branch of the Gê language family.

The Xavante habitat is open upland, with the distinctive vegetation known as *cerrado* that covers much of the Central Brazilian Plateau. Shrubs, savanna grasses, and low trees adapted for resistence to drought predominate, except along streams where the moisture supports gallery forests. Soil suitable for slash-and-burn agriculture is limited to the narrow strips of gallery forest. The rainy season lasts from the end of September through April, with 1100 to 1600 mm of rainfall over most of the region. In the dry season there is virtually no precipitation.

The Xavante first appear in recorded history in the eighteenth century. At that time, they lived east of the Araguaia River in what is now Goiás State. They probably formed one people with the Xerente, from whom they are now separated (Maybury-Lewis 1966). Gold strikes in Goiás in the 1730s started a rush of Portuguese prospectors into the region who soon became embroiled in fighting with the Indians. The efforts of the colonial governors to pacify hostile Indian groups were for a time more or less successful. Some of the Xavante–Xerente were sendentarized in large settlements where they were expected to engage in farming. But overcrowding, disease, and ultimately starvation reduced their numbers. By the end of the century most of the settlements were abandoned and the surviving Indians resumed hostilities against the settlers. At some time toward the middle of the nineteenth century, a group, or several groups, of Xavante broke away from white contact and migrated southwest across the Araguaia River to the unexplored lands of eastern Mato Grosso. The Xerente remained in northern Goiás, hemmed-in by settlers. Their numbers are now reduced to a few hundred.

After the Xavante withdrew westward they remained implacably hostile to whites. They also fought the Bororo on the southwest, the Karajá to the north and east, and among themselves. Their numbers seem to have increased rapidly, as there were estimated to be about 5000 Xavante when they were contacted in the 1940s (Ismael Leitão, personal communication). During these years the Xavante may have become more nomadic than they were previously. According to accounts collected from informants by the missionaries Giaccaria and Heide (1972), when a village was attacked the inhabitants might flee and spend several seasons wandering in the *cerrado* without planting crops or building a village.

During the 1940s the Brazilian government made continuous efforts to contact and pacify the Xavante. Most of the members of the first expedition were ambushed and killed by a Xavante group led by Apowẽ, an outstanding headman who was still alive and active at the time of my fieldwork. However, it was also Apowẽ's group that made the first friendly contact and accepted the government's presents. When Maybury-Lewis studied this group in 1958 they had built a village near the Indian post but spent little time there.

Maybury-Lewis reported that at the time of his study the Xavante grew only maize, beans, and pumpkins. Horticulture was clearly subordinated to their foraging activities based on hunting and the collection of wild roots.

> They used to spend only three weeks, or at most a month, of the year at their plantations, devoting roughly a week to clearing and planting, a week to harvesting the maize, and a week to harvesting the beans and pumpkins. In between these visits they would either remain at their village (which was usually at least a day's journey away from the plantations) or would go on trek. Someone occasionally went to look at the plantations and see that they did not get too overgrown, otherwise the crops received no care.
>
> The harvests were thought of less as providing the essentials for the life of the community than as bonuses to be used for celebrations (Maybury-Lewis 1967:48).

At the time of Maybury-Lewis's fieldwork, Xavante territory was still largely unoccupied by whites. The routes that the Xavante followed on trek were determined by the presence of game and by the localized distribution of specific plant resources, such as ripening palm fruits, cane for arrow shafts, a certain wood used for bow staves, or seeds to manufacture ceremonial regalia. They also planned their movements to avoid other Xavante groups with which they were on bad terms (Maybury-Lewis 1967:56).

Between the time of Maybury-Lewis's research and my fieldwork with the same group of Xavante, their life changed radically. Hunting and gathering activities are now restricted to the area of a reservation. This tract is surrounded by Brazilian cattle ranches and farms where upland rice is grown on a large scale, making use of the relatively barren *cerrado* soils by means of modern machinery, fertilizers, and special tillage techniques. The Xavante now go on trek for only a few weeks before the planting season. In the late 1960s they took up the growing of rice, which is now their major crop. Some Xavante families are growing a small surplus of rice for sale on local markets, and the men work occasionally for wages on nearby farms. As there has been little missionary influence on the Xavante group at Pimentel Barbosa, their traditional social and ceremonial life is largely intact.

Subsistence Practices

HUNTING

The Xavante take little interest in the care of domestic animals. Some families own a few pigs and chickens. There is a small herd of cattle on the reservation, cared for by a Brazilian employee of the government Indian post. When the community does some collective task, such as clearing the grass from the airstrip, the government agent has a steer killed, and the meat distributed. But almost all the animal protein in the Xavante diet comes from hunting and fishing (see Figure 12.1 for variation in diet in relation to seasonal distribution of subsistence activities).

The Xavante use several different hunting techniques: individual hunting; collective hunting expeditions lasting several days; and game drives, using fire, which are held in the dry season.[2] Time allocation observations showed that Xavante men spent an average of 1.06 hours/day in hunting (Table 12.1).

Individual hunting is the most common form of hunting, though not the most productive. One man, or two or three men, may decide to go out for a day's hunting. Or, if they have flashlight batteries, they may go to the gardens at night to wait for paca (*Cuniculus paca*). Practically all Xavante men today own .22-calibre rifles. I found that only about 21% of such hunts were successful (that is, some game was bagged) and the average share of dressed meat/hunter was 1.7 kg (Table 12.2). However, both the frequency and success rate of individual hunting may be underestimated. Men always carry their guns and are on the alert for game, whether going to their gardens or on some other errand. Several times I have seen men returning from work on neighboring farms come into the village carrying a deer or peccary they had shot on the way (Figure 12.2).

The Xavante say that collective hunting brings better results, and my observations bear this out. The success rate of collective hunts (four or more hunters) was 67%, and the average share of each hunter/day (some hunts last several days) was 4.7 kg of dressed meat (Table 12.2). This is probably because the most commonly bagged game animal is white-lipped peccary (*Tayassu pecari*) which runs in large herds. If he comes across a herd of white-lips, the individual hunter is lucky to shoot one or two, but with a number of hunters present some can head off the herd, giving everyone a chance to shoot more. On two occasions during my stay, a large herd of peccary was spotted close by the village. When the alarm was given, every man in the village seized his gun or, if it was not at hand, bow and arrows or a club, and gave chase. On one of these occasions, 22 peccaries were killed, and on the other, 20 (Figure 12.3). On the other hand, it may happen that most of the men in the village go out to search for a herd of peccary whose location has been reported and fail to find it, returning empty-handed at the end of the day.

On hunting expeditions, a number of men go out hunting for several days and usually return with a large amount of smoked meat. Dry-season game drives, which usually last only 1 day, use the technique of surrounding game by setting fire to the dry undergrowth. As a result, by the end of the dry season a large part of the reservation has been burned over.

Collective hunting may be most advantageous where there is high dependence on herd animals, and in relatively open country such as the *cerrado* where there is visibility over considerable distance. If collective

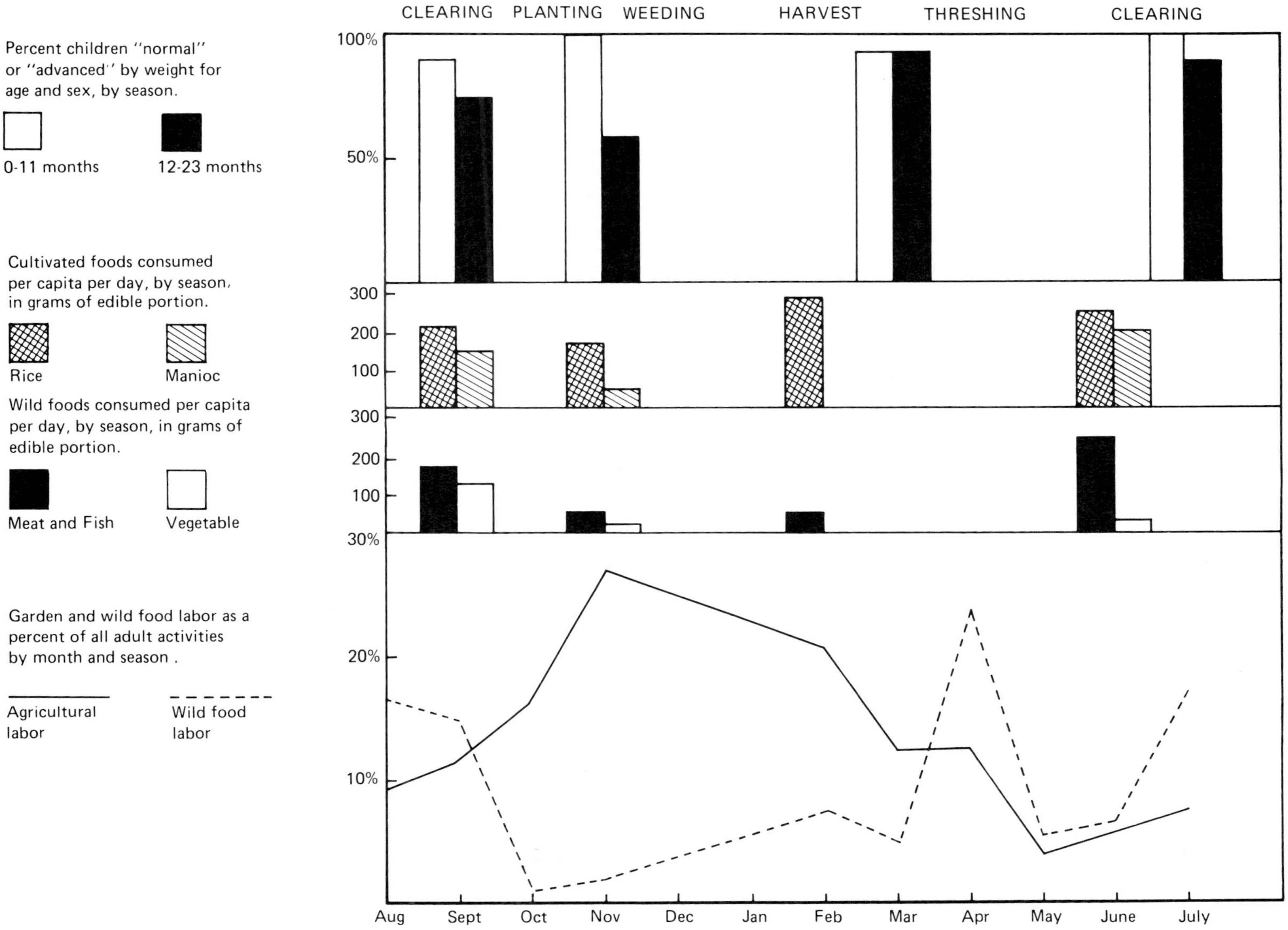

Figure 12.1 *Comparison of variation in children's weight over the agricultural year with seasonal variation in consumption of wild and cultivated foods and investment of adult labor time in wild-food procurement and agricultural tasks.*

Table 12.1

Xavante Adult Activities by Sex and by Season[a]

Activity	July–Sept. (garden clearing)		Oct.–Dec. (planting and weeding)		Jan.–March (Harvest)		April–June (postharvest)		Annual mean		Mean (both sexes)
	M	F	M	F	M	F	M	F	M	F	
Hunting	1.67	0.0	0.50	0.0	0.95	0.0	1.12	0.0	1.06	0.0	0.53
Fishing	0.91	0.31	0.0	0.0	0.08	0.07	0.64	1.05	0.41	0.36	0.39
Gathering	0.24	1.37	0.0	0.14	0.0	0.06	0.0	0.59	0.06	0.54	0.30
Total wild food	2.82	1.68	0.50	0.14	1.03	0.13	1.76	1.64	1.53	0.90	1.22
Gardening	1.71	1.32	4.02	3.33	2.53	3.28	1.09	1.19	2.34	2.28	2.31
Domestic animal care	0.04	0.0	0.0	0.0	0.04	0.0	0.39	0.18	0.12	0.05	0.09
Market activities	0.29	0.24	0.63	0.17	1.01	0.48	1.41	0.52	0.84	0.35	0.60
Nonsubsistence work[b]	1.25	4.66	0.48	5.32	1.48	6.15	1.11	4.77	1.08	5.22	3.15
Nonwork[c]	7.88	6.08	8.36	5.03	7.92	3.96	8.22	5.70	8.10	5.19	6.65

[a]Number of hr/14-hr day, from 6:00 A.M. to 8:00 P.M.

[b]Includes child care, food preparation, housekeeping, manufacture of tools, gathering firewood, and business transactions within the community.

[a]Includes eating, ceremonial activities, hygiene, conversation, recreation (e.g., soccer game), sleeping, and sitting idly.

Table 12.2

Xavante Hunts[a]

Date	Number of hunters	Duration in days	Game bagged	Estimated share/ hunter/day (kg dressed meat)[b]
8/1/76	2	1	0	0
8/16/76	12	1	0	0
9/8/76	1	night	0	0
9/10/76	10	1	1 peccary[c]	1.32
9/16/76	30	1	1 deer[d]	0.49
10/16/76	11	6	8 peccary, 3 tapir[e]	6.93
11/14/76	1	night	1 paca[f]	4.40
11/18/76	1	1	0	0
11/18/76	1	1	0	0
11/18/76	1	1	0	0
11/21/76	2	night	0	0
2/4/77	30	1	0	0
2/5/77	30	1	0	0
2/7/77	10	2	4 tapir[e]	23.46
3/5/77	1	1	0	0
3/16/77	1	1	0	0
3/27/77	1	1	0	0
3/27/77	1	1	1 deer[d]	14.68
6/11/77	1	1	0	0
6/14/77	1	1	0	0
6/14/77	2	night	2 paca[f]	4.40
7/23/77	20	1	3 peccary,[c] 1 deer[d]	2.72
7/27/77	20	1	2 deer,[d] 1 tapir[e]	7.33

[a]Hunts listed are those that showed up on randomized time allocation visits (see fns 2 and 4). When any individual was reported to be out hunting, I returned later to ask the hunter how many other hunters were with him, how long they were out, and what game was killed. The list, therefore, represents a sample of total hunts. Average yield/hunter/day for all hunts is 2.81 kg. Average yield/hunter/hour is 0.40 kg.

[b]Estimated share per hunter in kg dressed meat is the estimated amount of dressed meat from the total bag (as many animals were reported, rather than observed directly, average weights are used) divided by the number of hunter days. The duration of an average day of hunting was estimated to be 7 hours.

[c]All peccary on this list are white-lipped (*Tayassu pecari*). On other occasions I saw collared peccary brought in, usually by individual hunters.

[d]*Mazama* sp.

[e]*Tapirus terrestrius.*

[f]*Cuniculus paca.*

hunting also produces large quantities of game at one time, it also requires that meat should be widely shared and that there should be some way of preserving the surplus of fresh meat. The Xavante technique of smoking and drying meat keeps it in reasonably edible condition for about a week, and I have never seen any indication that meat was wasted.

To study the distribution and consumption of meat in the village, I monitored two households for 3-day periods, each at four different seasons of the year—24 days of observation in all.[3] I found that wild food, es-

Figure 12.2 *Solitary hunter bringing in a peccary.*

pecially fish and game, was widely shared. The larger household of the two I studied had five resident hunters; the smaller household had only one hunter who did not bring any meat into the household during the study periods. Therefore, 100% of the meat consumed during those periods in the smaller household was contributed from some other household. In the larger household, 60% of the meat consumed was brought in by resident hunters, with the rest coming from some other household. Yet there was little difference between the two households in mean/capita/day

Figure 12.3 *Part of bag from a herd of peccary.*

consumption of meat and fish. This averaged 138 g/capita/day for the smaller household, and 121 g/capita/day for the larger.

Comparison of Xavante hunting yields with those obtained by other central Brazilian groups indicate that for the Xavante, hunting is still reasonably productive. The average yield/hour spent hunting is 0.40 kg dressed meat for the Xavante. The Mekranoti obtained an average of 0.69 kg, but the Bororo and the Kanela, who both live in degraded habitats, obtain averages of only 0.20 kg and 0.11 kg, respectively (Werner *et al.* 1979:207). But this situation may soon change. Tsidadzé of the Uhö (Peccary) lineage, told me that his father and elder classificatory brother, both now dead, were "Masters of Pigs." That is, as leading men of their lineage, they possessed magical techniques to call the peccary herds (see also Giaccaria and Heide 1972:111–112). The Master of Pigs was not permitted to kill peccary himself, but stood quietly to one side, stroking the back of the "Father of the Herd," the large boar that leads the herd, while the other hunters killed as many peccary as they needed for that occasion. Tsidadzé never witnessed such a hunt, and his kinsmen died without passing their knowledge on to him, but he suspected that *someone* in the village had it and was using it secretly. Otherwise, he reasoned, how could you account for the recent appearance, on two different occasions, of large herds of peccary right at the edge of the village? Unfortunately, there may be another reason: the increasing occupation by ranches and farms of land around the reservation may be driving game onto the reservation where it may soon be overhunted.

FISHING

The present Xavante village on Pimentel Barbosa reservation is located near a minor stream where small fish are caught. Serious fishing involves a 20–30 km trip to a larger tributary of the Rio das Mortes or to the river itself. It appears that fishing became more important in Xavante subsistence after they acquired fishhooks from the Indian Protection Service in the 1950s (Maybury-Lewis 1967:51). Both men and women fish—women's fishing is perhaps recent, as Maybury-Lewis noted that women did not fish (see Table 12.1).

The Xavante fish only from March through September, when the water level in the rivers and streams is falling. It is also the season when their time is least taken up with garden work. Local fishing is casual. A few women go off together for a day's fishing, taking their small children with them. Fishing in local streams is also popular with boys in the 8–12 age group who return with their catch and roast it in the ashes of the household cooking fire. All local fishing is with hook and line.

The composition of parties that go on longer fishing trips varies. A family, including all the children, may go to one of the larger rivers and camp there for a week or more. On one occasion, a party of young men and boys went on a fishing trip while most of the other men were engaged in a game drive. Another fishing party consisted entirely of women and children. When the members of these longer fishing trips returned to the village they brought with them baskets of smoked fish, caught by using fish poison, and this fish was widely distributed in the village.

GATHERING

When Maybury-Lewis studied the Xavante, the staple of their subsistence was wild roots (Maybury-Lewis 1967:43–44). They continue to collect these tubers occasionally, apparently for the sake of variety in their diet, for sweet manioc now provides them with a cultivated starchy staple. But they continue to collect palm nuts, other wild fruits, and *palmito* (hearts of palm) in large quantities. *Palmito* is eaten fresh, but may also be dried in the sun and pounded into four which is boiled to make a kind of thick soup, considered a suitable food for small children, or to make into loaves that are baked, like maize loaves, in an earth oven. Gathering is mainly, but not exclusively, a female activity (see Table 12.1).

AGRICULTURE

There is no reason to believe that the Xavante, or any other Central Brazilian group, lacked knowledge of agriculture in recent times. However, they varied considerably in the extent to which they practised it. In the seventeenth century, the Dutch, who occupied the northeast coast of Brazil at that time, were acquainted with Tapuya groups from the interior. The subsistence practices of these groups, as described by Dutch observers, were very similar to those of Central Brazilians as we know them today. They planted maize and other vegetables which they ate up after the harvest, leaving only seed for the next season's planting (Marcgrave 1942:282). They were nomadic during much of the year, and the description of Tapuya trekking camps and log races might as easily apply to the Xavante before they were sedentarized (Maybury-Lewis 1965).

The Xavante still plant an archaic variety of maize (*Zea mays*) known to botanists as "interlocked soft corn." This race of maize is found only around the western and southern margins of the Amazon Basin (Brieger *et al.* 1958). Xavante maize varies in kernel color from white and yellow to bright red and purplish black, with mixtures of these colors. The Xavante greatly esteem this variation, carefully saving a stock of seed from each color variety in separate gourds stoppered with beeswax. As in previous times, they consume their maize crop, saving only seed and a little more to make the maize loaves that they use as ceremonial prestations.

It is upland rice (*Oryza sativa*), an introduced crop, that has become the Xavante staple in the past 5–6 years. Their reasons for accepting and intensifying the cultivation of this crop may have as much to do with their need for a marketable product as it has with their food subsistence requirements. (For further discussion of this, see Gross *et al.* 1979 and Flowers *et al.* 1982.) The Xavante now grow sufficient rice to cover their needs from one harvest to the next, though stocks may run low before the next crop is in (Figures 12.4 and 12.5).

Gardens are planted, weeded, and harvested by individual families, and the produce is regarded as belonging to the family that clears and plants the garden. Networks for informal exchange of garden foods seem to be much more restricted than those through which game and fish are distributed throughout the village. My food consumption studies showed that members of the larger household that I monitored shared food almost exclusively among themselves. On only two occasions did a visitor eat rice, and no one brought a gift of vegetable food.

Figure 12.4 *Harvesting rice.*

Figure 12.5 *Stacking rice for storage.*

On the other hand, while monitoring the smaller household, I counted 47 instances of exchange with outsiders that involved vegetable food, such as cooked rice, manioc cakes, and palm nuts. On 16 of these occasions, the household gained food, as a gift of food was brought to the household, or a household member ate with another family. On 31 occasions, the household gave food, because a visitor ate with the family, or a gift of food was sent to another household. However, about 85% of this frequent food exchange was with people from the two adjacent households who were close kin as well as neighbors.

Although the arrival of game or fish in the village was regarded as public information, people were reluctant to reveal the amount, or even the existence, of cereal stores. I find the following comment in my field notes taken while monitoring food consumption in the smaller of the households in November:

> I was unable to weigh rice stores, as Otomoaba insisted that there were none left in the house. However, I was in the house later when Tsipatsé (the 10-year-old son) took rice, carefully hidden under empty sacks, out of the bottom of a barrel in order to hull it. Perhaps seven or eight kilos were left. I also learned that the family had some rice stored in a shelter in their field. At this time households in the village had very unequal stores of rice. A few had a surplus stored to sell, others had eaten theirs up. So those who had rice were reluctant to admit how much for fear of claims to share it by those who had none.

Nevertheless, during this time of shortage Otomoaba gave plates of cooked rice to children from neighboring households, apparently as freely as to her own children.

There is no doubt that at present the Xavante are highly dependent on agriculture. In terms of labor time averaged over the year, they devote nearly twice as much time to garden labor for food as to collecting wild food (Table 12.1). For the Xavante, dependence on agriculture is increasing as the land they control becomes more circumscribed and their population increases.

THE SEASONAL ROUND

During my year of residence in the Xavante village, I followed the seasonal variation in activities by random spot checks.[4] I began these observations as the dry season was ending, shortly before the gardens were burned over and planting began at the beginning of October. I ended the observations after the harvest, when people were clearing new gardens for the following agricultural year (see Figure 12.1 and Table 12.1).

In August, about half the families in the village left on a trek lasting several weeks. All those who remained were clearing gardens. In September all were back in the village, and the men organized several collective game drives. At the same time the women went out gathering wild vegetable foods, especially palm fruits and *palmito.*

From the end of September until March, people spent much of their work time in the gardens, preparing the soil, planting, later weeding, and finally harvesting. Maize and manioc were planted first, then rice. Families went to the fields as a group, taking the babies and young children with them. Each family had a shelter at the edge of its fields where the babies and young children rested and played. Older children ranged further, but those under the age of 12 had few responsibilities beyond running errands such as fetching a few gourds of water from the stream, scaring away birds when the crops were ripening, and seeing that the younger ones did not wander far.

The harvest started in January, when the maize crop was ripe. Harvesting rice was an activity that went on for several months, for the Xavante planted rice in different plots from early October to December so that manpower for harvesting would not be required all at once. It was mostly men who cut the rice and stacked it, with some help from the women (Figures 12.4 and 12.5). Then the women threshed rice and brought it to their houses little by little, as it was needed.

After the harvest, men again turned their attention to hunting. But some of them, especially young unmarried men, left the village for a week or more at a time to work for wages on nearby farms. The women, still busy with threshing rice, did little gathering at the end of the rainy season. From April through June, much of the wild food effort was spent on fishing. By July some families were clearing new gardens, but until the rains began more time was spent on wild food than on garden labor (see Figure 12.1).

Health and Growth

COMPARISON WITH EARLIER STUDIES

A series of physiological, demographic, and genetic studies carried out in Xavante communities between 1962 and 1964 by a team of anthropologists and physicians (Neel and Salzano 1967; Neel *et al.* 1964; Niswander *et al.* 1967; Salzano *et al.* 1967; Weinstein *et al.* 1967) gives us a base of comparison from which to measure certain changes that have taken place since the Xavante changed their subsistence. The Xavante group studied at São Domingos in 1962 (Neel *et al.* 1964) is substantially the same as that now living at Pimentel Barbosa.

At the time of study, the investigators found the general physical status of the Xavante to be excellent. They were particularly impressed by the "exuberant health and vitality" of the men. Their appearance contrasted with that of the women who "although in apparent good health and nutrition when young gave an impression of early aging" (Neel *et al.* 1964:110). Some of the specific physiological characteristics of the population that distinguished it from those living under more "civilized" conditions were: "the musculature, the keen vision, the lack of dental caries, the slow pulses, the low blood pressures" (Neel and Salzano 1967:566). There was no evidence of malnutrition among the children (Neel *et al.* 1964:109).

The team took anthropometric measurements of the adult Xavante at São Domingos and at Simões Lopes. Stature and weight measurements show that the Xavante are among the tallest and heaviest South American Indians (Comas 1971: Table 12.23). Comparison of the height and weight measurements that I made of adult Xavante in March 1977 with the earlier ones shows that they are very similar (Table 12.3), indicating that if subsistence changes are affecting growth, they have not, as yet, had much effect on achieved stature.

To estimate the growth rate of Xavante children, Niswander *et al.* (1967:492–494) plotted the heights and weights of 30 Xavante boys and 15 girls between the ages of 4 and 12 years (estimated ages) against the 10th, 50th, and 90th percentiles of the growth standards of the Harvard School of Public Health. Although sample sizes were small, it appeared that Xavante children of both sexes were comparable in weight but considerably shorter than Boston children. For comparative purposes, I plotted the heights and weights of 25 boys and 26 girls in the 4–12-year age group against the same curves. The measurements were taken in March 1977, after the harvest had begun. Allowing for variation due to small sample size, Xavante children in this age group appear to be growing at a rate similar to 15 years earlier.

In spite of the impression of "feral vitality and physical resiliance" that the Xavante made on the investigators (Neel and Salzano 1967:566), they had in 1962 already suffered from a series of postcontact epidemics. This was demonstrated by the presence of antibodies to such diseases as mea-

Table 12.3

Mean Stature and Weight of Xavante Adults (Estimated Age over 18)

	Men			Women		
	No.	$\bar{X}$ Ht. (cm)	$\bar{X}$ Wt. (kg)	No.	$\bar{X}$ Ht. (cm)	$\bar{X}$ Wt. (kg)
São Domingos[a] 1962	24	168.1	67.2	34	154.7	54.0
Simoẽs Lopes[b] 1964	42	170.2	69.8	39	156.3	57.9
Pimentel Barbosa 1977	40	168.9	66.0	48	153.1	57.7

[a]Data from Neel *et al.* (1964:65).

[b]Data from Niswander *et al.* (1967:492).

sles, whooping cough, and polio (Neel *et al.* 1964:117–124). The reproductive histories of Xavante women, which I am now in the process of analyzing, indicate that many of their children died in measles epidemics. There are two young men in the village, now about 18 years old, who have partial paralysis from polio contracted when they were young children. There were no cases of measles, whooping cough, or polio during my stay in the village.

At the beginning of the dry season, when days are hot but the temperature may drop at night to less than 4°C, people suffer constantly from respiratory infections which may lead to pneumonia, especially in the young and the elderly.

Although at the time of the team investigations no evidence of tuberculosis among the Xavante was found, at present it is probably their most serious health problem. At Pimentel Barbosa in 1977 there were at least five or six cases of active tuberculosis, which often strikes apparently strong and well-nourished young adults. One young hunter (Figure 12.2) became seriously ill a few months after the picture was taken; tuberculosis was diagnosed in Goiania and he was subsequently hospitalized several times with little improvement.

Some cases of fever were probably malarial, as they responded to administration of chloroquine, but malaria was more likely contracted on trips to the larger rivers than at the site of the village, where there were few mosquitoes. In 1962 some evidence for malaria was found, of which the most marked was enlarged spleens in a number of women (Neel *et al.* 1964:119). In addition, there were several cases of Brazilian pemphigus in the village when I was there. This is a disease that usually has a fatal outcome (Weinstein *et al.* 1967:538).

A comparison of the age and sex structure of the two Xavante villages studied in 1962–1964 with my 1977 census at Pimentel Barbosa (Table 12.4) indicates that the population is now younger than it was at that time, with 48.6% under the age of 15. This is probably due in part to the fact that there have been fewer deaths in recent years from such epidemic diseases as measles and whooping cough. The relatively small number (27%) in the 15–30-year age group may be a reflection of past epidemics.

The Indian agent at Pimentel Barbosa began keeping systematic birth and death records in November 1974. In the 30-month period between that date and July 1977, when I left the village, there were three adult deaths, and three deaths of children between the ages of 2 and 15. Two of these deaths were the result of accidents.

The records kept between 1974 and 1977 indicate a mortality rate of 27% for infants under the age of 2. Table 12.5 gives the sex and age at death of these infants. All deaths except four occurred under the age of 6 months, an age up to which all the infants I observed—including those who died—

Table 12.4

Age and Sex Structure of Xavante Villages in 1962–1964, and in 1977[a]

	Age range				
	0–14	15–30	30+	Unknown	Total
São Domingos (1962)[b]					
Male	29	35	3	17	84
Female	39	31	8	16	94
Total	68 (38.2%)	66 (37.1%)	11 (6.2%)	33 (18.5%)	178 (100%)
Sex ratio[c]	74.4	112.9	37.5	106.3	89.4
Simões Lopes (1964)[b]					
Male	60	41	16	7	124
Female	39	43	15	5	102
Total	99 (43.%)	84 (37.2%)	31 (13.7%)	12 (5.3%)	226 (100%)
Sex ratio[c]	153.8	95.2	106.7	140.0	121.6
São Marcos (1964)[b]					
Male	83	54	39	42	218
Female	61	53	42	18	174
Total	144 (36.7%)	107 (27.3%)	81 (20.7%)	60 (15.3%)	392 (100%)
Sex ratio[c]	136.1	101.9	92.9	233.2	125.3
Pimentel Barbosa (1977)					
Male	65	31	32	0	128
Female	55	35	29	0	119
Total	120 (48.6%)	66 (26.7%)	61 (24.7%)	0	247 (100%)
Sex ratio[c]	118.2	88.5	110.3	—	107.6

[a]Number of individuals by age–sex group.
[b]The 1962–1964 data are from Salzano *et al.* 1967:471.
[c]Males/100 females.

appeared to be growing and gaining weight well. Most of the infant deaths that I witnessed occurred after a short respiratory illness that culminated in pneumonia. This pattern appears to be quite different from that reported from many developing countries, where death rate is highest among toddlers, and is often associated with malnutrition which becomes evident at the time of weaning (see, for example, Eveleth and Tanner 1976:243; Jelliffe 1968; Malcolm 1970).

During the 30 months of record-keeping, there were 62 live births in the village, 36 male and 26 female. Of the 60 women between the ages of 15 and 50, 44 gave birth once or more during this period (three births were of twins). Of these, 14 gave birth twice, and one woman gave birth a third time. The mean birth interval for women whose babies lived at least a year was 20.9 months. For women whose babies died it was 16.6 months.

The Xavante still give the impression, as they did 15 years ago, of being a handsome and vigorous people. They have survived the initial shock of contact, and their numbers now seem to be increasing rapidly.

Table 12.5

Deaths of Children Age 0–24 Months, By Age at Death (November 1974–July 1977)[a]

0				Months																							
Weeks				1	2	3	4	5	6	7	8	9	10	11	12	13	14	15	16	17	18	19	20	21	22	23	24
0	1	2	3																								
F			M	M	F			M			M	F						M		F							
M				M	M			M																			
M					M																						
M																											
F																											

[a]M is male; F is female.

CHILDREN'S NUTRITION AND GROWTH

I weighed the children in the village who were under 2 years of age four times during the year of observation.[5] I found that from birth through the sixth month, babies gained well at all seasons (Table 12.6). Although I weighed no babies immediately after birth, eight were weighed within a week after birth. The mean weight of these babies was 3.33 kg. All except two weighed more than 3.20 kg. The exception was a pair of female twins, who, when first weighed, were 2.50 and 2.70 kg. Although small for age, these twins gained weight well, and when I left the village they were thriving at 10 months of age.

It has often been observed from studies in Third World countries that babies do well during the first few months—and in fact often grow more rapidly than in developed countries (Eveleth and Tanner 1976:241–242; McGregor *et al.* 1961, 1968)—but after 6 months they fall behind in relation to Western standards. This is probably because supplementary foods are not available or are not sufficient, and after 6 months of age breast milk alone is not sufficient to support rapid weight growth (Jelliffe 1968). I observed a similar effect among the Xavante children. The mean weight of eight boys between the ages of 3 and 5 months was 106% of standard (World Health Organization [WHO] 1978) weight for age. The mean weight of four girls (including the twins) weighed between the same ages was 99.3% of standard. Xavante children are not fully weaned until 3 years or older, even if there is a younger sibling. Supplementary food is introduced after 6 months of age; this consists of boiled rice, mashed-up sweet manioc or bananas, and bits of meat or fish. Powdered milk and commercial baby formulas are not found in the village.

Figure 12.6 shows the weight growth of individual children at different ages and different seasons. The weight growth of a child during the interval between two weighings, as a percentage of the standard weight change for a child of that age for the interval, is indicated on the scattergram by a different symbol for (1) the interval between the September and November weighings, (2) the interval between the November and March weighings, and (3) the interval between the March and July weighings. The age assigned to the child is that reached at the midpoint between two weighings.

The scattergram indicates that till up to 6 months of age there is little decline with age in rate of weight growth, and seasonal influence appears to be negligible. Sixty percent of the comparisons between weighings of babies in the 0–5-month age group showed a weight gain greater than 100% of standard.

From 6 through 11 months there is a gradual decline in rate of weight growth, still with little apparent seasonal effect. A certain amount of the fat stored during the early months of rapid weight growth is lost. The mean weight of four boys weighed between the ages of 9 and 11 months was

Table 12.6

Children's Weight for Age, at Different Seasons

Age group	September, 1976				November, 1976				March, 1977				July, 1977			
	N	Mean age (mo)	Mean wt (kg)	% of standard	*N*	Mean age (mo)	Mean wt (kg)	% of standard	*N*	Mean age (mo)	Mean wt (kg)	% of standard	*N*	Mean age (mo)	Mean wt (kg)	% of standard
0–11 mo																
Boys	6	2.6	5.75	103	7	3.8	6.42	101	9	6.9	7.85	97	5	4.3	5.74	94
Girls	5	3.8	4.66	86	5	3.2	4.92	96	5	4.5	6.06	97	8	5.8	6.29	95
Both sexes	11	3.1	5.26	95	12	3.5	5.79	99	14	6.1	7.21	97	13	5.2	6.08	95
12–23 mo																
Boys	6	17.0	9.37	83	6	19.2	9.70	83	6	21.6	10.74	89	4	16.4	9.73	88
Girls	9	15.9	8.80	85	8	17.7	8.70	81	9	20.4	10.11	90	3	21.3	10.70	94
Both sexes	15	16.4	9.00	84	14	18.3	9.13	82	15	20.9	10.35	89	7	18.5	10.14	90

[a] "Standard" is weight for age according to "*A Growth Chart for International Use in Maternal and Child Health Care*" (WHO 1978).

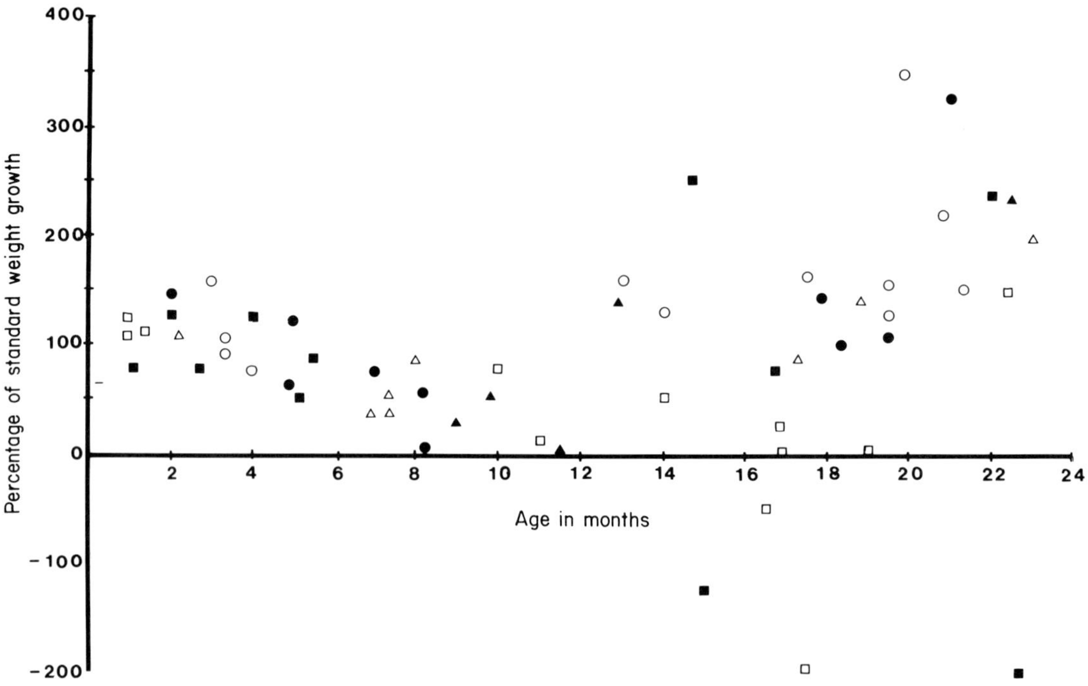

Figure 12.6 *Percentage of expected weight gain for age attained by individual children, at different seasons. Boys, from September to November (■); girls, from September to November (□); boys, from November to March (●); girls, from November to March (○); boys, from March to July (▲); girls, from March to July (△).*

92.7% of standard. The mean weight of six girls weighed between the same ages (again including the twins) was 85.5% of standard.

Table 12.6 shows that the mean percentage of standard weight for age of the 0–11-months age group varied little by season. The variation between a high of 99% of standard (in November) and a low for 95% of standard (in July) seems more closely related to the average age of the sample of babies observed at each weighing. Due to small sample size, the mean weight of the girls is depressed by the small-for-age twins, especially in September when they were born. For the 0–11 month age group, seasonal change in weight growth rate is not significant (by Wilcoxon signed-ranks matched-pairs test) at any season. This is probably because babies up to 1 year of age are primarily dependent for nourishment on mother's milk, which become increasingly inadequate as they get older.

After 12 months, as children become more dependent on the general supply of food in the village, a pronounced seasonal effect becomes apparent. Children in the 12–23-months age group dropped from a mean of 84% of standard weight for age in September, to 82% in November (see Table 12.6). This was the season when garden labor was most intense and when rice stocks were running low. Labor time spent in wild-food procurement reached its low in October (see Figure 12.1 and Table 12.1). All but 3 of the 12 children in this age group showed less than 100% of standard weight gain for age during the interval between the September and November weighings (see Figure 12.6). Two of the children gained no weight, and 4 showed a weight loss. The decline in percentage of standard weight for age from September to November is significant for this group (by Wilcoxon matched-pairs signed-ranks test, $p = .05$, one-tailed). It is probably not a coincidence that 2 of the 3 children who showed greater than standard weight gain for age belonged to families that had grown extra rice the previous season with the intention of selling it. Because transport problems had prevented them from selling all of this rice, they had a considerable amount stored at the Indian post, which they consumed as needed.

During the next interval between weighings, from November to March, weight growth for age of the children in the 12–23-month age group showed an almost completely opposite trend from the previous interval. *All* the children in this age group gained more than standard weight gain for age at this season. The increase in percentage of standard weight for

Figure 12.7 *Two-year-old girl, with mother.*

Table 12.7

Daily per Capita Consumption (g) of Representative Foods, by Season

	September	November	February	June
Rice	209.5	179.4	296.5	246.1
Manioc	153.0	54.0	0	215.0
Meat and fish	158.9	50.2	52.5	256.4
Wild vegetable foods	129.9	22.8	0	32.7

age is highly significant (by Wilcoxon matched-pairs signed-ranks test, $p = .001$, one-tailed). The average rate of weight gain for the group was 175% of standard weight gain for age, but the older children appeared to show the greatest relative weight gain (see Figure 12.6). By March, foods from the current harvest had been eaten in the village for about 2 months.

The same trend continued during the period from March to July. There are only five comparisons for weight growth of children in the 12–23-month age group for this interval, but four of these five children showed greater than 100% of standard weight gain (see Figure 12.6). Figure 12.7 shows a 26-month-old girl of normal weight with her mother.

These observations indicate that among children between the ages of 1 and 2 years, a lag in weight gain during the planting season was followed by a spurt after the harvest. One result of this pattern of seasonal changes in rate of weight growth would appear to be that for those children who reached their first birthday before the harvest, the deceleration in weight growth that started at about 6 months of age would continue until the harvest. Those who reached 12 months of age after the harvest, when garden foods were plentiful and people had time to collect wild foods as well, would have several months of accelerated weight growth before the next season of shortage, at a crucial stage of development. (For seasonal variation in per capita consumption of different types of foods, see Table 12.7). This might mean a considerable difference in the life chances of children born at different seasons of the year (McGregor *et al.* 1968).

Figure 12.1 summarizes, by month and seasonal activity, fluctuation in weight for age of children up to 2 years of age, variation by season in per capita consumption of wild and cultivated foods, and percentage of labor time devoted by adults to food production and to wild food procurement.

Discussion

A number of studies have been done on weight fluctuation associated with seasonal variation in food supply among hunter–gatherers (Lee 1979;

Truswell and Hansen 1976; Wilmsen 1978), and among subsistence agriculturalists (Chen *et al.* 1979; Hunter 1967; McGregor *et al.* 1961, 1968; Nurse 1975). These studies suggest some interesting comparisons with the Xavante experience.

The !Kung San, hunter–gatherers of the Kalahari Desert, are notably small and thin, and there has been considerable discussion among investigators who have studied them as to whether or not they can be considered undernourished. Truswell and Hansen (1976:190) believe that the small stature of the !Kung is probably due to "chronic or seasonal caloric insufficiency." They found that there was an increase in skinfold thicknesses from October (the dry season) to April–May (the rainy season). For adults the change was significant. This would imply considerable fluctuation in fat stores. However, with the exception of calories, the !Kung seem to have a well-balanced diet, with no evidence of qualitative nutritional deficiency (Metz *et al.* 1971; Truswell and Hansen 1976).

Wilmsen (1978) found seasonal weight fluctuations of up to 6% in the adults of a group of !Kung foragers. This degree of variation is as high as that reported for African agriculturalists under environmental stress (Hunter 1967) and considerably greater than the seasonal variation that Wilmsen found among the Herrero pastoralists that share the !Kung territory. Wilmsen believes that much of the variation in weight among the !Kung is related to seasonal variation in the food values of the wild plants that provide the staple of the !Kung diet. He points out that nuts (including the mongongo nut) and leguminous beans, foods that are rich in proteins, fats, and calories, ripen in quantity only at certain months of the year. At other times, the plant foods available may be mainly berries, fruits, and wild cucumbers, which contribute mostly vitamins and fiber. Wilmsen's data also showed that there was seasonal variation in quantity of meat consumed—from 15 g/capita/day in December to 220 g/capita/day in June. Births to !Kung women eating bush foods were more frequent in March–April. These babies would have been conceived in June–August when the calorie and protein intakes were highest and weights were at their peak. Births to !Kung women who were living among the Herrero, and therefore consuming milk and maize, were more evenly distributed through the year.

Wilmsen's findings would appear to support the idea that foragers are "more intimately associated with their environments [than are cultivators] and hunter–gatherer populations are, therefore, thought to be more immediately affected by environmental variables and to be more sensitive to perturbations in these variables" (Wilmsen 1978:65).

Lee, however, believes that Wilmsen's findings may have been misinterpreted. At the time of Wilmsen's study, in 1975–76, the !Kung at the /Xai/xai water hole, where Wilmsen did his study, had become heavily engaged in manufacturing handicrafts for sale. "What were the !Kung doing when they should have been out hunting and gathering?" Lee asks.

"We have to consider the possibility that they stayed at home precisely to manufacture the $2300 worth of crafts sold to Botswanacraft, a process that would take many hundreds of hours away from subsistence. . . . Instead of showing the range of variability in nutritional stress among foragers, the Wilmsen study shows how fragile these economies are in the face of mercantile penetration" (Lee 1979:440–441).

The success of the foraging strategy depends on scheduling activities so as to exploit each resource as it becomes available. Anything that interferes with scheduling tends to lower the efficiency of hunting and gathering. Once this occurs, foraging may not provide people with enough food, or it may become more costly in labor time; people may then turn to other strategies; they may, for example, increase their involvement in the market. In a comparative study of four native societies in Central Brazil (Gross *et al.* 1979), we found that the communities where people had to work hardest to get a living from slash-and-burn agriculture, because of circumscription and environmental degredation, were those where people invested most labor in the market.

Lee also made a study, in 1968 to 1969, of seasonal weight changes among the !Kung. The study included people living at a number of different water holes in the Dobe area. Lee found little seasonal fluctuation in adult weights. From July, the season of plenty, to January, when weights were lowest, adults lost not more than 1 or 2% of body weight. Lee concluded: "Measured objectively, the season of minimal weight among the !Kung is certainly not a hungry season by the standards of African agriculturalists" (Lee 1979:301).

However, the weight changes in a sample of 46 children weighed by Lee four times over the year did show a distinct pattern of seasonal changes in weight growth rate (Lee 1979:298–299). The children showed a substantial weight gain of 7.4% for boys and 8% for girls in the first 6 months of the year. From July to October, the boys gained 0.5% and the girls 1.6%. In the final quarter of the year there was an almost complete halt to weight growth, with the girls gaining 0.9% and the boys losing 0.6%. This indicates that 84% of the weight growth of the combined sample of boys and girls occurred during the first 6 months of the year. Children may be more sensitive than adults to seasonal variation in the food supply. Seasonal slowdown in weight growth, followed by recovery of weight, may help to explain the small adult stature of the !Kung.

We do not have evidence on seasonal changes in the diet and body weight of the Xavante during their nomadic years that would allow us to make direct comparisons with the !Kung. Certainly the Xavante are notably taller and heavier than the !Kung. A comparison of Truswell and Hansen's (1976:176) data on the growth of !Kung children with those collected by Niswander *et al.* (1967:494) on Xavante children, when the Xavante were still living largely by hunting and gathering, shows that between the ages of 4 and 12, !Kung children were consistently smaller and lighter than Xavante. How much of this difference is due to genetic factors and how much to environment, we do not know. The Xavante probably consumed a high-calorie diet. They had maize and beans from their gardens for 2 or 3 months of the year, and their wild vegetable staple was starchy roots (Maybury-Lewis 1967:44).

The fact that Xavante mobility was not as affected by change of season as that of the !Kung may be important in comparing the two groups. In the dry season of the Kalahari Desert, many water holes dry up and the !Kung congregate around those that have permanent water. Lee (1979:302 and 306) suggests that the greater weight loss of the people camped at the /Xai/xai water hole in January 1968, relative to other areas, was partly due to the higher population density there. Lee notes that the people who were still at /Xai/xai in January, and therefore weighed in the July–January weight comparison, included a disproportionate number of older and less mobile individuals who had not gone with the groups that made an early move out to the temporary water holes to exploit the new food resources that were becoming available as the wet season began. This again indicates the importance of mobility for hunter–gatherers.

In the *cerrado*, Xavante mobility was not restricted by season. In the dry season, if the camp site chosen was not near a stream, they could obtain water by digging a water hole (Maybury-Lewis 1967:58). The time spent at one place might vary from 1 day to 2 weeks, depending on the amenities of the site and on the game or other natural resources in the surrounding country. "There was little seasonal variation in Shavante life. They were able to go hunting and collecting as well in the rains as in the dry season and by these means they were assured of a constant and abundant food supply. There was only one proviso: that they should keep on the move" (Maybury-Lewis 1967:53).

A number of studies (e.g., Annegers 1973; Chen *et al.* 1979; Hunter 1967; Nurse 1975) provide evidence of seasonal food shortages and concommittent fluctuation in weight among subsistence agriculturalists. Hunter's (1967) study was carried out in a savanna region of West Africa characterized by low and unreliable rainfall, pockets of population density, and dependence on rainfall cultivation of cereal crops. He found that, on the average, adults lost 6% of their weight between the harvest and the peak of the hungry season, in June. The period of greatest food shortage, in the early part of the rainy season, was also the time when planting and weeding demanded maximum expenditure of calories in physical labor. "Some families, including pregnant women, nursing mothers and children, go without solid food of any sort for one or two days at a time, and men are at a loss as to how to keep up their strength for the daily round of weeding" (Hunter 1967:171). At this season people collected wild vegetable foods—fruits, nuts, and greens—but these were not enough to make up the food deficit. The pressure on this population was also indicated by the child mortality rate. Data collected from 214 mothers (of all ages) showed that 61% of all their offspring had died.

Nurse (1975) studied seasonal variation in two central African populations. He found that there was approximately a 3% loss of weight in adult men between October and March. These populations were apparently under less economic stress than that studied by Hunter, as livestock and hunting supplemented the vegetable protein in the diet. But at the beginning of the rains men ceased hunting because of the demands of agriculture.

Annegers (1973) points out that in West Africa seasonal food shortages are most likely to occur in the savanna regions where the agricultural season is limited by rainfall to a few months of the year. In these regions, cereal crops and seasonal hunting provide people with a relatively high-protein diet. But people must depend on the grain harvested at one season to supply them with food for the entire year. The harvest may vary according to abundance of rainfall and other factors. Storage of the crop may not be effective: after the onset of the rains part may be spoiled by mold and damp, and rodents and insects may eat part of the stored grain (Nurse 1975). Because the season of food shortages falls when people must work hardest in the fields, adults are favored in the distribution of the remaining food resources; small children are the sector of the population most affected by food shortage.

By contrast, in the more humid Guinean zones of West Africa, seasonal shortages are much less likely to occur because the variety of high-energy root crops grown allows harvesting throughout most of the year. But children are more likely to suffer from severe protein–calorie malnutrition which is often considered to be a result of weaning children on a diet based on low-protein roots. Annegers (1973:256) concludes that "The ecological circumstances that create the conditions under which seasonal food shortages may occur, and those circumstances responsible for severe endemic protein deficiency, are mutually exclusive."

A study of the growth of young children in a Gambian village (Billewicz 1967; McGregor *et al.* 1961, 1968; Thomson 1977) located in a region where there were pronounced wet and dry seasons, found evidence of seasonal variation in weight growth. The seasonal effect on weight growth became apparent after 6 months of age—rather younger than in Xavante children—and was most intense from 6 to 18 months. During the rainy season, children in this age group grew very slightly, or not at all, but there was rapid catch-up growth during the dry season.

Protein deficiency was not apparent in these children, who were weaned on a high-protein diet of millet, rice, and maize porridge. But the child mortality rate was high, with 43% of all children dying before the age of 7. The highest mortality was among children aged 9–14 months. Of the children in this age group that died, a significantly larger number was born in the rainy season, and died in the rainy season of the following year. The investigators did not ascribe these deaths primarily to undernutrition, even though the rainy season was the time of poor growth, because the children who died were not observed to be significantly smaller and lighter at weighing before their final illness than those who survived (Thomson 1977). Clinical cases of malnutrition that were observed among these children were usually secondary to infection.

During the rainy season, insect-borne diseases such as malaria and other infections were most prevalent in the village. The investigators suggested that the reason for the high death rate of children born in the rainy season was that they would reach the next rainy season at an age when their passive immunity to infection was wearing off and they had not yet acquired enough active immunity to resist the increased impact of the infections present in the rainy season. However, it would seem apparent that lowered resistence due to undernutrition must have put *all* the children at increased risk during the rainy season, and thereby contributed to the high death rate.

The studies in Gambia indicate that the ultimate effect on children of seasonal variation in nutritional status may depend on interaction with other factors. It is probably fortunate for Xavante children that the season when illnesses—mostly respiratory infections—are most common is at the beginning of the dry season when both meat and garden foods are abundant.

The way in which child care is reconciled with the agricultural labor required of mothers may also be important, as the season of low food supply usually coincides with planting and weeding the new crop. Thomson (1977) points out that in the Gambian village studied, women had exclusive responsibility for planting, weeding, and harvesting the rice crop. Mothers took their small babies to the fields with them, but usually left toddlers at home in the care of the elderly or of other siblings. Thus, when these children were most vulnerable they were receiving least maternal care. In another study, Chen *et al.* (1979) found that in a rice-growing region of rural Bangladesh, seasonal fluctuation in the weight growth of babies was apparent. Growth slowed in the months when there was highest demand for women's labor—in this case for harvesting and processing rice. Observations of breastfeeding women showed that they devoted less time to suckling their babies during the periods when they participated most intensely in crop-related activities.

Xavante women take as active a part as the men in subsistence farming; garden labor time is about equal for both sexes (see Table 12.1). But most of the time the entire family goes to the gardens as a group, so work seldom separates young children from their mothers. My spot checks showed that when the mothers of babies under 1 year old went to the gardens, they took their babies with them 87% of the time. Those who had toddlers 1 or 2 years old, and children 3–5 years old, had their children with them 75 and 76% of the time, respectively. When children were left in the village, it was almost always in the care of the maternal grandmother, who was normally resident in the same household as the mother.

These studies indicate that for the Xavante seasonal shortages are not so severe as those reported from Africa. Nor do demands on women's labor appear to be so heavy that they affect child care to the same extent as in Bangladesh. Xavante children get a good start in life, with rapid growth during the first few months, when they rely almost exclusively on mother's milk. Infant mortality is moderately high, but does not seem to be associated with malnutrition. Among toddlers, a seasonal deficit in weight growth seems to be largely made up after the harvest. However, as the dependence of the Xavante on agriculture increases, and as their growing population puts more pressure on resources, their position becomes more vulnerable. Eventually, fluctuations in their food supply could become as extreme as those experienced by African subsistence farmers who, like the Xavante, depend on rainfall cultivation of staple grain crops.

Conclusions

The Xavante at Pimentel Barbosa appear to be making a rapid adjustment to subsistence changes that seem inevitable—given that circumscription of their territory has deprived them of the mobility that made foraging, with minimal agriculture, a successful adaptation for them.

But there is little in the Xavante experience to support the hypothesis that the transition from subsistence based on wild food to agriculture would necessarily produce an increased and more reliable food supply. Fluctuation in food supplies appears to increase rather than to stabilize as this group attempts to reconcile the conflicting demands on labor for the collection of wild foods and the production of cultivated foods.

Observation of the process of adjustment to subsistence changes as it affects the Xavante and similar small groups may help us to greater understanding of the stresses to which populations were subjected during the prehistoric transition from wild food procurement to food production.

The Xavante experience may also have special implications for the transition to agriculture and for cultural evolution in South America, where there is a distinct dichotomy between regions where agriculture is based on seed crops, principally maize, and those where manioc and other root crops predominate. It indicates that certain pressures may have been specific to those populations that increased their reliance on seasonal grain crops.

When the agricultural cycle conflicted with hunting and fishing, a high-protein grain crop would permit such populations to give priority to the requirements of agriculture, but at the price of suffering from seasonal shortages. To prevent such shortages, it would be important to organize effective storage, and perhaps to control distribution. Populations whose primary dependence was on root crops, such as manioc, that can be stored in the ground and harvested year-round, but are low in protein, would be less subject to seasonal shortage of calories, but under more pressure to continue to hunt and fish to obtain animal protein.

Acknowledgments

I did the research for this paper while I was a fieldworker with the Human Ecology in Central Brazil Project, Daniel R. Gross, principal investigator. This project was supported by grants from the National Science Foundation and from the City University of New York. I should like to thank all the participants in this project for their constant support, cooperation, and criticism.

I should like to thank Warren DeBoer, Dan Gross, Ray Hames, Maddy Ritter, Sara Stinson, Bill Vickers, and Dennis Werner for reading and commenting on preliminary versions of this paper.

For help in the field, I am especially grateful to Basilio Da Sousa Barros and Ismael Leitão of the FUNAI post at Pimentel Barbosa. José Quintas and his wife Ana, directors of the Campus Avançado of the University of Brazil at Aragarças provided warm hospitality and assistance whenever I left the village for a few days of rest and to renew supplies. I would like to thank all the Xavante at Pimentel Barbosa, particularly the mothers who let me weigh their babies, and the families who watched patiently while I weighed their food before they ate it. Especially helpful were my friends and neighbors: Apowẽ, Tsurupredu, Warodi, Tsibupá, Wadzaê, Tsidadzê, Tsiriburã, Re'wai'õ, Wautó, and Otomoabá.

Notes

[1]Children in Europe and similar latitudes of the Southern Hemisphere have been observed to grow more rapidly in stature during spring and summer than during fall and winter. A study (Marshall and Swan 1971) that showed that in blind and partially sighed children the occurrence of maximum growth rate was not seasonal but was distributed throughout the year, suggested that in children with normal sight growth rate was influenced by some seasonal environmental factor. Hartung (1978) believes that this factor is probably the effect of increasing hours of daylight on the pineal gland. However, this effect would probably be minimal in children who, like the Xavante, live at latitudes near the equator where there is little seasonal variation in hours of daylight (Hartung 1978:280).

[2]To estimate hunting and fishing productivity, I used the time allocation spot checks (see fn 4). Whenever an individual in the household was reported to be out hunting or fishing, I returned later to find out what was killed or brought back.

[3]Two households were used for dietary studies: a small one consisting of one hunter, his wife, and three children, and a large extended family household which varied at the different study periods from 10 to 18 people, and from 1 to 5 hunters. I remained in the household I was observing throughout the day and weighed food as it was prepared.

Per capita estimates are based on the total consumption of each food in the household for the day, divided by the number of household members over the age of 1 year present on that day. As there are no set mealtimes, and food is constantly shared, there is no other way to arrive at individual consumption figures. The per capita consumption for each season was arrived at by averaging those recorded on each of the 6 days of observation. Food weights are given in edible portion (e.g., hulled rice, peeled manioc root, dressed meat).

With the exception of salt, which is now generally used in cooking rice, few purchased foods are consumed. When men earn cash by occasional labor on neighboring ranches, they may buy and bring back to their households a little coffee, manioc flour, sugar, or candy for the children.

[4]For the time-allocation study, I used the technique devised by A. Johnson (1975). I made 12 visits each week to as many households chosen at random, at randomly selected hours between 6:00 A.M. and 8:00 P.M. On entering the house, I took note of the activity in which each household member (or visitor) was engaged. Then I asked where each absent member was and what he was doing. Spot-checking with people on their return showed that these reports were generally accurate. The activity of each person was briefly described and coded. Provision was made for alternative analyses of simultaneous activities (such as cooking plus conversation). By the end of the year, about 5000 observations of individuals had been coded. For purposes of this study, a population of 198 was used because several families had their gardens so far from the village that they stayed there for weeks at a time. Their activities when they were in the village were not representative, as they were there on holiday, so to speak.

The analysis of these data is based on the assumption that the proportion of observations was the same as the proportion of time actually spent on that activity during the 14-hour day used. This procedure provides a more representative and reliable picture of time allocation than do studies of a few people over short periods.

[5]I weighed the children under the age of 2 in September, November, March, and July. Two anticipated weighings, in January and May, were missed because of my unforeseen absence from the village. I weighed the children on a standard baby scale, measured supine length, and the circumference of head, thorax, and upper arm. I had accurate birthdates of almost all this group, as the agent at the government Indian post had started keeping birth and death records some 18 months before my arrival.

Approximately 80% of the children falling within the age group were recorded at each weighing. A few could not be reached because their families were absent from the village, or for some other reason. The total sample was 41 children, of whom 10 were weighed four times, 12 three times, 7 twice, and 11 once.

The reference values of weight for age and sex used in comparisons are from *A Growth Chart for International Use in Maternal and Child Health Care* (WHO 1978). The expected weight gain for age and sex used is that of the 50th percentile. Children considered below normal weight are those below the 3d percentile for age and sex. Children considered advanced are those above the 50th percentile for age and sex (See Figure 12.1). My purpose in using standard tables was to provide me with a measure for comparisons between age groups and to show seasonal variation. This is not meant to imply that the 50th percentile of a standard table is necessarily a universally ideal weight for age.

IV
Settlement Pattern

CHAPTER 13

The Settlement Pattern of a Yąnomamö Population Bloc: A Behavioral Ecological Interpretation

Raymond B. Hames

Introduction

The factors that govern the settlement pattern of native Amazonians are a topic of widespread interest and debate. Generally, the size, mobility, and location of settlements have been interpreted as a means for adapting to the distribution of scarce subsistence resources (Gross 1975; Lathrap 1968; Meggers 1971). These interpretations assume that Amazonian populations are severely constrained by the distribution of strategic resources and, as a result, a settlement pattern is an adaptation designed to maintain a population's demographic equilibrium in relation to environmental constraints. There are several problems with this approach: (1) there is little evidence to indicate that Amazonian populations have saturated the environment such that there is little opportunity for growth without costly subsistence intensification which would threaten survival; (2) environmental constraints are narrowly conceived (i.e., merely the quality or density of one resource); and (3) the causal links between limiting resources and settlement pattern have not been empirically established. The aim of this chapter is to develop an integrated model of settlement and evaluate it with data collected on the Yąnomamö of Venezuela. The theoretical perspective I employ derives from evolutionary ecology (Pianka 1978) and behavioral ecology (Krebs and Davies 1978), fields of research which employ neo-Darwinian principles to model how group-living individuals adapt to the environment. In its reliance on optimization thinking, behavioral ecology deals with settlement pattern in much the same way as do locational geographers (Isard

ADAPTIVE RESPONSES
OF NATIVE AMAZONIANS

ISBN 0-12-321250-2

1972) and archaeologists (Plog and Hill 1972). In contradistinction, somewhat, to the above approaches, behavioral ecology considers reproductive as well as productive resources, employs a definition of adaptation consistent with neo-Darwinian theory, avoids complicated models based on homeostasis and/or group selection, and stresses the role that competitors (e.g., neighboring settlements) play in influencing settlement.

I begin with an outline of the components of a settlement pattern and a set of theoretical constructs which accounts for settlement variability. These constructs are then tested with data gained on a Yąnomamö population bloc (see Chagnon 1974 for a definition; and the following discussion here), and the fit between the data and the theory is discussed. In the conclusion, solid generalizations about Yąnomamö settlement pattern are formulated which may also characterize the settlement pattern of other Amazonian people and demonstrate the utility of evolutionary ecological models.

Components of Settlement Pattern

The ways communities map onto the environment as spatial and temporal entities is loosely defined as settlement pattern. Deciding where to settle, how long to stay, how large to be, and whether to forge connections with other settlements are behaviors which powerfully influence the efficiency of subsistence activities, physical well-being, and degree of impact on the environment. For horticultural Amazonian groups the regulation of village size, location, mobility, and regional integration commit individuals to relatively long-term environmental relations that are not easily or quickly modified without major costs. Because a settlement pattern covers a distinct number of phenomena, it is probably futile to attempt to develop a monolithic theory. Instead, I have chosen to break down settlement into its components and review some theories that seem particularly applicable to each component and integrate them at the level of behavioral ecological theory.

SETTLEMENT DURATION

In Amazonia, the great variability in the duration of a settlement or a camp is strongly tied to subsistence pattern. For example, the foraging Aché (Hill and Hawkes, this volume) abandon their camps daily while the Siriono (Holmberg 1950) and the Colombian Maku (Silverwood-Cope 1972), who gain the bulk of their subsistence from foraging, but also cultivate, abandon camps on a monthly basis. Of course, fully horticultural groups occupy their villages for a number of years. Nevertheless, some horticultural groups, such as the Mekranoti (Werner, this volume), Yąnomamö (Barandiaran 1967; Chagnon 1968a), Ye'kwana (Barandiaran 1962), Wayana (Butt-Colson 1972), and many Gê speakers (Gross, this volume) may spend several weeks to several months living in the forest on trek. Of the horticultural groups, those which inhabit the banks of large rivers and depend on fish for the bulk of their dietary protein appear to have the most permanent villages (e.g., Cubeo [Goldman 1963]; Boni and Akawaio [Butt-Colson 1972]; Shipibo [Bergman 1980]; and Cocamilla [Stocks, this volume]).

Resource depletion, structural decay of villages and environmental pollution, and warfare appear to be the most common causes of movement in Amazonia. Horticultural soil, game, fish, and wild plants are exploited at rates greater than their natural rates of recovery and through time this leads to diminishing rates of return per unit of effort (Johnson, this volume; Vickers 1980). Chagnon (1973) notes that when a Yąnomamö village occupies a new site, gardens are established on the perimeter. As gardens are abandoned in the process of swidden cultivation, they become increasingly distant from the village and they become less efficient to cultivate because of the increasing travel time to reach them (see also Butt-Colson 1972 for the Waiyana; and Johnson, this volume, for the Machiguenga). At about the same time, roof thatch and structural elements of village houses begin to decay and require near total rebuilding. The choice then is between rebuilding on the same site or moving the village to another site. The latter is frequently chosen because it has the additional benefit of shortening travel time to gardens. Chagnon (1973) refers to these moves, which range from 100 m to 1 km, as "micro-moves."

In Amazonia, game is a crucial resource which is harvested more rapidly than it is replaced reproductively. In a longitudinal study, Vickers (1980) has shown the rate of decline in hunting success is asymptotic through time; that is, it is severe in the first few years and then levels off. This decline forces hunters to extend the radii of their hunting area by exploiting successively more distant hunting zones (Hames 1980).

Gathered resources are energetically unimportant in most horticulturally based Amazonian societies (although they may have other important nutritional virtues), probably composing no more than 5–7% of total food intake (Dufour, this volume; Lizot 1977). No studies have been made to show how quickly they are depleted. However, Silberbauer's (1972) research on the depletion of gathered foods with increasing camp duration among the G/wi bushmen undoubtedly applies to Amazonians.

The degree to which humans deplete fishing resources in Amazonia is extremely problematical. Historic accounts (summarized in Meggers 1971) attest that extremely large settlements were found along the Amazon and its major tributaries, which may suggest that fish biomass suffers some depletion but never to levels which would force village abandonment. Stocks (this volume) notes that the Cocamilla who reside in large villages by Amazonian standards (i.e., more than 300 people) can, with modern technology and market incentives, deplete resources. But, when released from market conditions, they are able to satisfy their fishing needs easily with the maintenance of an equilibrium between fish and humans (see also Butt-Colson 1972; Bergman 1980).

In an important study of seven Amazonian societies, Carneiro (1970a) demonstrates a strong relationship between degree of dependence on fishing and settlement duration. Carneiro argues that fish in large riverine habitats represent an abundant and inexhaustible resource (under aboriginal conditions of exploitation) for meeting protein needs.

Depletion of food resources is a powerful incentive for movement (Butt-Colson 1972). If resource depletion were the only consideration, then one would predict that movement would occur when the overall ratio of output to input in subsistence declines. However, this argument rests on two weak assumptions: (1) a relatively homogeneous environment, that is, one area of the forest is as good as another; and (2) efficient subsistence is the only adaptive pressure. Moran (1980) has argued that the distribution of good horticultural soils is erratic in Amazonia, making movement risky. Below, I cite several cases when Yąnomamö relocated villages only to discover that the soil in the area was inadequate. On two occasions, the Ye'kwana, neighbors of the Yąnomamö, faced great privation after establishing new villages in areas with unforseen soil problems (Hames in press a). Lizot (1974) gives several more Yąnomamö examples and Stocks (this volume) and Hurault (1968) discuss the patchy distribution of fishing areas in relation to settlement. In general, there is some element of risk in settlement relocation even though native Amazonians regularly assess alternative areas while foraging (Butt-Colson 1972).

Factors other than subsistence directly affect the utility of staying or moving in the forest. In areas of endemic warfare, sites must be defensible (Harner 1973) or located away from more powerful enemies (Butt-Colson 1972; Chagnon 1968a). The presence of locally abundant pests such as chiggers (*Pulgans penetrans*), leaf-cutter ants (*Atta* sp.), and gnats (*Simulidae*) (Butt-Colson 1972; Chagnon 1968a) make some areas unfit for settlement or cause their abandonment before resource depletion sets in. These environmental factors and others (e.g., dependable water supply: Butt-Colson [1972]; Chagnon [1973]) determine whether alternative sites are superior to the present site or may themselves determine movement before subsistence declines in efficiency.

Some researchers (Butt-Colson 1972; Chagnon 1968a, 1973; Gross, this volume) have pointed out that warfare is a frequent motivation for movement. Small villages or villages with ineffective alliances are forced from their sites due to the raiding activities of more powerful villages. The goal of this movement is to put as much distance as possible between one's self and an enemy or to become hidden in the forest. The economic costs of such long-distance movements are enormous because of the difficulties involved in clearing a new village and garden sites, and having to return frequently to the abandoned distant village to harvest crops and planting stock (Chagnon 1974). Evidence indicates that this added stress increases mortality (Chagnon 1974). An alternative to immediate reestablishment of a new village is a fusion with an ally, which allows a base from which a weak village can set up a new village less stressfully. Regardless of whether warfare is a result of competition over mates (Chagnon 1968a; 1979b) or over scarce game resources (E. Ross 1978), warfare must be considered an immediate contingency to which villages adapt their settlement strategy.

Finally, there appear to be a number of social causes for movement in Amazonia. Most prominent among these are the death of a village headman (Arvelo-Jimenez 1971; Carneiro 1957; Murphy 1960) and internal disputes leading to fissioning (Arvelo-Jimenez 1971; Butt-Colson 1972; Chagnon 1968a; Gross, this volume). How such factors may be related to the more frequent environmental and warfare motives is presently unknown.

VILLAGE SIZE

The environmental limitations on village size in Amazonia have been debated for a long time. Initially, Steward (1949) and Meggers (1954, 1971) suggested that the quality of the soil for horticulture limited village size. As proof for this relationship, they showed that fluvial villages using silt-laden and nutrient-rich *várzea* soils tended to be much larger than interfluvial villages using nutrient-poor *terra firme* soils. In the first modern cultural ecological research in Amazonia, Carneiro (1957, 1961) argued that the soil–village size relationship was doubtful by showing quantitatively that interfluvial villages could maintain large and permanent populations without degrading the soil. Carneiro's findings were widely accepted (for an exception see Roosevelt 1980), but later protein was introduced as a limiting factor in village size. Lathrap (1968), Carneiro (1970a), and Gross (1975) argued that fluvial populations were larger than interfluvial populations because the former had access to greater quantities of animal protein, especially fish, than interfluvial populations who had meager fish resources and who had to rely on game for the bulk of animal protein intake.

While it is obvious that resource density potentially limits a village population, it tells us little else. Implicit in the environmental limitation formulation is an unintentionally ethnocentric assumption that humans attempt to aggregate in as large a group as possible and that the only factor which limits this propensity is the density or quality of resources. A more productive approach to village size is simply to invert the usual question and ask why humans are found in groups larger than the nuclear or extended family. We would then be available to concern ourselves with the adaptive value of group living and would have no need to assume that people have a natural desire to aggregate.

Alexander (1974) observes that groups larger than a mated pair and offspring evolve in response to the following factors: (1) a superabundance of local resources; (2) a scarcity of home sites; (3) predation; and (4) resources which are foraged more efficiently in cooperative groups. He notes that the first two factors, a superabundance of resources and a scarcity of home sites, do not commonly lead to the formation of groups with a high degree of sociality, because frequent, cooperative interaction patterns are not required. Predation (warfare) and foraging efficiency considerations, on the other hand, lead to greater sociality because they presuppose high degrees of interaction, communication, and coordination between individuals composing groups. Implicit in this model is the idea that a group would grow no larger than the optimal size to accomplish these tasks. Because home sites probably have rarely been important to group formation in humans and because superabundant resources is the traditional explanation of group formation used by Amazonian human ecologists, I will focus on warfare and economies of scale as determinants of village size in more detail.

For the Yąnomamö, Chagnon (1968a, 1974) has consistently argued that village size is determined by the intensity of warfare. Warfare is endemic throughout the Yąnomamö tribal distribution and a village must not have fewer than 40 individuals or it will not be able to fend off attacks or make successful counterattacks. Where warfare is intense, villages are large, and where it is infrequent, villages are small (see Gross 1979 on the Gê).

Depending on the environment and the modes of exploiting it, certain tasks can be made more efficient or effective through cooperation that allows an economy of scale to emerge, of which there are three main types. First, certain tasks cannot be done or become extremely inefficient to do (e.g., building a house) without the cooperation of two or more individuals. In the case of North American plains buffalo-hunting or subarctic caribou-hunting, overall hunting efficiency can be increased through information sharing and cooperative game drives. Second, certain tasks, such as harvesting and planting grain crops, require what Erasmus (1956) calls "peak labor loads." That is, the labor required for a task must be expended in a short period of time but the labor needed is greater than an individual or family can muster. Third, it may be advantageous to minimize the variance of intake for a particular food resource, especially when success rates for individuals attempting to acquire it are highly variable. For example, whale hunting by Point Barrow Eskimos is a very chancy business, with many whaling crews going an entire season without making a catch (Chance 1966). This variance in hunting success and hence food intake may be minimized through sharing a variable and unpredictable resource. In fact, certain food-getting strategies may not be possible, due to high variance, without widespread patterns of sharing. This factor alone may explain why in many societies large game animals, which are captured at high rates of variance, are frequently shared widely and why gathered foods tend not to be shared (See E. Smith 1981 for an excellent review of group formation among hunter–gatherers in terms of foraging efficiency). Leeds (1961, 1969) notes that the basic Amazonian economic pattern of root crop horticulture and hunting, fishing, and gathering are mainly individualistic pursuits little enhanced by cooperation. He concludes that this economic pattern does not require large settlements and structural complexity.

Another factor is also responsible for group formation. In environments that subject individuals to high levels of trauma or disease, it would be advantageous to live in a group so that others could render aid. For example, Nelson (1973) points out that Kutchin hunters hunt in pairs so that if one falls through lake ice or suffers some other problems, the other will be able to render aid.

Concentrated or superabundant resources, various economies of scale, and warfare are the factors which may be expected to account for settlement size variability. Furthermore, several of these factors may interact. For example, the presence of intense warfare may lead to the formation of settlements that are well above the optimal size for foraging efficiency. Where such conflicting factors interact, one would expect that group size would be set primarily by the factor that was most adaptive for all the members of a settlement.

COOPERATION IN SUBSISTENCE ACTIVITIES

Few Amazonian researchers have attempted to analyze the frequency and the causes of economic cooperation in subsistence activities (for an exception, see Leeds 1961, 1969). Whether or not cooperation enhances economic efficiency is unknown precisely. Here I review some of the literature on Amazonian subsistence to determine if cooperation is frequent and if economies of scale arise.

Gardening. In most Amazonian societies the tasks of clearing, burning, planting, weeding, and harvesting are accomplished by nuclear or extended family units (e.g., see Carneiro this volume; Johnson this volume; E. Ross 1976; Vickers 1976). Uncommonly, as among the Ye'kwana, clearing gardens is done communally and other garden tasks are also frequently cooperative, but on a lesser scale (Hames, in press b). Peak labor loads or economies of scale, as defined by Erasmus (1956), do not seem to occur in any of these activities. Where cooperation in garden making occurs, its adaptiveness may be in spreading risk of crop failure across a number of different gardens (Hames, in press b). In any event, communal gardening would not lead to the formation of villages of more than three or four cooperating families.

Fishing. While gardening adaptations show little variability in Amazonia, fishing is done in diverse ways. Hook-and-line, arrow, and trap fishing with baskets are mostly solitary or 2-person pursuits requiring no cooperation. Some forms of net fishing (Stocks this volume) may require the cooperation of 3 or 4 people. But fishing with poison (*barbasco* in Spanish-speaking countries and *timbo* in Portuguese-speaking countries) is done in groups of up to 40 people (Hames 1978), whereas fishing using damming techniques described by Beckerman (this volume) regularly requires 6 or more people. Damming and poison fishing regularly employ a number of people performing a variety of distinct tasks. It is also possible that individuals are merely sharing a common concentrated resource, which is harvested on a regular basis or, as Beckerman hypothesizes (this volume), fishermen actively share information on the location of relatively unpredictable fish resources which leads them to fish in groups. In either case, and especially the latter, formation of larger than average settlements may result where fish play a crucial role in the diet.

Gathering. Of all subsistence activities, gathering is the least cooperative. Picking fruit, digging for tubers, and climbing trees for honey are solitary activities (although frequently done in groups). A possible exception is the gathering and processing of *moriche* palm starch (*Mauritia flexuosa*) among groups such as the Warao (Suarez 1968). However, this use of *moriche* is rare in Amazonia.

Hunting. The work of Hill and Hawkes (this volume) on economies of scale in hunting represents the only precise data we have on the efficiency of group hunting in the lowlands. They show that for certain game, such as coati (*Nasua nasua*) and capuchin monkey (*Cebus apella*), small-group hunting leads to more of these species being taken per hunter when encountered, but as hunting group size increases, the efficiency of each hunter declines rapidly.

Throughout Amazonia it is consistently reported that white-lipped peccaries (*Tayassu pecari*) are hunted in teams which encompass nearly every able-bodied hunter in a settlement (E. Ross 1978). White-lipped peccaries are located frequently by a solitary hunter who returns to the village immediately to notify other hunters. As soon as possible, a hunting party is organized and led to the site and not uncommonly hunters may take up to 25 of these 40-kg animals (e.g., Murphy 1960; Wagley 1977). It may be that peccaries are hunted in groups because if a lone hunter were to return with a kill or kills, he would effectively deprive the other hunters in a village from going after this superabundant, prized resource. White-lipped peccaries are unpredictable in time and place (Kiltie 1980) and a lone hunter may sacrifice his short-term efficiency in hunting to maximize the group's hunting efficiency by reducing its search costs. Over the long term this is not maladaptive for the hunter who notifies others of the herd's presence because he can expect that he will be notified of the existence of a herd in the future which will minimize his search time and thereby maximize his hunting efficiency. Because white-lipped peccaries are an important resource for many groups, such as the Siona–Secoya (Vickers 1980), Ye'kwana and Yąnomamö (Hames 1979), Jívaro (E. Ross 1978), and Aché (Hill and Hawkes this volume), their hunting may be an important cause of group formation in certain parts of Amazonia (for a different view of peccaries see Berlin and Berlin in this volume).

For the important subsistence activities of hunting, fishing, gardening, and gathering, economies of scale seem to operate occasionally and when they do, they may not require much more than the personnel of four or five nuclear families. However, another economic advantage for group-living, an economy of sharing, must not be dismissed. On a seasonal or yearly basis, the efficiency and success of individual economic tasks may be high but the variance may be even greater. The most common example comes from hunting, which appears to be rather efficient in supplying dietary protein (Chagnon and Hames 1979) but varies on a daily basis due to relatively high failure rates (compared to other subsistence activities) and high variability in the size of game animals taken (E. Ross 1980). Therefore, group-living with widespread sharing of game (common in Amazonia) may be designed to minimize variance in food intake. This idea could also be extended to other subsistence activities and one might generalize that group-living may be an adaptation to buffer against the effects of garden failure (for examples see Lizot 1974; Hames in press b), inability to work due to sickness (Chagnon and Melancon 1981), or poor luck in hunting and fishing. However, the only attempt to test quantiatively this proposition, with respect to hunting, indicates that variance in hunting becomes sufficiently low when three or four hunters share their catches, leading to a settlement size of 12–15 people (Hill and Hawkes, this volume).

Settlement size tends to be greatest along major rivers and smallest in the deep forest (Lathrap 1968; Meggers 1971; Steward and Faron 1959). It appears that large size in riverine areas may be the result of superabundant resources such as fish and riverine game (Lathrap 1968) or perhaps rich soils (Meggers 1971; Roosevelt 1980) or intensification schemes that require many people (Roosevelt 1980). Smaller village size (25–100 people) in the interfluvial areas is probably the result of a more even distribution of resources and the lack of opportunities for an economy of scale in exploiting resources in groups. Warfare also contributes to settlement size but environmental factors probably limit the size of settlements when diminishing returns in subsistence activities begin to cost more than the advantages of protection by being large.

LOCATION

From an evolutionary ecological perspective, the location or position of a settlement is an adaptation to the distribution of resources, and competitors and/or predators in a habitat. One of the major ways a population can successfully adapt to the environment is to position itself so as to minimize the distance traveled to gain resources and to maximize the distance from competitors. In habitats with low population densities, the role that competitors play in influencing location is small and the distribution of resources is critical. As the density of the habitat increases, the role of competing settlements must be adjusted so that neighbors depress resource acquisition efficiency as little as possible. Resource depression through increased competition can be lessened by relocation, by increased time allocation to resource accrual ("scramble competition," Wilson 1971:186), or by territoriality and warfare ("direct competition," Wilson 1971:186). Thus, in relatively saturated niches, groups must reconcile the pulls of maximizing resource efficiency and minimizing competition.

It is necessary to distinguish between two different types or locations or areas in which a settlement is located: the domestic and extradomestic (see Butt-Colson 1972 for a similar distinction). The domestic area can be defined roughly as a central place; that is, the location where people return after foraging or gardening to sleep, socialize, process food, care for children, and perform a variety of maintenance activities. Outside the domestic area is the extradomestic area, where most resource acquisition activities occur. The differing qualities of each area interact to locate a settlement. In Amazonia, domestic areas must be well drained (Butt-Colson 1972; Hames 1978), have an easily accessible supply of water (Chagnon 1974), and, in areas of intense warfare, provide a vantage point to detect the approach of enemies (Harner 1973). For strongly riverine groups, adequate ports and quiet stretches of water are also important (Bergman 1980).

Generally, the qualities of extradomestic areas are far more important in determining settlement location. The character of resources and the location of other settlements (competitors) are the chief variables. In a simple model of location relative to resources, Horn (1968) shows that resource density and resource predictability are key variables to account for settlement location. When resources are predictable, settlements will be evenly dispersed, and when they are unpredictable settlements will tend to aggregate. (Resource density determines the actual size of settlements.) The goal of this locational strategy is to minimize travel time to resources from the settlement and to maximize distance from competitors. This model has been applied successfully by Heffly (1981) to Athabaskan hunter–gatherers.

The shape of the extradomestic area is determined by the density of competitors (Covich (1976) shows that in low-density environments, foraging areas will tend to be circular because this shape minimizes distance traveled from the domestic area to any resource location. This foraging pattern leads to discrete foraging areas with no overlap. As population density increases, foraging areas eventually assume a hexagonal shape, permitting the maximum number of areas that can be accommodated in a saturated environment. This hexagonal shape has been demonstrated for Australian aborigine foraging territories (Birdsell 1958) but appears to be polygonal for the Siona–Secoya (Vickers this volume).

Throughout Amazonia, with few exceptions (e.g., the Aché; Hill and Hawkes, this volume), gardening is the most important source of food energy, followed by hunting and/or fishing, with gathering least important. Because gardening contributes between 70 and 95% of dietary calories (Johnson and Montgomery 1976; Lizot 1977; Werner *et al.* 1979), the quality of land for cultivation, as determined by nutrient status, drainage, and so forth, is a powerful determinant of settlement location (Butt-Colson 1972; Chagnon 1973). In addition, what the so-called "protein debate" has made abundantly clear is that anywhere from 60 to 80% of all dietary protein derives from hunting and/or fishing (Lizot 1977; Ross 1976; Vickers 1976; Werner *et al.* 1979) and, as a result, settlement location to some degree is determined by an area's potential for the exploitation of fish and game resources. Work by Butt-Colson (1972), Bergman (1980), and Stocks (this volume) indicates that the distribution of oxbow lakes and other rich fishing areas importantly determines settlement location for riverine populations. Finally, arguments for gathered resources such as firewood have also been made (A. Johnson 1974). If we assume that survival and reproduction in Amazonia are determined by the efficient acquisition of all limiting resources, it is then evident that an optimal settlement location will be a compromise of all these resources (Vickers 1980). However, some resources are more limiting than others in that declining efficiency in their acquisition through time will govern settlement location more powerfully than others (Gross this volume).

It is useful to consider neighboring settlements as competitors that influence the location of a settlement. If competition may be defined as "the active demand made by two organisms [or settlements] on a common resource which is actually or potentially limiting" (Wilson 1971:185), then a village should be located to minimize the effects of competition because this will decrease resource accrual efficiency. This may be difficult or easy to accomplish, depending on population density and availability of virgin territory. Few studies on unacculturated native populations have addressed this problem in any detail (see Meggitt 1977 for an exception) and the sole study in Amazonia (Hames 1980) must be considered tentative. While neighboring settlements may depress resource accrual efficiency, they may also be beneficial as political allies or trading partners.

In certain contexts, such as endemic warfare, neighbors may benefit from one another's close proximity. Various studies of hunter–gatherers (e.g., Damas 1972; Helm 1968; J. Smith 1978) indicate that settlements maintain close communication and proximity at various times of the year in order to enhance trade, share information on the location of unpredictable but dense resources, and cooperate in resource acquisition. Given a relatively homogenous distribution of terrestrial resources in Amazonia, close proximity would seem to have no advantage in subsistence activities such as intervillage cooperative hunting or information sharing. In fact, close proximity would seem to reduce the efficiency of subsistence activities such as hunting (E. Ross 1978). Where warfare is endemic, Chagnon (1968a, 1974) has argued that villages who are allies tend to aggregate somewhat so that they can more easily mobilize against a common enemy or can quickly seek refuge with one another. Although close proximity does reduce subsistence efficiency, the benefits of surviving any enemy attack more than make up for it.

Closely related to village proximity is the phenomenon of intervillage visiting. In many native Amazonian societies, people spend weeks, months, or longer periods of time visiting kin in other villages. (Because of this practice, some ethnographic reports speak of the average village size of their main village of study [see Beckerman this volume; Dufour this volume]). For the Yąnomamö and many other groups the motivation behind visiting seems either social or religious. For example, during the dry season, the Yąnomamö hold mortuary ceremonies (*reahu*) by inviting neighboring settlements to mourn their dead. Although a great deal of time is spent in ceremonial activities, political leaders spend even more time assessing enemy strength, evaluating potential allies, planning raids, giving pledges of mutual support and solidarity, contracting marriages, and engaging in trade. It should be clear that village proximity, which facilitates frequent visiting and enhances the potential for rapid intervillage political action, is a kind of adaptation to the social environment.

In summary, the distribution and predictability of resources and the distribution of competing settlements will determine the location of a settlement. A settlement will optimize its location by minimizing its distance to resources and maximizing its distance from competitors. But, in certain instances, a settlement may attempt to draw nearer to a competing neighbor and thereby decrease its resource accrual efficiency, if the benefits of proximity (e.g., protection from enemies) outweigh the costs.

In the above discussion of some evolutionary ecological theory relating to settlement pattern, it is assumed that individuals attempt to maximize their inclusive fitness by influencing settlement size, duration, and location. Importantly, it should be noted that it is impossible to optimize simultaneously all dimensions of settlement because necessary resources are scarce and an efficient adaptation to one may lead to a inefficient adaptation to another. Thus, settlement pattern is a series of compromises. It is assumed that environmental constraints (largely resource density, variety, and predictability, and the presence of competitors) and technological factors (subsistence techniques and technology) are the variables that govern settlement pattern. An assumption not made here but commonly made in Amazonian human ecology is that the environment that native peoples exploit is relatively saturated or nearly maximally populated, and, as a result, many aspects of settlement are designed conservatively to prevent a population from overexploiting or degrading the environment in which they live (e.g., Denevan 1976; Lathrap 1968; Meggers 1971). Given that the Yąnomamö are growing in population and expanding geographically, this assumption cannot be credibly made. This does not mean that they are unconstrained environmentally but rather that their present adaptive abilities are flexible and capable of intensification without rapidly increasing rates of diminishing returns. In the following section, data are presented on the settlement pattern of one Yąnomamö population bloc as it relates to some of the issues just discussed. While it is usually easy to formulate optimization models, acquiring data to test them is not. As a result, the data presented and the discussion of them are not tests of optimization models but rather are an extended plausibility argument for their utility and a stimulus for further research which I hope to carry out soon.

The Haiyamo Population Bloc

METHODS

The data I am about to present on Yąnomamö settlement pattern derives primarily from informant accounts of past movements, causes of movements, fissions, fusions, and intervillage political relations. These data were gained by visiting all current villages (Figure 13.1) and interviewing those individuals, whenever possible, who witnessed particular events. In the case that living witnesses are unavailable, secondary accounts had to be used. While in the field, data gained from an informant were systematically cross-checked with other informants in his village and in other villages. In general, there is an inverse relationship between the accuracy and richness of informant accounts and the age of the events described. Descriptions of current villages are based largely on firsthand observations. Readers are referred to Chagnon (1974) for a detailed description of the

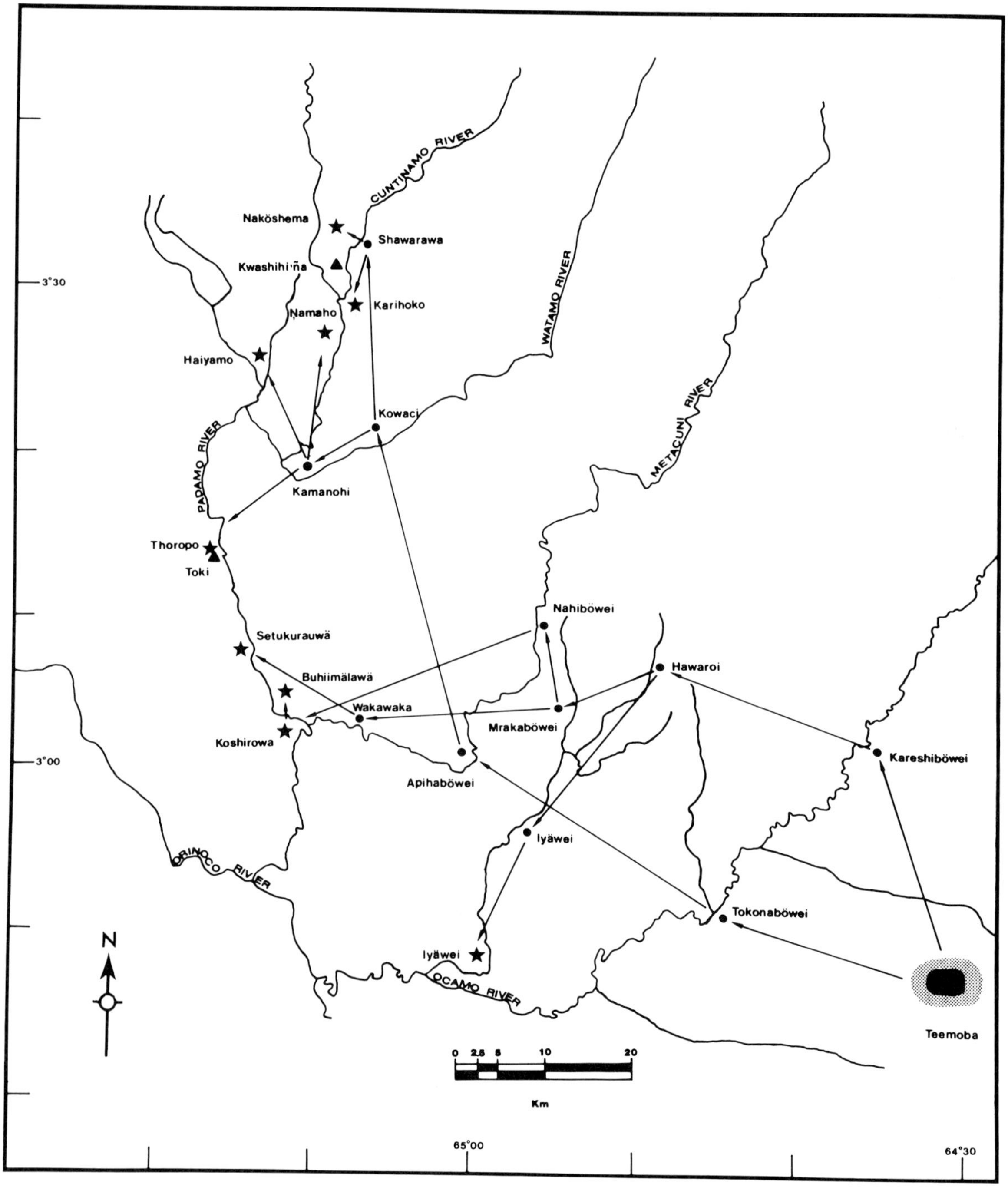

Figure 13.1 *Haiyamo population bloc settlement history (1920–1976). Dots indicate former Yąnomamö village sites, arrows show direction of movement, stars indicate current (1976) Yąnomamö villages, and trianges show Ye'kwana villages.*

methodological procedures followed and some of the problems of data verification and collection.

Crucial to the description of Yąnomamö settlement dynamics is the concept of a population bloc. A population bloc refers to a group of villages that share a common historical origin. A population bloc is the result of successive fissions and movements and growth of a single parental village through time (Chagnon 1974). To facilitate description, population blocs will be broken down into sub-blocs that represent groups of villages that have fissioned recently from a parental village, or that reside in the same general area.

MACROMOVEMENT HISTORY

In the upper-middle to lower segments of the Padamo River there are eight villages with a combined population of 412 (as of July 1976) whose ancestors largely originated from a single village, Teemoba, near the Maniviche River prior to 1920 (Figure 13.1). (An additional village, Iyawei, located near the Salesian Mission of Ocamo at the confluence of the Ocamo and Orinoco Rivers, is also issue of the ancestral village of Teemoba but will not be considered in this history.) These eight villages can be conveniently divided into three sub-blocs: (1) Wakawaka, which includes the villages of Koshirowa, Buhiimälawä, and Setukurauwä; (2) Kamanohi, which includes the villages of Toropo, Haiyamo, and Namaho; and (3) Shawarawa, which includes the villages of Naköshema and Karihoko (Figure 13.1). These sub-block names refer to former village sites where all members of each sub-bloc lived together before fissioning. Although these sub-bloc names are not autodenominations they are sometimes used by Yąnomamö outside the sub-bloc.

For the most part, villages of the Haiyamo population bloc are strung out along the Padamo River. Distance between villages ranges from 4 to 24 km with a mean of nearly 11 km. Average village population is 52, and ranges from 23 to 92 (Table 13.1). Although the residents of a single village share a common *shabono* (a circular compartmentalized lean-to), some villages have a more dispersed settlement structure. For example, the village of Toropo is divided into two roughly equal-sized houses separated by a 15-minute walk, and some families in Naköshema and Haiyamo have separate houses, in which they spend most of their time, 1–2 km from the main village. This pattern is similar to one described by Smole (1976:62) for the Barifiri population bloc in the Parima Highlands. Even though such families or groups of families live apart from the main settlement, they consider themselves and are considered by others to be members of the main village.

The members of the Haiyamo population bloc were pushed out of the

Table 13.1

Haiyamo Bloc and Neighboring Village Populations

Village	Population
Haiyamo bloc	
Koshirowa	92
Namaho	78
Haiyamo[a]	60
Karihoko	47
Setukurauwä	46
Toropo[b]	36
Buhiimälawä	36
Naköshema[c]	23
Bloc total	418
Bloc mean	52
Neighboring villages	
Iyawei	110
Rabai	28
Witokaya	86

[a]Village divided into two population segments of 20 and 16 people and separated by a 15-minute walk.
[b]Includes a small settlement of eight people about 1 km downstream.
[c]Includes a small settlement of nine people about 2 km to the north.

Ocamo River basin in the 1930s as the result of raiding from more powerful villages in the area. In the late 1930s they arrived in the mid-Padamo basin, which at that time was virtually uninhabited. The Haiyamo bloc represents nearly the westernmost extension of the Yą̧nomamö population and here their settlement pattern has been somewhat altered by contact with Ye'kwana inhabitants (a linguistically unrelated culture belonging to the Carib language family) and Protestant missionaries (New Tribes). Below, I will describe the history of macromoves of the Haiyamo population bloc which is schematically portrayed in Figure 13.1 and summarized in Table 13.2.

The village of Teemoba near the Manaviche River is the oldest place where senior members of all current Padamo River Yą̧nomamö villages agree that their ancestors once lived. Teemoba was occupied from about 1915 to 1920. During that time Teemoba villagers warred with ancestors of the present village of Mahekodo (see Lizot 1976:36 for the current location of Mahekodo) whose members, in a series of raids, killed approximately 10 Teemoba males and abducted a number of nubile females. These raids forced them to refuge temporarily in an area known as Heawi, and they then set up a new village in 1928, 26 km to the east at Kareshiböwei. Soon after their move to Kareshibowei, lessening of external threats and some internal dissension led to a fission. The ancestors of the present inhabitants

Table 13.2

Summary of Haiyamo Settlement History (Macromoves)

Old site	New site	Date	Distance (km)	Cause of movement
Kamanohi and Shawara sub-blocs				
Teemoba[a]	Tokonaböwei	1920	24	Raids from more powerful enemies
Tokonaböwei[a]	Apihaböwei	1930	31	Abduction of women and raids
Apihaböwei[a]	Modoshema	1940	42	Witchcraft and fear of raids
Modoshema[b]	Kowaci	1941	6	Poor garden site.
Kowaci[b]	Modoshema	1945	6	Intravillage conflict, leading to fission.
Modoshema[b]	Shawarawa	1952	20	Fear of raids and asked to leave by Ye'kwana.
Shawarawa[b]	Naköshema	1965	5	Peaceful fission from Karihoko.
Shawarawa[b]	Karihoko	1965	5	Peaceful fission from Naköshema.
Kowaci[c]	Kamanohi	1955	9	Fear of raids and lack of garden land.
Kamanohi[c]	Namaho	1958	14	Fission due to bickering and tension.
Kamanohi[c]	Haiyamo	1967	16	Wanted to be near to Ye'kwana for trade.
Kamanohi[c]	Toropo	1962	13	Living with Ye'kwana begins.
Shanama[c]	Toki	1970	5	Living with Ye'kwana.
Wakawaka sub-blocs				
Teemoba	Kareshi	1920	26	Raids from more powerful enemies
Kareshi	Hawaroi[d]	1928	27	Raids from more powerful enemies
Hawaroi	Mrakaböwei	1940	15	Fear of raids and abduction of women
Mrakaböwei	Nahiböwei[e]	1947	9	Fission due to constant bickering
Nahiböwei	Koshirowa[f]	1960	30	Raids from more powerful enemies and missionary influence
Mrakaböwei	Wakawaka	1949	25	Fear of raids and desire to be nearer allies
Wakawaka	Setukurauwä	1961	17	Sorcery and desire to be nearer allies and mission
Setukurauwä	Baiimi[g]	1976	?	On trek (*heniyomou*) to become lost to enemies
Koshirowa	Buhiimälawä	1970	7	Fission due to sorcery

[a]Ancestors to Shawarawa and Kamanohi sub-blocs living together.
[b]Ancestors of Shawarawa sub-bloc after fission from Kamanohi sub-bloc.
[c]Members of Kamanohi sub-bloc after fission with Shawarawa.
[d]Iyawei fissions in about 1935.
[e]Koshirowa leaves ancestors of present-day Setukurauwä and Buhiimälawä.
[f]Some members have moved in with Buhiimälawä and Setukurauwä.
[g]Some of Setukurauwä have joined Buhiimälawä for protection against Koshirowa.

of Haiyamo, Toropo, Namaho, Shawarawa, and Karihoko (see Figure 13.1) moved to a new garden site, Dokonaböwei, about 3 km distant, while the current ancestors of Koshirowa, Buhiimälawä, and Setukurauwä set up a new village structure at the head of their gardens, about 1 km distant. The fission was described by informants in terms of "we didn't have to be so big (populous)" or "we were too big."

THE WAKAWAKA SUB-BLOC

Constant demands for women and raids from Yąnomamö villages in the east caused the inhabitants of Kareshiböwei to move slowly approximately 27 km distant to the new site of Hawaroi between 1928 and 1930. They lived peacefully here for a number of years and the ancestors of the current members of the Iyäwei fissioned, relocating about 8 to 10 km to the southeast. For a time they had friendly relations with the nearby village of Auwei (today, Witokaya; see Lizot 1976:36 for the location) but these relations became embittered after a chest-pounding duel at a *reahu* (mortuary ceremony) got out of hand, causing several serious injuries (rocks were held in the fists). The duel apparently was related to a lack of reciprocity in the exchange of women in marriage.

As a result of poor political relations with Witokaya, the inhabitants of Hawaroi temporarily fused with their former co-residents at Dokonaböwei and then established a new village called Mrakaböwei high up on the Iyäwei-ko River about 51 km from Hawaroi. Although this river drains into the Ocamo, it is quite near the Metacuni River of the Padamo River basin. While at Mrakaböwei in about 1947, the ancestors of the current villages of Koshirowa and Buhiimälawä fissioned from the ancestors of the current village of Setukurauwä and settled in Nahiböwei about 8 km to the north. About 2 years later, those who remained at Mrakaböwei moved about 25 km west to Wakawaka.

The causes of the fission at Mrakaböwei are complex and subsequent bitter relations between Koshirowa and Buhiimälawä and Setukurauwä still exist today. Shaboa, the leader of the Koshirowa faction, believed that the Setukurauwä headman, Kurewä, was magically killing his agnates and affines and was much too aggressive toward their traditional enemies of Witokaya and Hakokoi. As a result, they slowly moved to the site of Nahiböwei about one-half day's walk upstream. The fission was not clean in that some personnel from both villages alternately lived with ancestors of the current villages of Koshirowa, Setukurauwä, and Buhiimälawä. At Nahiböwei they attempted to create an alliance with their traditional enemies, the more powerful allied villages of Auwei and Witokaya to the south. They invited members of Auwei and Witokaya to a *reahu* to accomplish the alliance but it ended in a bloody club fight which seriously injured several Witokaya men (the Koshirowa had refused to cede two sisters to a Witokaya man). Soon after, a large Witokaya and Auwei force returned and killed several Nahiböwei men and abducted the two sisters. As a result, in 1960 the Nahiböwei decided to move about 30 km away to their current village of Koshirowa at the confluence of the Metacuni and Padamo Rivers. There, a New Tribes missionary family had set up a small airstrip and a house, had moored a houseboat, and had tried for a number of years to lure the Nahiböwei. In 1960 the invitation was taken, due to their untenable political position.

The Koshirowa have lived at this site since 1961. As a result of the New Tribes mission's presence, traditional religious activities have all but stopped, *reahus* are held within the village instead of the usual practice of intervillage *reahus,* and most social contacts with neighboring related villages have ceased. In 1972, however, several people of one of the village's minor lineages died of malaria. The headman's lineage was blamed for the act, which caused a segment of the village to fission and to set up a new village, called Buhiimälawä, about 7 km upstream. In March 1976, 12 members of Koshirowa who were close kin to the fissioned Buhiimälawä village joined Buhiimälawä for the same reason, but this time they implicated the missionary in the witchcraft.

After the ancestors of the current members of Koshirowa and Buhiimälawä fissioned from the ancestors of the current members of Setukurauwä at Mrakaböwei and Wakawaka in the 1940s, the Setukurauwä people remained in the area for only a few years. During that time they lived in fear of *aworare* magic aimed at their children from a group of villages, collectively known as Kobari, high up on the Metacuni River. Moreover, they were weakened militarily after the fission and lived in fear of attacks from Witokaya and Auwei to the south. During that time they reestablished good relations with a number of villages occupying the middle course of the Padamo whose ancestors had co-resided with them at Teemoba in 1920 before fissioning. As a result, in 1961 they decided to leave Wakawaka and move about 17 km northwest, establishing the village of Setukurauwä near the banks of the Padamo. In July 1976 they were planning to fuse with Buhiimälawä in anticipation of attack from Koshirowa.

THE MIDDLE-PADAMO SUB-BLOCKS: KAMANOHI AND SHAWARAWA

As mentioned previously, the ancestors of the current villages of Toropo, Haiyamo, Namaho, Naköshema, and Karihoko fissioned from the Koshirowa sub-bloc around 1920 at Teemoba and settled at Dokonaböwei. They left Dokonaböwei in 1930 because demands for women were followed by raids from Madodoi to the southeast. They settled about 30 km further west at Apihaböwei from about 1930 to 1940. They were again threatened by Madodoi and Madodoi's new ally, Hakokoi, which forced them to flee about 40 km away to Modoshema, a small affluent of the Yatamo River that drains into the Padamo. They stayed there for about 2 years and left to establish a new village at Kowaci, about a half-day's walk away on the Yatamo. They left Modoshema because of the area's poor garden soil and fear of sorcery by the village of Yawarawä, located several days' walk to the north.

The village of Kowaci is critical to the history of the Kamanohi and Shawarawa sub-blocs. The village had a population of about 150 and was divided into two factions. One was headed by Kumauwä and the other by Oikomo, whose descendants dominate the political structure of the present villages of Toropo, Haiyamo, and Namaho, and Naköshema and Karihoko, respectively. Rivalry between Oikomo and Kumauwä was intense and it ultimately led to the shooting death of Kumauwä by Oikomo and his brothers. Evidently, his murder was well planned and the supporters of Oikomo immediately fissioned from Kowaci and reoccupied the old village at Modoshema upstream in 1945. This village was only several kilometers from the Ye'kwana village of Monatta'ña and Oikomo quickly arranged an alliance with the Ye'kwana to protect them from Kowaci by marrying two of his daughters to the Ye'kwana headman, Pedro Santiño. He also quickly allied himself with Yawarawä, a former enemy village.

In about 1955 the Ye'kawa grew tired of the Yąnomamö presence at Monatta'ña and asked Oikomo and his people to leave because they did not want to have any Ye'kwana mistaken for Yąnomamö by a Kowaci raiding party. Oikomo decided to move nearer another Ye'kwana group at Yudinama'ña to maintain the economic and political advantages of living near Ye'kwana and also to be nearer to the Yąnomamö of Yawarawä and Moyenawä, who were allied with them against the Kowaci. This new village was named Shawarawa and was located about 20 km from Monatta'ña.

At Shawarawa they were able to forge strong alliances with Yawarawä and especially with the village of Rabai (which was co-residing with the Ye'kwana village of Modeshiha'ña on the upper Padamo) through the exchange of women and frequent *reahus.* For about the next 15 years, political relations between Shawarawa and Kowaci were rather stable with little mutual raiding and some intermigration between them. A split occurred between Naköshema and Karihoko, both sons of Oikomo, over whether more peaceful relations should be established between themselves and Kowaci. Karihoko's faction wanted more peaceful relations so his people could descend the Padamo River to trade with missionaries without being attacked by the Kowaci. As a result, in 1965 Karihoko's faction fissioned peacefully from Shawarawa and moved about 5 km away.

Later, the village of Shawarawa was moved closer to the Ye'kwana village of Yudinama'ña, which had relocated several years earlier about 6 km north of the confluence of the Padamo and Cutinamo Rivers at Kwashi-ihi'ña. Currently, relations between Shawarawa and descendents of the Kowaci are very strained (more on this follows).

THE KAMANOHI SUB-BLOC

The present villages of Toropo, Haiyamo, and Namaho, which compose the Kamanohi sub-bloc (see Figure 13.1), maintain the most stable and powerful intervillage alliance in the Padamo region. Each village is dominated by a core of agnatic kinsmen who are descendents of the headman, Kumauwa, who was murdered by Oikomo at Kowaci. As a result of this murder, Oikomo's faction fled and his descendents now form the core of the current villages of Naköshema and Karihoko. This incident explains Kamanohi's solidarity and consuming hatred for the people of Karihoko and Naköshema.

After the 1945 murder of their headman, Oikomo, the Kowaci continued to live in Kowaci for about 10 years. As a result of diminishing horticultural land and threats from Yawarwä and Shawarawa to the north, they moved about 9 km downstream to Kamanohi, near the confluence of the Padamo and Yatamo Rivers. A dispute of unknown cause led the ancestors of the present village of Namaho to found a new village a short distance from Kamanohi for about 2 years and then set up the permanent present village of Namaho south of the confluence of the Padamo and Cuntinamo, 14 km distant, in 1958. The dispute was serious enough for the Namaho to re-establish friendly relations with Shawarawa, cemented by the exchange of women. But relations between Namaho and Shawarawa soon deteriorated as the Shawarawa headman, Naköshema, demanded more women and threatened violence. This led to the fusion of Namaho with their former co-residents of Kowaci at a new site of Toropo for a short period of refuge in 1963. In 1965 they returned to their old village site of Namaho when they felt they could withstand the military threat of Shawarawa (Shawarawa at this time had fissioned with one village, Karihoko, wanting to maintain friendly relations with members of the Kamanohi sub-bloc). The Namaho have continued to occupy the same general area ever since.

After the current members of the village of Namaho left Kamanohi, 3 years after it was founded in 1958, the current members of Haiyamo and Toropo moved near the confluence of the Yatamo and Padamo Rivers in order to be nearer to the Ye'kwana village of Nosamo'ña (today, Toki, see Figure 13.1) to gain more frequent contact with the Venezuelan malaria service (*Malarialogia*) and to gain trade goods more easily. In about 1964, relations between two factions in the village whose members compose the core of the current villages of Haiyamo and Toropo became strained because of constant bickering and being "too large." As a result, the Haiyamo faction moved about 10 minutes upstream on the Yatamo, occupying the abandoned village site of Modoshema in 1966. They never intended to stay at Modoshema long, and by 1967 they had begun to establish a new village near the confluence of the Yamapo and the Padamo, about 20 km away, which they called *Yamapo Ora Pabeha* ("mouth of the Yamapo River"). This group is known as Haiyamo and they have been living in the same general area since 1967 (as of 1976).

After the Haiyamo fissioned from Toropo in 1964, the remaining people began to establish closer ties with the Ye'kwana village of Nosamo'nã. But in 1966 the Ye'kwana left the immediate area to settle on the Padamo at Shanama'nã. One faction of the village decided to follow the Ye'kwana and co-reside with them at Shanama'ña. In 1967 the remaining faction at Toropo decided to rejoin their relatives at Shanama'ña because a measles epidemic struck and they could gain medical attention from New Tribes missionaries who were living with the Ye'kwana at Shanama'ña. In 1970 the Ye'kwana abandoned Shanama'ña because of flooding problems in the village and gardens and moved to their present site of Toki. The Yąnomamö of Toropo followed and they remain at Toki today.

CURRENT SITUATION

Relations between the Shawarawa and Kamanohi sub-blocs and within the Wakawaka sub-bloc were very tense when I left the field. In August of 1975, members of the Kamanohi sub-bloc invited those of Naköshema and Karihoko to a club fight at Namaho. Karihoko declined to attend, saying they had nothing to settle and desired to maintain good relations with Kamanohi. In addition, a sizable contingent from Iyäwei was also present. The fight was held and five people were killed. Members of each sub-bloc gave radically different versions of how the club fight led to arrows and shotguns, but each agreed that two men from Iyäwei and three from Naköshema (including the headman) were killed. After the killings, members of the Kamanohi sub-bloc and their ally, Iyäwei, mounted a raid against Naköshema which was to take them 3–4 days to the north where the Naköshema were refuging with their allies, Kobari (the Naköshema had abandoned their village site). The raiding party failed to locate Kobari and returned home. The Ye'kwana of Toki and Kwashiihi'ña have attempted to play the role of peacemakers because different warring factions are associated with each of their villages. They have relayed messages between the factions, confiscated shotguns from the Yąnomamö, and threatened to kill any Yąnomamö attempting a raid against the Yąnomamö who reside with them.

The villages of Buhiimälawä and Setukurauwä are pitted against Koshirowa. The core members of Buhiimälawä had lived with Koshirowa as late as 1970 but in 1975, 12 residents of Koshirowa who had kin ties with Buhiimälawä left Koshirowa. In both cases, those who left Koshirowa accused the core patrilineal groups of causing malaria. The co-residence of these two historically antagonistic groups for such a long period of time was probably a result of mission influence. Residents of Buhiimälawä and those who fissioned from Koshirowa in 1976 found Koshirowa attractive because of the missionaries stationed there. In any event, members of Buhiimälawä and Setukurauwä currently live in fear of Koshirowa. This has caused members of Setukurauwä to begin establishing a new village about 8 to 10 km inland and to the west for protection against Koshirowa. In addition, Toropo is considering joining them because the Ye'kwana no longer want them as co-residents because of Toropo's recent conflicts with the Shawarawa sub-bloc. Finally, some members of Buhiimälawä are planning to join Setukurauwä because of their fear of Koshirowa.

MICROMOVES

In the preceding section, movements of long distance, or macromoves, were found to stem usually from political causes, and rarely from factors relating to subsistence, according to informant accounts. However, macromoves do not exhaust all village movements. Yąnomamö villages are moved as little as 50 m–1 km every 4 or 5 years, largely because of the ecology of shifting cultivation. Chagnon (1973) has termed such movements "micro-moves." Factors such as nutrient quality, drainage, extent of soil, and topography limit the efficiency of gardening. It is possible that a decline in the efficiency of hunting and gathering may play a contributory role because the efficiency of these activities does decline with increasing time of residence (Hames 1980; A. Johnson 1974; Vickers 1980). Nevertheless, when informants were questioned closely about the role of wild-resource depletion as a cause of micromoves, they admitted depletion did occur but never cited it as a primary cause of a micromove.

Data on micromoves were collected in much the same way as data on macromoves; however, data on micromoves are much less accurate because micromoves are rather mundane events to the Yąnomamö and do not leave the lasting impressions that macromoves do. As a result, the data present a false trend of increasing frequency of micromoves in recent years.

When settlers occupy a new site, they first construct temporary shelters and clear the area for a garden. Permanent houses are then built directly next to the cleared garden, frequently no more than 30–40 m distant. Through time, gardens are abandoned and become successively more distant from the village, increasing the walking time needed to accomplish the daily tasks of weeding, planting, harvesting, or monitoring. As the gardens become more distant through time, roof thatch begins to rot and the village may become infested with cockroaches (*Blatta* sp.), chiggers (*Pulgans penetrans*), crickets (*Gryllus domesticus*), and other pests. When travel time to gardens, leaky roofs, and pests reach irksome proportions, the Yąnomamö simply decide to build a new village near the head (or, using the Yąnomamö metaphor, the *bei ka hushibo,* or "nose") of their newest gardens. Thus, in a single stroke they shorten travel time to gardens and create a more enjoyable domestic scene. This is the most basic and common process behind a micromove.

Table 13.3 lists the causes of micromoves for all members of the Haiyamo bloc from 1920 to 1976. It is obvious that informants attribute gardening-related problems as the most common cause for movement. It is probable that all subsistence activities decline in efficiency with increasing length of occupation. Although precise information is unavailable, micromoves undoubtedly increase the efficiency in gardening by shortening travel time. However, it is improbable that other subsistence activities, particularly hunting, are much enhanced by micromoves due to a shortening of travel time or movement away from depleted areas. Earlier, I showed that the maximum distance the Yąnomamö hunters of Toropo travel to hunt is 11 km (Hames 1980). If we generously assume that the average micromove is 1 km, and assume that a Yąnomamö hunting territory is roughly circular, an additional 23 km^2 of new hunting area will become available at no increase in travel time. Because game resources are radially depleted by the Yąnomamö, this new area will have the highest game concentrations but

Table 13.3

Causes of Micromoves for the Yaiyamo Population Bloc

Cause of move	*N*	%
Garden land too distant	12	35
Garden land of good topography used up	6	18
Well-drained soils used up	4	12
Deteriorated village structures	3	9
Poor soil[a]	3	9
Unattractive village site[b]	2	6
High levels of insect pests	1	3
More attractive stream nearby	1	3
Too many large and hard trees for easy garden-making	1	3
Unpleasant, ugly, or tiresome site	1	3
	34	100

[a]Crops yielded poorly on these soils, forcing a quick move.
[b]A muddy village site that took a long time to dry out during the rainy season.

will represent only 6% of the total hunting area. Therefore, in all probability micromoves do enhance hunting success (and perhaps gathering and fishing in much the same way), but not greatly.

Micromoves significantly enhance garden efficiency and this can be shown by comparing two segments of the Toropo population. Toropo is divided into two segments—Toropo 1 and Toropo 2—separated by a 14–17-minute walk. Toropo 1 was 6 years old when I arrived, whereas Toropo 2 moved to its current location 5 months prior to my arrival. The people of Toropo 1 were cultivating a single large, subdivided patch of forest which required a round-trip walk averaging 52 mintues, a result of 6 years residence in the same site. Toropo 2 had two of their gardens less than 20 m from their house, a third which cost them 7 minutes of round-trip travel, and one more costing 12 minutes. Given the rate at which each of these four gardens were worked, I estimate that round-trip travel time for Toropo 2 was 8 minutes, compared to 52 minutes for Toropo 1. Because data on clearing, weeding, planting, and harvesting were collected by instantaneous scan sampling (Altman 1974) and timing of travel was done by a focal-person method, only a rough estimate can be made of the average savings in total garden labor time from a micromove. Further complications arise when the ratio of travel time to in-garden labor time varies for each garden task. For example, harvesting of plantains takes no more than a few minutes, which would mean that travel time is more than in-garden labor for this task, while weeding, planting, and clearing can occupy the better part of the day, making travel time much less than in-garden labor time. Finally, in-garden labor time appears to increase as travel time increases. With these cautions in mind, I estimate that micromoves can decrease total garden labor by at least 14–18 minutes per day per adult, and frequently much more.

Although efficient gardening is the primary cause of micromoves, the political environment also plays a role. The risk of abduction for women and ambush for men increases as gardens become more distant from the village. Ambushes and abduction frequently occur early in the morning when people travel to their gardens to defecate or to harvest (Chagnon 1968a). This risk is of special concern for villages that are chronically raided. Although recent Haiyamo population bloc micromoves seem not to have been influenced by this risk, it was cited as a factor for movement more than 20 years ago. In some informant accounts, it was difficult to determine whether exposure to ambush and abduction or increases in gardening costs through long travel time was the more important cause of movement.

THE ALLIANCE SYSTEM

Data gained on the frequency of Yanomamo macromoves and micromoves indicate them to be relatively mobile village horticulturalists. But even while settled in a village, they engage in a great deal of intervillage visiting, for a variety of reasons, which gives their settlement pattern yet another dimension of structure. Along with intervillage marriage, intervillage visiting strongly reflects patterns of political alliance in a region. Quantitative data on intervillage visiting were gained through instantaneous scan sampling (Altmann 1974) or spot-checks (A. Johnson 1975). A total of 10,835 observations were made on all village residents during a 216-day period and 1618 observations were coded as visits to neighboring villages (Table 13.4). This means that the Yąnomamö of Toropo spent nearly 15% of their time, or 30 out of 216 days, visiting other villages. Eighty-three percent of these visits were to Iyäwei, Namaho, and Haiyamo, the most important political allies of the Toropo. Reasons for visits varied from purely social visits to kin (29%) to *reahus* (66%) which reaffirm political ties. Raiding was listed as a kind of "visit" and during my field stay it represented 3% of all Toropo visits, or about 1.5% of an adult male's total time budget. The main significance of the data on intervillage visiting is to drive home the fact that Yąnomamö settlement pattern cannot be understood except at the regional or intervillage level.

Table 13.4

Visiting Patterns of the Toropo Yąnomamö

Village visits	Frequency	% of all visits
Village visited		
Iyawei	824	50.9
Namaho	301	18.6
Haiyamo	219	13.5
Setukurauwa	120	07.4
Witokaya	75	04.6
Rabai	41	02.5
Kwashiihi'ña[a]	26	01.6
Koshirowa	4	00.6
Other	8	00.5
	1618	100
Reason for visit		
Social[b]	464	28.6
Reahu	1063	65.6
Joint raid[c]	52	03.2
Shamanic cure	29	01.7
Other	10	00.6
	1618	100

[a]Ye'kwana village (for location see Figure 13.1).

[b]An omnibus category which includes social visits to kin, trade, and some wage labor for the Salesian missionaries at Ocamo.

[c]Participation with Haiyamo, Iyawei, and Namaho in a raid against Naköshema and Kobari.

Discussion

In the beginning of this chapter I described some of the environmental factors that possibly account for variability in village movement, location, and size. I then described tbe data I had gained on these aspects of settlement pattern and the environmental factors that appear to account for them. It is clear that, given the nature of the data collected, I am unable to make any decisive tests of most of the hypotheses presented. Nevertheless, I am in a position to make some empirical generalizations about the nature of Yąnomamö settlement pattern in the region studied that have implications for Yąnomamö and Amazonian settlement pattern and that can be verified by future research.

Village movement, and hence village duration, is largely governed by local depletion of garden soil and by intertribal raiding. Macromoves, or movement induced by intertribal raiding, have caused the eight villages that currently compose the Haiyamo population bloc to move, over approximately 55 years, between 80 and 115 km (a mean of 99 km) in a northwesterly direction from their ancestral site of Deemoba (Figure 13.1). The mean distance of each macromove was 16.7 km, ranging from 5 to 42 km. These movements occurred an average of once every 9.2 years (see Table 13.2), with the shortest interval between macromoves being 1 year, and the longest, 16 years.

During its 55-year history, the villages of the Haiyamo population bloc made 18 macromoves, but the number of micromoves was almost double that amount at 34, a rate of one micromove every 4.8 years. Chagnon (1974) has pointed out that some villages are powerful enough not to have to make macromoves and they thus exploit the same general area through a series of micromoves for periods up to 40 years. In one sense, micromoves are not really moves at all because these moves are most frequently less than 1 km and, although new areas of forest are open to cultivation, most hunting, fishing, and gathering activities essentially occur in the same places.

Figure 13.1 shows that the eight villages of the Haiyamo population bloc have a strongly riverine orientation. With the exceptions of Buhiimälawä, Naköshema, and Karihoko, all are located on bluffs overlooking the Padamo River. And even those which are not located on bluffs are within a 1–2-k walk from a major river. This contrasts somewhat with villages in the Shamatari bloc, which tend to live in the interfluves (Chagnon 1974). The ancestral homeland of the Yąnomamö is the Parima highlands, an area devoid of large and navigable rivers which are easily fished. As the Yąnomamö grew and moved to lower elevations, they encountered larger and quieter rivers, which allowed them to exploit fish more intensively after learning the art of dugout canoe-making from the Ye'kwana Indians. This riverine orientation is also caused by a desire to have easy contact with missionaries, traders, and malaria eradication workers who use major waterways.

Intervillage spacing varies considerably among the Yąnomamö. As pointed out by Chagnon (1974), where intervillage warfare is intense, villages are widely spaced, and where it is moderate, villages are closely spaced. To support this correlation he shows that the average distance separating villages of the Namoweiteri bloc is 48 km and the rate of male mortality in warfare is 24%, compared to the Shamatari whose villages are 80 km apart and whose male mortality rate in warfare is 30% (Chagnon 1974:129, 160–161). Intervillage distance for the Haiyamo bloc ranges from 6.6 to 23.3 km, with a mean of 11.7 km. Although my demographic data are not as accurate as Chagnon's, I estimate male mortality in warfare to be no more than 12%. This estimate seems to support Chagnon's observation on the relationship between warfare intensity and intervillage distance. In other parts of the Yąnomamö tribal distribution there are few quantitative data on distance between villages and warfare intensity. Data gathered by Smole in the Parima Highlands seem to indicate that only 11–12 km separate most villages (Smole 1976:56, 76) and Smole has argued that Yąnomamö warfare in his area of study is much less intense than described by Chagnon for the Shamatari and Namoweiteri blocs (Smole 1976:232, n. 75).

While the Yąnomamö attempt to maximize the distance between themselves and enemies, they also attempt to draw as close as necessary to allies (Chagnon 1974, 1975; Lizot 1978b). Map data (Figure 13.1), descriptions of current political relations, and visiting patterns (Table 13.4) for the Haiyamo population support this empirical generalization. Deviations are almost entirely a result of missionary and Ye'kwana influence.

Crucial to any description of a locational system is a determination of whether local groups are territorial or have home-range systems (Brown 1975). Because warfare is a form which territoriality frequently takes, this question is all the more important for the Yąnomamö. Gross (1975) has argued that Yąnomamö warfare may lead to population redistribution, either to lessen pressure on game resources or to enhance hunting efficiency. In a similar but more complicated vein, E. Ross (1978) argues that territoriality is the ultimate cause of Yąnomamö warfare in particular and Amazonian warfare in general, and that its main effect is to create no-man's lands to prevent overexploitation of game (see also Durham 1976).

Brown's (1964) behavioral ecological model of territoriality postulates that territoriality will evolve when the benefits accruing to exclusive use of a territory are greater than the costs of defending it (see Dyson-Hudson and Smith 1978 for ethnographic applications). Operationally, territoriality occurs when resources are dense and predictable. On formal grounds, given that terrestrial game resources are sparse and somewhat unpredictable in Amazonia (Gomez-Pompa 1973), there is little reason to suppose that Yąnomamö warfare is a result of competition over scarce protein resources through maintenance of a territorial system.

On empirical grounds, disconfirmation of the protein theory of warfare is more difficult because of continual modification of its predictions (e.g., compare Harris 1974:vi to Harris 1979b), or because of predictions which are excessively vague, trivial, or which can be more easily accounted for by simpler models (Harris 1980:221). On the basis of long-term research into Yąnomamö warfare and demography, Chagnon has argued that Yąnomamö warfare has nothing to do with competition over scarce productive resources (Chagnon 1968a, 1974), and researchers, with the exception of Lizot (1978b:507), who have made in-depth studies of Yąnomamö hunting have found the Yąnomamö to be nonterritorial in respect to hunting areas (Hames 1980; Smole 1976;78–79). A key to testing the protein model would be a demonstration that warfare does not lead either to an eventual increase in protein consumption or to a reduction in hunting time that is more than sufficient to offset the costs of warfare (e.g., a 20–30% rate of adult male mortality). This direct test has never been made and would be extremely difficult to do.

An indirect way of evaluating the hypothesis (e.g., see Gross 1975) would be to observe the direction of a victorious village's movement after the defeat of an enemy. For the Toropo Yąnomamö, I (Hames 1980) documented that slightly better hunting areas exist midway between villages as a result of radial depletion of game (rates of return increase with distance from the village). The only way a victorious village could either increase its protein intake or lower the cost of protein acquisition after driving an enemy out would be to move to an area approximately midway between its current location and the defeated victim's former location. This hypothesis assumes that differences in game density are largely a result of hunting pressure.

I could not use most of the early historical data (prior to 1955) on the Padamo population bloc to evaluate the relationship between a victor's and a loser's movements because I was unable to identify precisely the location of all nonbloc neighboring villages. But more recent data (from about 1955 to 1976) can be used because the locations of all neighboring villages are known with some precision. As described earlier, and shown in Figure 13.1, the inhabitants of Kowaci fought and fissioned a number of times until they formed two separate villages (Shawarawa and Kamanohi) about 30 km apart. It is probably safe to conclude that hunting efficiency and/or game consumption may have improved for both groups as a result of this fission. In about 1965 Shawarawa fissioned peacefully into Naköshema and Karihoko. However, Karihoko most probably moved to an area where hunting was more difficult because the village was located nearly midway between (about 6 km from each) the Ye'kwana village of Kwashii'ña and the Yąnomamö village of Namaho (Figure 13.1). Naköshema probably gained little from its move because it remained about the same distance from the Ye'kwana village of Kwashii'ña. The important thing about these movements is that, even though there existed virgin hunting territory to the northeast and east of Shawarawa into which both Naköshema and Karihoko could have moved, one village moved into an area where hunting was probably more difficult and the other into an area where hunting was no better than in its previous location.

The fissioning patterns of Kamanohi are similar to Shawarawa. Starting in 1958 and ending in 1962, the creation of three villages (Haiyamo, Namaho, and Toropo) peacefully dissolved Kamanohi. By doing so, pressure on resources was probably lessened for each village. (The same may be said of the fissioning pattern of Koshirowa that led to the formation of Buhiimälawä and Sedukurawä [Figure 13.1].) When I left the field in July of 1976, all three villages were engaged in bitter hostilities against Naköshema, forcing Naköshema out of the area. The cost to one of these villages (Namaho) was the death of two men. But, as far as I could determine, their gains have been nil because they do not hunt in any of the areas formerly hunted by Shawarawa. Furthermore, the village of Karihoko (a former enemy) most directly depresses Namaho's potential game resources and would have been an ideal target if the function of warfare were to lessen the cost and/or to increase the consumption of terrestrial protein. Finally, if protein stress were a problem for any of the villages in the area, then eastern or western movement into virgin areas apparently would be a cheaper alternative than loss of human life.

The role of the Ye'kwana cannot be left out of the Padamo River Yąnomamö settlement pattern. The Ye'kwana villages of Toki'ña and Kwashii'ña have satellite Yąnomamö villages associated with them (Hames 1980). In other areas of Ye'kwana–Yąnomamö contact (Chagnon *et al.* 1970; Ramos 1978; Taylor 1976), Yąnomamö are frequently used as laborers by Ye'kwana. Yąnomamö live near Ye'kwana in the Padamo in order to gain easy access to trade goods from the Ye'kwana, and goods and medical attention from visitors to the Ye'kwana (missionaries, medical personnel, and environmental sanitation personnel). Yąnomamö who live near Ye'kwana villages are of little economic value to the Ye'kwana because little exchange occurs. Their main benefit is as a source of wives for the Ye'kwana (Chagnon *et al.* 1970; Hames 1978). Yąnomamö presence undoubtedly depresses Ye'kwana hunting efficiency (Hames 1979, 1980) and, if they wished, by using their overwhelming military superiority (Arvelo-Jimenez 1971; Chagnon *et al.* 1970; Hames 1978), the Ye'kwana could simply demand that the Yąnomamö leave the vicinity. But this expulsion has never occurred.

The data indicate that village fissioning probably leads to better hunting efficiency for the departed segment because they move into an area of higher game abundance. But the immediate costs of movement are high because an extraordinary amount of time and energy must be spent in constructing a new village, clearing new gardens, and transporting heavy planting stock (e.g., plantain ratoons). The benefit for the segment that remains is an immediate lowering of pressure on all resources. Although these are obvious effects of fissioning, one cannot conclude that fissioning is designed primarily to achieve these effects. An important research question that needs to be answered in quantitative terms is: Which village segment gains the most economically—those members who remain or those who depart? As villages grow large, not only does the efficiency of resource exploitation decline, but the frequency of intravillage competition for mates increases (Chagnon 1975, 1979a). Fissioning reduces the problems of mate competition, especially for the politically weak segments who are forced to leave. Therefore, village fissioning reduces resource and mate competition and the difficulties associated with each. The economic consequences of warfare are more difficult to assess. Warfare between nonadjacent villages may force the loser out of the area, but no economic benefits will accrue to the winner because the winner will not be able to exploit the loser's territory or to relive any pressure on local resources. This result is clearly the case in the conflict between Naköshema and Namaho. If Yąnomamö warfare were a matter of competition over scarce game resources, then minimally one would expect fighting to occur between neighbors, as in the case of Highland New Guinea warfare (Meggitt 1977). In this situation winners would have something to gain.

Groups that rely heavily on fish for dietary protein are the best candidates for the development of territoriality. In some parts of Amazonia, fish are highly concentrated in oxbow lakes (Stocks this volume) and other favored habitats, such as back swamps (Bergman 1980). Fish tend to be highly predictable resources owing to their vulnerability in the dry season and regularity of movement (Goulding 1980) and rivers are easy to monitor for poachers. Thus, they represent in particular habitats a dense and predictable resource that appears economically defendable, according to Brown's model (Brown 1964). In keeping with the model, reports of territoriality over fishing areas have been made by Goldman (1963) for the Cubeo, Butt-Colson (1972) for the Akawaio, Dreyfus (1972) for Takanoans, and Hurault (1968) for the Boni and Wayana. All of these groups depend on fishing for their main source of dietary protein. (But see Vickers [this volume] for an account of territoriality for the Siona–Secoya who depend on game for their protein supply.)

Because the important resources upon which the Yąnomamö depend are relatively evenly and thinly distributed, one would expect the Yąnomamö to live in small and evenly dispersed villages. Large village size in the Yąnomamö habitat creates high levels of intravillage competition, leading to rapid resource depression and lowered rates of return in subsistence activities (Horn 1968). However, another environmental factor, warfare, leads to the creation of villages large enough to defend themselves or to engage in offensive activities (Alexander 1974; Chagnon 1968a). These two factors pull in opposite directions: (1) strain toward small villages, which enhances resource accrual efficiency; and (2) pressure toward large villages to adapt to chronic warfare. The interaction of these two factors results in optimal village size.

Yąnomamö village size varies considerably on a regional basis. Villages of the Shamatari and Namoweiteri population blocs average 158 and 134 people, respectively (Chagnon 1974:136, a sample of 10 villages), whereas two population blocs of what Lizot refers to as the "Central Yanomamo" average 59 people per village (Lizot 1978b:512, a sample of 35 villages). Mean village size of the Haiyamo bloc is 52 (Table 13.1) and is 44 for the Shiitari bloc of the upper-middle section of the Ocamo River (Fredlund, personal communication, a sample of 7 villages). For the rest of the Yąnomamö tribal distribution, mean village size appears to range between 40 and 60 people (Migliazza 1972; Smole 1976). Chagnon (1968a, 1972, 1974) has argued that village size increases with the intensity of warfare. With male mortality as a measure of warfare intensity, Chagnon shows that the Namoweiteri have a mean village size of 134 people and male mortality is 24%, whereas the Shamatari have a mean size of 158 people and a mortality rate of 30%. The data I have on Haiyamo male mortality in warfare are much less accurate and refined than the data collected by Chagnon, but male mortality seems not to have been greater than about 12%. Other researchers who have worked in the Parima or in Brazil among the Yąnomamö indicate that warfare is less intense than Chagnon described for the Shamatari and Namoweiteri populations (Ramos 1979; Ramos and Albert 1976; Smole 1976; Taylor 1976), and the villages they studied ranged from 40 to 60 persons or even smaller (e.g., Saffirio and Hames n.d.).

Village fissioning is a regular aspect of Yąnomamö settlement pattern. According to the Yąnomamö, many but not all fissions are attributable to intravillage conflicts (see Chagnon 1968a, 1974, for the structure of village fissioning). Although political reasons are given for village fissioning, there may be a strong underlying economic and ecological dynamic (Chagnon 1975). As I pointed out above, the only large advantage to large village size among the Yąnomamö and other interfluvial populations is in warfare, and where warfare is low, village size is small. Village size beyond the optimal for exploiting resources will lead to intravillage competition (in the economic sense) for agricultural land and wild resources, which either lowers consumption or increases labor costs, or both. The simplest solution to this problem is to leave areas of high resource competition and to move into areas of low resource competition. Because resources, particularly garden land, are evenly distributed, those who split are in a sense making a micromove to improve resource accrual efficiency. Characterizing fissions as a kind of micromove is plausible when one realizes that some fissions are often the same distance as micromoves.

One of the most remarkable characteristics of the Yąnomamö has been their rate of population growth, at least until very recently. Chagnon (1974) shows that the Shamatari and Namoweiteri populations have sustained a growth rate of about 1.5% over the past 40 or 50 years, and Lizot (1978b) makes nearly the same estimate for two central Yąnomamö blocs between 1900 and 1950. (Research indicates that Yąnomamö population growth has slowed in some areas [Chagnon 1979a] or has been negative [Lizot 1976] in recent years because of increased incidences of malaria and white-introduced diseases such as influenza and measles [Chagnon and Melancon 1981]). Again, my data on the Haiyamo bloc, though inferior to those collected by Lizot and Chagnon, suggest that its rate of growth is about the same as documented elsewhere for the Yąnomamö. In 1950, James Barker estimated the population of Wakawaka (now Koshirowa, Buhiimälawä, and Setukurauwä) and Kowaci (now Toropo, Haiyamo, and Namaho) to be 100 each (Barker 1953), but he omitted the village of Shawarawä (today Karihoko and Nokoshema), which I estimate had a population of about 70. Today, after a period of 25 years, the population of these villages is 412. Although I suspect that Barker's estimates are a bit low, the data indicate a growth rate comparable to rates documented by Chagnon and Lizot.

Table 13.5

Some Estimates of Yanomamö Population Density in Various Regions

Location	Area (km^2)	Population	Density (persons/km^2)	Source
Central Parima highlands	—	—	2.31	Migliazza (1972:18)
Southwest Parima headwaters	1036	800	0.77	Migliazza (1972:18)
Upper Buuta and Mahekodo rivers	2331	1000	0.42	Migliazza (1972:19)
Namoweiteri (mid-Orinoco)	—	—	1.44	Chagnon (1974:127)
Shamatari (mid-Mavaca River)	—	—	0.67	Chagnon (1974:127)
Mid-Padamo[a]	2201	590	0.27	Chagnon & Hames (1979:911)
Central Yanomamɨ	6082	2068	0.34	Lizot (1978:503)
Barifiri	647	84	0.18	Smole (1976:77)
Serra do Surucucus	20,106	3800	0.19	Ramos and Taylor (1979:108)
Upper Cauburi River	5439	250	0.04	Migliazza (1972:19)
Paragua and Uraicaa headwaters	6216	250	0.04	Migliazza (1972:19)
Total[b]	44059	8842	0.20	

[a]Includes Ye'kwana residents of the area.

[b]Total and average calculated from sources which contained data on area and population only.

This remarkable rate of population growth by the Yąnomamö is the best evidence that they are not severely limited by subsistence resources. The aboriginal homeland of the Yąnomamö population lies in the Parima Highlands of Venezuela (Chagnon 1974; Lizot 1978b). Their present pattern of expansion south, southwesterly, and southeasterly was probably made possible by a demographic vacuum caused by the decimation of Carib and Arawakan populations along the Orinoco and its major tributaries, due to white-introduced diseases (see Civrieux 1970; Schomburgk 1840), for some early accounts of Yąnomamö expansion). Table 13.5 presents available data on Yąnomamö population density. Some of the data are gross estimates and will be modified with future research; nevertheless, they indicate that Yąnomamö population density is only 0.2 people/km^2, which is identical to the estimate made by Denevan (1970:82) for Amazonian *terra firme* populations. Although population density figures are not very useful without ecological and economic data, the data on subsistence that we do have (Hames 1979, 1980; Lizot 1978a; Saffirio and Hames n.d.) argue against suggestions that Yąnomamö warfare is attributable to competition over scarce subsistence resources or that increasing environmental resistance (Pianka 1978:113–120 is likely to slow the rate of growth in the near future.

Conclusion

Sahlins (1964) perceptively pointed out that a culture adapts not only to the physical and biological components of its environment, but also to its human components. He argued that neighboring settlements are important factors to which a settlement must adjust. Despite this observation, the trend in cultural ecology has been to focus on various forms of subsistence adaptation while paying scant attention to the effects of neighboring settlements. I have tried to demonstrate here that Yąnomamö settlement pattern can be understood only in terms of both human and nonhuman factors in the environment. This process of adaptation occurs at the intravillage level where, for example, co-residents can either cooperate in mutually beneficial ways to exploit the environment more efficiently or unite at the intervillage level to defend their women against abduction by enemy villagers.

Another trend in cultural ecology has been the assumption that adaptation, competition, and limiting factors relate to subsistence or material objects, which enhance survival or system stability. This overemphasis on survival runs counter to an evolutionary ecological concept of adaptation. Successful adaptation to the environment is measured in terms of reproductive success. Obviously, one must survive to adulthood to reproduce but survival without reproduction (or enhancement of kin's reproduction [Hamilton 1964]) is not adaptive. Studies of Yąnomamö labor and diet (Hames 1979, 1980; Lizot 1978a, 1978b; Saffirio and Hames n.d.; Smole 1976) and health (Neel 1970) indicate the Yąnomamö survive quite well. Moreover, they reproduce even more successfully (Chagnon 1975; Neel and Weiss 1975). Data on Yąnomamö labor, diet, health, and reproduction strongly indicate they are encountering little environmental resistance. Yąnomamö adaptive performance is made possible by living in a low-density environment and by having a settlement pattern that distributes them over time and space, allowing them to take advantage of environmental opportunities, and reducing the hazards of competition from other villages.

One of the key issues in Amazonian research has been the debate over the causes of Yąnomamö warfare. It has been argued that Yąnomamö warfare is an outcome of competition over a limiting resource. I agree. Evolutionary ecology defines a limiting resource as any factor in the environment which, if not gained in sufficient quantity, will cause a decline in fitness. But the resource over which the Yąnomamö compete is not the kind that indirectly leads to fitness, such as animal protein, but is something much more direct: namely, mates (Chagnon 1979a, 1979b). Cultural ecology made a great theoretical leap forward when it admitted that the stomach was an adaptive constraint and now, if it goes a little lower and admits reproduction, even more progress will be made (Chagnon and Hames 1980).

Acknowledgments

Funds for fieldwork upon which this chapter is based came from the National Institute of Mental Health, Grant No. NIMH 5 R01 MH 26008-SSR (N. A. Chagnon, principal investigator); and a generous grant in the form of a postdoctoral fellowship from the Harry Frank Guggenheim Foundation gave me time to write it up. I would like to thank Ilene Levin Hames and Bill Vickers for their careful criticism of the manuscript and John Haddidian for the rendering of Figure 13.1.

CHAPTER 14

Village Movement in Relation to Resources in Amazonia

Daniel R. Gross

Introduction

This chapter has two purposes. The first will be to review the debate concerning the role of resources in determining settlement patterns in Amazonia.[1] The second will be to present new data that bear on this subject. I shall record my point of view from the outset: the extensive human use of soils, plants, and animals leads to steadily rising labor inputs for each unit produced, especially for food. Under extensive management systems, the increased labor time cost of subsistence production provides a strong incentive for groups to keep their settlement density low and to move about frequently (5–20-year intervals). This argument has been advanced with regard to animal protein capture (Gross 1975), garden productivity (Meggars 1954), and other resources (A. Johnson 1977), and it has drawn much critical response.

Since 1975 there has been a lively debate and a healthy amount of new data generated concerning the relationship between settlement and animal protein capture(Beckerman 1979, 1980; Chagnon 1980; Chagnon and Hames 1979; Hames 1980; A. Johnson 1980; Lizot 1977; Nietschmann 1980; Roosevelt 1980; E. Ross 1978, 1980; Spath 1981; Vickers 1975, 1980; Werner *et al.* 1979). I was recently accorded the opportunity to respond to some criticisms and to react to the new data (Gross 1982), so I shall not deal at length with this topic here. Some of the evidence has been negative with respect to the protein hypothesis, whereas other data support my views.

ADAPTIVE RESPONSES
OF NATIVE AMAZONIANS

ISBN 0-12-321250-2

Some new ways of evaluating the question have also emerged, for example, optimal foraging models.

Although strides have been made, there still are issues requiring clarification and data to be collected. One problem that has arisen is that the "environmental limitations" viewpoint (or simply the "environmental" position) has been confused with a number of other positions. Some impassioned arguments have been made against assertions that were never made. The ecological approach has been referred to as deterministic, reductionist, and incapable of dealing with all aspects of society. I will therefore devote a few pages here to clarify what I see as the central theoretical and methodological issues.

Some of the objections to arguments based on environmental limitations raise the issue of determinism. To argue that protein or agricultural productivity is limiting on population density and the permanence of settlements *is* deterministic. Any explanation is deterministic when it is constructed in such a fashion that the phenomenon to be explained can be logically deduced from the explanation, that is, from statements setting forth the necessary and sufficient conditions of the phenomenon. In the present analysis, the dependent variables are the size, density, and permanence of native Amazonian settlements. The form of the explanation is a combination of general probabilistic statements plus a set of statements describing specific conditions that jointly permit the general phenomenon to be deduced. In other words, the phenomenon is seen to be a particular case of a more general set of regularities.

Reductionism means different things to different people. An argument that seeks to understand a phenomenon as a specific case of a more general class of phenomena is one sense. The dispersion of Amazonian villages is understandable in terms of general ecological law concerning energy flow and nutrient cycling that apply to other species populations as well as to humans. This does not, in my view, transform human behavior generically to animal behavior, nor will the general laws to which we appeal explain all we might want to know about human settlement patterns. The only claim is to explain part of the variation in settlement patterns in terms of a general principle. Anthropology is not in danger of being swallowed up by ecology any more than psychology is going to be consumed by neurology.

Another issue that has been raised is "monocausal explanation." Some critics argue that to explain sociocultural differences and similarities in terms of the availability of protein amounts to single-factor explanation, and they state a preference for multifactorial explanations (e.g., Vickers 1980). Looking for the influence of one variable on a phenomenon is not to deny the influence of other variables. I am sure that factors other than resource availability affect settlement patterns in Amazonia and I would be happy to see them investigated. It is possible to acknowledge the complexity of a phenomenon and still to attempt to account for part of the variation in terms of a single variable or a small number of variables. The investigator need not decide in advance how many variables are at work, but can test rival hypotheses utilizing greater or smaller numbers of variables. Problems arise only if two hypotheses appear to account equally well for a phenomenon. There is a convention in science that states that, in such cases, the simpler hypothesis is to be preferred.

A final issue that disturbs some critics of the environmental position is that it cannot explain the specific content of culture in Amazonia. This objection presumes that it is the investigator's purpose to explain everything and that adequate explanations will be "complete." My explanation addresses itself to a limited set of questions and does not purport to explain such matters as cosmology, kinship nomenclature, nonkin sodalities, household organization, etc.

Lest I appear to be claiming too little for my approach, let me state that I focused on settlement patterns because I think they are intermediate variables that have an important influence on other features of culture. Settlement patterns influence both the elaborateness and complexity of social organization. By elaborateness, I refer to the "horizontal" proliferation of social units and kinds of formalized social ties that do not involve hierarchical power relationships. To illustrate, the size and permanence of settlements interacting with other variables, such as fluctuation in resource base and the nature and incidence of warfare, affects the presence or absence of nonkin sodalities that link settlements, kin groups, and other units of the society (see Gross 1979; Ritter 1980). By complexity, I mean social institutions that involve hierarchical power relations, ranging from centralized decision making to socioeconomic stratification.

But, to repeat, the specific content of cultural institutions is not explained by the ecological requirements they fulfill. The content is the result of a selection process in which specific features of an existing cultural repertoire are conserved or amplified, while others are not. Some features may be conserved although they are adaptively neutral. These may be determined through an investigation of the unique historical influences to which a society was exposed. It is somewhat distressing, however, to see attempts at explaining some aspect of sociocultural variation compared to detailed ethnohistorical accounts. Even the best examples of ethnohistory do not achieve completeness in the sense of thoroughly covering a conventional inventory of "notes and queries."

Turning now to Amazonia, some critics have approached the environmental limitation position by arguing, in effect, that the phenomenon that the environmental limitation hypothesis seeks to explain is an artifact of a set of historically specific events. If this were so, the environmental explanation would be undermined because it would no longer be relevant. Beckerman (1979) suggests that preconquest Amazonia was more densely populated than most estimates imply. He bases this claim mainly on reports from early travellers into the Amazon Valley and he discounts the testimony of later observers because, he suggests, massive depopulation from disease and massacre occurred very early. Beckerman cites evidence relating to the very high death toll from introduced disease and warfare early in the sixteenth century.

The major problem with Beckerman's argument is that the evidence he cites refers to only a small part of the region. Much of the information that supports the claim of high preconquest population densities comes from observers who travelled along the major rivers, primarily the Solimões and the Amazon. Virtually all students of the Amazon agree that there were densely populated native groups with complex structures living along the major streams of the region, and there is a fair amount of archeological evidence to support this (Hilbert 1962a, 1962b; Meggers and Evans 1957; Simões 1974). The test of Beckerman's suggestion must come from upland areas away from the major rivers. Unfortunately, to date, there has been very little systematic archaeological or ethnohistorical research on population size in these areas, which constitute 95% of the land area of Amazonia. The remainder of Beckerman's approach consists of an attempt to demonstrate that there were many alternative sources of protein in the tropical forests other than fish and game. I have discussed this evidence elsewhere (Gross 1982).

Central Brazilian Settlement Pattern

In this section, I shall present some data based on recent research among four native groups of Central Brazil that relates to the issues just presented. The groups are the Xavante (also spelled "Shavante"), Bororo, Timbira and Kayapó all of which are part of the so-called Gê–Bororo language stock of central Brazil. They are already well known ethnographically (Bamberger 1974; J. C. Crocker 1969a, 1969b; W. Crocker 1972, 1974, 1979; Dreyfus 1963; Maybury-Lewis 1967, 1979, Melatti 1972, 1973; Nimuendajú 1939, 1942, 1946; Vidal 1977; Viertler 1976). The research conducted by myself and my associates focused mainly on relationships between these groups and the environment (Flowers this volume; Flowers *et al.* 1982; Gross *et al.* 1979; Werner this volume; Werner *et al.* 1979). I shall present some data relating to the movement of village locations by one population unit of each of the four ethnic groups just named. The data were obtained by four ethnographers working as a team, each of whom resided in a single village for approximately 12 months from early 1976 to early 1977. Most of the data come from informants's accounts, and a small part of it has been obtained from the work of other ethnographers.

The reader is referred to some of the literature cited above for ethnographic and ecological information about the four societies in our study. All four groups are relatively unacculturated Indian groups who speak a native language (exclusively in the case of the Mekranoti Kayapó and most of the Xavante at Pimentel Barbosa). The population and other details about each group are given on Table 14.1 and their location is indicated on Figure 14.1. Of the four groups, the Kanela (the Timbira group studied) and the Bororo at Posto Gomes Carneiro have been in direct contact with Brazilians the longest, are the most closely circumscribed by Brazilian settlements, are the most heavily involved in commercial exchange, and have placed the greatest pressure on their resources. Game is scarce around both the Bororo and Kanela villages, but the Bororo exploit the relatively plentiful fish in the São Lourenço River adjacent to their village. The Xavante and the Kayapó are relatively less acculturated and commercialized, live in relative isolation from Brazilian settlements, and rely more heavily on wild game resources. All four groups rely on swidden horticulture for most of their food and on forest products for many of their tools, utensils, and most house construction materials. All four groups wear native or modified western clothing, and all four actively practice indigenous ceremonialism and other religious activities. Although all have been exposed to missionaries, none of them has any converts to Christianity.

It is a commonplace in lowland South America that native villages move about frequently. Some observers take it for granted that these migrations are due primarily to the local exhaustion of resources (e.g., Meggers 1971), whereas others focus on the internal and external social relations of village

Table 14.1

Basic Data on Four Central Brazilian Village Communities[a]

Community name	FUNAI[b] post name	Language (family)	Size of reservation (ha)	Habitat description	Principal crops
Mekranoti	Mekranoti	Kayapó (Gê)	Not delimited	Evergreen forest	Manioc, sweet potato, banana, maize
Xavante	Pimentel Barbosa	Xavante (Gê)	205,000	Cerrado	Rice, manioc maize
Bororo	Gomes Carneiro	Bororo (Macro-Gê)	19,000	Cerrado/pantanal	Rice, maize, manioc, beans
Kanela	Ramkokamekra	Timbira (Gê)	225,000	Cerrado	Manioc, sweet potato, rice, maize, yam, beans

[a]See also Flowers *et al.* [1982]; Gross *et al.* [1979]; Werner *et al.* [1979].
[b]Fundação Nacional do Indio (National Indian Foundation).

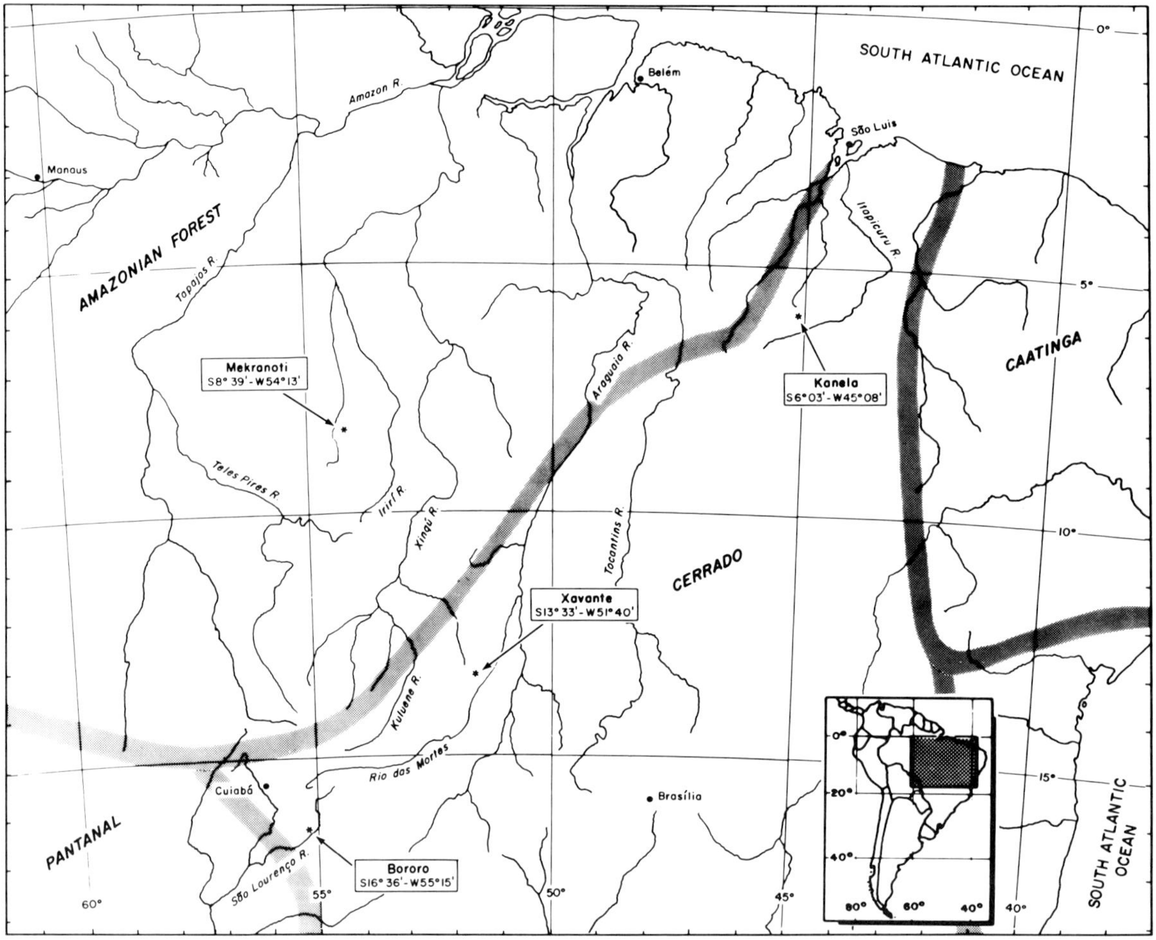

Figure 14.1 *The study area. Shaded lines indicate boundaries between major ecological regions. (Gross* et al. *1979; copyright 1979 by the American Association for the Advancement of Science.)*

populations in order to understand these movements (e.g., Chagnon 1973). In a few instances, investigators see these two approaches as contradictory. I do not see any necessary conflict between an approach that stresses the social and motivational aspects of village movement and one that looks for explanations in the rates at which resources are depleted. Until now, there have been very few systematic studies of this question (but see DeBoer 1981; Meggers and Evans 1979) in spite of its obvious importance to a number of leading issues in South American studies.

The reasons for this neglect are not hard to find. It is no easy task to follow the movement through space and time of small, native villages in the vastness of Amazonia. The ephemeral nature of villages and of the very materials of which they are constructed make them difficult sites for archeological studies. Ethnologists, on the other hand, find it difficult to trace the wanderings of villages for more than a few dozen years into the past. Groups vary in the way they designate villages: some use place-names but have no way of referring to the population aggregate at a given settlement location. For others, the situation is the opposite. Sometimes groups are referred to under a specific name, which may be that of a moiety or a men's society, but which includes many persons who are not members of its namesake unit.

Along with the relatively coordinated movements of entire villages or sizeable segments of villages, there is also the uncoordinated trickle of individuals from place to place. Groups split into two or three smaller units, the segments joining other groups, only to separate from them later, reconstituting an earlier population unit, perhaps at another location. Informant accounts may vary widely, and often the field worker is obliged to choose among conflicting accounts without a firm basis for doing so. Written documents are available for some areas, but they are subject to similar sources of error. All of these problems were encountered in coordinated field studies among four village societies in central Brazil.

For the present study, I am taking the data relevant to a single, more-or-less continuous village population reconstructed from about 1900 to the present (Tables 14.2 and 14.3). Recognizing differences in the habitats, culture, language, social morphology, and other features of the four populations, I am primiarly concerned here with differences that may be related to the degree of environmental circumscription and habitat degradation. In each case, the village units moved a dozen or so times from 1900 to the present, split into two or more subgroups, often converging, fissioning again, only to reunite once more. Where subgroups continued to maintain their autonomy, I attempted to follow their itineraries as well.

Table 14.3 presents data on village moves. Any migration over any distance involving 20 people or more, and that involved the construction of houses for year-round use (i.e. not as a temporary shelter such as those built on hunting trips), was considered to be a village move. In a few cases, the criterion of building new houses was dropped, as when 20 or more people were absorbed into a preexisting village, moving into houses already constructed.

The codings are based on informants' firsthand accounts of moves that took place in the past. For each move, at least one informant's opinion was solicited as to why a move had occurred. Where possible, more than one informant was consulted. Often, informants gave more than one reason for a move. Given the remoteness in time of these events, the unreliability of informants' accounts and overt conflicts of opinion, how are we to determine the true reasons for the movement of a village? It is possible to

Table 14.2

Reasons Given by Informants for Village Moves since 1900 in Four Central Brazilian Villages

	Mekranoti		Xavante		Bororo		Kanela	
Reason for move	*N*	%	*N*	%	*N*	%	*N*	%
Unknown[a]	11	31	5	21	6	50	6	24
Internal dispute[b]	9	25	2	8	0	0	5	20
External attack[c]	4	11	6	25	2	17	2	8
Illness or epidemic[d]	5	14	0	0	2	17	5	20
Government or missionary inducement[e]	2	6	4	17	1	8	4	16
Trade, marketing, presents[f]	3	8	6	25	0	0	0	0
Resource scarcity[g]	0	0	0	0	1	8	2	8
Decrepitude[h]	2	6	1	4	0	0	1	4
	36	100	24	100	12	100	25	100

[a]Reason for move could not be determined.
[b]Arguments, fights, murders, witchcraft accusations, etc. within the village and among villagers.
[c]Armed attack or the threat of one by outsiders (Indians or not) against the village.
[d]Outbreak of serious illness among several people or belief that the place was unhealthful.
[e]Forced settlement, "attraction" by FUNAI, provision of medical care, protection or other service from a government, military or missionary outpost or coerced settlement.
[f]The offer or actual transfer of gifts, payments for native products, employment, etc. to the Indians by outsiders.
[g]Shortages or difficulty in acquiring game, food, garden land, firewood or other native resource.
[h]Presence of vermin, deterioration of houses, or their loss to fire.

suppose that many moves took place when several different individual or factional motivations converged. Even where the villagers' opinions are unanimous that, say, sorcery was the cause of a factional split, we need not accept the opinions of the actors. We may doubt that sorcery was practiced or that it had the effects it was alleged to have had. It is also possible that the actors deliberately or unconsciously conceal the causes of a move. In the Xavante village studied by Nancy Flowers, for example, villagers frequently disclaimed knowledge of the whereabouts of a certain clan formerly represented in the village. She suspects that all the members of the

Table 14.3

Type of Move since 1900 in Four Central Brazilian Villages, According to Informants

	Mekranoti		Xavante		Bororo		Kanela	
Kind of move	*N*	%	*N*	%	*N*	%	*N*	%
Fission	9	25	9	38	4	33	9	36
Fusion	9	25	5	21	2	17	7	28
Entire village moved	18	50	10	42	6	50	9	36
	36	100	24	100	12	100	25	100

clan were killed or driven off by the leading village faction. In sum, we need not seek unanimity of motivation among informants.

Upon inspecting the data, perhaps the most remarkable feature is how comparable the cases are, even when we compare the extremes on a scale of acculturation (Gross *et al.* 1979). The Kayapó village known as Mekranoti is the least acculturated of the four, situated in an isolated area of tropical forest west of the Xingu River. The Mekranoti practiced active warfare until the late 1960s and have been in permanent contact only since 1966. The Kanela, on the other hand, live in the cerrado region with its open, dry forest and scrub and have been in permanent contact with Brazilians for well over a century, and are involved in the regional economy. Each group moved several times since 1900: the Mekranoti, at least 35 times; the Kanela, 25 times (Table 14.2). The proportion of moves attributed to internal disputes is about the same for both groups. Other percentage values for different motives are roughly comparable. Thus, in spite of the differences among the groups, the data suggest that the same kind of actor considerations lie behind village migrations.

Another salient point is the importance of conflict (both internal and external) as a motive for village movement. Under "internal disputes" I included physical violence or the threat of it, sorcery, and adultery accusations. External fighting was coded whenever a village was raided by another village, or when a move was precipitated by fear of such an attack. Attacks by Brazilians were also included in this category. The combined frequency of these two motives is over a third for the Mekranoti and Xavante and somewhat less for the Bororo and Kanela (Table 14.2). But they constitute the single most important reason for moving in all four villages. People moved because they feared violence: kidnapping, injury, and death. The finality of death or serious injury may play an important part in convincing people of the necessity to move, in contrast to the less dramatic effects of resource depletion. To some extent, this argument also holds for epidemic diseases that figure importantly in reasons given for village moves (Table 14.2).

Our data reveal what informants viewed as the most important reasons for their actual village moves. What can the data tell us about possible environmental effects on decisions to move? Elsewhere (Gross *et al.* 1979; Werner *et al.* 1975), my coworkers and I suggest that impending scarcities of certain goods (e.g., game, garden produce) are telegraphed to villagers by falling output–input ratios. People become aware that they must expend more effort or time to secure a constant amount of food or other goods. If there is such recognition, why do not informants give it frequently as a motivation for moving? There are two possibilities: (1) resource depletion rarely reaches a point where it is noticed by Amazonian villagers, or (2) falling output ratios are "translated" into more compelling reasons such as potential illness, injury, or death through a social process.

In two of the groups in our study, resource depletion is never mentioned as a primary reason to explain a village migration, whereas in the other two (Kanela and Bororo), resource depletion is mentioned in fewer than 10% of the instances (Table 14.2). Does this mean that these groups rarely, if ever, approach the limits of their resources?

There is some evidence that all four groups are using up local resources faster than they can be replenished. This process is more pronounced in the Bororo and Kanela cases, where the settlements have been circumscribed by the Brazilian national frontier over several decades. We showed (Gross *et al.* 1979) that the Kanela and Bororo areas have lower forest biomass, lower soil nutrient content levels, lower crop productivity, lower food availability, and greater dependence on the external market. It may also be significant that Kanela informants give depletion of gardens as the most important *general* reason for moving the village, although they rarely mention it in specific instances. But depletion is also measurable in the areas surrounding the two groups that have been in contact for much less time. I shall present some data concerning the change in soil nutrient levels in successive years of swidden use.

Another important contrast emerges in the average distance of the moves of each village or village segment. Until the past 10 years, the Xavante and Mekranoti travelled from 50 to 250 km on each migration. The migrations undertaken by the Kanela and Bororo groups were much shorter, ranging from 2 to 20 km. In some cases, a move amounted to nothing more than shifting a village site a few hundred meters from the old one in order to expand its size or to abandon old, vermin-infested dwellings (see Chagnon 1973; Hames, this volume). To the extent that these moves were conditioned by depletion of resources, it appears that the range over which the Kanela and Bororo could move was far smaller than for the Xavante and the Mekranoti.

This is consistent with historical data concerning the annual treks that these and other central Brazilian groups practiced. These were foraging expeditions lasting several months (Baldus 1970; Nimuendaju 1939, 1946; Wagley 1977; Werner this volume). Entire villages would decamp, often splitting into smaller units, leaving behind the gardens and central villages, and travelling through unsettled areas in search of wild food, arrow cane, and so forth. The Xavante ranged over hundreds of kilometers of territory at least until the late 1950s (Maybury-Lewis 1967). The Mekranoti, as of 1976, trekked over long distances (Werner 1978, this volume). The Bororo and Kanela, who once trekked over long distances as well, have been virtually surrounded by Brazilian settlements for over a century and have abandoned their trekking practices. The distances over which each group could range reflects the degree of circumscription that group underwent and therefore the degree to which intensified use of local resources was necessary.

The final aspect of settlement movement that the data reveal is the frequency with which people return to a given location. All the groups in the study repeatedly returned to a particular location, although houses might not be constructed in precisely the same spot. Of the 36 village movements in this century recorded for the Mekranoti, 10 of them refer to sites occupied previously at least once. There are five such revisited sites and each of them was occupied at least twice since 1900, some three times, and one four times. The picture is similar for other groups, but it is particularly striking to find this pattern for the Mekranoti because they have wandered throughout this century over an area roughly the size of Belgium, virtually unoccupied by other groups. The Xavante also repeatedly returned to particular locations to build their base villages, up until about 20 years ago. Thus, even those groups having effective use of large expanses of territory focus their efforts on a relatively few sites. It is plausible that these locations are those offering the greatest abundance of resources, or areas that offer a range of demanded resources (e.g., water, freedom from disease, good hunting, good garden land, firewood, building materials, medicinal and decorative plants, etc.).

To summarize, the Kanela and Bororo have lived in a far more restricted habitat throughout this century in contrast to the Xavante and the Mekranoti villages we studied. Their village migrations are far shorter than those of the Xavante and Mekranoti. They are the only groups for which any mention of environmental depletion occurs in informants' accounts of migrations, although even for them this motivation is rarely mentioned. This may be due to the fact that, in a given situation of choice, the depletion of resources may be far less compelling and dramatic than an enemy attack or a local feud. It also seems plausible to assume that a given move may be strongly conditioned or caused by resource depletion whereas, at the same time, informants view it as having been precipitated by warfare or sorcery.

Garden Soil Depletion

In this section, I shall present some data pertaining to soil nutrient and other values gathered in the same study to which I have just referred. The objective of this part of our work was to observe the change in swidden soil nutrient values and pH, after the swidden was burned over and after it had been cultivated for 1 or 2 years in relation to the forest surrounding the swidden. Soil nutrient and pH values themselves do not necessarily reflect the productivity or longevity of a swidden. In the tropics, soil values are perhaps less important indicators of agricultural potential than the vegetation supported by a given quadrat. This is because nutrients tend to be stored in living biomass to a greater degree in the tropics than in temperate zones (Nye and Greenland 1960).

When people prepare a swidden in a tropical forest, they modify the habitat by felling and burning the vegetation. The ashes and other materials enter the soil and become available to cultivated plants, weeds, and regrowth. Our soil tests were conducted on "sequences" of adjacent patches consisting of forest, recently burned-over swiddens, second-year gardens and third-year gardens. The forest may have been primary forest or secondary growth, but in no case had it been used within the past 20 years for cultivation.[2] It cannot be determined from these data whether

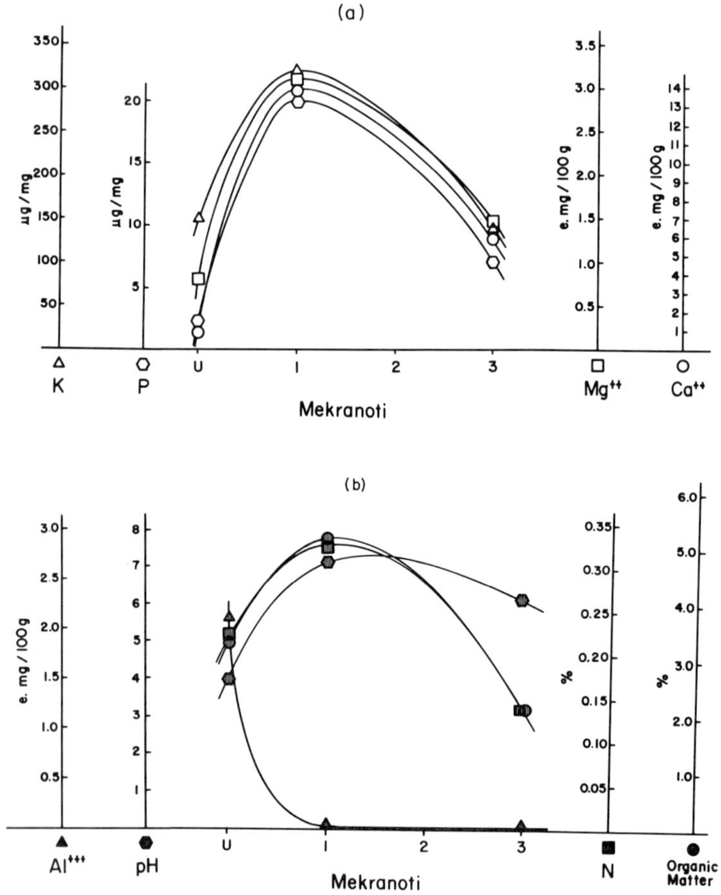

Figure 14.2 *Soil nutrient and pH plots for Mekranoti. (a) K, P, Mg^{++}, and Ca^{++}. (b) Al^{++}, pH, N, and organic matter. Symbols along abscissa: (u) unburned forest; (1) garden used for 1 year; (2) garden used for 2 years; (3) garden used for 3 years.*

swiddens are fallowed because of weed competition, decline in yields, animal predation, or other reasons. However, our data show a coincidence in the return of nutrient levels to preclearing values, declining crop yields, and the decision to fallow a garden. In general, annual crops are not replanted after the third year of cultivation in any of our four study areas (see Flowers *et al.* [1982] for a discussion of cropping practices in the four groups. Gross *et al.* [1979] provides garden yield data for the principal crops cultivated by the Mekranoti, Xavante, Bororo and Kanela). Inspection of figures 14.2 through 14.5 and Table 14.4 suggest some interesting processes at work. Generally speaking, burning-over felled vegetation appears to increase the levels of K, P, Mg^{++}, Ca^{++}. Change in N values is not consistent across the four cases, although theoretically it could be

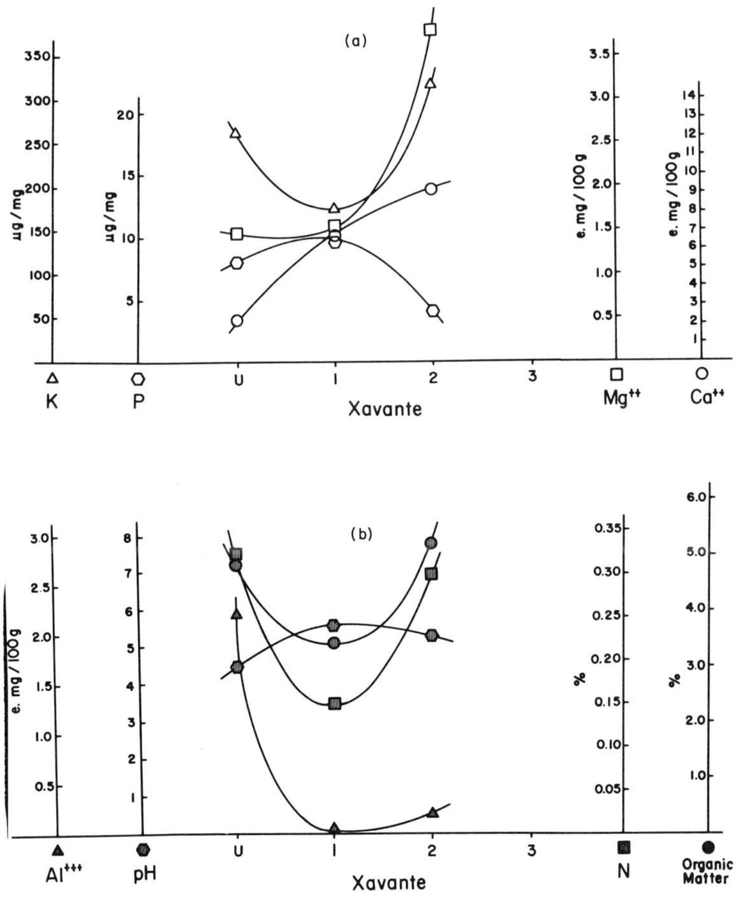

Figure 14.3 *Soil nutrient and pH plots for Xavante. (a) K, P, Mg^{++}, and Ca^{++}. (b) Al^{+++}, pH, N, and organic matter.*

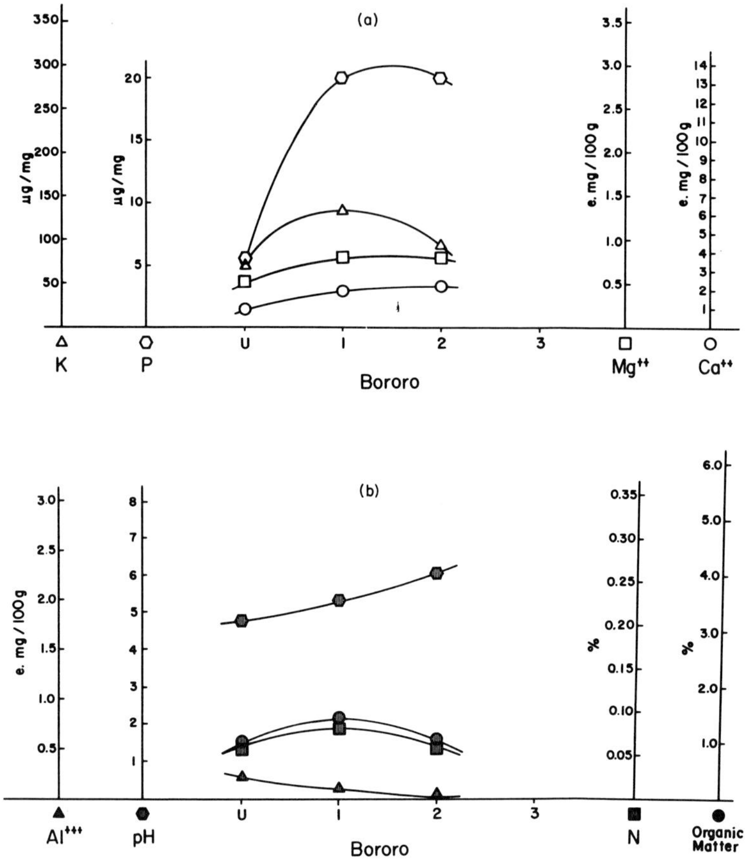

Figure 14.4 *Soil nutrient and pH plots for Bororo. (a) K, P, Mg^{++}, and Ca^{++}. (b) Al^{+++}, pH, N, and organic matter.*

expected to decline as a result of the heat of burning (Nye and Greenland 1960). The introduction of Ca^{++} and Mg^{++} to the soil in the ashes of burned litter is probably responsible for raising pH values from highly acidic, characteristic of tropical soils to nearly neutral. The same two elements are probably also responsible for the reduction in available Al^{+++} which is highly toxic to cultivated plants. Organic matter rises in every case except the Xavante. This may be because the test for organic matter (OM) is sensitive to the C added to the soil in burning. The Xavante exception may be due to the fact that this group, unlike the others, practices annual burning of a wide area of land (including future garden plots) in order to drive game.

Thus, clearing and burning gardens appears to enhance soil qualities desirable for successful cultivation by increasing nutrient availability, de-

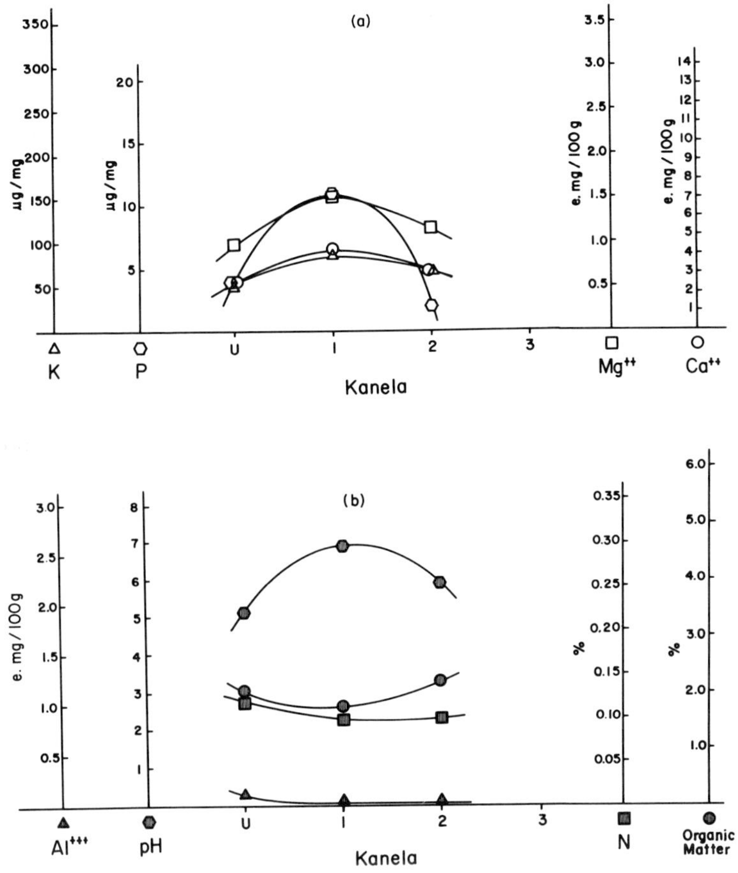

Figure 14.5 *Soil nutrient and pH plots for Kanela. (a) K, P, Mg^{++} and Ca^{++}. (b) Al^{+++}, pH, N, and organic matter.*

creasing toxic aluminum, and raising pH, as well as by exposing the plants to sunlight and by eliminating competitors. It can be tentatively concluded that there is no absolute level of fertility at which gardens will be planted or fallowed by Amazonian cultivators. The nutrient levels *after burning* at the Bororo and Kanela sites are still below the before-burning levels at the Xavante and Mekranoti garden sites (Figures 14.2–14.5). The fact that plant biomass is lower at the Kanela and Bororo sites is reflected in the smaller overall changes (flatter curves in Figures 14.4 and 14.5) in soil values after burning and cultivation. In other words, there are fewer nutrients available for release to the soil after burning.

Thus the Bororo and Kanela are obliged to work longer hours for lower yields than the Xavante and the Mekranoti (see Table 14.5). The Kanela achieve higher yields in Calories/hectare because they rely more heavily on

Table 14.4

Soil Nutrient Values at Three Stages of Swidden Growth at Four Central Brazilian Sites Cultivated by Native Peoples[a]

	K	P	Mg^{++}	Ca^{++}	Al^{+++}	pH	N	OM
Mekranoti								
Unburned	148	2	0.80	0.95	2.10	4.00	0.22	3.25
1st year	316	20	3.12	14.00	0.00	7.20	0.33	5.12
3rd year	136	7	1.44	6.00	0.00	6.20	0.14	2.12
Xavante								
Unburned	260	8	1.50	2.25	2.20	4.50	0.32	4.75
1st year	172	10	1.54	6.80	0.00	5.60	0.15	3.35
2nd year	316	4	3.80	9.20	0.20	5.35	0.30	4.90
Bororo								
Unburned	68	5	0.50	0.85	0.20	4.80	0.06	0.95
1st year	132	2	0.78	1.90	0.10	5.30	0.08	1.39
2nd year	92	20	0.78	2.15	0.00	6.10	0.06	1.01
Kanela								
Unburned	56	4	1.00	2.55	0.10	5.20	0.12	1.99
1st year	84	11	1.56	4.25	0.10	6.70	0.10	1.72
2nd year	68	2	1.18	3.25	0.00	6.00	0.10	2.18

[a]See note 2. K, and P are expressed as μg/mg; Mg^{++}, C^{++}, and Al^{+++} are expressed as equivalent mg/100 g.

manioc, which makes lower demands on the soil but has a high caloric yield. For any particular crop, however, Kanela yields are lower than at the Xavante and Mekranoti gardens we measured (Table 14.5). It is likely, based on their former behavior, that the Kanela and Bororo would have moved their villages to locations where larger stands of undisturbed forest, suitable for swiddens, are located. But they are prevented from doing so by the boundaries of their reservations. Their adaptation to this state of affairs has been to disperse their gardens widely over the land available to them, apparently taking advantage of every favorable spot available.

The relatively sharp peaks in the Mekranoti curves may reflect a rela-

Table 14.5

Garden Yields, Maize Yields, and Garden Labor Yields of Food Energy (kcal) for Four Central Brazilian Village Communities

	Garden yield (all crops) (10^6 kcal/ha)	Maize yield (10^6 kcal/ha)	Garden labor yield (all crops) (10^3 kcal/hr)
Mekranoti	28.84	12.39	17.6
Xavante	7.80	10.81	7.1
Bororo	5.01	2.81	1.5
Kanela	15.59	0.59	5.1

tively undegraded habitat. The U-shaped curves shown in the Xavante sequences (Figure 14.3) are not easy to explain. Unlike the other three cases, some nutrient values (Mg^{++}, K, Ca^{++}, N), actually rise between the first and second years of cultivation in the Xavante garden. Again, this may be related to the annual burning that is unique to the Xavante. In any case, the Xavante themselves believe that second-year gardens give better yields than, first-year gardens and Flowers's yield data (personal communication) support this belief. Also, the Xavante are the only one of the four groups to plant maize in the second and third years of cultivation. This suggests that the management and habitat conditions prevalent in the Xavante area are conducive to this more demanding crop, maize, than in the other three sites where only root crops are cultivated in second- and third-year swiddens. Unfortunately, the sequences from the Xavante site did not include a third-year garden that would have revealed the tendency of nutrient levels over the entire life of the swidden. Inspection of the graphs from the other three groups (Figures 14.2, 14.4., 14.5), however, reveals a tendency for the soil values to return to levels close to those that prevailed prior to cultivation in the unburned plot.

Meggers (1954) suggests that limited agricultural potential was responsible for the relatively small size and the impermanence of Amazonian settlements. Carneiro (1960) proposed a formula for calculating the carrying capacity of swiddeners. Using data collected in the upper Xingu basin, Carneiro (1961) rebutted Meggers. He calculated that the Kuikuru, who relied on manioc for 85% of their diet could sustain a village of 2000 inhabitants indefinitely. Carneiro's study was very influential even though it reported data for a single location in Amazonia, in an area which is not typical of many other areas in the region. Later studies suggest that Carneiro's estimate of cultivable land surrounding the village may be too high, even for the Xingu (Gregor 1977:13–17). Carneiro's projection is based on a single, starchy root crop deficient in many important nutrients (Gross 1975; Spath 1981).

In a comprehensive review of the problem, Roosevelt (1980) concluded that Carneiro's estimates of agricultural carrying capacity in Amazonia were not representative of Amazonia in general. She argues that even the flood plains of rivers could not support dense, permanent populations unless maize, a more nutritious crop, were substituted for manioc as the principal staple. Our results from central Brazil appear consistent with Roosevelt's point of view. They suggest that maintenance of soil fertility is highly dependent on the elephantine process of forest succession. Groups that exceed these rates are condemned to ever-higher labor inputs for a constant amount of food (Gross *et al.* 1979). There are alternative strategies, crops, and techniques (A. Johnson 1977), but no study has yet suggested that upland horticulturalists can long exceed the limits imposed by Amazonian nutrient cycles. It would appear that natives implicitly recognize these limits by frequently moving their villages and by not allowing them to grow too large. People from outside the region need to be constantly reminded of this lesson. The recent collapse of Daniel Ludwig's Jari Enterprises joins a long list of well-financed schemes to turn Amazonia into a breadbasket.

Thus I find myself drawn to a position that seeks to recognize the broad limits to human population growth and settlement density imposed by environmental factors. I do not deny that any particular case is complex and that many different variables combine to produce the particular settlement pattern of a given group. Nevertheless, the yields per unit labor of human-powered subsistence systems in Amazonia seem to make it difficult for native groups to establish large, dense, permanent settlements, except along the banks of major rivers.

The frequent movement of settlements, and their tendency to fission upon reaching a size of 300 or more, appears to be the result of these limitations. Meggers and Evans (1979) estimate that the Taruma of the upper Essequibo (Guyana) moved their village sites on the average of every 17.2 years until they became extinct in 1925. DeBoer (1981) suggests that Shipibo on the upper Ucayali maintained buffer zones between their frequently moved villages, a practice that could be highly beneficial in permitting game and floral resources to regenerate (but see Stocks this volume). The same function could be ascribed to the Jívaro settlement pattern described by Harner (1972), principally as an adaptation to the frequent raiding in this society.

Meggers and Evans urge that ethnologists begin to delve more deeply into reasons for moving villages. This, as we have seen, is a difficult task. It should be kept in mind that the native model of village movement and an ecological explanation need not be in conflict. The stress by natives on life-threatening factors in their decisions to move may be a way of translating environmental imperatives into social ones. But simply documenting the frequency of movements and the duration of settlements and the natives' explanations of them will not be sufficient to verify this hypothesis. There must also be measurements taken of the population density and resource availability in terms of actual rates of capture. Only then will we be able to state with some confidence that such practices as warfare, internal disputes, sorcery, and so forth are pivotal aspects of a system whose leverage is provided by subsistence.

Conclusion

The debate will no doubt continue, and undoubtedly new and better data will continue to be generated. Up until now, however, virtually all the information concerning Amazonian habitats and human interactions with them have been collected in relatively small areas, usually delineated by the customary range used by the inhabitants of a single settlement. In some cases, this is an artifact of the ethnographic reality: a "tribe" or language group consists of one or two small villages. For example, there are a number of studies of such "peoples" as the Kalapalo, Kuikuru, Waurá, Mehinaku, Bakairi, Kamaiurá, and others of the headwaters of the Xingu River in Brazil. Because, in that region today, village and language group are largely coterminous, it is easy to lose sight of the fact that these groups all participate in a regional system of exchange involving mutual feasting, marriage exchanges, migrations, trade with Brazilians, witchcraft accusations and occasional executions, and, until recently, raiding (Galvão 1953; Murphy and Quain 1966; Oberg 1953; Zarur 1975). Judging from maps available in the literature, the radius of the sphere of subsistence activities for each village overlaps considerably with those of other villages. Similar situations exist in the Brazil–Venezuela region occupied by the Yąnomamö, Yanoama, Yanam, and related groups (Chagnon 1973; Hames 1980; Spielman *et al.* 1974), in the northwest Amazon (Jackson 1976); and no doubt in other regions as well. If we take into consideration regional exchange and overlapping resource spheres in areas with a substantial non-Indian presence, we shall find that virtually all of Amazonia consists of a series of regional exchange systems (see also Morey and Morey 1977).

The meaning of this for Amazonian cultural ecology is that data concerning production and consumption from a single village location are not likely to yield the information we require about the potential of Amazonian environments for human use. To the extent that there is local specialization, for example, we will find some villages producing more of certain goods than are required for local consumption (see Silverwood-Cope 1972 for such an account). The others will, in compensation, produce less. Thus, the pressure on any particular resource will not be a function simply of the requirements of the local villagers. Overlapping spheres of activity means that some areas may be subject to exploitation for, say, hunting by more than one village population. Thus, the pressure on resources for such areas must be calculated in terms of the combined intensity of exploitation. If this is so, attempts to estimate resource availability in terms of an area delimited by the circumference of a circle drawn around a single village are misleading.

Yet another factor that may confound attempts to assess the impact of human activities on resources around a single village is the variability of Amazonian environments. As we come to know more and more about Amazonian soils, flora, and topography, we become increasingly aware of microregional variability of great magnitude (see Moran 1980). Furley (1980) has recently argued that maps on a scale of 1 : 100,000 are virtually useless for environmental planning in Amazonia. Thus, the response of resources to human activities at a given location at a given moment in time is not necessarily a valid sample of response over the entire range of the population that that village typifies. Finally, we must take into consideration the impact of all the various kinds of movements of people from place to place including macro- and micromoves (Chagnon 1973), "treks" (Maybury-Lewis 1967; Nimuendajú 1939; Werner 1978), rotation of gardening, hunting, and gathering sites (Hames 1980), fission and fusion (Carneiro n.d.; Chagnon 1968a), offensive raids and defense, marriage exchange (Jackson 1976), intervillage feasting, and temporary labor migration. The effect of these movements of people will be to redistribute the pressure on resources.

All of these factors multiply the difficulties faced by the investigator who seeks to assess the impact of subsistence and other activities on resources for a single village. The time has come, it would appear, to adopt a regional perspective in Amazonia. This implies abandoning, for a time, the single village or village cluster as the unit of research. The problem of delimiting a region for research purposes will be a thorny one and cannot be resolved here, but it may be useful to consider such parameters as trade networks, relative floristic homogeneity, and annual migratory ranges as tentative markers of regional limits. In any case, it is not necessary to treat regions as closed systems, but only to presume that resources within them are finite. I would also urge that investigators overcome the common inclination of anthropologists to exclude the activities and impact of nonindigenous groups in regional systems. Although the activities and impact of Indians and non-Indians might be kept separate for the purposes of analysis, there is no justification for ignoring the impact of any group simultaneously exploiting a common resource with another.

Acknowledgments

Support for the field research and subsequent analysis of data was provided by grants from the National Science Foundation (Grants BNS 76–03378 and BNS 78–25295) and from the PSC/BHE (Professional Staff Congress/Board of Higher Education) Faculty Research Award Program of the City University of New York (Grants 11479E and 12182). D. Gross was principal investigator on each of these projects. The field studies were carried out by N. Flowers (Xavante), F. Leoi (Bororo), M. Ritter (Kanela), D. Werner (Kayapó), and George Eiten (botanical consultant). The author thanks all of his coworkers for their tireless devotion. For assistance in various forms in the research, analysis, or writeup, he also thanks Silêde Gross, Gentil Martins Dias, Roque de Barros Laraia, Delvair Montagner Melatti, Roberto Cardoso de Oliveira, Dora Cerruti Rodrigues, Josefa Bispo dos Santos, Renate Brigitte Vertler, George Zarur, and all of the villagers who made us welcome and assisted us in our investigations in Central Brazil.

Notes

[1]Amazonia is defined here as the area drained by the Amazon and Orinoco rivers, including the lower eastern slopes of the Andes, the Central Brazilian Shield, and the Guayanas.

[2]The soil tests reported on here were conducted in the following manner. Ten shallow samples (0–10 cm) of approximately 1 liter each were collected from spots scattered over a 0.1 ha plot. Duff and loose pieces of burnt wood were scraped away before digging into the soil. These were carefully mixed together and approximately 1 kg was dried, tagged, and sent to various laboratories for analysis. The results reported are from the soils laboratory of the Instituto Agronômico de Campinas through the courtesy of Dr. Alcyr Nascimento. The tests performed provided values for potassium (K), phosphorus (P), magnesium (Mg^{++}), calcium (Ca^{++}), aluminum (Al^{+++}), acidity (pH), nitrogen (N), and organic matter (OM). The values for each swidden sequence are plotted on Figures 14.2–14.5. There are two plots for each sequence, the first for Al^{+++}, pH, N and OM, the second showing K, P, Mg^{++}, and Ca^{++}.

For each of the four field sites, the fieldworkers located one or more sequences consisting of adjacent plots representing unused forest (or fallowed land), recently burned-over swiddens, second- and third-year swiddens, as well as regrowth. The complete data for soils will be published elsewhere. Adjacent plots (within 100 m of each other) were selected in order to control for other sources of variability, such as humidity, substrate differences, etc. Because each of the groups planted mainly in gallery forests along streams, the sequences are generally located in a straight line, parallel to and equidistant from the stream bed. The method of selection of sampling sites made it possible to analyze the data as if they were longitudinal, reflecting changes in soil values over time. Because of the possibility of microregional variation, data from different garden locations were not averaged, thus the data on each plot represent the values from a single sequence from each of the four field sites (see Figure 14.1). When more than one sequence was measured, the one chosen was that displaying the highest overall nutrient values.

CHAPTER 15

The Territorial Dimensions of Siona–Secoya and Encabellado Adaptation

William T. Vickers

Introduction

Although there has been considerable emphasis on environmental factors in the interpretation of Amazonian societies through the years, it is only recently that detailed ecological studies of specific communities have begun to appear (such as the papers in this volume). This is a healthy trend because, as Vayda and Rappaport have argued (1968), the appropriate unit of cultural ecological analysis is the community and its associated environment (cf. Gross, this volume).[1] Purported analyses at the tribal or culture area levels usually rely on general observations of a nonquantitative nature and this relative lack of specificity in the description of man–land relationships has contributed to a lack of intersubjectivity among scholars concerning the significance of various factors as determinants of cultural patterns (see, for example, Eric Ross 1978 and his commentators).

Even though ecological anthropologists have now begun to focus their efforts on adaptation at the community and even individual (e.g., the evolutionary ecologists and optimal foraging theorists) levels, little attention has been given to the spatial parameters of Amazonian adaptation. Questions such as (1) To what extent does territoriality exist? (2) What are the determinants of territoriality? and (3) How does territoriality relate to other aspects of culture (e.g., settlement pattern, sociopolitical organization, and warfare)? have largely gone unasked and unanswered by anthropologists (but see Hames, this volume). Lathrap (1970) emphasizes the distinction between the *várzea* floodplain and the *terra firme* uplands, but

ADAPTIVE RESPONSES
OF NATIVE AMAZONIANS

ISBN 0-12-321250-2

his analysis is at the macrolevel of the Amazon Basin as a whole. Jane Bennett Ross (1980) discusses the role of buffer zones between settled areas as sanctuaries where animals breed and replenish faunal populations, but her analysis, while valuable, is essentially general and nonquantitative.

Most ethnographic maps of Amazonia have provided culture area or linguistic distributions on a regional scale, whereas the maps provided by ethnologists studying specific communities have shown settlement sites as dots on otherwise undifferentiated landscapes. Few anthropologists have provided maps that indicate precisely the extent to which specific Amazonian communities range over the landscape, or whether formal or informal boundaries delimit home ranges or territorial divisions. In this regard, anthropologists can learn from the analyses provided by certain cultural geographers. Smole (1976:77), for example, gives a clear definition of the territorial base of a Venezuelan Yanoama community. He does not, however, carry his analysis to the point of estimating the resources to be found within the indicated area. Nietschmann's work on the Miskito Indians of Nicaragua (1973), although falling outside of Amazonia, provides detailed analyses of the spatial distributions of resources and resource utilization. He concludes that the Miskito are not presently limited by the availability of agricultural land, but that they are degrading the faunal resources (particularly the green sea turtles that are their preferred prey). This analysis focuses primarily on declining yields and does not include estimates of animal population levels.

The lack of emphasis on territorial spaces and their associated resources has hampered the interpretation of human ecological data from Amazonia. Much of the debate on protein availability, for example, has centered on the grams of protein intake per day, or the amount of game taken per hour (see, for example, Chagnon and Hames 1979; Gross 1975). While these data are very important, the emphasis on such statistics does not adequately consider the extent of the resource base from which the kills are being made, nor the impact of the hunting activity on this base. In order to advance the level of understanding on the protein question, it is imperative that we begin to take a look at the sizes and characteristics of the hunting and fishing areas available to Amazonian natives, including the populations and reproductive dynamics of the significant fauna. Only when this is done can game yields and protein intake be evaluated in the proper context, and only then will it be possible to begin to reach conclusions about the relative scarcity or abundance of protein and its significance in the overall adaptive system. Other resources, such as soils and flora, likewise need to be inventoried and plotted in the spatial dimension (e.g., see Bodley and Benson's [1979] treatment of palms in the Shipibo community of San Francisco).

The present paper focuses on the spatial organization of the Siona–Secoya community of San Pablo in northeastern Ecuador. Its purpose is to reveal how the Siona–Secoya carry out their subsistence activities within a defined space, and how the size of this space is conditioned by environmental and technological factors. Furthermore, the analysis relates these findings to theories of territoriality and proposes a territorial model of aboriginal western Tucanoan communities (the linguistic grouping to which the Siona–Secoya belong) using data from the journals of eighteenth century Jesuit missionaries. Although the latter aspect involves some ethnohistorical comparison, this chapter is not primarily an exercise in ethnohistory. If it were so, it would not meet the criterion of this volume, which is to present ecological analyses generated from empirical field research. The empirical aspect of this chapter is to present an analysis of the land-use patterns of modern San Pablo in order to delineate the limits of resource utilization for the community. The subsequent ethnohistorical analysis is an attempt to determine the extent to which the modern pattern can be used as a model for aboriginal territoriality among various western Tucanoan groups. This is an important issue, for if we can begin to understand the limits of territoriality for specific Amazonian populations, we will have made an important step toward providing a scale for assessing resource availabilities and the cultural significance of resource use by humans. These issues bear directly on the questions of settlement pattern, social and political organization (including warfare), and ideology and religion (particularly with regard to witchcraft and witchcraft accusations).

The Modern Settlement

San Pablo is a community of Siona and Secoya Indians located on the Aguarico River in northeastern Ecuador (see frontis map). The Siona and Secoya are related members of the western branch of the Tucanoan linguistic family, and are the survivors of a once-numerous group that was called the Encabellado by the seventeenth-century and eighteenth-century Spanish. The aboriginal Encabellado occupied an extensive area of approximately 82,000 km^2 between the Napo and Putumayo rivers. Steward estimates that their population at the time of the conquest was 16,000, with a density of 0.2 individuals/km^2 (1949:663). The modern population probably does not exceed a thousand individuals in various settlements in Ecuador, Colombia, and Peru. The Aguarico River (a northern tributary of the Napo) was part of the traditional Encabellado homeland, and is one of the areas where western Tucanoan peoples survive today. (Other areas that still have survivors include the Angusilla, Yubineto, and Santa María Rivers in Peru, and various sites along the Putumayo River in Colombia.) Specific multifamily settlement sites within the Aguarico Basin include the Eno

River (a southern tributary of the Aguarico), the Cuyabeno River (a northern tributary of the Aguarico), and Campo Eno, San Pablo, and Sewaya (all located on the banks of the Aguarico proper). In addition, there are a number of dispersed households located at various points along the Aguarico. The estimated population of the Siona–Secoya residing in the environs of San Pablo was 297 in 1980 (excluding the Cuyabeno River group).

San Pablo is the largest Siona–Secoya settlement in Ecuador and serves as something of a headquarters for the people because it is where the president of the *comuna* resides, and has a school, a native paramedic, and a jungle airstrip that serves as a link to the Summer Institute of Linguistics (SIL) basecamp at Limoncocha on the Napo River.[2] The office of president and the legal status of *comuna* are recent innovations associated with the Ecuadorian government. Traditional settlements have headmen, who are referred to as *ĩnti ba'iki* ("this one who lives"), as their principal leaders. A number of the Siona–Secoya who are not permanent residents of San Pablo nevertheless maintain houses there because they periodically come to the village for visitation and other purposes. Frequent movement is one of the more salient aspects of Siona–Secoya culture, whether it be for hunting and fishing, the collection of craft materials, or the periodic relocation of settlement sites. San Pablo has a long but intermittent history as a Siona settlement site, with the Secoya being more recent immigrants from the Santa María and Angusilla rivers to the east. In general, Siona and Secoya settlements tend to go through cycles of establishment, growth, decline, and then relocation. When people move, they usually reestablish their communities at or near named locations where previous settlements once existed. Thus the pattern of village or household movement is not random, but represents cyclical movements to and from specific locations within the larger regional context.

The most recent settlement of San Pablo began in early 1973 when various households began moving there from the Cuyabeno River settlement of Puerto Bolivar. By mid-1974, no fewer than 24 households had been established there and the population approached 200. By 1978, 7 households (consisting primarily of Sionas) fissioned and occupied a site by the name of Camp Eno on the Aguarico (about 7 km upriver from San Pablo). At about the same time, 10 other households (consisting primarily of Secoyas) established pioneer gardens and shelters downriver from San Pablo. The explanations given for these moves fit a rather typical tropical forest pattern, including the expression of traditional antagonisms between certain households, the desire to avoid illness (which the people associate with the more densely settled villages), and the search for lesser populated areas in which to practice horticulture, hunting, and fishing. This waxing and waning of the population of a particular settlement is an entirely normal aspect of Siona–Secoya adaptation and can be understood primarily in ecological terms, even though the expressed reasons for movements often include social and psychological motivations (Vickers 1976, 1980; see Gross this volume for similar findings among the Gê). I will return to this issue in subsequent sections of this chapter. Despite the ongoing process of settlement fusion and fission, the Siona and Secoya do maintain a sense of community in that they communicate with one another through visitation and encounters made when travelling along the rivers and through the forest.

Patterns of Resource Use

Like most native Amazonian peoples, the Siona–Secoya practice a mixed subsistence economy involving shifting cultivation, hunting, fishing, and collecting. Each of these subsistence modes focuses on differing resource categories, and each presents differing behavioral, ecological, and technological features. Although the gross caloric contributions of the various modes vary, each provides dietary components that are complementary and that enhance the overall nutritional status of the people. Shifting cultivation is by far the largest supplier of calories to the diet. Because I have described the dynamics of Siona–Secoya shifting cultivation elsewhere (Vickers 1976, 1979), I will not dwell on them here. It is sufficient to mention that dietary surveys and garden yield analyses indicate that approximately 72% of the calories in the diet come from cultivated plants. Gardens are typically located adjacent to or near settlement sites, or along rivers and streams so that the produce can be easily transported by canoe to the house sites. The Siona–Secoya select their garden sites with care and rely on their judgments concerning soil and drainage conditions (Vickers 1976:59–66). Gardens have primary cultivation cycles that last approximately 3 years. Peach palm (*Bactris gasipaes*) and some other fruit trees remain in "abandoned" gardens and the people return to harvest these tree crops over a period of years.

Twenty-six ha were cleared for gardens at San Pablo during the first 2 years (1973–1974) of the most recent occupation of that site. The mean population during this period was 132 individuals, so the mean garden area per individual was approximately 0.2 ha. Actually, the primary unit of garden production and consumption is not the individual, but the household or a related cluster of households. As of January 1974 there were 16 such productive units, with a mean area of land under cultivation of 1.6 hectares. Given the approximately 3-year cropping cycle of Siona–Secoya shifting cultivation, it seems unlikely that a community the size of San Pablo would ever have more than 40 ha of land under production at any given moment. By 1980, the population of the Siona–Secoya at San Pablo and in the surrounding area had grown to 297, but it is doubtful that more than 60 ha were under cultivation at this time. This is but 0.05% of the area depicted in Figure 15.1.

The fact that so little land is under cultivation at any given time reinforces the often-expressed view that the availability of land for gardens is not a limiting factor in aboriginal Amazonian adaptation (e.g., Carneiro 1961). However, there are factors that constrain the access to and use of land in Amazonia. For example, by 1979 some of the people at San Pablo were complaining that there was "no more land" for gardens despite the fact that literally thousands of hectares of forest lay beyond the village and its ring of cultivated plots. What these people were saying, in effect, was that the nearby land had now been incorporated into several roughly concentric bands of current and abandoned gardens, and that any new gardens made beyond this zone would require uncomfortably high travel and transportation costs (see Hames this volume). A number of people stated that they would prefer to move their house sites to new locations along the river rather than to commute ever-increasing distances to their gardens. So even at San Pablo, where there is a relative abundance of good soils and well-drained land, many of the people did not see the possibility of pushing new gardens ever deeper into the forest as a desirable land-use option.

The availability of land that is suitable for cultivation also varies considerably from site to site. In the Cuyabeno River drainage, most of the land consists of swampy depressions and arable land is extremely rare, being limited to small and scattered embankments adjacent to the channel. In the Cuyabeno region, the amount of well-drained land constitutes a very real limitation on the size and permanence of human settlements. At San Pablo, almost all gardens are made with 1 or 2 km of the house sites, but on the Cuyabeno the distances are much larger (up to 10–15 km) because the people must travel by canoe to their gardens made on the dispersed embankments. Hence the land requirements for gardening can be very significant despite the commonly held assumption that Amazonian lands are abundant.

Wild resources are incorporated into the Siona–Secoya diet by hunting, fishing, and collecting. After horticulture, hunting is the largest supplier of calories to the diets of the people residing at San Pablo (18%). It is also the largest supplier of dietary protein (75%), and provides significant amounts of fats. Fishing provides about 6% of the protein at San Pablo where most of this activity is conducted during the dry season when the Aguarico's water levels are low (November–January). According to informants at the Cuyabeno River settlement, fishing in their area surpasses hunting as a supplier of protein because of the year-round availability of many varieties of small- and medium-sized fish and the relative difficulty of terrestrial hunting in the swampy environs of the region (Vickers 1976:123–126).

Collecting is the least productive of the subsistence modes in terms of its total caloric yields, but meets specific needs by providing fresh fruits that are probably high in vitamins, and palm fruits that have high oil (i.e., fat) contents. Collecting is also important as a source of food during migrations and at new village sites where gardens have not yet reached the production stage.

The hunting, fishing, and collecting modes of subsistence require that the people move well beyond the cleared perimeters and gardens of established settlements. By necessity, the Siona–Secoya must fan out into the forest and travel up and down the waterways in order to locate and kill, catch, and collect the dispersed feral resources. Such trips may occupy a few hours, an entire day, or periods of days and even weeks. The Siona–Secoya enjoy these activities and whole families depart on several expeditions each year to exploit wild resources. The most notable seasonal event of this type is the turtle and turtle-egg quest that takes place in November and early December when river turtles (*Podocnemis expansa*) lay their eggs on the beaches of the Cuyabeno and lower Aguarico. Another seasonal quest occurs when *Mauritia* palms fruit in September, but this involves shorter trips to the forest around San Pablo.

The movements to and from the village for hunting, fishing, and collecting activities may be accomplished by walking, by canoe, or by a combination of these two modes of travel. The characteristics of these modes of travel determine to a large degree the range or extent to which the Siona–Secoya can exploit their surrounding environment for feral resources. The movements associated with hunting, fishing, and collecting are not haphazard but follow named trails and streams (see Figure 15.1). Of course, when game or specific plant resources are spotted off these routes, people leave them to pursue their objectives. The trails and streams actually form a complex network of interconnected paths that allow many variations of departure and return. Figure 15.1 shows how this network radiates from the settlement of San Pablo. This map is based on knowledge gained as I accompanied Siona–Secoya hunters and collectors on various trips, plus interviews that I conducted with informants. Data concerning the map were cross-checked with Elías Piaguaje (the extremely articulate and intelligent president of Comuna San Pablo) who provided a sketch map of the local trails based on his own knowledge of the routes.

Several points concerning Figure 15.1 deserve explanation. Firstly, the suffixes *-siaya*, *-ya*, or *-ña* denote rivers or streams in both the closely related Siona and Secoya dialects of western Tucanoan. For example, the Shushufindi River is *gãtɨsiaya* ("cane river"), the Aguarico is *haiya* ("big river"), and various smaller streams are *wa'iya* ("meat river"), *neaña* ("black river"), and *ne'eña* ("*Mauritia*-palm river"). Trails are designated by the term *ma'a*, and are usually named after a stream that they either parallel or lead to, or after some other feature of the local landscape. Hence *napo*

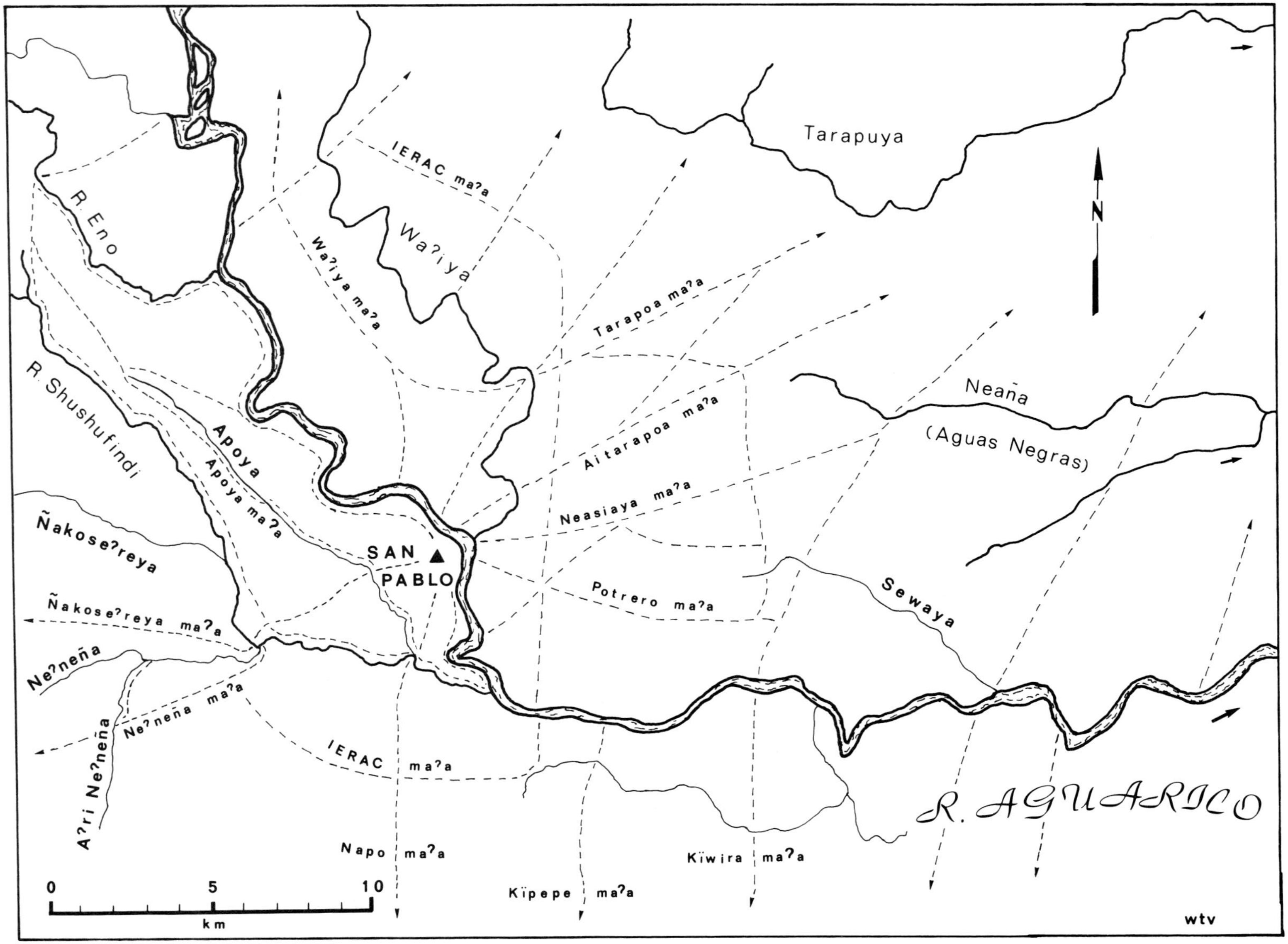

Figure 15.1 *Territory and hunting trails around the Siona–Secoya community of San Pablo in eastern Ecuador.*

ma'a is the trail that leads to the Napo River, *kɨwɨra ma'a* ("whirlpool trail") is the path that begins near a large whirlpool on the Aguarico River, *potrero ma'a* ("pasture trail") is the path that crosses a pasture, and *IERAC ma'a* is a land surveying trail which was cleared to facilitate the work of the *Instituto Ecuatoriano de Reforma Agraria y Colonización* (IERAC).

Taken as a whole, this network of streams and trails provides multiple access routes to the feral resources located in the vicinity of San Pablo. The rectangular area shown in Figure 15.1 measures 29.1 by 39.4 km, or 1149.1 km^2. I estimate that approximately 95% of the subsistence activities of the people residing at San Pablo take place within this area. The remaining subsistence activities that reach beyond this area consist primarily of the seasonal hunting, fishing, and turtling trips to the lower Aguarico and Cuyabeno rivers.

The land area shown in Figure 15.1 is not exploited uniformly because greater pressure is placed on resources in those areas nearest the human settlement. The Siona–Secoya themsevles use a set of about 17 toponyms to refer to specific zones within this region. In another publication dealing with optimal-diet breadth theory (Hames and Vickers 1982), I classify the San Pablo hunting grounds into *near, intermediate,* and *distant* hunting zones (whose mean distances from the settlement center are 5, 11, and 25 km, respectively).

I characterize the hunts used to exploit these zones as (1) morning or evening hunts (lasting from a few minutes to several hours, but less than half a day); (2) day-hunts (lasting for the better part of a day); (3) overnight hunts (extending into a second day); and (4) expedition hunts (lasting from 3 days to several weeks). The vast majority of Siona–Secoya hunting and fishing trips are of the 1-day variety and occur within "near" or "intermediate" zones (Hames and Vickers 1982). During 1973 to 1975, the average hunt lasted 7.6 hours and extended to a distance of 8.5 km from the settlement (with a round-trip distance of approximately 17 km). This distance can be viewed as the radius of the area of heaviest resource exploitation; this area covers 227 km^2, or 20% of the total area depicted in Figure 15.1. The relationship of this differential pressure on resources to settlement dynamics will be discussed in subsequent sections of this chapter.

Historical Data on Western Tucanoan Settlement Patterns

The lands occupied by the western Tucanoan peoples at the time of European contact were located between 1°N and 4°S, and between 73 and 77°W in what is presently part of Ecuador, Colombia, and Peru. This area is to the east of the Andean foothills and its elevations range from 300 to 100 m above sea level from west to east. As indicated previously, Steward estimates that the native population was 16,000 in an area of 82,000 km^2, or 0.2 persons/km^2 (1949:663). In order to analyze the man–land relations in detail, however, it is necessary to focus on the historical data from a somewhat smaller portion of the western Tucanoan area that centers on the confluence of the Aguarico and Napo rivers. The reason for this is that this is the area for which reasonably detailed accounts of native local groups are available. The primary sources of information are the journals of Jesuit missionaries from the seventeenth and eighteenth centuries (Chantre y Herrera [1901] provides a compilation from these sources). Figure 15.2 depicts the locations of eighteenth-century Jesuit missions among the Encabellado, based on my analysis of the historical materials.

Jesuit accounts of the 1730s and 1740s identify 15 Encabellado local groups in an area estimated to cover 16,864 km^2 along the Aguarico, Cuyabeno, Napo, and Santa María rivers. If Steward's estimation of aboriginal western Tucanoan population density (0.2 persons/km^2) is correct, the population for this region would have been about 2912, or about 194 persons/local group. This size is consistent with that of modern Siona–Secoya communities such as San Pablo, and is well within the range reported for Amazonian peoples. I believe that the estimate of approximately 194 persons/local community is within the realm of probability, but I doubt that all of the households of each community would have been located in a single nucleated settlement. A pattern of patrilocal, patrilineal extended households located at different sites, but still within "commuting distance" of each other would be more consistent with the known practices of the Siona–Secoya. Nevertheless, each of these local communities were under the influence of a prominent headman–shaman, or *ĩnti ba'ikɨ*. The oldest male of any household or extended household was and is a de facto headman for the members of his household and performs or attempts to perform shamanistic rituals during times of crisis (e.g., curing illnesses). But not all household heads have equal abilities or charisma, and the more able and knowledgeable headmen gain influence over many households. Such household groupings are associated with a particular river or a section of a large river and are usually related through consanguine or affinal ties. The influence of such prominent headmen–shamans does not represent total authority; the fact that most heads of households are potential leaders in their own right contributes to the easy fissionability of western Tucanoan settlements.

The Jesuit accounts clearly indicate that when the missionaries attempted to form nucleated settlements combining the followers of various headmen, they met with considerable resistance owing to the historic pattern of sorcery accusations between communities (Vickers 1981b). As a consequence, the first Encabellado missions founded under Jesuit influ-

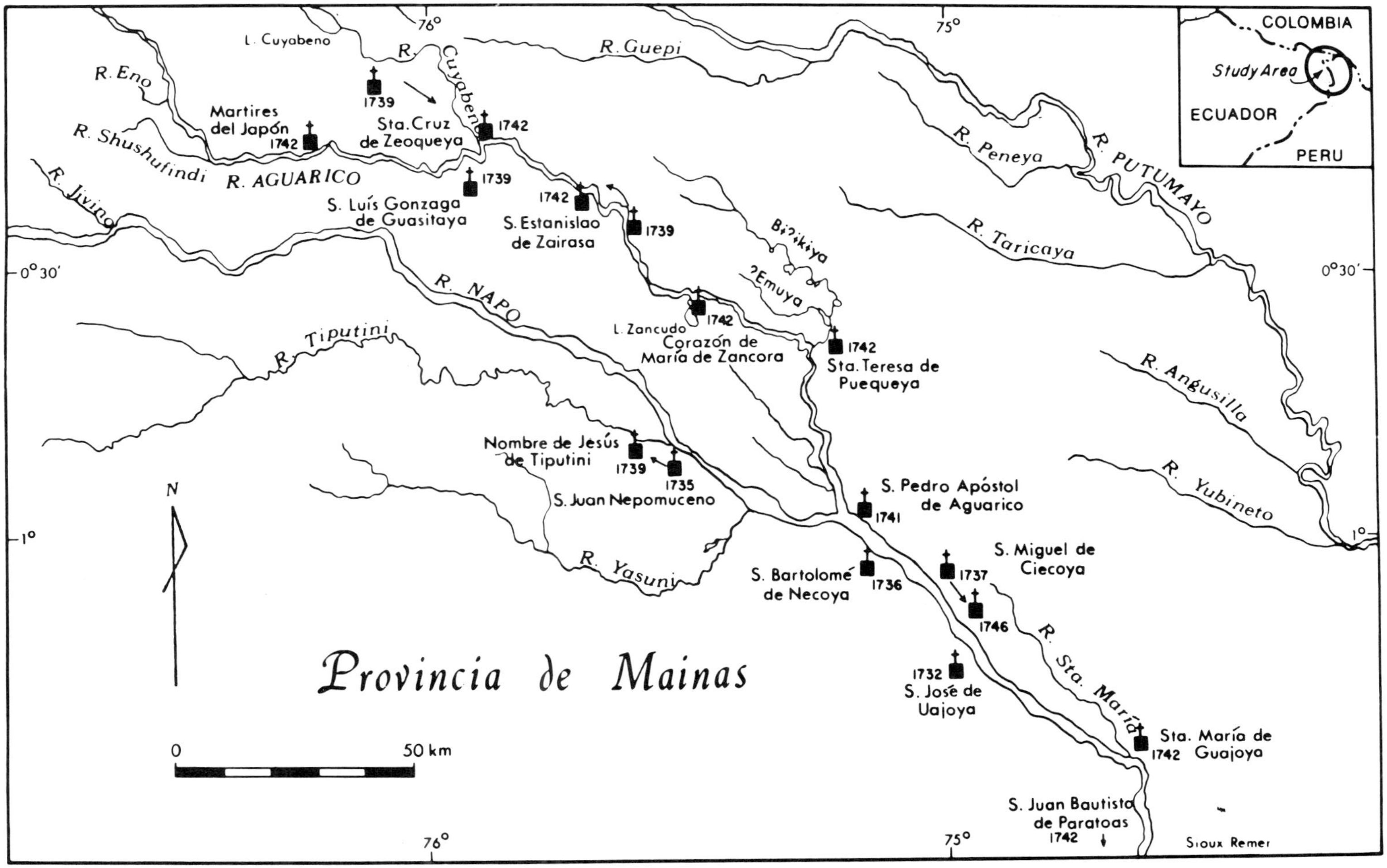

Figure 15.2 *Eighteenth-century Jesuit missions among the Encabellado.*

ence were not true reductions, but were traditional communities under individual headmen that were simply shifted to sites closer to the main rivers. In the Jesuit view, this was to be a temporary arrangement that would place the Indians at sites easily accessible by canoe until they could be cajoled into forming a single large reduction at the confluence of the Aguarico and Napo rivers. The name of this reduction, which was never fully consolidated, was San Pedro Apóstol de Aguarico.

The documentation of these temporary mission locations provides a useful key for the analysis of aboriginal western Tucanoan territoriality. The mean distance between adjacent mission sites is 38 km. In order to estimate the territorial size of each local community, I take half of the distance between the missions as the radius of the average territory, which yields an average size of 1124 km^2. This estimate is remarkably close to the 1149 km^2 area exploited by the modern Siona–Secoya settlement of San Pablo (Figure 15.1). If simple circles with 19-km radii were to be drawn around the Jesuit mission sites, a degree of overlapping would occur between certain missions. Modern Siona–Secoya territories are not circular, but tend to form semirectangular or polygonal blocks along the main rivers (Figure 15.1 represents such a block). This is in part due to the fact that a major hunting and collecting pattern is to travel up or down the rivers by canoe to a trail that then leads inland. (One might expect a more circular area of exploitation if the people simply walk outward from a central point.) Although there is some uncertainty over the presence or significance of canoes among the aboriginal western Tucanoans (Steward 1948:742), the Jesuit accounts and mission locations suggest a pattern of land use that is consistent with that of today's surviving communities.

Figure 15.3 depicts a hypothetical model of Encabellado territorial divisions in which the boundaries are depicted in linear form and are placed so as to include streams, rivers, and resource areas associated with specific settlement sites known to modern informants. Many twentieth-century settlement sites of the Siona–Secoya share the same approximate locations as the eighteenth-century Jesuit missions. For example, San Pablo occupies a territory in the area inhabited by Zapua and his people who formed the mission of Martires del Japón (founded in 1742). Likewise, the modern Cuyabeno River settlement is in the area of the Hijo de Mumu group who formed the mission Santa Cruz de Zeoqueya (founded in 1739). During the 1970s, a Secoya extended household occupied a site on the Aguarico called Caño Negro that is in the area inhabited by Zairasa's group in the eighteenth century, associated with the mission San Estanislao. Similar groups of twentieth-century Sionas and Secoyas had sites at *Sõncora* (corrupted to Zancudo in the local Spanish), Lagarto Cocha (*Pi'ikɨya*), Pantoja, and the Santa María (or *Wahoya*) River that match up with the eighteenth-century missions of Corazón de María de Zoncora, Santa Teresa de Puequeya, San Pedro Apóstol de Aguarico, and Santa María de Guajoya. Overall, it seems

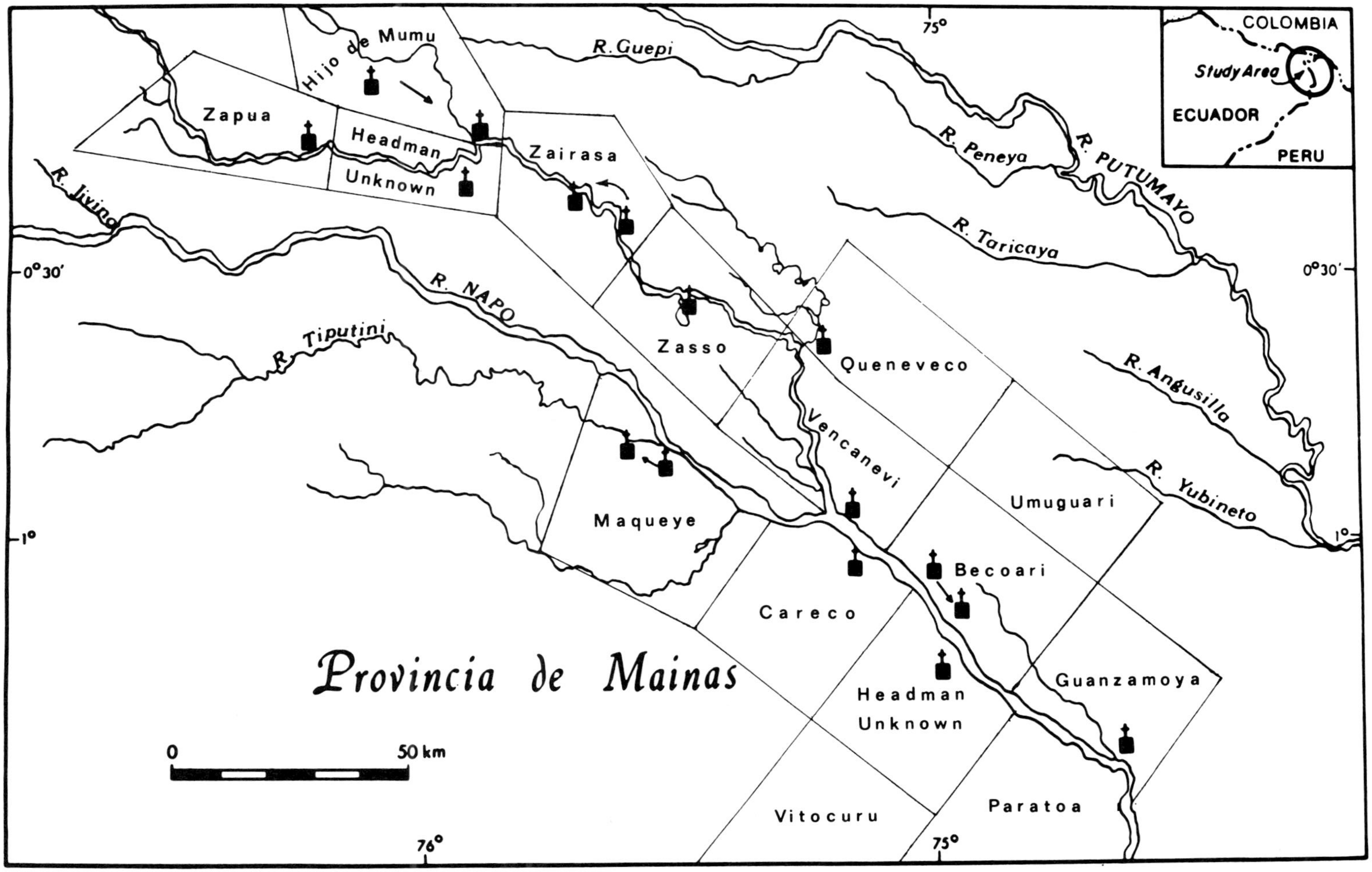

Figure 15.3 *Hypothetical reconstruction of the territorial divisions of eighteenth-century Encabellado communities (named after their headmen).*

reasonable to suggest that the linear territorial boundaries shown in Figure 15.3 are somewhat arbitrary, but that the general loci and approximate size of the suggested territories are supported by both historical documents and modern observations. The tendency for territories to take polygonal shapes is also consistent with the findings of ecologists (see Wilson [1975:270–271] for a discussion of this phenomenon).

Resource Allocation and Maintenance

To this point, I have described the land base of a modern Siona–Secoya community and made an ethnohistorical reconstruction of the land bases of 15 eighteenth-century Encabellado communities. As noted in the preceding section, there appears to be a high degree of congruence in the sizes of the areas associated with the native communities of these two periods, despite the differing methods used in the respective analyses. Although some might attribute this congruence to chance, I believe that it derives from certain fundamental technoenvironmental similarities that characterize both the modern Siona–Secoya and their Encabellado ancestors. Although more than three centuries separate these two populations, they share the same general environment and similar subsistence technologies. While it is true that the Siona–Secoya of today have steel tools, shotguns, nylon fishing line, steel fishhooks, and some postconquest cultivars (of which plantains, *Musa* ×*paradisiaca,* L., are the most significant), their major modes of subsistence (i.e., shifting cultivation, hunting, fishing, and collecting) are the same as those of the Encabellado. The vast majority of their important cultivars are aboriginal and the wild resources they exploit are the very same ones that were used by their ancestors. Hence the broad outlines of the subsistence economies of the two periods are essentially similar. Of course, the native population has been much reduced from its aboriginal level, but the surviving communities exploit the lands around their settlements in much the same manner as before. The decline in population does not mean that these communities can expand their territorial bases indefinitely, because the methods of transportation are still very real constraints on foraging behavior. Also, other peoples, such as Quichua Indians from the Napo River and Ecuadorian colonists, have filled in the voids left by the decline in the native western Tucanoan population.

There are two very real sets of constraints that influence the range of sizes for territories. In the first instance, the territory must be large enough to include the resources that are necessary for a subsistence system based on shifting horticulture, hunting, fishing, and collecting. The relative density of resources and the level of technology powerfully influence the density of the human population that can be supported in a given area. Within Amazonia, variations in such factors are clearly related to differences in settlement size and permanence, and in the levels of sociopolitical organization attained (see the introductory chapter of this volume). In the second instance, the territory should not be too large because after a certain point expansion does not improve the cost–benefit equation of subsistence and may even lower it (Wilson 1975:268). The concept of "economic defendability" applies here; an overly large territory requires too many energetic costs to control and does not significantly enhance the efficiency of subsistence when compared to a somewhat smaller, but adequate, territory (Brown 1964). Of course, this discussion refers to a situation in which several factors, such as the levels of technology, are being held constant. Larger territories make sense when differing resource and technological conditions allow denser populations and more complex forms of sociopolitical organization. The chiefdoms of the Amazonian *várzea* (floodplain) and the conquest state of the Inca in the Andes are but two examples from South America where larger territories (encompassing many village-level units) and hierarchical sociopolitical institutions evolved as a consequence of more intensive production systems operating under differing environmental conditions.

In the introduction to this chapter, I argued that it is important to identify community home ranges or territories and to inventory their resources because such data allow one to analyze the significance of the rates of resource utilization by the members of the community. This is an area in which most anthropological and even cultural ecological studies are deficient. The kind of resource data found in most such studies are qualitative; that is, they indicate that this or that species of animal, vegetation, or soil type is present in an area, and make a few descriptive comments about their characteristics and significance to man. Quantitative data are generally harder to come by because their collection requires more sophisticated research designs and techniques, which many anthropologists are unfamiliar with, or which may be precluded by constraints on time, funding, transportation, the physical characteristics of the environment, or other impediments. Despite these kinds of problems, quantitative analyses on Amazonian human adaptation are increasing (as the chapters in this volume attest). What I would like to do now is to provide examples of the kind of resource base–resource utilization analysis that I have been touting in this chapter. I have already described the methods used to identify the territorial areas exploited by modern San Pablo and the various eighteenth-century Encabellado communities. The next step is to make a qualitative and quantitative inventory of the resources associated with these territories. Ideally, such an inventory would include all of the resources that are judged to be significant in the adaptation of the people (e.g., cultivable land, flora, fauna, etc.), as well as other environmental factors that are relevant to the adaptive process (e.g., climatic conditions). Unfortunately, such an extensive analysis is beyond the scope of this chapter, so I will focus on two important sets of resources to illustrate the method. The resources to be discussed are cultivable land and game animals.

CULTIVABLE LAND

During 1973 to 1975, I collected soil samples from gardens and forest at San Pablo and also at the Siona–Secoya community on the Cuyabeno River. These samples were analyzed by the *Instituto Nacional de Investigaciones Agropecuarias* (INIAP) of Ecuador following methods developed by the International Soil Fertility Evaluation and Improvement Project of North Carolina State University (Vickers 1976:256–260). The samples from both areas were dark brown to brownish black in color, with the Cuyabeno samples being slightly more acidic (pH 4.9–5.8) than the San Pablo samples (pH 6.1–6.3). The samples from both areas showed relatively low levels of nitrogen and phosphorous, but high levels of potassium, calcium, and magnesium. These soils were from or near cultivated areas and were demonstrably adequate to support the traditional shifting cultivation practices of the Siona–Secoya (see Figure 15.4). However they did not represent the full range of soil types from the two areas nor did they reveal the distribution of such types over the landscape.

Fortunately, the data gap on soil type distributions has been addressed by scientists from the Ecuadorian *Ministerio de Agricultura y Ganadería* (MAG) and the French *Office de la Recherche Scientifique et Technique Outre-Mer* (ORSTOM). The scientists compiled a detailed (scale 1:250,000) edaphic map of eastern Ecuador (MAG–ORSTOM 1980). The MAG–ORSTOM map shows that the 1149 ha area around San Pablo (Figure 15.1) has three major soil categories. The best soils were formed by alluvial deposits over an ancient clay bed and consist of matter of volcanic origin mixed with sand and pebbles. These soils, which are technically classified as Vitrandepts and Dystrandepts, are loose, well drained, and relatively fertile. They occur on both sides of the Aguarico River, but are predominant in the western third of the area depicted in Figure 15.1. Overall, they make up 44.6% of the soils in the study area. These are the soils from which I drew my San Pablo samples.

The northeastern quarter of Figure 15.1 is characterized by ancient depressions that were excavated by river action and which are now partially refilled by decanted clays and organic residuals. These low and swampy areas have soils that are classified as Tropaquepts, Tropaquents, and Tropofibrists, and are not suitable for cultivation in their natural waterlogged state. However, there are a few alluvial dikes in the area with sufficient elevation to allow cultivation. My Cuyabeno samples were drawn from

Figure 15.4 *Siona–Secoya youth planting maize (Zea mays* L.*) in a burned garden plot at San Pablo.*

such dikes. The aforementioned swampy zones comprise 31.2% of the area in Figure 15.1. Finally, 10.7% of the area consists of rounded hills which are composed of highly weathered and compacted red-clay soils (Oxic Dystropepts). These soils have high levels of aluminum toxicity and inferior agricultural potential. They are found primarily in the southeastern corner and along the northern fringe of Figure 15.1. An additional 12.3% of the area consists of miscellaneous soil types that are not suitable for cultivation.

Overall, the MAG–ORSTOM map indicates that the San Pablo area has some of the best soils in Amazonian Ecuador. Nevertheless, more than half of the land area is unsuitable for cultivation due to poor drainage characteristics or fertility. The Siona–Secoya of San Pablo are fully aware of these characteristics and make their gardens on the more fertile alluvial terraces and dikes (see Vickers [1976:61–62] for the native classification of soil types). The 44.6% of the area that comprises the best soils cover 507 km^2. As discussed earlier, the Siona–Secoya are unlikely to have more than 60 ha, or 0.1% of these good soils under cultivation at any given moment. If we consider a new clearing rate of 20 ha/year and a 20-year fallow cycle for "abandoned" gardens, the cultivable land requirement for the San Pablo community grows to 400 ha, or 0.8% of the best soils areas. However, the Siona–Secoya are unlikely to make their gardens at sites that are far removed from the major rivers or streams, regardless of the soil quality, because of the added transportation costs, so another adjustment needs to be made in the estimate of the "prime" area for cultivation. If we limit the prime area to well-drained alluvial soils that occur within 1 km of a navigable stream, the estimate of 507 km^2 is reduced to 122 km^2, and the 400 ha land requirement rises to 3.3% of the total.

These calculations can lead to but one conclusion; cultivable land is not a major limiting factor for the Siona–Secoya in the San Pablo area (given their present population level). However, it does seem to limit the ability of all of these people to live in a single nucleated settlement because such a residence pattern would force them to travel distances to their gardens that they consider excessive. This analysis is an example of how it is possible to assess the availability of a given resource at the level of a specific community. If the focus of the analysis had been the Siona–Secoya community on the Cuyabeno River, the conclusions would have been very different because the overwhelmingly dominant soil types there are the heavily weathered, compacted, and toxic Oxic Dystropepts and the swampy Tropaquepts, Tropaquents, and Tropofibrists. The few alluvial dikes that are suitable for cultivation are used intensively, but most are less than a hectare in size and their limited occurrence constitutes a powerful limiting factor on cultivation and the subsequent human population that the area can support. Hence, a given environmental variable such as cultivable land may constitute a limiting factor in one community but may not in another.

GAME

The question of game availability is closely connected to the debate concerning whether protein is the primary limiting factor on aboriginal adaptation to Amazonia (see the discussion of this issue in the introductory chapter of this volume). During the early phases of this debate, the principal indicators that were employed were the mean daily intake per individual and the hunting yield per unit of time. As more anthropologists have returned from the field with quantitative data on these factors (e.g.,

Aspelin 1975; Beckerman 1980; Chagnon and Hames 1979; Hames 1979; Hawkes and Hill this volume; Vickers 1975, 1979; Yost and Kelley this volume), it has become increasingly clear that most Amazonian peoples are adept hunters whose success rates are higher than had been presumed, and who have surprisingly high protein intakes by world standards. However, such factors are only a part of the resource availability–resource utilization equation. Hunting yields and protein intakes that are more than adequate when human populations are low may not be so under different demographic conditions. And there are likely to be highly significant variations in the faunal inventories of differing Amazonian habitats.

Again, it is desirable to get a measure of the resource base, which in this case consists of the faunal populations of the area under study. As with a number of other ethnologists who have worked in Amazonia, I have made (in 1973 to 1975, and in 1979) empirical observations consisting of records of hunting, fishing, collecting, and gardening yields, plus dietary surveys. Although my hunting yield data do give an indication of the numbers of various species that were killed, I did not conduct actual censuses of animals in the wild. The reason for this is that animal censusing is not something that can be done casually by untrained people. It is a complex task that is best completed by professional zoologists who have some knowledge of the habits of the species to be studied. Unfortunately, there are very few published animal censuses that meet the needs of anthropologists who are attempting to study the human ecology of Amazonia. One of the most widely-cited studies used to support the idea of protein scarcity (Fittkau and Klinge 1973) is clearly deficient because it is really a study of plant biomass and makes impressionistic estimates about animal biomass that are not based on actual censuses of faunal populations. Furthermore, the study was conducted near the Amazonian metropolis of Manaus where one could hardly expect to find a representative number of game animals after centuries of postconquest settlement. In one of the best censuses of Neotropical mammals, Eisenberg and Thorington (1973) estimate that the biomass of Barro Colorado Island, Panama is 5317 kg/km^2. Although Barro Colorado has some distinctive attributes (see Hames and Vickers 1982:361), Eisenberg and Thorington conclude that their findings on its biomass levels are consistent with the findings from studies in other tropical regions of the world (1973:152). They also cite a study from Surinam by Walsh and Gannon (1967) that indicates a mammalian biomass of 3051 kg/km^2. In a previous publication (Vickers 1979), I calculated that the mammalian *game* biomass (i.e., only those species that are actually hunted by the Siona–Secoya) in the San Pablo area is 1882 kg/km^2 (prior to settlement and the application of hunting pressure). In order to make a conservative estimate, however, I reduced this to 1500 kg/km^2. If the total Surinam mammalian biomass estimate of 3051 kg/km^2 is actually more representative of Amazonian conditions, my estimate for game mammals in the San Pablo area could be further reduced to approximately 1080 kg/km^2. These figures indicate that the 1149 km^2 area (Figure 15.1) around San Pablo should contain between 1.2 and 2.2 million kg of game mammal biomass, and between 3.5 and 6.1 million kg of total mammalian biomass (excluding aquatic and volant species).

My research of 1973 to 1975 in the then-new settlement of San Pablo, indicates that the annual kill of game mammals is 17,333 kg. This represents a 1–2% kill of the estimated 1.2–2.2 million kg game mammal biomass in the 1149 ha territory. However, as I have discussed earlier, the area of heaviest hunting pressure occurs in the 225 km^2 area centering on the settlement of San Pablo. If hunting were limited to this area, the annual kill would represent 5–9% of its game mammal biomass.

Of course, the annual kill need not remain at a static level; as hunting pressure continues over a period of years the yield per unit of hunting time declines due to the depletion of game populations (Vickers 1980). Also, any increase in the human population would place more hunting pressure on the game resources.

Gross (1975:529) estimates that up to 20% of the game biomass could be killed each year "without depletion" because this rate would allow population replenishment at natural rates of reproduction. Although Gross is not specific as to how he derived this general estimate, it does seem to be based on the known gestation periods for specific Amazonian game animals (which are very short for rodents such as the agouti [*Dasyprocta*] and quite long for a large ungulate such as the tapir [*Tapirus terrestris*]).

All of these estimates must be considered provisional until we have better studies of Amazonian fauna, but with these best current estimates it appears that the game mammal resources of San Pablo are not deficient, given the current levels of the human population and their exploitation of the resource base. However, if a population of Amazonian hunters remains in a given village for an extended period of time, the efficiency of hunting (caloric yield relative to caloric expenditure) does decline (see Vickers [1980] for a discussion concerning the relationship between settlement age and hunting yields).

The data on soil and game resources that I have been discussing indicate that both play a role in Siona–Secoya migration patterns even though neither is limiting in an absolute sense. Insofar as can be determined, the Siona–Secoya have never lived in permanent villages; the usual pattern is for them to relocate their settlements at intervals ranging from 5 to 20 years. Such relocations undoubtedly have adaptive significance despite the fact that they are sometimes explained as resulting from interpersonal squabbles or even boredom with a given site (Vickers 1980). As discussed earlier, the soils in the immediate vicinity of San Pablo are excellent and relatively extensive by Amazonian standards, yet some households have begun to relocate from that village, citing the scarcity of land for gardens. In reality the people do not want to stay in their present houses and solve their land problem by leapfrogging the existing band of gardens because this would increase both their travel time in walking to the new gardens and their energetic costs of carrying produce back to the settlement. Hence they prefer moving to a new section of the river where they can make their new gardens close by their new houses. They likewise recognize that such a relocation in an area that has not been inhabited for some time will enhance their hunting yields. Although I have not made a detailed discussion of other resources in this chapter, a complete analysis should also consider fish and other aquatic resources, floral resources (for food, and building and craft materials), and any other resources or conditions that can be demonstrated to be significant to human adaptation. At San Pablo, for example, people are complaining that the area has been depleted of the palms that are used for thatch and cedar trees (*Cedrela odorata*), which are the preferred wood for making canoes.

The periodic migrations of the Siona–Secoya are not unpredictable moves through an unknown landscape. Life history materials indicate that the vast majority of settlement relocations occur within a definable space (usually a particular river drainage or section of one of the larger rivers, such as the Aguarico or Napo), and that new settlements are usually established on previously inhabited sites. The analysis of locational data from San Pablo and from the eighteenth-century Jesuit chronicles indicates that communities were and are composed of populations within the range of 100 to 300 individuals, with a mean territorial size of about 1100 to 1150 km^2. When the people settle at a given site, they exploit the surrounding area (approximately 225 km^2 in size) most intensively for their varied subsistence activities, with the outlying areas being rested (fallowed). After a period of years the people relocate their settlement somewhere within the rested area. This model of Siona–Secoya territoriality suggests that only about one-fifth of the territory receives intensive exploitation at any given time. The pattern of migration within the territory represents a means of both exploiting and maintaining the resource base. The motive for migration need not be the total depletion of a local area; rather, a move after several years at one site tends to enhance the cost–benefit ratio of the various subsistence activities because an area with relatively depleted resources is traded for one with relatively abundant resources. The five-to-one ratio of relatively unexploited land to heavily exploited land gives a theoretical ability to relocate in up to five areas within the territory before returning to the original site. In actual practice, such a mechanical rotation cycle need not be practiced because there is enough slack in the system to allow periodic movements to and from fewer than five preferred sites.

The Question of Territorial Defense

In assessing the spatial relationships between organisms and their habitats, ecologists differentiate between "home ranges" which are not defended and "territories" which are (Odum 1971:209). In this chapter, I have referred to the areas exploited by the Siona–Secoya and the Encabellado as territories. In this section I will give a brief discussion of why the territorial designation is appropriate.

Warfare has been documented in hundreds of lowland South American societies. Many observers have been concerned with the manifestations of warfare in specific groups, such as Chagnon's (1968a) description of the Yąnomamö as "the fierce people," the taking of human trophies such as Jívaro shrunken heads (Harner 1972), cannibalism (Staden 1928), and Mundurucú long-distance raiding (Murphy 1960). A wide variety of structuralist and mentalist hypotheses have been proposed to account for the witchcraft and warfare phenomenon in Amazonia. In the summary article on tropical forest warfare in the Handbook of South American Indians Lowie argues that "revenge seems to have been the foremost motive for warfare, but the Parintintin fought mainly for sport, and the Tupinamba to gain prestige and to acquire victims to be eaten. The craving for glory also figured largely, as indicated by the use of trophies (Lowie 1948:34)." Other prominent anthropologists take similar positions concerning Amazonian warfare. Harner attributes Jívaro war to the desire for heads and women (1972:182–183). Goldman states that the Cubeo practice war "to seize women and [for] revenge (1963:162)." Metraux sees the warfare pattern as the result of "murders, kidnappings, [and] sorcery suspicions (1949:384–385)." Such explanations are largely based on the manifest explanations of the native actors.

In contrast, Robert Murphy's analysis of Mundurucú warfare does not rest on the manifest explanations of the Indians themselves. He argues that the famous long-distance raids of the Mundurucú reduce internal tensions produced by a structural conflict that is caused by a matrilocal residence rule and a patrilineal descent rule (1960). Therefore Murphy's explanation is primarily a structural-functional one.

In discussing Yąnomamö warfare, Chagnon states that "the conflicts are not initiated or perpetuated with territorial gain as an objective or consequence . . . aggressiveness and warfare are admired, but their commitment to this way of life cannot be explained in terms of land shortage (1968b:110)." Instead, Chagnon sees their warfare as arising from sexual competition for women. This view can be subsumed under the evolutionary biological concept of "inclusive fitness" which holds that males compete for sexual rights to females in order to maximize the representation of their genes in suceeding generations. But evolutionary biologists also recognize the existence of resource competition as well as sexual competition (e.g. Wilson 1975:243). They subsume resource competition within their theory by pointing out that access to resources also enhances an individual's inclusive fitness because an organism must derive energy from its environment if it is to survive to reproduce. However, scholars such as Chagnon and Hames (this volume) view sexual competition as having primacy over resource competition as an explanation for Yąnomamö warfare.

Although I do not deny the existence of sexual competition and the fact that it is often related to agonistic behavior, I do not feel that sexual competition alone is an adequate explanation for organized human warfare. In my view, the warfare patterns characteristic of modern nation-states, ancient city-states, and horticultural societies such as those of the Amazon are much more obviously linked to resource competition and the forms of territoriality that derive from resource competition than to sexual competition. However, it must be emphasized that territoriality and resource competition are flexible responses to specific demographic and environmental conditions and are not inevitable. When human population densities are very low or resources low or unpredictable, they may be minimal (e.g., see Lee and DeVore 1968:12, 156–158; Dyson-Hudson and Smith 1978).

Other scholars also suggest economic or ecological hypotheses to account for Amazonian warfare (e.g., Lathrap 1970). Morey and Marwitt (1975) attempt to show examples of warfare conducted for rational economic gain (i.e., warfare in which the actors consciously pursue territorial acquisition, and raiding for slaves, crops, and other kinds of loot). They are able to document some cases of this type, but have difficulty dealing with conflict in which the motivation is ascribed to nonmaterial motives and/or in which there is no expropriation of territory or property. As a consequence of this difficulty, they conclude that the warfare patterns of the Jívaro and Tucano fail to support their model of economically motivated warfare. In the discussion that follows I will explain why universal property expropriations are not essential for a model of warfare based on ecological principles.

Steward classifies the western Tucanoans (including the Siona–Secoya and their Encabellado ancestors) as marginal fighters: "The Western Tucanoans were not notorious warriors . . . the special enemy of the Encabellado were the neighboring Awishira, a bellicose people who continually raided them for loot and women (1948:746)." And under the heading of "Some Atypical Peoples of the Montaña" Steward and Faron write, "In warfare, they are the victims of more aggressive tribes on the larger rivers, rather than genuine warriors themselves (1959:346)."

Despite these characterizations by Steward and Faron, historical accounts indicate the presence of a rich witchcraft and warfare complex among the Encabellado that was analogous to those of other Amazonian societies. This aboriginal complex included such features as sorcery accusations, raiding, defensive works, the taking of trophy heads, cannibalism, and the formation of strategic alliances between groups. I will document some of the events of this nature here because they are relevant to my interpretation that links Amazonian warfare to resource competition and territoriality.

In 1635, Captain Juan de Palacios and a group of soldiers and Franciscans descended the Aguarico and attempted to found a mission at a site 18 leagues above the mouth of the river. Palacios and some of his men were killed by Encabellado residing in this region (Juan de Velasco 1946, cited in Espinosa Perez 1955:11–14). Diego de Brieda and Andres de Toledo escaped downriver and journeyed all the way to Pará at the mouth of the Amazon. The arrival of the Spanish friars astounded the Portuguese who subsequently sent Pedro Teixeira on a return expedition to establish Portuguese dominion in the west. Seventy soldiers and several hundred Indians departed upriver with Teixeira on October 26, 1637. When the expedition reached the mouth of the Aguarico Teixeira ordered his main body to encamp there and travelled on to Quito with a small body of men. The Encabellado did not welcome this intrusion into their territory and attacked the Portuguese camp. In retaliation, the expedition burned the Encabellado settlements in the vicinity of the camp (Reis 1931:39).

In 1654 padre Raimundo de Santa Cruz entered Encabellado territory in search of a route to connect the missions of the Marañon and Quito. His party lost its way and five Jívaros were sent ashore to seek directions. They approached an Encabellado settlement where all but one were captured and beheaded (Chantre y Herrera 1901:164–165; Rodríguez 1684:189–193; cited in Espinosa Perez 1955:12, 17).

In the 1700s, Jesuits attempted to missionize the Encabellado. They found the Indians living in scattered settlements. As discussed previously, the Jesuit strategy was to unite the Encabellado in reductions along the banks of the Napo and Aguarico. The headmen of the various local groups were attracted to the iron tools of the missionaries but refused to join in unified settlements with other Encabellado groups. Fear of neighboring sorcerers was often given as the reason for resisting resettlement in the reductions, and compromises were effected by which the headmen brought their groups to locations along the river banks, but remained in autonomous settlements that were within or adjacent to their traditional territories.

At least 15 Encabellado missions were founded along the Aguarico and Napo between 1732 and 1742 (see Figure 15.2). The Jesuits never gave up their master plan of forming large settlements, but they could not make it work. The Indians either left the new settlements or revolted. On January 4, 1744 padre Francisco Real was assassinated at San Miguel de Ciecoya by the headman Curazaba, and seven Encabellado settlements were deserted as the word of his death spread. By 1769 the Indians of all but two missions had returned to their usual haunts.

The Jesuit journals indicate that the pattern of indigenous warfare continued even during their presence. For example, the Awishira attacked the Encabellado at San Juan Paratoas in 1745 (Chantre y Herrera 1901:400–405), and the Encabellado insisted on retribution because "it was their custom" (p. 403). Padre Martín Iriarte implored the Encabellado to refrain from killing the Awishira, but indicated that they might take prisoners. Iriarte and a lay helper named Salvador Sanchez accompanied the Encabellado on the raiding expedition. The Encabellado party journeyed down the Napo to the mouth of the Aushiri River and then trekked inland to a location near the Curaray. Here they surprised the Awishira and captured one woman. This raid was considered to be successful because "after this they did not discover signs [of the enemy] . . . in the forests which they crossed in search of game" (p. 405).

In the nineteenth century, there were no active missions among the western Tucanoans of the Aguarico and Napo rivers and contacts with them were sporadic. During this period outsiders referred to them as the Piojé. Simson (1886:193) states that they were still considered to be *aucas,* a generic term denoting savage status. It is evident from the oral traditions of the surviving Siona–Secoya that warfare continued. Some generations ago (the exact date is unknown) a western Tucanoan group known as the Tetetes was driven off the middle Cuyabeno River by other Tucanoans (probably Siona) from the Aguarico River. The actual raid consisted of a typically Amazonian early morning sneak attack in which a number of Tetetes were reported to have been killed. The motive for the attack is said to have been a sorcery accusation that occurred when one of the Siona began smoking tobacco during a visit that the Tetetes made to the Aguarico people. Following the raid on their settlement, the Tetetes retreated to the forests near the headwaters of the Cuyabeno and Paca Yacu Rivers; as of 1968, three surviving Tetetes were known to be living in this area. Sionas subsequently colonized the region from which the Tetetes had been displaced.

Langdon (1974) describes a series of sorcery killings of Siona shamans on the Putumayo River in the twentieth century. These events are corroborated by Siona migrants to the Aguarico River whom I interviewed during 1973 to 1975. The last reported murder of a shaman by raiders from another village took place in Peru circa 1972. Today the overt manifestations of the traditional witchcraft–warfare complex continue (see Figure 15.5), but the rate of raiding has declined due to depopulation and the restraining influences of missionaries and civil authorities.

All of these examples demonstrate that witchcraft and warfare among the western Tucanoans are qualitatively similar to the witchcraft and warfare patterns reported for other Amazonian societies. Furthermore, I have

Figure 15.5 *Siona shaman treating a female victim of sorcery.*

documented that these activities among the western Tucanoans are linked to settlement pattern and territoriality (i.e., the defense of the home range). Are we to believe that in some Amazonian cultures the witchcraft–warfare complex derives primarily from social structural causes, while in others it is essentially caused by a desire for revenge, and in yet others results from a competition for women? Or does a more fundamental process underlie these phenomena? The agonistic behaviors associated with witchcraft and warfare among the western Tucanoans can be understood as complementary aspects of resource competition and territoriality, and I believe that this model can be applied to many other Amazonian societies as well. Although structural considerations, revenge motivation, and sexual competition all exist to one degree or another and are not mutually exclusive factors, resource competition appears to be a more parsimonious and general explanation for warfare in Amazonian societies, as well as for warfare in more intensive agricultural and industrial societies.

It should be emphasized that resource competition is not synonymous with the narrowly defined concept of warfare for rational economic gain as proposed by Morey and Marwitt (discussed previously). In their model, property or goods must change hands if warfare is to have economic significance. This approach to Amazonian warfare is too simplistic. Property or land do not have to change hands in order for warfare to have material consequences. Rather, the agonistic behavior associated with witchcraft and warfare serves to establish, advertise and defend the resource bases of Amazonian (and other) populations. That is, it tends to allocate and maintain the resources that are necessary for the survival of the people residing in specific communities. Territorial gains and losses do occur, but they need not occur continuously nor in all cases. When neighboring societies have roughly equivalent populations and levels of technology and social organization, all-out conquest is likely to be the exception rather than the rule. (Descriptions of highland New Guinea warfare by Rappaport [1968] and Heider [1979] corroborate this view.)

Conclusion

In this paper I have attempted to define the spatial dimensions of a particular Siona–Secoya community and to compare my findings with a reconstruction of Encabellado territoriality in the eighteenth century. The data that have been analyzed suggest considerable spatial concordances for the two periods and I have argued that the observed similarities derive from similar resource inventories and subsistence technologies. I have also proposed that the ecological concept of territoriality is applicable to aboriginal man–land relationships in Amazonia, and that warfare and witchcraft are associated with resource competition and territorial behavior. I have also argued that future studies of Amazonian human adaptation should investigate the spatial parameters of specific communities in order to advance our understanding of resource availabilities and exploitation strategies. Such studies should provide powerful empirical keys to the explanation of Amazonian cultural patterns.

Acknowledgments

My 1973–1974 fieldwork among the Siona–Secoya in Ecuador was funded by the Doherty Foundation, while the 1974–1975 period was supported by the National Institute of Mental Health (Grant # 1 Fol MH 58552-01). Subsequent research from June to August 1979 was made possible by a grant from the Florida International University Foundation, Inc., and the College of Arts and Sciences and Latin American and Caribbean Center of Florida International University. The *Instituto Nacional de Antropología e Historia* of Ecuador (INAH) provided my host-country affiliation. I would like to thank Mark B. Rosenberg, Anthony P. Maingot, James A. Mau, Raymond B. Hames, James A. Yost, and Edite Vickers for their assistance during various phases of the fieldwork and in the preparation of this paper. This work would not have been possible without the cooperation of the Siona and Secoya peoples of the Aguarico River region of eastern Ecuador.

Notes

[1]Gross (this volume, p. 448) argues for a regional approach that "implies abandoning . . . the single village or village cluster as the unit of research." Although I have no quarrel with the desirability of regional analyses, I believe that high-quality community studies must form the building blocks for larger regional studies. Furthermore, the methods of community-level data collection and analysis have not yet been developed to their full potential.

[2]On May 22, 1981 Ecuador's president, Jaime Roldós Aguilar, issued a decree to terminate the work of the Summer Institute of Linguistics (SIL) in the country. As of June 1982 this decree is being implemented on a step-by-step basis.

References

Aguirre, M.
1941 La Compañía de Jesús en Venezuela. Caracas: Monte Albán.

Albuquerque, M. de
1969 A Mandioca na Amazônia. Belem: Supertêndencia de Desenvolvimiento da Amazônia.

Alexander, R.
1974 The Evolution of Social Behavior Annual Review of Ecology and Systematics 5:325–383.

Alexander, R. and G. Borgia
1978 Group Selection, Altruism and the Levels of Organization of Life. Annual Review of Ecology and Systematics 9:449–474.

Altmann, J.
1974 The Observational Study of Behavior. Behaviour 48:1–41.

Annegers, J.
1973 Seasonal Food Shortages in West Africa. Ecology of Food and Nutrition 2:251–257.

Århem, K.
1976 Fishing and Hunting among the Makuna. Arstryck (Annals): 27–44. Göteborg, Sweden.

Arvelo-Jiménez, N.
1971 Political Relations in a Tribal Society. Ph.D. dissertation. Department of Anthropology, Cornell University.

Aspelin, P.
1975 External Articulation and Domestic Production: The Artifact Trade of the Mamainde of Northwestern Mato Grosso, Brazil. Latin American Studies Program Dissertation Series No. 58. Cornell University.

1976 Nambicuara Economic Dualism: Levi-Strauss in the Garden Once Again. Bijdragen tot de Taal, Land-en Volkenkunde 132:135–160.

1979 Food Distribution and Social Bonding among the Mamaindê of Mato Grosso, Brazil. Journal of Anthropological Research 35(3):309–327.

Atran, S.
1981 Natural Classification. Social Science Information 20(1):37–91

Axelrod, A.
1964 Nutrition in Relation to Acquired Immunity. *In* Modern Nutrition in Health and Disease, 3rd ed., M. Wohl and M. Goodland, eds. pp. 94–112. Philadelphia: Lea and Febiger.

Ayres, M. and F. Salzano
1972 Health Status of Brazilian Cayapó Indians. Tropical and Geographical Medicine 24(2):178–185.

Baldus, H.
1970 Tapirapé: Tribo Tupí no Brasil Central. São Paulo: Companhia Editora Nacional.

Balikci, A.
1968 The Netsilik Eskimos: Adaptive Processes. *In* Man the Hunter, R. Lee and I. DeVore, eds. pp. 78–82. Chicago: Aldine.

Bamburger, J.
1968 The Adequacy of Kayapó Ecological Adjustment. Proceedings, International Congress of Americanists, Vol. 3, pp. 373–379. Stuttgart.
1974 Naming and the Transmission of Status in a Central Brazilian Society. Ethnology 13(4):533–560.
1979 Exit and Voice in Central Brazil: The Politics of Flight in Kayapó Society. *In* Dialectical Societies: The Gê and Bororo of Central Brazil, D. Maybury-Lewis, ed. pp. 147–178. Cambridge: Harvard University Press.

Banner, H.
1961 O Indio Kayapó em seu Acampamento. Boletim do Museu Paraense Emilio Goeldi, Belem, Para, 13:1–51.

Barandiaran, D. de
1962 Actividades Vitaes de Subsistencia de los Indios Yekuana o Makiritare. Antropologica 11:1–29.
1967 Agricultura y Recolección entre los Indios Sanema–Yanoama o el Hacha de Piedra y la Psicologia Paleolítica de los Mismos. Antropologica 19:24–50.

Barker, J.
1953 Memoria sobre la Cultura de los Guaika. Boletín Indigenista Venezolana 1(3–4):433–489.

Barrett, O.
1928 The Tropical Crops. New York: Macmillan.

Bates, D. and S. Lees
1979 The Myth of Population Regulation. *In* Evolutionary Biology and Human Social Behavior, N. Chagnon and W. Irons, eds. pp. 273–289. North Scituate, Massachusetts. Duxbury Press.

Bates, H.
1864 The Naturalist on the River Amazons. 2nd ed. London: John Murray.

Beckerman, S.
1975 The Cultural Energetics of the Barí (Motilones Bravos) of Northern Columbia. Ph.D. dissertation. Department of Anthropology, University of New Mexico, Albuquerque.
1978 Comment on Ross. Current Anthropology 19:17–18.
1979 The Abundance of Protein in Amazonia: A Reply to Gross. American Anthropologist 81(3):533–560.
1980 Fishing and Hunting by the Barí of Colombia. *In* Working Papers on South American Indians, Vol. 2, R. Hames, ed. pp. 67–111. Bennington, Vermont: Bennington College.

Beebe, W.
1925 Studies of a Tropical Jungle: One Quarter of a Square Mile of Jungle at Kartabo, British Guiana. Zoologica 6(1):5–193.

Behrens, C.
1981 Time Allocation and Meat Procurement among the Shipibo Indians of Eastern Peru. Human Ecology 9:189–220.

Bergman, R.
1974 Shipibo Subsistence in the Upper Amazon Rainforest of Peru. Ph.D. dissertation. Department of Geography, University of Wisconsin, Madison.
1980 Amazon Economics: The Simplicity of Shipibo Indian Wealth. Ann Arbor: University Microfilms International.

Berlin, B.
1972 Speculations on the Growth of Ethnobotanical Nomenclature. Journal of Language in Society 1:51–86.
1973 Folk Systematics in Relation to Biological Classification and Nomenclature. Annual Review of Ecology and Systematics 4:259–271.
1976 The Concept of Rank in Ethnobiological Classification: Some Evidence from Aguaruna Folk Botany. American Ethnologist 3:381–399.

Berlin, B. and E. Berlin
1978 Etnobiología, Subsistencia, y Nutrición en una Sociedad de la Selva Tropical: Los Aguaruna (Jíbaro). *In* Salud y Nutrición en Sociedades Nativas, A. Chirif, ed. pp. 13–47. Lima: Centro de Investigatión y Promoción Amazónica.

Berlin, B., J. Boster, and J. O'Neill
1981 The Perceptual Basis of Ethnobiological Classification: Data from Aguaruna Ornithology. Journal of Ethnobiology 1:95–108.

Berlin, B., D. Breedlove, and P. Raven
1973 General Principles of Classification and Nomenclature in Folk Biology. American Anthropologist 75:214–242.
1974 Principles of Tzeltal Plant Classification: An Introduction to the Botanical Ethnography of a Mayan Speaking Community in Highland Chiapas. New York: Academic Press.

Berlin, B. and R. McDiarmid
Ms. Theoretical Implications of Jívaro Lizard Classification. Department of Anthropology, University of California, Berkeley.

Berlin, B., and J. Patton
1979 La Clasificación de los Mamíferos de los Aguaruna, Amazonas, Perú. Berkeley: Language Research Laboratory.

Berlin, B. and C. Swift
Ms. The Aquatic Resources of the Huambisa Jívaro. Department of Anthropology, University of California, Berkeley.

Berlin, E.
In press Social Implications of Dietary Patterns in Three Amazonian Communities. *In* Energy in Tropical Ecosystems, D. Cattle and K. Schwerin, eds.

Berlin, E. and E. Markell
1977 An Assessment of the Nutritional and Health Status of an Aguaruna Jivaro Community, Amazonas, Peru. Ecology of Food and Nutrition 6:69–81.

Bidou, P.
1976 Les Fils de L'Anaconda Céleste (Les Tatuyo): Etude de la Structure Socio-Politique. Unpublished Ph.D. dissertation for Troisième Cycle, University of Paris.

Billewicz, W.
1967 A Note on Body Weight Measurements and Seasonal Variation. Human Biology 39:241–250.

Birdsell, J.
1958 On Population Structure in Generalized Hunting and Gathering Populations. Evolution 12:189–205.

Black, F. L., W. J. Hierholzer, D. P. Black, S. H. Lamm, and L. Lucas
1977 Nutritional Status of Brazilian Kayapo Indians. Human Biology 49(2):139–154.

Blake, E.
1977 Manual of Neotropical Birds, Vol. 1. Chicago: University of Chicago Press.
Blakeslee, J.
Ms. Lluvia en Limonchoca, Provincia de Napo, Ecuador. December 20, 1960 to January 10, 1980. Quito, Ecuador.
Blurton-Jones, N. and R. Sibly
1978 Testing the Adaptiveness of Culturally Determined Behavior: Do Bushmen Women Maximize their Reproductive Success by Spacing Births Widely and Foraging Seldom? *In* Human Behavior and Adaptation, Society for the Study of Human Biology, Symposium 18. V. Reynolds and N. Blurton-Jones, eds. pp. 135–157. London: Taylor and Francis.
Bodley J. and F. Benson
1979 Cultural Ecology of Amazonian Palms. Laboratory of Anthropology, Reports of Investigation No. 56. Washington State University, Pullman.
Boserup, E.
1965 The Conditions of Agricultural Growth. Chicago: Aldine.
Boster, J.
In press A Comparison of the Diversity of Jivaroan Gardens with that of the Tropical Forest. Human Ecology.
Bourliere, G.
1963 Observations on the Ecology of Some Large African Mammals. *In* African Ecology and Human Evolution, F. Howell and F. Bourliere, eds. pp. 43–54. Chicago: Aldine.
Brand, D.
1943 Tapioca from a Brazilian Root. *In* Agriculture in the Americas, Vol. 3. pp. 93–96. United States Department of Agriculture, Washington, D.C.: U.S. Government Printing Office.
Brieger, F., J. Gurgel, E. Paterniani, A. Blumenschein, and M. Alleoni
1958 Races of Maize in Brazil and other Eastern South American Countries. National Science Foundation, National Research Council Publication 593. Washington, D.C.
Brown, J.
1964 The Evolution of Diversity in Avian Territorial Systems. Wilson Bulletin 6:160–169.
1975 The Evolution of Behavior. New York: Norton.
Brown, M.
1980 Una Paz Incierta: Un Estudio Etnográfico Sobre los Aguaruna del Alto Río Mayo, Perú. Lima: Centro Amazónico de Antropología y Aplicación Práctica.
Bulmer, R.
1970 Which Came First, the Chicken or the Egg-head? *In* Echanges et Communications, J. Pouillon and P. Miranda, eds. pp. 1069–1091. Paris: Mouton.
1974 Folk Biology in the New Guinea Highlands. Social Science Information 13:9–28.
Bulmer, R. and J. Menzies
1972–1973 Karam Classification of Marsupials and Rodents. Journal of the Polynesian Society 81(4):477–479, 82(1):86–107.
Bulmer, R. and M. Tyler
1968 Karam Classification of Frogs. Journal of the Polynesian Society 77:333–385.
Butt-Colson, A.
1972 Land Use and Social Organization of Tropical Forest Peoples of the Guianas. *In* Human Ecology in the Tropics, J. Garlick and R. Keay, eds. pp. 33–63. London: Pergamon Press.
Butzer, K.
1971 The Significance of Agricultural Dispersal into Europe and Northern Africa. *In* Prehistoric Agriculture, Stuart Streuver, ed. pp. 313–334. Garden City, New York: Natural History Press.

Campos, R.
1977 Producción de Pesca y Caza en una Comunidad Shipibo en el Río Pisqui. Amazonía Peruana 2(2):34–56.

Cañadas, L.
1977 Mapa Bioclimático: Ecuador. Ministerio de Agricultura y Ganadería, Dirección de Planificación, Regionalización. Quito, Ecuador: ORSTOM.

Carneiro, R.
1957 Subsistence in Social Structure: An Ecological Study of the Kuikuru Indians. Ph.D. dissertation. Department of Anthropology, University of Michigan, Ann Arbor. University Microfilms, Ann Arbor.
1960 Slash-and-Burn Agriculture: A Closer Look at its Implications for Settlement Patterns. *In* Men and Cultures: Selected Papers of the International Congress of Anthropological and Ethnological Sciences, A. Wallace, ed. pp. 229–234. Philadelphia: University of Pennsylvania Press.
1961 Slash-and-Burn Cultivation Among the Kuikuru and its Implications for Cultural Development in the Amazon Basin. *In* The Evolution of Horticultural Systems in Native South America, Causes and Consequences: A Symposium, J. Wilbert, ed. pp. 46–67. Antropológica. Supplement Publication No. 2.
1970a The Transition from Hunting to Horticulture in the Amazon Basin. Proceedings of the VIII International Congress of Anthropological and Ethnological Sciences, Tokyo-Kyoto, Vol. 3, pp. 244–248.
1970b A Theory of the Origin of the State. Science 169:733–738.
1970c Hunting and Hunting Magic Among the Amahuaca of the Peruvian Montaña. Ethnology 9:331–341.
1974 On the Use of the Stone Axe by the Amahuaca Indians of Eastern Peru. Ethnologische Zeitschrift Zürich 1:107–122.
1977 Recent Observations on Shamanism and Witchcraft Among the Kuikuru Indians of Central Brazil. Annals of the New York Academy of Sciences 293:215–228.
1978 The Knowledge and Use of Rain Forest Trees by the Kuikuru Indians of Central Brazil. *In* The Nature and Status of Ethnobotany, R. Ford, ed. pp. 210–216. Museum of Anthropology, University of Michigan, Anthropological Papers, No. 67.
1979a Tree Felling with the Stone Axe: An Experiment Carried out Among the Yanomamö Indians of Southern Venezuela. *In* Ethnoarchaeology: Implications of Ethnography for Archaeology, C. Kramer, ed. pp. 21–58. New York: Columbia University Press.
1979b Forest Clearance Among the Yanomamö, Observations and Implications. Antropólogica 52:39–76.

Carneiro, R.
Ms. Village Splitting as a Function of Village Size. American Museum of Natural History, New York.

Cassidy, C.
1980 Nutrition and Health in Agriculturalists and Hunter–Gatherers: A Case Study of Two Prehistoric Populations. *In* Nutritional Anthropology, N. Jerome, R. Kandel, and G. Pelto, eds., pp. 117–146. Pleasantville, New York. Redgrave.

Chagnon, N.
1968a Yąnomamö: The Fierce People. New York: Holt, Rinehart and Winston. (See also revised 2nd ed., 1977).
1968b Yąnomamö Social Organization and Warfare. *In* Anthropology of Armed Conflict and Aggression. M. Fried, M. Harris, and R. Murphy, eds. pp. 109–159. Garden City, New York: Natural History Press.
1973 The Culture-ecology of Shifting (Pioneering) Cultivation among the Yąnomamö Indians. *In* Peoples and Cultures of Native South America, D. Gross, ed. pp. 126–144. New York: Doubleday and Natural History Press.

1974 Studying the Yąnomamö. New York: Holt, Rinehart and Winston.
1975 Genealogy, Solidarity and Relatedness: Limits to Local Group Size and Patterns of Fissioning in an Expanding Population. Yearbook of Physical Anthropology 19:95–110.
1975 Response to Marvin Harris' "Protein Theory of Warfare." Psychology Today 8:6–7.
1979a Is Reproductive Success Equal in Egalitarian Society? *In* Evolutionary Biology and Human Social Behavior, N. Chagnon and W. Irons, eds. pp. 373–401. North Sciutate, Massachusetts: Duxbury Press.
1979b Mate Competition, Favoring Close Kin, and Village Fissioning among the Yąnomamö Indians. *In* Evolutionary Biology and Human Social Behavior, N. Chagnon and W. Irons, eds. pp. 86–131. North Sciutate, Massachusetts: Duxbury Press.
1980 Highland New Guinea Models in the South American Lowlands. *In* Working Papers on South American Indians, Vol. 2, R. Hames, ed. pp. 111–130, Bennington, Vermont: Bennington College.

Chagnon, N. and R. Hames
1979 Protein Deficiency and Tribal Warfare in Amazonia: New Data. Science 20(3):910–913.

Chagnon, N. and R. Hames
1980 La "Hipótesis Proteica" y la Adaptación Indígena a la Cuenca del Amazonas: una Revisión Crítica de los Datos y la Teoría. Interciencia 5(6):346–358.

Chagnon, N. and W. Irons, eds.
1979 Evolutionary Biology and Human Social Behavior: An Anthropological Perspective. North Scituate, Massachusetts: Duxbury Press.

Chagnon, N. and T. Melancon
1981 Reproduction, Numbers of Kin and Epidemics in Tribal Populations: A Case Study. Paper presented to the IUSSP Conference Center for Population Studies, Harvard University, 9–11 February 1981.

Chagnon, N., J. Neel, L. Weitkamp, H. Gershowitz, and M. Ayres
1970 The Influence of Cultural Factors on the Demography and Pattern of Gene Flows from the Makiritare (Ye'kwana) to the Yąnomamö Indians. American Journal of Physical Anthropology 32:339–349.

Chance, N.
1966 The Eskimo of North Alaska. New York: Holt, Rinehart and Winston.

Chantre y Herrera, J.
1901 Historia de las Misiones de la Compañía de Jesús en el Marañón Español. Madrid: Imprenta de A. Avrial.

Charnov. E.
1973 Optimal Foraging: Some Theoretical Explorations. Ph.D. dissertation. Department of Biology, University of Washington.
1976 Optimal Foraging: the Marginal Value Theorem. Theoretical Population Biology 9:129–136.

Charnov, E., G. Hyatt, and G. Orians
1976 Ecological Implications of Resource Depression. The American Naturalist 110:247–259.

Charnov, E. and G. Orians
1973 Optimal Foraging: Some Theoretical Explanations. Department of Biology, University of Utah, Salt Lake City.

Chen, L., A. Chowdhury, and S. Huffman
1979 Seasonal Dimensions of Energy Protein Malnutrition in Rural Bangladesh: The Role of Agriculture, Dietary Practices, and Infection. Ecology of Food and Nutrition 8:175–187.

Chiarelli, A.
1972 Taxonomic Atlas of Living Primates. New York: Academic Press.
Chirif, A. and C. Mora
1977 Atlas de Comunidades Nativas. Sistema Nacional de Apoya a la Movilización Social. Lima: Dirección de Organizaciones Rural.
Chodkiewicz, L.
1980 Gasoline and Proteins. Paper presented at the 79th Annual Meeting of the American Anthropological Association, Washington, D.C., December 3–7, 1980
Chumap Lucia, A. and M. García-Rendueles
1979 "Duik Muun . . .": Universo Mítico de los Aguaruna. Lima: Centro Amazónico de Antropología y Aplicación Práctica. Serie Antropológica II.
Civrieux, M. de
1970 Watunna: Mitología Makiritare. Caracas: Monte Avila Editores.
Clastres, P.
1968 Ethnographie des Indiens Guayaki (Paraguay–Brésil). Journal de la Société des Américanistes de Paris 57:9–61.
1972 The Guayakí. *In* Hunters and Gatherers Today, M. Bicchieri, ed. pp. 138–174. New York: Holt, Rinehart and Winston.
Cochrane, T. and P. Sanchez
In press Land Resources of the Amazon Basin. *In* Land Use and Agricultural Resources in the Amazon Basin, S. Hecht, and G. Nores, eds. CIAT.
Comas, J.
1971 Anthropometric Studies in Latin American Indian Populations. *In* The Ongoing Evolution of Latin American Populations, F. Salzano, ed. pp. 333–404. Springfield, Illinois: C. C. Thomas.
Cook, W.
1909 Through the Wildernesses of Brazil by Horse, Canoe and Float. New York, American Tract Society (4).
Cooper, J.
1942 Areal and Temporal Aspects of Aboriginal South American Culture. Primitive Man 15:1–38.
Coursey, D.
1973 Cassava as Food: Toxicity and Technology. Chronic Cassava Toxicity: Proceedings of an Interdisciplinary Workshop, London, England, 29–30 January 1973, pp. 27–36. Ottawa: International Development Research Centre Monograph.
Covich, A.
1976 Analyzing Shapes of Foraging Areas: Some Ecological and Economic Theories. Annual Review of Ecology and Systematics 7:235–257.
Crocker, J.
1969a Men's House Associates among the Eastern Bororo. Southwestern Journal of Anthropology 25:236–260.
1969b Reciprocity and Hierarchy among the Eastern Bororo. Man 4:184–194.
Crocker, W.
1972 The Non-Adaptation of a Savanna Indian Tribe (Canella, Brazil) to Forced Forest Relocation: An Analysis of Factors. Anais. I Seminario de Estudos Brasileiros. São Paulo: Setembro.
1974 Extramarital Sexual Practices of the Ramkokamekra–Canela Indians: An Analysis of Sociocultural Factors. *In* Native South Americans: Ethnology of the Least-Known Continent, P. Lyon, ed. pp. 184–194. Boston: Little, Brown.
1979 Canela Kinship and the Question of Matrilineality. *In* Anthropological Perspectives, M. Margolis, and W. Carter, eds. pp. 225–249. New York: Columbia University Press.

Damas, D.
1972 The Copper Eskimo. *In* Hunters and Gatherers Today, M. Bicchieri, ed. pp. 3–50. New York: Holt, Rinehart and Winston.
Da Matta, R.
1973 A Reconsideration of Apinayé Social Morphology. *In* Peoples and Cultures of Native South America, D. Gross, ed. pp. 277–291. Garden City, New York: Natural History Press.
1976 Um Mundo Dividido: A Estrutura Social dos Indios Apinayé. Petropolis: Editora Vozes, Ltda.
Dart, R. A.
1953 The Predatory Transition from Ape to Man. International Anthropological and Linguistic Review 1:201–217.
Davis, S.
1977 Victims of the Miracle. New York: Cambridge University Press.
DeBoer, W.
1981 Buffer Zones in the Cultural Ecology of Aboriginal Amazonia: an Ethnohistorical Approach. American Antiquity 48:364–377.
Denevan, W.
1966a A Cultural–Ecological View of the Former Aboriginal Settlement in the Amazon Basin. The Professional Geographer 18(6):346–351.
1966b The Aboriginal Cultural Geography of the Llanos de Mojos of Bolivia. Ibero-Americana No. 48: 1–185. Berkeley: University of California Publications.
1970 The Aboriginal Population of Western Amazonia in Relation to Habitat and Subsistence. Revista Geográfica 72:61–86.
1976 The Native Population of the Americas in 1492. Madison: University of Wisconsin Press.
Diamond, J.
1966 Zoological Classification Systems of a Primitive People. Science 151:1102–1104.
Divale, W. and M. Harris
1976 Population, Warfare, and the Male Supremacist Complex. American Anthropologist 78:521–538.
Dobyns, H.
1966 Estimating Aboriginal American Population: An Appraisal with a New Hemispheric Estimate. Current Anthropology 7:395–416.
Dole, G.
1978 The Use of Manioc Among the Kuikuru: Some Interpretations. *In* The Nature and Status of Ethnobotany, R. Ford, ed. pp. 217–247. Museum of Anthropology, University of Michigan, Anthropological Papers, No. 67.
Dreyfus, S.
1963 Les Kayapó du Nord, Etat de Pará, Brésil: Contribution a l'Etude des Indiens Gê. Paris: Mouton & Co.
1972 Introduction. Journal de la Société des Américanistes 61:9–16.
Dricot, J. and Ch. Dricott-D'Ans
1977 Influence des Transformations Socio-économiques sur l'état de Nutrition de Indiens Machiguenga (Amazonie Péruvienne): Aspects Méthodologiques. Biometric Humaine 12 (2,3,4):77–89.
Drico-D'Ans, C. and J. Dricot
1978 Influence de l'Acculturation sur la Situation Nutritionnelle en Amazonie Péruvienne. Annales de la Société Belge de Médecine Tropicale 5811(1):39–48.
Dufour, D.
1979 Insects in the Diet of Indians in the Northwest Amazon. Paper presented at the 1979 Annual Meeting of the American Association of Physical Anthropology, Buffalo, New York.
1981 Household Variation in Energy Flow in a Population of Forest Horticulturalists.

Ph.D. dissertation. Department of Anthropology, State University of New York at Binghamton.
Ms. The Composition of Some Vaupés Foods. University of Colorado, Boulder, Colorado.

Dufour, D. and J. Zarucchi
1979 *Monopteryx angustifolia* and *Erisma Japura:* Their use by Indigenous Peoples in the Northwestern Amazon. Botanical Museum Leaflets 27(3–4):69–91.

Duke, J., D. Aulik, and T. Plowman
1975 Nutritional Value of Coca. Botanical Museum Leaflets 24(6):113–119.

Durham, W.
1976 Resource Competition and Human Aggression, Part 1: A Review of Primitive Warfare. Quarterly Review of Biology 51:381–415.

Dwyer, P.
1976 An Analysis of Rofaifo Mammal Taxonomy. American Ethnologist 3:425–445.

Dyson-Hudson, R. Smith, and E. Smith
1978 Human Territoriality: An Ecological Reassessment. American Anthropologist 80:21–41.

The Economist (International Edition)
1979 Will Cowboys Wipe Indians Out? June 30, 1979, pp. 66–67.

Eisenburg, J. and R. Thorington, Jr.
1973 A Preliminary Analysis of a Neotropical Mammal Fauna. Biotropica 5(3):15–161.

Emlen, J.
1966 The Role of Time and Energy in Food Preference. American Naturalist 100:611–617.
1973 Ecology: an Evolutionary Approach. Reading, Massachusetts: Addison-Wesley.

Emmons, L.
Ms. The Mammals of Ecuador. Quito, Ecuador.

Erasmus, C.
1956 Culture Structure and Process: The Occurrence and Disappearance of Reciprocal Farm Labor. Southwestern Journal of Anthropology 12:444–469.

Espinosa Perez, L.
1955 Contribuciones Lingüísticas y Etnográficas sobre Algunos Pueblos Indígenas del Amazonas Peruano. Madrid: Consejo Superior de Investigaciones Científicas, Instituto Bernardo Sahagún.

Estioko-Griffin, A. A., and P. B. Griffin
1981 Woman the Hunter. *In* Woman the Gatherer, Frances Dahlberg, ed. pp. 121–151. New Haven: Yale University Press.

Eveleth, P. and J. Tanner
1976 Worldwide Variation in Human Growth. International Biological Programme 8. Cambridge: Cambridge University Press.

Fagundes-Neto, U., R. G. Baruzzi, J. Wehba, W. S. Sivestrini, M. B. Morais, and M. Cainelli
1981 Observations of the Alto Xingu Indians (Central Brazil) with Special Reference to the Nutritional Evaluation in Children. American Journal of Clinical Nutrition 34:2229–2235.

Falesi, I.
1974 Soils of the Brazilian Amazon. *In* Man in the Amazon, C. Wagley, ed. pp. 201–229. Gainesville: University Presses of Florida.

FAO/WHO (United Nations Food and Agricultural Organization and World Health Organization)
1973 Energy and Protein Requirements: Report of a Joint FAO/WHO Ad Hoc Expert Committee. WHO Technical Report, series no. 522. Geneva: WHO.

Feit, H.
1973 The Ethno-ecology of the Waswanipi Cree: or How Hunters Can Handle Their

Resources. *In* Cultural Ecology, Bruce Cox, ed. pp. 115–125. Toronto: McClelland and Stewart.

Feld, Steven

1979 Sound and Sentiment: Birds, Weeping, Poetics, and Song in Kaluli Expression. Ph.D. dissertation. Department of Anthropology, University of Indiana, Bloomington.

Fittkau, E.

1970 Role of Caimans in the Nutrient Regime of Mouth Lakes Affluents. Biotropica 2:138–142.

1973 Crocodiles and the "Nutrient" Metabolism of Amazon Waters. Amazonia 4:103–133.

Fittkau, E., U. Irmler, W. Junk, F. Reiss, and G. Schmidt

1975 Productivity, Biomass, and Population Dynamics in Amazonian Water Bodies. *In* Tropical Ecological Systems: Trends in Terrestrial and Aquatic Research, F. Golley and E. Medina, eds. pp. 289–311. New York: Springer Verlag.

Fittkau, E. and H. Klinge

1973 On Biomass and Trophic Structure of the Central Amazon Rain Forest Ecosystem. Biotropica 5:1–14.

Flannery, K.

1968 Archaeological Systems Theory and Early Mesoamerica. *In* Anthropological Archaeology in the Americas, B. Meggers, ed. pp. 67–87. Washington D.C.: The Anthropological Society of Washington.

Fleming, T.

1971 Population Ecology of Three Species of Neotropical Rodents. Miscellaneous Publications of the Museum of Zoology 143:1–77. Ann Arbor: University of Michigan.

Flowers, N., D. Gross, M. Ritter, and D. Werner

1982 Variation in Swidden Practices in Four Central Brazilian Societies. Human Ecology 10(2):203–217.

Freeman, L., Jr.

1968 A Theoretical Framework for Interpreting Archaeological Materials. *In* Man the Hunter, R. Lee and I. DeVore, eds. pp. 262–267. Aldine: Chicago.

Frikel, P.

1969 Migração, Guerra e Sobrevivencia Suiá. Revista de Antropologia 17:106–136.

Furley, P.

1980 Development Planning in Rondonia Based on Natural Renewable Resource Surveys. *In* Land, People and Planning in Contemporary Amazonia, Occasional Publications, No. 3, F. Barbira-Scazzocchio, ed. pp. 37–45. Cambridge: Centre of Latin American Studies, Cambridge University.

Galvão, E.

1953 Cultura e Sistema de Parentesco ods Tribus do Alto Rio Xingu. Boletim do Museu Nacional (Antropologia). Rio de Janeiro. n.s. 14:1–56.

1959 Aculturação Indígena no Rio Negro. Boletim do Museu Paraense Emilio Goeldi: n.s. Antropologia 7:1–60.

1964 Encontro de Sociedades Tribal e Nacional no Rio Negro. Actas y Memorias, XXXV Congreso Internacional de Americanistas 3:329–340.

Galvão, E. and A. de Oliveira

1973 A Situação Atual dos Baniwa (Alto Rio Negro). O Museu Goeldi no Sesquicentenario. Belem: Publicações Avulsas.

Giaccaria, B. and A. Heide

1972 Xavante (Akwẽ Uptabi: Povo Autêntico). São Paulo: Editorial Dom Bosco.

Gilmore, R.

1950 Fauna and Ethnozoology of South America. *In* Handbook of South American

Indians, Vol. 6, J. Steward, ed. pp. 345–564. Washington, D.C.: U.S. Government Printing Office.

Glanville, E. and R. Geerdink

1970 Skinfold Thickness, Body Measurements and Age Changes in Trio and Wajana Indians of Surinam. American Journal of Physical Anthropology 32:455–461.

Goldman, I.

1963 The Cubeo. University of Illinois Studies in Anthropology No. 2. Urbana: University of Illinois Press.

Gomez-Pompa, A.

1973 Recovery of Tropical Ecosystems. *In* Fragile Ecosystems, E. Farnsworth and F. Golley, eds. pp. 113–138. New York: Springer-Verlag.

Gonzalez, O.

1977 Los Indígenas de Las Riberas del Guainía–Río Negro. Paper presented at the 7 Cipecam.

1980 Mitología Guarequena. Caracas: Monte Avila.

Good, K.

1978 Comment on Ross. Current Anthropology 19(1)20–21.

Goulding, M.

1980 The Fishes and the Forest: Explorations in Amazonian Natural History. Berkeley: University of California Press.

Gourou, P.

1953 The Tropical World. E. Laborde, transl. London: Longmans, Green, and Co.

Gregor, T.

1977 Mehinaku: The Drama of Daily Life in a Brazilian Indian Village. Chicago: University of Chicago Press.

Gross, D.

1975 Protein Capture and Cultural Development in the Amazon Basin. American Anthropologist 77:526–549.

1979 A New Approach to Central Brazilian Social Organization. *In* Contemporary Brazil, M. Margolis and W. Carter, eds. pp. 321–342. New York: Columbia University Press.

1982 Proteína y Cultura en la Amazonía: Una Segunda Revisión. Amazonía Peruana 3(6):85–97.

Gross, D., G. Eiten, N. Flowers, F. Leoi, M. Ritter, and D. Werner

1979 Ecology and Acculturation Among Native Peoples of Central Brazil. Science 206:1043–1050.

Gross, D. and B. Underwood

1971 Technological Change and Caloric Costs: Sisal Agriculture in Northeast Brazil. American Anthropologist 73:725–740.

Guallart, J.

1964 Los Jíbaros del Alto Marañón. América Indígena 24:315–331.

1976 Peces, Pescadores y Pesqueras entre los Aguaruna de Alto Marañón. Biota 10:372–396.

Gudeman, S.

1978 The Demise of a Rural Economy. London: Routledge and Kegan Paul.

Hames, R.

1978 A Behavioral Account of the Division of Labor among the Ye'kwana of Southern Venezuela. Ph.D. dissertation. Department of Anthropology, University of California, Santa Barbara.

1979 A Comparison of the Efficiencies of the Shotgun and the Bow in Neotropical Forest Hunting. Human Ecology 7(3):219–252.

1980 Game Depletion and Hunting Zone Rotation Among the Ye'kwana and Yąnomamö of Amazonas, Venezuela. *In* Working Papers on South American

Indians, Vol. 2., R. Hames, ed. pp. 24–62. Bennington, Vermont: Bennington College.
In press a Monoculture, Polyculture, and Polyvariety in Tropical Forest Swidden Cultivation. Human Ecology.
In press b Garden Labor Exchange and Exchange Balance among the Ye'kwana. Ethology and Sociobiology.
Ms. The Wild Resource Quest in Amazonia: Time, Efficiency, and Utility. *In* Neither Hell nor Paradise, E. Moran, ed.

Hames, R. and W. Vickers
1982 Optimal Foraging Theory as a Model to Explain Variability in Amazonian Hunting. American Ethnologist 9(2)358–378.

Hampton, I.
1960 Energy Expenditure and Caloric Intake of an Undergraduate Expedition. Nutrition 14:174–188.

Harako, R.
1976 Mbuti as Hunters: A Study of the Economic Anthropology of the Mbuti Pigmies. Kyoto University African Studies 10:37–99.
1981 The Cultural Ecology of Hunting Behavior among Mbuti Pigmies in the Ituri Forest, Zaire. *In* Omnivorous Primates, R. S. O. Harding and G. Teleki, eds., pp. 499–555. New York: Columbia University Press.

Harding, R. S. O., and G. Teleki
1981 Omnivorous Primates. New York: Columbia University Press.

Harner, M.
1972 The Jívaro: People of the Sacred Waterfalls. Garden City, New York: Doubleday and Natural History Press.
1977 The Ecological Basis for Aztec Sacrifice. American Ethnologist 4:117–135.

Harris, M.
1974 Cows, Pigs, Wars and Witches: The Riddles of Culture. New York: Random House.
1975 Culture, People and Nature. New York: Crowell.
1977 Cannibals and Kings. New York: Random House.
1979a Our Pound of Flesh. Natural History 88(7):36–38.
1979b Cultural Materialism. New York: Random House.
1980 Culture, People, and Nature. 3rd edition. New York: Harper and Row.

Hart, J. A.
1978 From Subsistence to Market: A Case Study of the Mbuti Net Hunters. Human Ecology 6:325–353.

Hartung, John
1978 Light, Puberty, and Aggression: A Proximal Mechanism Hypothesis. Human Ecology 6(3):273–297.

Hawkes, K., K. Hill, and J. O'Connell
1982 Why Hunters Gather: Optimal Foraging and the Aché of Eastern Paraguay. American Ethnologist 9(2):379–391.

Hawkes, K. and J. O'Connell
1981 Affluent Hunters? Some Comments in Light of the Alyawara Case. American Anthropologist 83:622–626.

Hawkes, J., J. O'Connell, K. Hill, and E. Charnov
1982 How Much is Enough?: Hunters and "Limited Needs". Paper presented at "Human Sociobiology: New Research and Theory." Symposium at Northwestern University, November 12–14, 1982.

Hayden, B.
1981 Subsistence and Ecological Adaptations of Modern Hunter–Gatherers. *In* Om-

nivorous Primates, R. S. O. Harding and G. Teleki, eds., pp. 344–421. New York: Columbia University Press.

Hays, T.
1982 Utilitarian/adaptationist Explanations of Folk Biological Classification: Some Cautionary Notes. Journal of Ethnobiology 2(1):89–94.

Hecht, S.
1981 Deforestation in the Amazon Basin: Magnitude, Dynamics and Soil Resource Effects. Studies in Third World Societies 13:61–108.

Heaton, H., ed.
1934 The Discovery of the Amazon According to the Account of Friar Gaspar de Carvajal and Other Documents. Translated from the Spanish of J. Toribio Medina by B. Lee. New York: American Geographical Society.

Heffley, S.
1981 The Relationship between Northern Athabaskan Settlement Patterns and Resource Distribution: An Application of Horn's Model. *In* Hunter–Gatherer Foraging Strategies, B. Winterhalder and E. Smith, eds. pp. 126–147. Chicago: University of Chicago Press.

Heider, K. G.
1979 Grand Valley Dani: Peaceful Warriors. New York: Holt, Rinehart and Winston.

Helm, J.
1968 The Nature of Dogrib Socioterritorial Groups. *In* Man the Hunter, R. Lee and I. DeVore, eds. pp. 118–125. Chicago: Aldine.

Herrera, R.
1979 Nutrient Distribution and Cycling in an Amazon Caatinga Forest on Spodosols in S. Venezuela. Ph.D. dissertation. Department of Soil Science, University of Reading, England.

Herrera, R., C. Jordan, H. Klinge, and E. Medina
1978 Amazon Ecosystems: Their Structure and Functioning with Particular Emphasis on Nutrients. Interciencia 3:223–231.

Herrera, R., C. Jordan, E. Medina, and H. Klinge
1981 How Human Activities Disturb the Nutrient Cycles of A Tropical Rainforest in Amazonia. Ambio 10:109–114.

Hilbert, P.
1962a New Stratigraphic Evidence of Culture Change on the Middle Amazon (Solimões). Vienna: Atken des Internationalen Amerikanistenkongress.
1962b Preliminary Results of Archeological Research on the Japurá River, Middle Amazon. Vienna: Akten des Internationalen Amerikanistenkongress.

Hill, J.
Ms. Music, Myth, and Communication in Wakuénai Society. Ph.D. dissertation. Department of Anthropology, Indiana University.

Hill, K.
Ms. The Aché of Eastern Paraguay: Current Conditions and Recent History. Department of Anthropology University of Utah, Salt Lake City.

Hipsley, E. H. and N. Kirk
1965 Studies of Dietary Intake and Expenditure of Energy by New Guineans. South Pacific Commission Technical Paper No. 147. Noumea.

Holmberg, A.
1950 Nomads of the Long Bow: the Siriono of Eastern Bolivia. Smithsonian Institution, Publications of the Institute of Social Anthropology, No. 10. Washington, D.C.: U.S. Government Printing Office.

Holmes, R.
1981 Diet, Nutritional Status and Acculturation Among the Curripaco Indians of the

Amazon Territory. M.A. thesis. Department of Anthropology, Instituto Venezolano de Investigaciones Científicas, Caracas.

Horn, H.
1968 The Adaptive Significance of Colonial Nesting in the Brewer's Blackbird (*Euphagus cyanocephalus*). Ecology 49:682–694.

Holvey, D., ed.
1972 Merck Manual. 12th edition. Rahway, New Jersey: Merck, Sharp and Dohme Research Laboratories.

Hugh-Jones, C.
1979 From the Milk River: Spatial and Temporal Processes in Northwest Amazonia. London: Cambridge University Press.

Hugh-Jones, S.
1979 The Palm and the Pleiades: Initiation and Cosmology in Northwest Amazonia. London: Cambridge University Press.

Hunn, E.
1975 A Measure of the Degree of Correspondence of a Folk to a Scientific Classification. American Ethnologist 2:309–327.
1976 Toward a Perceptual Model of Folk Biological Classification. American Ethnologist 3:508–524.
1977 Tzeltal Folk Zoology: The Classification of Discontinuities in Nature. New York: Academic Press.
In press The Utilitarian Factor in Folk Biological Classification. American Anthropologist.

Hunter, J. M.
1967 Seasonal Hunger in a Part of the West African Savanna: A Survey of Body Weights in Nangodi, North-East Ghana. Transactions of the Institute of British Geographers 41:167–185.

Hurault, J.
1968 Les Indiens Wayana de la Gutane Français: Structure Sociale et Coutume Familiale. Paris: ORSTOM.

INCAP, ICCND
1961 Food Composition Table for Use in Latin America. Bethesda, Maryland: Interdepartmental Committee on Nutrition for National Defense.

INN (Instituto Nacional de Nutrición)
1967 Tabla de Composición de Alimentos Colombianos. Tercera Edición. Bogotá:Siglo XX.

Irons, W.
1980 Is Yomat Social Behavior Adaptive? *In* Sociobiology: Beyond Nature/Nurture? G. Barlow and J. Silverberg, eds. pp. 417–463. Boulder, Colorado: Westview Press.

Isard, W.
1972 Location and Space Economy, Cambridge: Massachusetts Institute of Technology Press.

Jackson, J.
1976 Vaupés Marriage: A Network System in an Undifferentiated Region of the Northwest Amazon. *In* Social Systems, Vol. 2, Regional Analysis, C. Smith, ed. pp. 65–93. New York: Academic Press.

Jelliffe, D.
1966 Assessment of the Nutritional Status of the Community. WHO Monograph Series 53. Geneva: World Health Organization.
1968 Infant Nutrition in the Sub-Tropics and the Tropics. WHO Monograph Series 29. Geneva: World Health Organization.

Johnson, A.
1974 Carrying Capacity in Amazonia: Problems in Theory and Method. Paper read at

the 73rd Annual Meeting of the American Anthropological Association, November 19–24, 1974, Mexico City.

1975 Time Allocation in a Machiguenga Community. Ethnology 14:301–310.

1977 The Energy Costs of Technology and the Changing Environment: a Machiguenga case. *In* Material Culture, H. Letchman and R. Merrill, eds. pp. 155–167. St. Paul, Minnesota: West Publishing Co.

1978 In Search of the Affluent Society. Human Nature 1(9):50–60.

1980 The Limits of Formalism in Agricultural Decision Research. *In* Agricultural Decision Making, P. Bartlett, ed. pp. 138–158. New York: Academic Press.

1982 Reductionism in Cultural Ecology: The Amazon Case. Current Anthropology 23(4):413–428.

Johnson, O.

1978 Domestic Organization and Interpersonal Relations Among the Machiguenga Indians of the Peruvian Amazon. Ph.D. dissertation. Department of Anthropology, Columbia University, New York.

Johnson, O. and A. Johnson

1976 Male/Female Relations and the Organization of Work in a Machiguenga Community. American Ethnologist 2:634–648.

Jordan, C.

1979 Nutrient Dynamics of a Tropical Rain Forest Ecosystem. Annual Report to the National Science Foundation. Institute of Ecology, University of Georgia.

Jordan, C. and C. Uhl

1978 Biomass of a "Tierra Firme" Forest of the Amazon Basin. Oecologia Plantarum 13(4):387–400.

Kensinger, K.

1975 The Cashinahua of Eastern Peru. Providence, Rhode Island. Haffenreffer Museum, Brown University.

Key, M. R.

1979 The Grouping of South American Indian Languages. Tübingen: Gunter Narr Verlag.

Kiltie, R.

1980 More on Amazon Culture Ecology. Current Anthropology 21:541–544.

Klinge, H.

1978 Studies on the Ecology of Amazon Caatinga Forest in Southern Venezuela. Acta Científica Venezolana 29(4):258–62.

Klinge, H., and E. Medina

1979 Rio Negro Caatingas and Campinas, Amazonas States of Venezuela and Brazil. *In* Heathlands and Related Shrublands of the World, Vol. A, Descriptive Studies, R. L. Specht, ed. pp. 483–488. Amsterdam: Elsevier Scientific.

Klinge, H., E. Medina and R. Herrera

1977 Studies on the Ecology of Amazon Caatinga Forest in Southern Venezuela. Acta Científica Venezolana 28:270–276.

Kloos, P.

1971 The Maroni River Caribs of Surinam. Assen. The Netherlands: Van Gorcum & Comp. n.v.

Kracke, W.

1978 Force and Persuasion: Leadership in an Amazonian Society. Chicago: University of Chicago Press.

Kramer, D., and W. Nowell

1980 Central Place Foraging in the Eastern Chipmunk *Tamias striatus*. Animal Behavior 28:772–778.

Krebs, J. and N. Davies, eds.
1978 Behavioural Ecology: an Evolutionary Approach. Sunderland, Massachusetts: Sinaver.

Langdon, E.
1974 The Siona Medical System: Beliefs and Behavior. Ph.D. dissertation. Department of Anthropology, Tulane University.

LaPointe, J.
1970 Residence Patterns and Wayana Social Organization. Ph.D. dissertation. Department of Anthropology, Columbia University, New York.

Larrick, J., J. Yost, J. Kaplan, G. King and J. Mayhall
1979 Patterns of Health and Disease Among the Waorani Indians of Eastern Ecuador. Medical Anthropology 3(2):147–189.

Larson, M.
1966 Vocabulario Aguaruna. Lima, Peru: Instituto Lingüístico de Verano.

Lathrap, D.
1968 The "Hunting" Economies of the Tropical Forest Zone of South America: an Attempt at Historical Perspective. *In* Man the Hunter, R. Lee and I. DeVore, eds. pp. 18–23. Chicago: Aldine.
1970 The Upper Amazon. New York: Praeger.

Layrisse, M. and Z. Layrisse
1980 Biomedical Observations on the Warao. *In* Demographic and Biological Studies of the Warao Indians, J. Wilbert and M. Layrisse, eds. pp. 160–165. University of California: UCLA Latin American Center Publications.

Lee, R.
1968 What Hunters Do for a Living, or, How to Make Out on Scarce Resources. *In* Man the Hunter, R. Lee and I. DeVore, eds. pp. 30–43. Chicago: Aldine.
1979 The !Kung San: Men, Women, and Work in a Foraging Society. Cambridge: Cambridge University Press.

Lee, R. and I. DeVore, eds.
1968 Man the Hunter. Chicago: Aldine.

Leeds, A.
1961 Yaruro Incipient Tropical Forest Horticulture. *In* The Evolution of Horticultural Systems in Native South America, Causes and Consequences: A Symposium, J. Wilbert, ed. pp. 13–46. Antropologica Supplement Publication No. 2. Caracas.
1969 Ecological Determinants of Chieftainship among the Yaruro Indians of Venezuela. *In* Environment and Cultural Behavior, A. Vayda, ed. pp. 377–394. Garden City, New York: The Natural History Press.

Leung, W., and M. Flores
1961 Food Composition Table for Use in Latin America. Bethesda: INCAP–ICNND.

Lévi-Strauss, C.
1948 La Vie Familiale et Sociale des Indiens Nambikwara. Paris: Journal de la Société des Américanistes.
1966 The Savage Mind. Chicago: University of Chicago Press.

Linares, O.
1976 "Garden Hunting" in the American Tropics. Human Ecology 4:331–349.

Lizot, J.
1974 El Río de Los Periquitos: Breve Relación de un Viaje entre Los Yanomamɨ del Río Siapa. Antropologica 37:3–23.
1976 The Yanomamɨ in the Face of Ethnocide. Document 22. Copenhagen: International Work Group for Indigenous Affairs.
1977 Population, Resources, et Guerre chez les Yanomamɨ. Libre 2:111–145.
1978a L'Economie Primitive. Libre 4:69–113.
1978b Population, Resources, and Warfare among the Yanomamɨ. Man 12:497–517.

Lowenstein, F.
1973 Some Considerations of Biological Adaptation by Aboriginal Man to the Tropical Forest. *In* Tropical Forest Ecosystems in Africa and South America, B. Meggers, E. Ayensu, and D. Duckworth, eds. pp. 293–310. Washington, D.C.: Smithsonian Institution Press.

Lowie, R.
1948 The Tropical Forests: An Introduction. *In* The Tropical Forest Tribes, J. Steward, ed. pp. 1–56. Handbook of South American Indians. Vol. 3. Washington, D.C.: U.S. Government Printing Office.
1949 Social and Political Organization of the Tropical Forest and Marginal Tribes. *In* Handbook of South American Indians. Vol. 5. J. Steward, ed. pp. 313–350. Washington, D.C.: U.S. Government Printing Office.

MacArthur, R.
1972a The Economics of Consumer Choice. *In* Geographical Ecology, R. MacArthur, ed. pp. 59–69. New York: Harper and Row.
1972b Geographical Ecology. New York: Harper & Row.

MacArthur, R. and E. Pianka
1966 On the Optimal Use of a Patchy Environment. American Naturalist 100:603–609.

McCay, B.
1981 Optimal Foragers or Political Actors? Ecological Analyses of a New Jersey Fishery. American Ethnologist 8:356–382.

McGregor, I., W. Billewicz, and A. Thomson
1961 Growth and Mortality in Children in an African Village. British Medical Journal, December 23, 1961, pp. 1661–1666.

McGregor, I., A. Rahman, B. Thompson, W. Billewicz, and A. Thomson
1968 The Growth of Young Children in A Gambian Village. Transactions of the Royal Society of Tropical Medicine and Hygiene 62(3):341–352.

McNair, J.
In press Optimal Giving-Up Times and the Marginal Value Theorem. American Naturalist.

MAG–ORSTOM
1980 Mapa Morfo-Edafológico Provisional, Nor-Oriente, Hoha No. 2. Quito, Ecuador: Ministerio de Agricultura y Ganadería and Office de la Recherche Scientifique et Technique Outre-Mer.

Majnep, I. and R. Bulmer
1977 Birds of My Kalam Country. Auckland: Auckland University Press.

Malcolm, L.
1970 Growth, Malnutrition, and Mortality of the Infant and Toddler in the Asai Valley of the New Guinea Highlands. The American Journal of Clinical Nutrition 23(8):1090–1095.

Malier, G.
1967 Hydrobiology in the Amazon Region. Atlas do Simpósia Sobre a Biota Amazónica. 3 (Limnologia):1–7.

Marcgrave, J.
1942 [1648]Historia Natural do Brasil. São Paulo: Imprensa Oficial do Estado.

Marques, J., M. Santos, N. Villa Nova, and E. Salati
1977 Precipitable Water and Water Vapor Flux Between Belem and Manaus. Acta Amazônica 7(3):11–20.

Marshall, W. A., and A. V. Swan
1971 Seasonal Variation in Growth Rates of Normal and Blind Children. Human Biology 43:502–516.

Matos Arvelo, M.
1912 Vida Indiana. Barcelona: Casa Editorial Maucci.

Maybury-Lewis, D.
1965 Some Crucial Distinctions in Central Brazilian Ethnology. Anthropos 60:340–358.
1966 On Martius' Distinction Between Shavante and Sherente. Revista do Museu Paulista n.s. 16:263–288.
1967 Akwẽ–Shavante Society. Oxford: Clarendon Press.

Maybury-Lewis, D., ed.
1979 Dialectical Societies: The Gê and Bororo of Central Brazil. Cambridge, Massachusetts: Harvard University Press.

Mayr, E.
1969 Principles of Systematic Zoology. New York: McGraw-Hill.

Meehan, B.
1977 Hunters by the Seashore. Journal of Human Evolution 6:363–370.

Meggers, B.
1954 Environmental Limitation on the Development of Culture. American Anthropologist 56:801–824.
1957 Environment and Culture in the Amazon Basin: An Appraisal of the Theory of Environmental Limitations. *In* Studies in Human Ecology, Social Science Monograph, A. Palerm, ed. No. 3. pp. 71–89. Washington, D.C.: Panamerican Union.
1971 Amazonia: Man and Culture in a Counterfeit Paradise. Chicago: Aldine.
1975 Application of the Biological Model of Diversification to Cultural Distributions in Tropical Lowland South America. Biotropica 7:141–161.

Meggers, B. and C. Evans, Jr.
1957 Archeological Investigations at the Mouth of the Amazon. Bureau of American Ethnology, Bulletin No. 167. Washington, D.C.: U.S. Government Printing Office.
1979 An Experimental Reconstruction of Taruma Village Succession and Some Implications. *In* Brazil: Anthropological Perspectives, M. Margolis and W. Carter, eds. pp. 39–60. New York: Columbia University Press.

Meggitt, M.
1977 Blood is their Argument: Warfare among the Mae Enga Tribesmen of the New Guinea Highlands. Palo Alto, California: Mayfield.

Melatti, J.
1972 O Messianismo Kraho. São Paulo: Editora Herder e Editora da Universidade de São Paulo.
1973 O Sistema de Parentesco dos Indios Kraho. Brasília: Fundacão Universidade de Brasília, Departamento de Ciencias Sociais (Serie "Antropologia," No. 3).

Metz, J., D. Hart, and H. Harpending
1971 Iron, Folate, and Vitamin B_{12} Nutrition in a Hunter–Gatherer People: A Study of the !Kung Bushmen. American Journal of Clinical Nutrition 24:229–242.

Métraux, A.
1946 The Caingang. *In* The Marginal Tribes, J. Steward, ed. pp. 445–475. Handbook of South American Indians, Vol. 1. Washington, D.C.: U.S. Government Printing Office.
1949 Warfare, Cannibalism, and Human Trophies. *In* The Comparative Anthropology of South American Indians, J. Steward, ed. pp. 383–409. Handbook of South American Indians. Vol. 5. Washington, D.C.: U.S. Government Printing Office.

Métraux, A. and H. Baldus
1963 The Guayakí. *In* The Marginal Tribes, J. Steward, ed. pp. 435–444. Handbook of South American Indians. Vol. 1. New York: Cooper Square Publishers.

Migliazza, E.
1972 Yanomama Grammar and Intelligibility. Ph.D. dissertation. Department of Anthropology, Indiana University.

Montgomery, E. and A. Johnson
1976 Machiguenga Energy Expenditure. Ecology of Food and Nutrition 6:97–105.

Moran, E.
1974 The Adaptive System of the Amazonian Caboclo. *In* Man in the Amazon, C. Wagley, ed. pp. 34–68. Gainesville: University of Florida Press.
1980 Mobility and Resource Use in Amazonia. *In* Land, People and Planning in Contemporary Amazonia, F. Barbira-Scazzocchio, ed. pp. 46–57. Centre of Latin American Studies, Occasional Publications No. 3. Cambridge: Cambridge University Press.
1981 Developing the Amazon. Bloomington: Indiana University Press.

Moran E., ed.
In press The Ecosystem Concept in Anthropology. Boulder, Colorado: Westview Press.

Moran, E. and J. Hill
In press Subsidy and Resource Use in the Rio Negro Basin. *In* Affluence and Cultural Survival, R. Salisbury, ed. Minneapolis, Minnesota: West Publ. Co.

Morey, N. and R. Morey
1975 Relaciones Comerciales en el Pasado en los Llanos de Colombia e Venezuela. Montalbán 4:533–564.

Morey, R.
1970 Warfare Patterns of the Colombian Guahibo. Actas y Memorias del Congreso Internacional de Americanistas, Vol. 4, pp 59–68. Lima: Instituto de Estudios Peruanos.

Morey, R. and J. Marwitt
1975 Ecology, Economy and Warfare in Lowland South America. *In* War: Its Causes and Correlates, M. Nettleship, R. Givens, and A. Nettleship, eds. pp. 439–450. The Hague: Mouton.
1978 Ecology, Economy and Warfare in Lowland South America. *In* Advances in Andean Archaeology, D. Browman, ed. pp. 247–258. The Hague: Mouton

Murdock, G.
1968 The Current Status of the World's Hunting and Gathering Peoples. *In* Man the Hunter, R. Lee and I. DeVore eds. pp. 13–20. Chicago: Aldine.

Murphy, R.
1960 Headhunter's Heritage. Berkeley: University of California Press.

Murphy, R. and Y. Murphy
1974 Women of the Forest. New York: Columbia University Press.

Murphy, R. and B. Quain
1966 The Trumai Indians of Brazil. The American Ethnological Society Monograph No. 24. Seattle: The University of Washington Press.

Myers, N.
1979 The Sinking Ark. Oxford: Pergamon Press.

Neel, J.
1970 Lessons from a "Primitive" People. Science 170:815–822.

Neel, J. and F. Salzano
1967 Further Studies on the Xavante Indians. X. Some Hypotheses–Generalizations Resulting from These Studies. American Journal of Human Genetics 19:554–574.

Neel, J., F. Salzano, P. Junqueira, F. Keiter and D. Maybury-Lewis
1964 Studies on the Xavante Indians of the Brazilian Mato Grosso. American Journal of Human Genetics 16(1):52–140.

Neel, J. and K. Weiss
1975 The Genetic Structure of a Tribal Population, the Yanomama Indians. XIII. Biodemographic Studies. American Journal of Physical Anthropology 42(1):25–51.

Nelson, R.
1973 Hunters of the Northern Forest. Chicago: University of Chicago Press.

Nicholaides, J. *et al.*
In press Crop Production Systems in the Amazon Basin. *In* The Dilemma of Amazonian Development, E. Moran, ed. Boulder: Westview Press.

Nietschmann, B.
1973 Between Land and Water: The Subsistence Ecology of the Miskito Indians, Eastern Nicaragua. New York: Seminar Press.
1978 Comment on Ross. Current Anthropology 19:23–24.
1980 The Limits to Protein. *In* Working Papers on South American Indians, Vol. 2, R. Hames, ed. pp. 131–137. Bennington, Vermont: Bennington College.
Nimuendajú, C.
1939 The Apinayé. Anthropological Series No. 8. Washington, D.C.: The Catholic University of America.
1942 The Sherente. Los Angeles: Publications of the Frederick Webb Hodge Anniversary Fund.
1946 The Eastern Timbira. Publications in American Archaeology and Ethnology, No. 41. Berkeley: University of California.
1950 Reconhecimento dos Rias Içana, Ayari y Vaupés. Journal de la Société des Américanistes n.s. 39:125–82.
Niswander, J., F. Ketter, and J. Neel
1967 Further Studies on the Xavante Indians. II. Some Anthropometric, Dermatoglyphic and Non-Quantitative Morphological Traits of the Xavante of Simões Lopes. American Journal of Human Genetics 19:490–501.
Norgan, N. G., A. Ferro-Luzzi, and J. V. G. A. Durnin
1974 The Energy and Nutrient Intake and Energy Expenditure of 204 New Guinea Adults. Philosophical Transactions of the Royal Society (London) 268(893):309–348.
Nurse, G.
1975 Seasonal Hunger Among the Ngoni and Ntumba of Central Malawi. Africa 45(1):1–11.
Nye, P. and D. Greenland
1960 The Soil Under Shifting Cultivation. Commonwealth Bureau of Soils, Technical Communication No. 51. Farnham Royal, Berkshire: Commonwealth Bureau of Soils.
Oberg, Kalervo
1955 Indian Tribes of Northern Mato Grosso, Brazil. Institute of Social Anthropology Publication No. 15. Washington, D.C.: Smithsonian Institution.
O'Connell, J. and K. Hawkes
1981 Alyawara Plant Use and Optimal Foraging Theory. In Hunter–Gatherer Foraging Strategies, B. Winterhalder and E. Smith, eds. pp. 78–93. Chicago: University of Chicago Press.
Odum, E.
1969 The Strategy of Ecosystem Development. Science 164: 262–270.
1971 Fundamentals of Ecology. 3rd edition. Philadelphia: W. B. Saunders Co.
Odum, H.
1971 Environment, Power, and Society. New York: Wiley.
Odum, H. and R. Pigeon
1970 A Tropical Rainforest. Springfield, Virginia: National Technical Information Service.
Oke, O.
1973 The Mode of Cyanide Detoxification. Chronic Cassava Toxicity: Proceedings of an Interdisciplinary Workshop, London, England, 29–30 January 1973, pp. 97–104. Ottawa: International Development Research Centre Monograph.
Oliveira, A. de
1975 A Terminologia de Parentesco Baniwa. Boletim do Museu Paraense Emilio Goeldi, n.s. 56:1–37.

Orians, G. and N. Pearson
1979 On the Theory of Central Place Foraging. *In* Analysis of Ecological Systems, D. Horn, G. Stairs, and R. Mitchell, eds. pp. 43–60. Columbus: Ohio State University Press.
Osuntokun, B.
1973 Ataxic Neuropathy Associated with High Cassava Diets in West Africa. Chronic Toxicity: Proceedings of an Interdisciplinary Workshop, London, England, 29–30 January, pp. 127–138. Ottawa: International Development Research Centre Monograph.
Osuntokun, B., G. Monekosso, and J. Wilson
1969 Relationship of a Degenerative Tropical Neuropathy to Diet: Report of a Field Survey. British Medical Journal 2:547–550.
Oyama, M. and H. Takehara
1967 Revised Standard Soil Color Charts. Tokyo: International Soil Science Society.
Patterson, T.
1971 The Emergence of Food Production in Central Peru. *In* Prehistoric Agriculture, Stuart Streuver, ed. pp. 181–207. Garden City, New York: Natural History Press.
Patton, J., B. Berlin, and E. Berlin
In press Aboriginal Perspectives of a Mammal Community in Amazonian Peru: Knowledge and Utilization Patterns among the Aguaruna Jívaro. *In* Mammalian Diversity in South America, M. A. Mares and H. H. Genoways, eds. Pittsburgh: University of Pittsburgh Press.
Pearson, N.
1974 Optimal Foraging Theory. Quantitative Science Paper No. 39, Center for Quantitative Science in Forestry, Fisheries, and Wildlife, University of Washington.
Pianka, E.
1970 On *r*- and *K*- Selection. American Naturalist 104:592–597.
1978 Evolutionary Ecology. 2nd edition. New York: Harper and Row.
Pinton, S.
1965 La Maison Barí. Journal de la Société des Américanistes 54(2):247–333.
Pires, J.
1978 The Forest Ecosystems of the Brazilian Amazon: Description, Functioning and Research Needs. *In* Tropical Forest Ecosystems: A State-of-Knowledge Report, pp. 607–627. Paris: UNESCO.
Pires, J. and G. Prance
1977 The Amazon Forest: A Natural Heritage to be Preserved. *In* Extinction is Forever, G. Prance and T. Elias, eds. pp. 158–194. Bronx: The New York Botanical Gardens.
Plog, F. and J. Hill
1972 The Southwestern Archaeological Group: Revisions of the Research Design. Prescott College Anthropological Reports 3:7–20.
Posey, D.
1982 Keepers of the Forest. Garden 6(1):18–24.
Prance, G.
1978 The Origin and Evolution of the Amazon Flora. Interciencia 3(4):200–205.
Prentice, A. M., R. G. Whitehead, S. B. Roberts, and A. A. Paul
1981 Long Term Energy Balance in Child Bearing Gambian Women. American Journal of Clinical Nutrition 34:2790–2799.
Primov, G.
1979 Urban Growth and Migration in Amazonia: The Case of Territorio Federal Amazonas (Venezuela). Paper presented at XLIII International Congress of Americanists. Vancouver, Canada.

PRORADAM (Proyecto Radargramétrico del Amazonas)
1979 La Amazonía Colombiana y Sus Recursos. Bogotá: República de Colombia.
Pulliam, H.
1974 On the Theory of Optimal Diets. American Naturalist 108:59–74.
Pyke, G., H. Pulliam, and E. Charnov
1977 Optimal Foraging: A Selective Review of Theories and Tests. Quarterly Review of Biology 52:137–154.
Radcliffe-Brown, A. R.
1940 On Social Structure. Journal of the Royal Anthropological Institute 70 (1):1–12.
Ramos, A.
1978 El Indio y los Otros: la Visión Multiétnica de un Indio Maiongong. América Indígena 37(1):89–112.
1979 On Women's Status in Yanoama Societies. Current Anthropology 20:185–187.
Ramos, A. and B. Albert
1976 Yanoama Descent and Affinity: The Sanuma/Yanomama Contrast. Actes du XLIII Congrès International des Américanistes 2:71–90.
Ramos, A. and K. Taylor
1979 The Yanomama in Brazil 1979. IWGIA Document No. 37. Copenhagen: ARC/IWGIA/SI.
Rappaport, Roy A.
1968 Pigs for the Ancestors: Ritual in the Ecology of a New Guinea People. New Haven, Connecticut: Yale University Press.
Reichel-Dolmatoff, G.
1971 Amazonian Cosmos. Chicago: University of Chicago Press.
1976 Cosmology as Ecological Analysis: a View from the Rain Forest. Man 11:307–318.
Reis, A.
1931 Historia do Amazonas. Manaus, Brazil: Edição do Instituto Biologico.
Ritter, M.
1980 The Conditions Favoring Age-Set Organization. Journal of Anthropological Research 36(1):87–104.
Roberts, T. R.
1973 Ecology of Fishes in the Amazon and Congo Basins. *In* Tropical Forest Ecosystems in Africa and South America: A Comparative Review, B. J. Meggers, E. S. Ayensu, and W. D. Duckworth, eds. pp. 239–254. Washington, D.C.: Smithsonian Institution Press.
Rodríguez, M.
1684 El Marañón y Amazonas. Madrid.
Roosevelt, A.
1980 Parmana: Prehistoric Maize and Manioc Subsistence Along the Amazon and Orinoco. New York: Academic Press.
Ross, E.
1976 The Achuara Jívaro: Cultural Adaptation in the Upper Amazon. Ph.D. dissertation. Department of Anthropology, Columbia University.
1978 Food Taboos, Diet, and Hunting Strategy: The Adaptation to Animals in Amazon Cultural Ecology. Current Anthropology 19:1–36.
1980 Reply. Current Anthropology 21(4): 544–546.
Ross, J.
1980 Ecology and the Problem of Tribe: A Critique of the Hobbesian Model of Preindustrial Warfare. *In* Beyond the Myths of Culture: Essays in Cultural Materialism, E. Ross, ed. pp. 33–60. New York: Academic Press.
Ruddle, K. and R. Chesterfield
1977 Education for Traditional Food Procurement in the Orinoco Delta. Ibero-Americana No. 53. Los Angeles: University of California Press.

Saake, W.
1968 Mythen über Inapirikuliden Kulturheros der Baniwa. Zeitschrift für Ethnologie. Band 93, Heft 1–2.

Saffirio, J. and R. Hames
In press The Forest and the Highway: A Comparative Study of the Effects of Road Construction and Manufactures of the Yanomama Indians of Roraima, Brazil. *In* Working Papers on South American Indians and Cultural Survival Occasional Papers. Bennington, Vermont: Bennington College.

Sahlins, M.
1964 Culture and Environment: The Study of Cultural Ecology. *In* Horizons in Anthropology, S. Tax, ed. pp. 132–147. Chicago: Aldine.
1972 Stone Age Economics. Chicago: Aldine.

Salati, E., J. Marques, and L. Molion
1978 Origem e Distribuição das Chuvas N Amazonia. Interciencia 3(4):200–205.

Salisbury, R.
In press Affluence and Cultural Survival. St. Paul, Minnesota: West Publishing Co.

Salzano, F., J. Neel, and D. Maybury-Lewis
1967 Further Studies on the Xavante Indians. I. Demographic Data on Two Additional Villages: Genetic Structure of the Tribe. American Journal of Human Genetics 19:463–489.

Sanchez, P.
1976 Properties and Management of Soils in the Tropics. New York: John Wiley and Sons.

Sanders, W. and B. Price
1968 Mesoamerica: The Evolution of a Civilization. New York: Random House.

Sauer, C.
1950 Cultivated Plants of South and Central America. *In* Physical Anthropology, Linguistics, and Cultural Geography of South American Indians, J. Steward, ed. pp. 487–543. Handbook of South American Indians. Vol. 6. Washington, D.C.: U.S. Government Printing Office.

Scazzocchio, F., ed.
1980 Land, People and Planning in Contemporary Amazonia. Latin American Studies Centre Publications No. 4. Cambridge: Cambridge University Press.

Schoener, T.
1971 Theory of Feeding Strategies. Annual Review of Ecology and Systematics 11:369–404.

Schomburgk, R.
1840 Report of the Third Expedition into the Interior of British Guyana Comprising the Journey to the Sources of the Essequibo, to the Caruma Mountains, and to Fort San Joaquim, on the Rio Branco, in 1837–38. Journal of the Royal Geographical Society 10:191–247.
1922 Richard Schomburgk's Travels in British Guiana 1840–1844. 2 Vols. W. Roth, transl. Georgetown, British Guiana: Daily Chronicle Office.

Schuster, M.
1956 Die Soziologie der Waika. International Congress of Americanists, Proceedings, 32.

Scudder, T.
1975 The Ecology of the Gwembe Tonga. Kariba Studies Vol. II. Manchester: Manchester University Press.

Seigler, D. S. and J. T. Pereira
1981 Modernized Preparation of Cassava in the Llanos Orientales of Venezuela. Economic Botany 35(3):356–362.

Silberbauer, G.
1972 The G/wi Bushmen. *In* Hunters and Gatherers Today, M. Bicchieri, ed. pp. 271–326 New York: Holt, Rinehart and Winston.

Silverwood-Cope, P.
1972 A Contribution to the Ethnography of the Conombian Maku. Ph.D. dissertation. Department of Anthropology, Cambridge University.

Simões, M.
1974 Contribuição a Arqueologia do Baixo Rio Negro. Programa Nacional de Pesquisas Arqueológicas: Resultados Preliminares do Quinto Ano, 1969–1970. Museu Paraense Emilio Goeldi, Publicações Avulsas 26:165–188.

Simson, A.
1886 Travels in the Wilds of Ecuador and the Exploration of the Putumayo River. London: Sampson Low, Marston, Searle, and Livingston.

Sioli, H.
1975a Tropical Rivers as Expressions of their Terrestrial Environments. *In* Tropical Ecological Systems: Trends in Terrestrial and Aquatic Research, F. Golley and E. Medina, eds. pp. 275–288. New York: Springer-Verlag.
1975b Amazonian Tributaries and Drainage Basins. *In* Coupling of Land and Water Systems, A. Hasler, ed. pp. 199–213. New York: Springer-Verlag.

Siskind, J.
1973 Tropical Forest Hunters and the Economy of Sex. *In* Peoples and Cultures of Native South America, D. Gross, ed. pp. 226–241. Garden City, New York: Natural History Press.

Siverts, H.
1972 Tribal Survival in the Alto Marañón: The Aguaruna Case. Copenhagen: Work Group for Indigenous Affairs.

Smith, C.
1974 Regional Analysis. 2 vols. New York: Academic Press.

Smith, E.
1981 The Application of Optimal Foraging Theory to the Analysis of Hunter–Gatherer Group Size. *In* Hunter–Gatherer Foraging Strategies, B. Winterhalder and E. Smith, eds. pp. 84–108. Chicago: University of Chicago Press.

Smith, J.
1978 Economic Uncertainty in an Aboriginal Affluent Society: Caribou and Caribou Eater Chippewyan Adaptive Strategies. Arctic Anthropology 15(1):68–88.

Smith, N.
1979 Aquatic Turtles of Amazonia: an Endangered Resource. Biological Conservation 16(3):165–176.
1980 Anthrosols and Human Carrying Capacity in Amazonia. Annals of the Association of American Geographers 70(4):553–66.
1981a Man, Fishes, and the Amazon. New York: Columbia University Press.
1981b Colonization Lessons from a Tropical Forest. Science 214:755–761.
1981c Caimans, Capybaras, Otters, Manatees and Man in Amazonia. Biological Conservation 19(3):177–187.

Smole, W.
1976 The Yanoama Indians: a Cultural Geography. Austin: University of Texas Press.

Sourdat, M. and E. Custode
1977 Reconocimiento Morfográfico y Edafológico de la Amazónia Ecuatoriana, Zona Nor-Oriental. Dirección de Planificación, Regionalización. Quito, Ecuador: MAG–ORSTOM.

Spath, C.
1971 The Toxicity of Manioc as a Factor in the Settlement Patterns of Lowland South America. Paper presented at the 70th Annual Meeting of the American An-

thropological Association. New York, November 1970.
1981 Getting to the Meat of the Problem: Some Comments on Protein as a Limiting Factor in Amazonia. American Anthropologist 83:377–379.

Spielman, R., E. Migliazza and J. Neel
1974 Regional Linguistic and Genetic Differences among Yanomama Indians. Science 184:637–644.

Staden, Hans
1928 Hans Staden: The True History of His Captivity. Edited and translated by Malcolm Letts. London: Routledge and Kegan Paul.

Steinen, K. von den
1897 Unter den Naturvölkern Zentral-Brasiliens. Berlin: Geographische Verlagshandlung Dietrich Reimer.
1940 Entre os Aborígenes do Brasil Central. E. Schaden, transl. São Paulo: Departamento de Cultura.

Sternberg, H.
1975 The Amazon River of Brazil. Wiesbaden: Franz Steiner Verlag.

Steward, J.
1938 Basin–Plateau Aboriginal Sociopolitical Groups. Bureau of American Ethnology Bulletin. Washington, D.C.: U.S. Government Printing Office.
1944–1949 (editor) Handbook of South American Indians, Vols. 1–7. Bureau of American Ethnology Bulletin. Washington, D.C. U.S. Government Printing Office.
1946 The Marginal Tribes. Handbook of South American Indians. Vol. 1. Washington, D.C.: U.S. Government Printing Office.
1948 The Tropical Forest Tribes. Handbook of South American Indians, Vol. 3. Washington, D.C.: U.S. Government Printing Office.
1949 South American Cultures: An Interpretative Summary. *In* The Comparative Ethnology of South American Indians, J. Steward, ed. pp. 669–672. Handbook of South American Indians, Vol. 5. Washington, D.C.: U.S. Government Printing Office.
1955 Theory of Culture Change. Urbana: University of Illinois Press.

Steward, J. and L. Faron
1959 Native Peoples of South America. New York: McGraw-Hill.

Stini, W.
1980 Human Adaptability to Nutritional Stress. *In* Nutrition, Food, and Man, P. B. Pearson and J. R. Greenwell, eds. pp. 124–140. Tucson: University of Arizona Press.

Stocks, A.
1976 Notas Sobre Los Autóctonos Tupi del Perú. Amazonía Peruana 1:59–72.
1978 The Invisible Indians: A History and Analysis of the Relations of the Cocamilla Indians of Loreto, Peru, and the State. Ph.D. dissertation. Department of Anthropology, University of Florida.
1981 Los Nativos Invisibles: Etnohistoria y Realidad Actual de los Cocamilla del Río Huallaga. Lima: Centro Amazónico de Antropología y Aplicación Práctica.
In press Native Enclaves in the Upper Amazon: A Case of Regional Non-integration. Ethnohistory.

Suarez, M.
1968 Los Warao: Indígenas del Delta del Orinoco. Caracas: Instituto de Investigaciones Científicas Venezolano.

Symons, D.
1979 The Evolution of Human Sexuality. Oxford: Oxford University Press.

Tanno, T.
1976 The Mbuti Net Hunters in the Itwi Forest, Eastern Zaire. Kyoto University African Studies 10:101–136.

White, L.
1959 The Evolution of Culture. New York: McGraw-Hill.
Whitten, N., Jr.
1978 Ecological Imagery and Cultural Adaptability. American Anthropologist 80(4):836–859.
Wilbert, J.
1976 To Become a Maker of Canoes: An Essay in Warao Enculturation. *In* Enculturation in Latin America: An Anthology, J. Wilbert, ed. pp. 303–358. Los Angeles: University of California Press.
1979 Geography and Telluric Lore of the Orinoco Delta. Journal of Latin American Lore 5(1):129–250.
Williams, G.
1966 Adaptation and Natural Selection. Princeton: Princeton University Press.
Wilmsen, E.
1978 Seasonal Effects of Dietary Intake on Kalahari San. Federation Proceedings 37:65–72.
1982 Studies in Diet, Nutrition, and Fertility among a Group of !Kung Bushmen in Botswana. Social Science Information 21:95–126.
Wilson, E.
1971 Competitive and Aggressive Behavior. *In* Man and Beast: Comparative Social Behavior, J. Eisenberg and W. Dillon, eds. pp. 181–218. Washington, D.C.: Smithsonian Institute Press.
1975 Sociobiology: The New Synthesis. Cambridge: The Belknap Press of Harvard University Press.
Winterhalder, B.
1981a Optimal Foraging Strategies and Hunter–Gatherer Research in Anthropology: Theory and Models. *In* Hunter–Gatherer Foraging Strategies, B. Winterhalder and E. Smith, eds. pp. 13–35. Chicago: University of Chicago Press.
1981b Foraging Strategies in the Boreal Environment: An Analysis of Cree Hunting and Gathering. *In* Hunter–Gatherer Foraging Strategies, B. Winterhalder and E. Smith, eds. pp. 66–98. Chicago: University of Chicago Press.
1982 Stalking the Fundamental Theorem: Observations on a Nutritious if Dangerous Quarry. Paper presented at "Optimization in Anthropology and Ecology." Symposium held at University of Nebraska, Lincoln, April 22–23, 1982.
World Health Organization (WHO)
1978 A Growth Chart for International Use in Maternal and Child Health Care. Geneva: World Health Organization.
Wright, R.
1981 Baniwa History and Religion. Ph.D. dissertation. Department of Anthropology, Stanford University.
Wrigley, G.
1969 Tropical Agriculture. New York: Praeger.
Wynne-Edwards, V.
1962 Animal Dispersion in Relation to Social Behavior. Edinburgh: Boyd G. Oliver.
Yellen, J.
1977 Archaeological Approaches to the Present: Models for Reconstructing the Past. New York: Academic Press.
Yost, J.
1981a People of the Forest: The Waorani. *In* Ecuador in the Shadow of the Volcanoes, G. Ligabue, ed. pp. 95–115. Centro Studi Richerche Ligabue. Venice: Ediciones Libri Mundi.
1981b Twenty Years of Contact: the Mechanisms of Change in Wao ("Auca") Culture. *In* Cultural Transformations and Ethnicity in Modern Ecuador, N. Whitten, Jr., ed. pp. 677–704. Urbana: University of Illinois Press.

Zarur, G.
1975 Parentesco, Ritual e Economia no Alto Xingu. Brasília: Fundação Nacional do Indio.
1979 Ecological Need and Cultural Choice in Central Brazil. Current Anthropology 20:649–653.

Index

STUDIES IN ANTHROPOLOGY

Under the Consulting Editorship of E. A. Hammel,
UNIVERSITY OF CALIFORNIA, BERKELEY

Andrei Simić, THE PEASANT URBANITES: A Study of Rural-Urban Mobility in Serbia

John U. Ogbu, THE NEXT GENERATION: An Ethnography of Education in an Urban Neighborhood

Bennett Dyke and Jean Walters MacCluer (Eds.), COMPUTER SIMULATION IN HUMAN POPULATION STUDIES

Robbins Burling, THE PASSAGE OF POWER: Studies in Political Succession

Piotr Sztompka, SYSTEM AND FUNCTION: Toward a Theory of Society

William G. Lockwood, EUROPEAN MOSLEMS: Economy and Ethnicity in Western Bosnia

Günter Golde, CATHOLICS AND PROTESTANTS: Agricultural Modernization in Two German Villages

Peggy Reeves Sanday (Ed.), ANTHROPOLOGY AND THE PUBLIC INTEREST: Fieldwork and Theory

Carol A. Smith (Ed.), REGIONAL ANALYSIS, Volume I: Economic Systems, and Volume II: Social Systems

Raymond D. Fogelson and Richard N. Adams (Eds.), THE ANTHROPOLOGY OF POWER: Ethnographic Studies from Asia, Oceania, and the New World

Frank Henderson Stewart, FUNDAMENTALS OF AGE-GROUP SYSTEMS

Larissa Adler Lomnitz, NETWORKS AND MARGINALITY: Life in a Mexican Shantytown

Benjamin S. Orlove, ALPACAS, SHEEP, AND MEN: The Wool Export Economy and Regional Society in Southern Peru

Harriet Ngubane, BODY AND MIND IN ZULU MEDICINE: An Ethnography of Health and Disease in Nyuswa-Zulu Thought and Practice

George M. Foster, Thayer Scudder, Elizabeth Colson, and Robert Van Kemper (Eds.), LONG-TERM FIELD RESEARCH IN SOCIAL ANTHROPOLOGY

R. H. Hook (Ed.), FANTASY AND SYMBOL: Studies in Anthropological Interpretation

Richard Tapper, PASTURE AND POLITICS: Economics, Conflict and Ritual Among Shahsevan Nomads of Northwestern Iran

George Bond, Walton Johnson, and Sheila S. Walker (Eds.), AFRICAN CHRISTIANITY: Patterns of Religious Continuity

Philip Burnham, OPPORTUNITY AND CONSTRAINT IN A SAVANNA SOCIETY: The Gbaya of Meiganga, Cameroon

John Comaroff (Ed.), THE MEANING OF MARRIAGE PAYMENTS

Michael H. Agar, THE PROFESSIONAL STRANGER: An Informal Introduction to Ethnography

Robert J. Thornton, SPACE, TIME, AND CULTURE AMONG THE IRAQW OF TANZANIA

Linda S. Cordell and Stephen Beckerman (Eds.), THE VERSATILITY OF KINSHIP

Peggy F. Barlett (Ed.), AGRICULTURAL DECISION MAKING: Anthropological Contributions to Rural Development

Thayer Scudder and Elizabeth Colson, SECONDARY EDUCATION AND THE FORMATION OF AN ELITE: The Impact of Education on Gwembe District, Zambia

Eric B. Ross (Ed.), BEYOND THE MYTHS OF CULTURE: Essays in Cultural Materialism

Gerald D. Berreman (Ed.), SOCIAL INEQUALITY: Comparative and Developmental Approaches

Karla O. Poewe, MATRILINEAL IDEOLOGY

J. D. Lewis-Williams, BELIEVING AND SEEING: Symbolic Meaning in Southern San Rock Paintings

David Riches, NORTHERN NOMADIC HUNTER-GATHERERS: A Humanistic Approach

George A. Collier, Renato I. Rosaldo, and John D. Wirth (Eds.), THE INCA AND AZTEC STATES, 1400-1800: Anthropology and History

Joseph Shepher, INCEST: A Biosocial View

Raymond B. Hames and William T. Vickers (Eds.), ADAPTIVE RESPONSES OF NATIVE AMAZONIANS

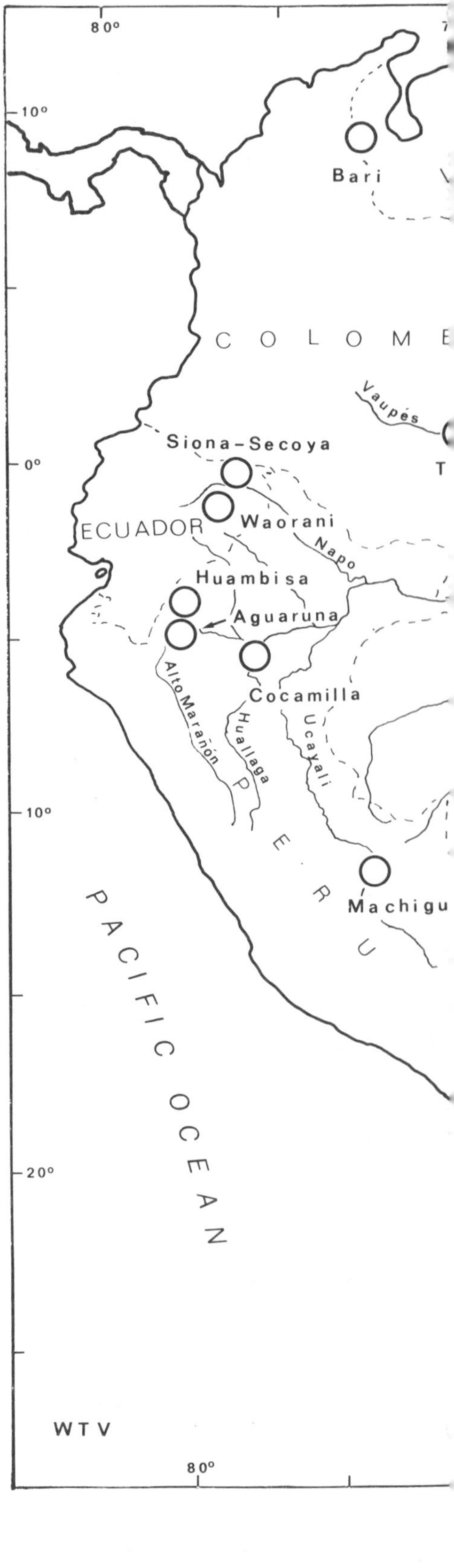

Locations of the lowland South American Indian societies that are the subjects of this volume.